Science and Society

Science and Society

An Anthology for Readers and Writers

Edited by
Catherine Nelson-McDermott
Laura Buzzard
Don LePan

broadview press

Library and Archives Canada Cataloguing in Publication

    Science and society : an anthology for readers and writers / edited by Catherine Nelson-McDermott, Laura Buzzard, Don LePan.

Includes writings previously published in Broadview anthology of
    expository prose.
Includes bibliographical references.
ISBN 978-1-55481-192-2 (bound)

    1. Technical writing. 2. Science. 3. Science—Moral and ethical aspects. I. LePan, Don, 1954-, editor of compilation II. Buzzard, Laura, editor of compilation III. Nelson-McDermott, Catherine A. (Catherine Ann), 1962-, editor of compilation

Q171.S3468 2014   808.06'65   C2013-908515-7

Broadview Press is an independent, international publishing house, incorporated in 1985.

We welcome comments and suggestions regarding any aspect of our publications—please feel free to contact us at the addresses below or at broadview@broadviewpress.com.

North America	PO Box 1243, Peterborough, Ontario K9J 7H5, Canada 555 Riverwalk Parkway, Tonawanda, NY 14150, USA Tel: (705) 743-8990; Fax: (705) 743-8353 email: customerservice@broadviewpress.com
UK, Europe, Central Asia, *Middle East, Africa, India,* *and Southeast Asia*	Eurospan Group, 3 Henrietta St., London WC2E 8LU, UK Tel: 44 (0) 1767 604972; Fax: 44 (0) 1767 601640 email: eurospan@turpin-distribution.com
Australia and New Zealand	NewSouth Books, c/o TL Distribution 15-23 Helles Ave., Moorebank, NSW 2170, Australia Tel: (02) 8778 9999; Fax: (02) 8778 9944 email: orders@tldistribution.com.au

www.broadviewpress.com

Developmental Editor: Laura Buzzard

Broadview Press acknowledges the financial support of the Government of Canada through the Canada Book Fund for our publishing activities.

PRINTED IN CANADA

Contents

Preface

There are fewer books available on the subject of writing about science than there are on writing in almost any given discipline in the humanities and social sciences. Perhaps the assumption has been that writing is relatively more important in disciplines such as history and philosophy, while, in the sciences, "the facts speak for themselves." This book aims, by contrast, to provide a real-world sense of the degree to which writing matters in the sciences, and of the *ways* in which this writing matters. The anthology is also designed to encourage the sort of critical analysis that helps a reader to realize when science has gone wrong: when research has been badly designed, when bias of one sort or another has distorted the process, when the results have been represented in misleading ways.

As editors, we have kept in mind these twin concerns – quality of writing and quality of research – in choosing the selections for this anthology. Some of these articles have been chosen because they effectively and engagingly communicate the results of scientific research to the reader. But we have also chosen pieces that arguably represent bad science, bad writing, or both. Our hope is that the inclusion of both positive and negative examples will give readers a sense not only of what to aim for (or what to strenuously avoid) when writing about science, but also of what to look for when reading scientific material critically. We hope, then, that the material will help students to develop their own skeptical reading strategies, as well as to define their own writing styles.

The selections in this volume are organized in clusters on such politically or ethically controversial topics as climate change, IQ testing, corporate funding of medical research, and factory farming. Each cluster of materials offers a variety of approaches – both in terms of writing style and in terms of viewpoint. Together, the selections offer a glimpse of how diverse, how lively, and how provocative science writing can be.

Like similar collections, this anthology includes science writing aimed at a popular audience; unlike other such collections, it also includes a substantial number of articles from scientific journals, so that the reader has the opportunity to encounter scientific research first-hand. We have done our best to choose research articles that are relatively accessible to general readers, and we have frequently excerpted sections from articles, in order to remove material that is particularly dense or difficult. A few academic pieces, however, are provided in their entireties.

Our hope is that this anthology will be helpful for several sorts of university courses – including interdisciplinary "Science and Society" courses as well as courses with a primary focus on writing. Much of the material that may be of interest more to the latter than to the former has been included in appendices, where we hope it will be readily accessible for students in courses with a writing focus, but unobtrusive for those taking "Science and Society" courses.

This book began life as a research and writing composition class at the University of British Columbia; it is therefore designed around teaching academic science-based writing and scholarly critical inquiry. Six of the selections in these pages also derive from *The Broadview Anthology of Expository Prose* – a collection that is unusual among composition readers in that it includes numerous examples of scholarly writing, and numerous examples of writing about science.

Readers are encouraged to consult the companion website to the book (http://sites.broadviewpress.com/scienceandsociety), which includes additional readings and a wide variety of grammar and usage exercises (many of them interactive).[a] A separate site for instructors provides additional discussion questions and background information.

We welcome comments about the text from both instructors and students, and we also welcome suggestions for additions or changes in future editions; please email us at broadview@broadviewpress.com.

---

a   All new copies of this volume are sold with a card that provides an access code for the companion site. If you have a used copy, you can purchase access to the companion site for a modest fee through the Broadview Press website: www.broadviewpress.com.

Introduction

This introduction is short because we want you to read it.[a] It will give you an overview of the ways academics approach the scientific and experimental processes, and of the ideals, influences, and pressures that affect the writing up of experiments. It outlines some of the ways the principles of science and science writing can be corrupted, and it suggests ways to guard against being misled, either as a reader or as a scientist and writer.

Many overviews of science – and of writing about science – leave the reader with the impression that scientific inquiry follows a set of simple ideals. For instance, according to Lynn Quitman Troyka and Douglas Hesse (authors of a well-known introductory text on academic writing), scientists follow "the scientific method ... to make discoveries." Troyka and Hesse add that scientists "usually write to inform their audience about factual information" and to "explain cause and effect as systematically and objectively as possible" (97). This is a typical summary of the best aims of science and science writing.

Science and Society is designed to provide you with illustrations of science and of science writing that lives up to these aims. But it is also designed to provide you with a sense of the many ways in which the workings of science (and of writing about science) can go wrong. Sometimes these are purely a matter of poor communication strategies; with this in mind, *Science and Society* looks at different types of writing about science, examines the audiences they are aimed at, and explores the ways in which they function well or poorly.

Like any other field of human endeavor, science (and writing about science) can mislead, can become lost in blind alleys, can promulgate false information, can reach the wrong conclusions. Sometimes such things occur accidentally; from time to time they can also occur as a result of scientists being deliberately misleading. And they can occur, too, as a result of systemic factors. If, for example, all research into a particular area is being funded by companies with a vested interest in certain conclusions being reached, it is likely that the scientific objectivity of that research will be compromised, to

---

a   We would also like you to read the material in the two appendices, "How to Use Sources" and "Writing about Science: A Closer Look," in which we provide more detail about some of the practical elements of scientific writing, such as how not to plagiarize and how to structure a science article.

some degree at least. Therefore, an understanding of science and of scientific writing requires not only that readers be able to comprehend what they are reading (or writing about), but also that they have the tools to think critically about it. This anthology tries to help "street-proof" you for the sciences by focussing on both good and bad instances of the scientific communication process.

THE SCIENTIFIC PROCESS

If you've read about science in the news lately, you've probably seen reports along the lines of "Scientists Prove X!" In the case of headlines like this, it usually turns out that the study has been misrepresented by either the reporter or the scientist. In fact, the scientific process is a skeptical one, and scientists' conclusions tend to be tentative and heavily qualified. The core activity of this process involves investigating, through research, the extent to which hypotheses are either supported or disproved by evidence. Science may speak with near-complete certainty when it comes to certain hypotheses being shown to be false. For instance, the hypothesis that the sun and the planets revolve around the earth has been disproved, as has the hypothesis that bloodletting will cure a wide range of diseases, and we no longer believe that thinking occurs in the heart. But science as a field of study is not designed to provide certainty about positive conclusions. If there are many positive findings on the same subject, then this information is considered to be supportive of a theory, rather than absolute proof that this theory fully explains the subject in ways that cannot be challenged. Even about something as basic as gravity, science has not reached settled conclusions; the theories that have vied to explain its operation – notably, those of Isaac Newton and then of Albert Einstein – have more recently been joined by those of quantum field theory.

At the level of quantum field theory, the process of formulating and testing hypotheses is an almost purely theoretical one: hypotheses are tested through advanced mathematical calculation rather than in the physical world. This kind of approach often only yields practical results many years later, when other researchers have figured out how the theory might be applied in specific real-world circumstances. A joke told by Ewan Birney, associate director of the European Bioinformatics Institute, might illustrate this point:

> A group of wealthy investors wanted to be able to predict the outcome of a horse race. So they hired a group of biologists, a group of statisticians, and a group of physicists. Each group was given a year to research the issue. After one year, the groups all reported to the investors. The biologists said that they could genetically engineer an unbeatable racehorse, but it would take

200 years and $100bn. The statisticians reported next. They said that they could predict the outcome of any race, at a cost of $100m per race, and they would only be right 10% of the time. Finally, the physicists reported that they could also predict the outcome of any race, and that their process was cheap and simple. The investors listened eagerly to this proposal. The head physicist reported, "We have made several simplifying assumptions: first, let each horse be a perfect rolling sphere ..." ("Scientists Tell Us")

Birney suggests, "This is really the joke form of 'all models are wrong, some models are useful.'" When it comes to science, no concepts ever describe the world completely or with complete certainty, but good concepts can nonetheless produce functional knowledge and useful results. It is good to keep such cautions in mind as one encounters written descriptions of scientific ideas and experiments.

Luckily for most readers, many areas of scientific practice are less rarified than theoretical physics, and most scientific hypotheses may be tested by experimentation. As well, most scientific inquiry is not revolutionary in character; dramatic new developments are few, and far between. Instead, scientific inquiry tends to proceed incrementally, with only modest advancements the norm. Ideally, such inquiry is also supposed to be concerned with repeating (or trying to repeat) work first done by others. This last practice – replication – is vital to scientific inquiry. A great deal of the tedious detail that is included in scientific research papers is there to give other scientists enough information to replicate the experiment. A new scientific discovery should gain widespread support only once repetition of the experiment(s) produces the same results.

It is also useful to remember that the scientific process is a relatively recent innovation, rooted in the work of Francis Bacon, René Descartes, and other thinkers of the seventeenth century. It was not until the nineteenth century that the ways of science began to underpin a general understanding of how things work. The twentieth century, though, was one of widespread confidence that the scientific process was the best way to organize our perceptions of the world – and that science could and would solve most of the world's problems. In the twenty-first century, individuals may be a little more cautious in those beliefs, but a sense of confidence in the objectivity of the scientific process still underpins our relationship to the world.

WRITING ABOUT SCIENCE: BIASES, COMPLEXITY, AND ACCESS

Writing about science in the twenty-first century is inherently difficult in a number of ways. Many scientific topics are so complex that it is very difficult for scientists themselves, and more so for intermediaries such as journalists, to accurately report the results of experiments and studies to a general audience. Oversimplification and distortion often occur at this stage.

But this is only one of the ways in which accurate and ethical writing about science is difficult to achieve. One other has already been mentioned in passing: the bias toward dramatic discoveries. This bias may be most evident in the way scientific activity is reported, – or, more specifically, in the selection of what actually gets reported. Whereas a claim that a new drug might cure Alzheimer's disease or some form of cancer is likely to be headlined, subsequent studies in which researchers find they are, in fact, unable to replicate these results are likely to receive much less coverage (if, indeed, such reports are considered newsworthy at all).

This particular bias is not confined to reporting on scientific research in newspapers and blogs; it is also endemic to both science research and science publishing. There is a strong research and publication bias against negative studies and negative results. At its most flagrant, this kind of bias leads to researchers themselves suppressing results that don't support their hypotheses – a practice of which there are more examples than one would like to think. A slightly less obvious, but equally damaging trend is the pervasive publication bias toward positive results. Statistically speaking, negative results tend to be more reliable than positive ones.[a] Yet researchers are encouraged in many ways to aim for new discoveries rather than testing others' results. For one thing, scientific journals are still far more interested in publishing positive results (or initial results that support a study's hypothesis) than in publishing anything in which researchers repeat a study and find the same result(s). Studies that show mixed results or that simply do not support their stated hypotheses have routinely not been published at all.[b] It's hard to ignore the ways in which such practices undermine real scientific understanding.

In recent years, a number of scandals relating to publication bias and to skewed research results have heightened awareness of these problems, and some journals now deliberately publish more negative results than in the past (or have stated that they intend to do so). Registers for human trials have also been established in a number of countries so that negative trial results don't get entirely lost, even if their authors are not immediately able to find a publisher for them. The registers are often hard to work with, however; it

---

a   If you find this statement unlikely or troubling, you may be interested in a long article in *The Economist*. "Trouble at the Lab" covers contemporary issues regarding scientific research; along the way, the authors set out the reasons why research that identifies positive results tends to include many more false results than research that reports negative results does.

b   For a discussion of the damaging effects of this trend, see Alok Jha's "False Positives."

is frequently difficult to find a study's results. More broadly, the suppression of negative results continues to have serious consequences. Perhaps most notably, the suppression of negative clinical trial data for new drugs has led to some truly nasty experiences for humans taking commercially produced drugs. (You can read about a few well-known examples in this anthology, and many further examples exist – enough to provide considerable support to those who tout the advantages of older, cheaper drugs, which have proven histories and better-documented side-effects.)

Mistakes and misunderstandings can, of course, occur in all forms of science and in all forms of science writing; not all of them result from positive findings or publication biases. As noted, the results of almost any scientific study can be misunderstood in the larger community context, or elements can be misreported. Standard study statements about the caution required in interpreting results ("These results should be interpreted with caution, as the study size is smaller than optimal," "More research is necessary in this area," etc.) are frequently misrepresented, misunderstood, or ignored in the press or by hostile parties. And sometimes the existence of vigorous disagreement over a scientific issue is taken as evidence that it is not possible to come to any sound conclusions. Climate change studies are a good example here: even though there is a near-universal consensus among scientists that global warming is occurring and that humans are responsible for causing a substantial part of that warming, the presence of a tiny minority within the scientific community (and a much more vocal minority outside that community) who argue that "the jury is still out" on global warming has left some in the media more concerned about being "even-handed" than about reporting collective agreement. This approach relies on a false assertion of objectivity that the philosopher Thomas Nagel has called "the view from nowhere" (also the title of his 1986 book).

Another set of issues relating to scientific research and scientific writing concerns increased levels of competitiveness. Whereas modern academic research was in the 1950s still "a rarified pastime," it has grown to in many ways resemble an industry, and competition has become cut-throat in nature, as "the obligation to 'publish or perish' has come to rule" ("How Science Goes Wrong" 13). Strongly related to this increase in competitiveness has been a tremendous increase in the commercialization of research. In the early twentieth century, almost all reputable scientific and academic journals were published by various university groups, and the material in them was generally written by people employed by universities, often public universities. These journals would then be purchased by individuals and by university libraries for generally reasonable amounts (though they were certainly not inexpensive). The journal articles were indexed by academic groups that held an interest in the specific subject area, and these indexes were published and accessed in hard copy, usually at a university library. While a person needed to belong to a university to borrow journal issues, anyone could read these issues in the library itself.

In the middle decades of the twentieth century, however, that model began to change. Increasingly, publishing rights to scientific journals were acquired by private companies such as Springer-Verlag, Pergamon, and Elsevier – and the journals became increasingly more expensive, to the point where a year's subscription to an academic journal could cost many thousands of dollars. With the advent of the internet, prices became even higher, and restrictions on access became tighter. Starting with the dominance of the graphic-interface version of the internet (Web 2.0), commercial consortiums began to bundle together electronic libraries of scientific journals – and to charge libraries millions of dollars each to purchase access to these bundles of documents. As a result, access to the digital articles (and many journals came to be available only in digital form) became restricted to those who had university and college library accounts (or company subscriptions), or those able to pay large fees for often time-limited access to single articles. Given that certain of the journals in these exorbitantly expensive bundles were important for university libraries, funds available for the purchase of new journals in the sciences (or, for that matter, in the humanities and social sciences) shrank to ever smaller levels.

All of this led, in the very late twentieth and early twenty-first centuries, to a pushback from academics, many of whom objected to seeing their work (which had often been paid for out of public funds) sequestered behind a paywall. The arXiv site (which opened in August 1991) may have been the earliest site to offer free access to scientific research. Numerous other sites have followed, as an Open Access movement, enthusiastically championed by scientists, has gathered force; by 2000, there were 744 open-access scholarly journals, and by 2011 there were 6,714 (Laakso and Björk). The flagship set of open-access journals at PLOS (the Public Library of Science) is deserving of special mention; other important sites include the PubMed Central database, ResearchGate, and the Social Science Research Network.

The open access movement continues to gather momentum. In early December of 2013, for example, Randy Schekman, a moving force behind the open-access scientific journal *eLife*, announced that he would be boycotting a number of the best-known scientific journals – among them *Nature* and *Science*. Schekman described these "high impact" journals as resembling "fashion designers who create limited-edition handbags. They know that scarcity stokes demand." Such colorful criticism from a scientist who was, that same week, awarded the Nobel Prize for Physiology or Medicine suggests that open access publishing and open access advocacy have become important avenues of publication in a comparatively short period.

The open-access model has not been without its problems – not least of which has been the ways various elements keep trying to monetize it. Some publishing entities have sensed a golden opportunity in this model and have created journals that are not in fact genuine academic journals. Such "predatory publishers" will publish your article for $3,000 or more in a publication that may look like a scholarly journal but does not follow normal scholarly practices of peer review, does not have reputable editors, and does not have any academic recognition; they are, in fact, only there to take in money.[a] In other cases, seemingly scholarly journals may not charge the author a publication fee, but may instead be funded entirely by industry sources, an approach which, at the very least, encourages scientists to self-censor their research results. Even well-known and otherwise reputable publishers may have some publications in this category among their large lists of journals.[b] It is thus more important than ever to pay attention to the source of the research one reads about in a scientific journal – or in journalistic reports of scientific research. As a researcher seeking publication, you now have to investigate every journal you don't already know is rock-solid, especially since you usually can only publish an article's material once. As a reader, you must also be aware of these publication practices when judging the reliability of science writing. And as someone who relies on the application of science in the world around you, you are affected by these practices in other ways too; the sources of research, the ways in which research has been reviewed before publication, and the ways in which research and the publication of research have been funded all play a part in shaping the ideas you encounter; the medicine you consume; the buildings in which you live, study, or work; and the policies of your country's government.

WORKS CITED

ArXiv. Cornell University Library, 1991. Web. 2 Dec. 2013.

Beall, Jeffrey. "Beall's List of Predatory Publishers 2013." *Scholarly Open Access*. Wordpress, 4 Dec. 2012. Web. 2 Dec. 2013.

——. "Predatory Publishing." *Scientist*. LabX Media Group, 1 Aug. 2012. Web. 2 Dec. 2013.

Grant, Bob. "Elsevier Published 6 Fake Journals." *Scientist*. LabX Media Group, 7 May 2009. Web. 2 Dec. 2012.

---

a   For further information on the dangers of predatory publishing, see Jeffrey Beall's "Predatory Publishing" and "Beall's List of Predatory Publishers."

b   See, for example, Bob Grant's "Elsevier Published 6 Fake Journals."

"How Science Goes Wrong." *Economist*. Economist, 19 Oct. 2013. Web. 3 Jan. 2014.

Jha, Alok. "False Positives: Fraud and Misconduct Are Threatening Scientific Research." *Guardian*. Guardian, 13 Sept, 2012. Web. 2 Dec. 2013.

Laakso, Mikael, and Bo-Christer Björk. "Anatomy of Open Access Publishing: A Study of Longitudinal Development and Internal Structure." *BMC Medicine* 10.124 (2012): n. pag. Web. 2 Dec. 2012.

Nagel, Thomas. *The View from Nowhere*. New York: OUP, 1986. Print.

Schekman, Randy. "How Journals Like *Nature*, *Cell* and *Science* are Damaging Science." *Guardian*. Guardian, 9 Dec. 2013. Web. 3 Jan. 2014.

"Scientists Tell Us Their Favourite Jokes: 'An Electron and a Positron Walk into a Bar ...'" *Observer*. Guardian, 29 Dec. 2013. Web. 3 Jan. 2014.

"Trouble at the Lab." *Economist*. Economist, 19 Oct. 2013. Web. 3 Jan. 2013.

Troyka, Lynn Quitman, and Douglas Hesse. *Quick Access: Reference for Writers*. 3rd Canadian ed. Toronto: Pearson, 2007. Print.

A Note on the Texts

We have endeavored to reproduce the selections in this anthology as they originally appeared, apart from the silent correction of obvious spelling or typographical errors; the articles have, however, been formatted for consistency within the volume. In the case of excerpted selections, omissions are marked with ellipses. Except where otherwise indicated, all references have been included.

Please note that the explanatory footnotes did not appear in the original articles; the editors have added them where we felt information would be useful to students. Any endnotes or references that appear at the end of an article were written by that article's original authors. Our editors' explanatory footnotes are indicated with superscript letters, while original authors' endnotes are indicated with superscript numbers.

Acknowledgments

We would like to acknowledge the contributions of a number of people to the creation of this book. Our thanks go to Alysia Miller, Christine Bugler, and Luke McLeod for their assistance in the preparation of text and explanatory notes; thanks also to Alan McGreevy for helpful comments on matters of pedagogy and scientific accuracy, and to Martin Boyne for his keen editorial eye. We are grateful to Eileen Eckert for her excellent work in designing and typesetting the book, and to Maureen Okun for her contribution of the citation and documentation material, which formed the basis for our "How to Use Sources" appendix. Additionally, we would particularly like to thank those who gave us valuable advice as this book took shape: Gisèle M. Baxter, University of British Columbia; Christopher Bloss, Auburn University; Kathleen Gonso, Northeastern University; Heather Graves, University of Alberta; Adam Lawrence, Cape Breton University; Olga Menagarishvili, Georgia Institute of Technology; Cecilia Musselman, Northeastern University; and Carol Senf, Georgia Institute of Technology.

Catherine Nelson-McDermott also wishes to thank her many keen composition students for the fun and hard work over the years. You (and future students) are the reason this book exists.

And, Dad, this one's for you.

I

Human Geology:
Reading Planet Earth

one

JARED DIAMOND

Easter's End

This 1995 article, written by popular science writer Jared Diamond for Discover *magazine, outlines recent scientific discoveries regarding the history of Easter Island. The introduction to the article poses the following question: "In just a few centuries, the people of Easter Island wiped out their forest, drove their plants and animals to extinction, and saw their complex society spiral into chaos and cannibalism. Are we about to follow their lead?"*

☙

Among the most riveting mysteries of human history are those posed by vanished civilizations. Everyone who has seen the abandoned buildings of the Khmer, the Maya, or the Anasazi[a] is immediately moved to ask the same question: Why did the societies that erected those structures disappear?

Their vanishing touches us as the disappearance of other animals, even the dinosaurs, never can. No matter how exotic those lost civilizations seem, their framers were humans like us. Who is to say we won't succumb to the same fate? Perhaps someday New York's skyscrapers will stand derelict and overgrown with vegetation, like the temples at Angkor Wat and Tikal.[b]

Among all such vanished civilizations, that of the former Polynesian society on Easter Island remains unsurpassed in mystery and isolation. The mystery stems especially from the island's gigantic stone statues and its impoverished landscape, but it is enhanced by our associations with the specific people involved: Polynesians represent for us the ultimate in exotic romance, the background for many a child's, and an adult's, vision of para-

---

a *Khmer* Empire based in what is now Cambodia. Its capital in Angkor was abandoned in the fifteenth century; *Maya* Civilization in what is now southern Mexico, Guatemala, and Belize. Its major cities were abandoned by the end of the ninth century; *Anasazi* Ancient Pueblo peoples of what is now the southwestern United States. They abandoned their settlements in the twelfth and thirteenth centuries.

b *Angkor Wat* Complex of temples built by the Khmer in Angkor; *Tikal* Ruined Mayan city in Guatemala.

25

dise. My own interest in Easter was kindled over 30 years ago when I read Thor Heyerdahl's[a] fabulous accounts of his Kon-Tiki voyage.

But my interest has been revived recently by a much more exciting account, one not of heroic voyages but of painstaking research and analysis. My friend David Steadman, a paleontologist, has been working with a number of other researchers who are carrying out the first systematic excavations on Easter intended to identify the animals and plants that once lived there. Their work is contributing to a new interpretation of the island's history that makes it a tale not only of wonder but of warning as well.

5   Easter Island, with an area of only 64 square miles, is the world's most isolated scrap of habitable land. It lies in the Pacific Ocean more than 2,000 miles west of the nearest continent (South America), 1,400 miles from even the nearest habitable island (Pitcairn). Its subtropical location and latitude – at 27 degrees south, it is approximately as far below the equator as Houston is north of it – help give it a rather mild climate, while its volcanic origins make its soil fertile. In theory, this combination of blessings should have made Easter a miniature paradise, remote from problems that beset the rest of the world.

The island derives its name from its "discovery" by the Dutch explorer Jacob Roggeveen, on Easter (April 5) in 1722. Roggeveen's first impression was not of a paradise but of a wasteland: "We originally, from a further distance, have considered the said Easter Island as sandy; the reason for that is this, that we counted as sand the withered grass, hay, or other scorched and burnt vegetation, because its wasted appearance could give no other impression than of a singular poverty and barrenness."

The island Roggeveen saw was a grassland without a single tree or bush over ten feet high. Modern botanists have identified only 47 species of higher plants native to Easter, most of them grasses, sedges, and ferns. The list includes just two species of small trees and two of woody shrubs. With such flora, the islanders Roggeveen encountered had no source of real firewood to warm themselves during Easter's cool, wet, windy winters. Their native animals included nothing larger than insects, not even a single species of native bat, land bird, land snail, or lizard. For domestic animals, they had only chickens.

European visitors throughout the eighteenth and early nineteenth centuries estimated Easter's human population at about 2,000, a modest number considering the island's fertility. As Captain James Cook recognized during

---

a   *Thor Heyerdahl* Norwegian anthropologist famous for conducting water voyages in reconstructed historical vessels; his book *Kon-Tiki* (1950) recounts a trip by raft from Peru to Polynesia.

his brief visit in 1774, the islanders were Polynesians (a Tahitian man accompanying Cook was able to converse with them). Yet despite the Polynesians' well-deserved fame as a great seafaring people, the Easter Islanders who came out to Roggeveen's and Cook's ships did so by swimming or paddling canoes that Roggeveen described as "bad and frail." Their craft, he wrote, were "put together with manifold small planks and light inner timbers, which they cleverly stitched together with very fine twisted threads.... But as they lack the knowledge and particularly the materials for caulking and making tight the great number of seams of the canoes, these are accordingly very leaky, for which reason they are compelled to spend half the time in bailing." The canoes, only ten feet long, held at most two people, and only three or four canoes were observed on the entire island.

With such flimsy craft, Polynesians could never have colonized Easter from even the nearest island, nor could they have traveled far offshore to fish. The islanders Roggeveen met were totally isolated, unaware that other people existed. Investigators in all the years since his visit have discovered no trace of the islanders' having any outside contacts: not a single Easter Island rock or product has turned up elsewhere, nor has anything been found on the island that could have been brought by anyone other than the original settlers or the Europeans. Yet the people living on Easter claimed memories of visiting the uninhabited Sala y Gomez reef 260 miles away, far beyond the range of the leaky canoes seen by Roggeveen. How did the islanders' ancestors reach that reef from Easter, or reach Easter from anywhere else?

Easter Island's most famous feature is its huge stone statues, more than 10 200 of which once stood on massive stone platforms lining the coast. At least 700 more, in all stages of completion, were abandoned in quarries or on ancient roads between the quarries and the coast, as if the carvers and moving crews had thrown down their tools and walked off the job. Most of the erected statues were carved in a single quarry and then somehow transported as far as six miles – despite heights as great as 33 feet and weights up to 82 tons. The abandoned statues, meanwhile, were as much as 65 feet tall and weighed up to 270 tons. The stone platforms were equally gigantic: up to 500 feet long and 10 feet high, with facing slabs weighing up to 10 tons.

Roggeveen himself quickly recognized the problem the statues posed: "The stone images at first caused us to be struck with astonishment," he wrote, "because we could not comprehend how it was possible that these people, who are devoid of heavy thick timber for making any machines, as well as strong ropes, nevertheless had been able to erect such images." Roggeveen might have added that the islanders had no wheels, no draft animals, and no source of power except their own muscles. How did they transport the giant statues for miles, even before erecting them? To deepen the mystery, the statues were still standing in 1770, but by 1864 all of them had been pulled down, by the islanders themselves. Why then did they carve them in the first place? And why did they stop?

The statues imply a society very different from the one Roggeveen saw in 1722. Their sheer number and size suggest a population much larger than 2,000 people. What became of everyone? Furthermore, that society must have been highly organized. Easter's resources were scattered across the island: the best stone for the statues was quarried at Rano Raraku near Easter's northeast end; red stone, used for large crowns adorning some of the statues, was quarried at Puna Pau, inland in the southwest; stone carving tools came mostly from Aroi in the northwest. Meanwhile, the best farmland lay in the south and east, and the best fishing grounds on the north and west coasts. Extracting and redistributing all those goods required complex political organization. What happened to that organization, and how could it ever have arisen in such a barren landscape?

Easter Island's mysteries have spawned volumes of speculation for more than two and a half centuries. Many Europeans were incredulous that Polynesians – commonly characterized as "mere savages" – could have created the statues or the beautifully constructed stone platforms. In the 1950s, Heyerdahl argued that Polynesia must have been settled by advanced societies of American Indians, who in turn must have received civilization across the Atlantic from more advanced societies of the Old World. Heyerdahl's raft voyages aimed to prove the feasibility of such prehistoric transoceanic contacts. In the 1960s the Swiss writer Erich von Däniken, an ardent believer in Earth visits by extraterrestrial astronauts, went further, claiming that Easter's statues were the work of intelligent beings who owned ultramodern tools, became stranded on Easter, and were finally rescued.

Heyerdahl and von Däniken both brushed aside overwhelming evidence that the Easter Islanders were typical Polynesians derived from Asia rather than from the Americas and that their culture (including their statues) grew out of Polynesian culture. Their language was Polynesian, as Cook had already concluded. Specifically, they spoke an eastern Polynesian dialect related to Hawaiian and Marquesan, a dialect isolated since about A.D. 400, as estimated from slight differences in vocabulary. Their fishhooks and stone adzes resembled early Marquesan models. Last year[a] DNA extracted from 12 Easter Island skeletons was also shown to be Polynesian. The islanders grew bananas, taro, sweet potatoes, sugarcane, and paper mulberry – typical Poly-

---

a *Last year* I.e., in 1994.

nesian crops, mostly of Southeast Asian origin. Their sole domestic animal, the chicken, was also typically Polynesian and ultimately Asian, as were the rats that arrived as stowaways in the canoes of the first settlers.

What happened to those settlers? The fanciful theories of the past must give   15
way to evidence gathered by hardworking practitioners in three fields: archeology, pollen analysis, and paleontology.

Modern archeological excavations on Easter have continued since Heyerdahl's 1955 expedition. The earliest radiocarbon dates[a] associated with human activities are around A.D. 400 to 700, in reasonable agreement with the approximate settlement date of 400 estimated by linguists. The period of statue construction peaked around 1200 to 1500, with few if any statues erected thereafter. Densities of archeological sites suggest a large population; an estimate of 7,000 people is widely quoted by archeologists, but other estimates range up to 20,000, which does not seem implausible for an island of Easter's area and fertility.

Archeologists have also enlisted surviving islanders in experiments aimed at figuring out how the statues might have been carved and erected. Twenty people, using only stone chisels, could have carved even the largest completed statue within a year. Given enough timber and fiber for making ropes, teams of at most a few hundred people could have loaded the statues onto wooden sleds, dragged them over lubricated wooden tracks or rollers, and used logs as levers to maneuver them into a standing position. Rope could have been made from the fiber of a small native tree, related to the linden, called the hauhau. However, that tree is now extremely scarce on Easter, and hauling one statue would have required hundreds of yards of rope. Did Easter's now barren landscape once support the necessary trees?

That question can be answered by the technique of pollen analysis, which involves boring out a column of sediment from a swamp or pond, with the most recent deposits at the top and relatively more ancient deposits at the bottom. The absolute age of each layer can be dated by radiocarbon methods. Then begins the hard work: examining tens of thousands of pollen grains under a microscope, counting them, and identifying the plant species that produced each one by comparing the grains with modern pollen from known plant species. For Easter Island, the bleary-eyed scientists who performed that task were John Flenley, now at Massey University in New Zealand, and Sarah King of the University of Hull in England.

---

a   *radiocarbon dates*   Estimates of the age of an object or material, as measured by comparing the relative proportions of two isotopes (stable carbon-12 and unstable carbon-14) that it contains. The ratio between the isotopes changes as the carbon-14 steadily decays.

Flenley and King's heroic efforts were rewarded by the striking new picture that emerged of Easter's prehistoric landscape. For at least 30,000 years before human arrival and during the early years of Polynesian settlement, Easter was not a wasteland at all. Instead, a subtropical forest of trees and woody bushes towered over a ground layer of shrubs, herbs, ferns, and grasses. In the forest grew tree daisies, the rope-yielding hauhau tree, and the toromiro tree, which furnishes a dense, mesquite-like firewood. The most common tree in the forest was a species of palm now absent on Easter but formerly so abundant that the bottom strata[a] of the sediment column were packed with its pollen. The Easter Island palm was closely related to the still-surviving Chilean wine palm, which grows up to 82 feet tall and 6 feet in diameter. The tall, un-branched trunks of the Easter Island palm would have been ideal for transporting and erecting statues and constructing large canoes. The palm would also have been a valuable food source, since its Chilean relative yields edible nuts as well as sap from which Chileans make sugar, syrup, honey, and wine.

20   What did the first settlers of Easter Island eat when they were not glutting themselves on the local equivalent of maple syrup? Recent excavations by David Steadman, of the New York State Museum at Albany, have yielded a picture of Easter's original animal world as surprising as Flenley and King's picture of its plant world. Steadman's expectations for Easter were conditioned by his experiences elsewhere in Polynesia, where fish are overwhelmingly the main food at archeological sites, typically accounting for more than 90 percent of the bones in ancient Polynesian garbage heaps. Easter, though, is too cool for the coral reefs beloved by fish, and its cliff-girded coastline permits shallow-water fishing in only a few places. Less than a quarter of the bones in its early garbage heaps (from the period 900 to 1300) belonged to fish; instead, nearly one-third of all bones came from porpoises.

Nowhere else in Polynesia do porpoises account for even 1 percent of discarded food bones. But most other Polynesian islands offered animal food in the form of birds and mammals, such as New Zealand's now extinct giant moas and Hawaii's now extinct flightless geese. Most other islanders also had domestic pigs and dogs. On Easter, porpoises would have been the largest animal available – other than humans. The porpoise species identified at Easter, the common dolphin, weighs up to 165 pounds. It generally lives out at sea, so it could not have been hunted by line fishing or spearfishing from shore. Instead, it must have been harpooned far offshore, in big seaworthy canoes built from the extinct palm tree.

---

a   *strata*   I.e., layers of sediment with qualities that allow differentiation between each layer.

In addition to porpoise meat, Steadman found, the early Polynesian settlers were feasting on seabirds. For those birds, Easter's remoteness and lack of predators made it an ideal haven as a breeding site, at least until humans arrived. Among the prodigious numbers of seabirds that bred on Easter were albatross, boobies, frigate birds, fulmars, petrels, prions, shearwaters, storm petrels, terns, and tropic birds. With at least 25 nesting species, Easter was the richest seabird breeding site in Polynesia and probably in the whole Pacific.

Land birds as well went into early Easter Island cooking pots. Steadman identified bones of at least six species, including barn owls, herons, parrots, and rail. Bird stew would have been seasoned with meat from large numbers of rats, which the Polynesian colonists inadvertently brought with them; Easter Island is the sole known Polynesian island where rat bones outnumber fish bones at archeological sites. (In case you're squeamish and consider rats inedible, I still recall recipes for creamed laboratory rat that my British biologist friends used to supplement their diet during their years of wartime food rationing.)

Porpoises, seabirds, land birds, and rats did not complete the list of meat sources formerly available on Easter. A few bones hint at the possibility of breeding seal colonies as well. All these delicacies were cooked in ovens fired by wood from the island's forests.

Such evidence lets us imagine the island onto which Easter's first Polynesian 25 colonists stepped ashore some 1,600 years ago, after a long canoe voyage from eastern Polynesia. They found themselves in a pristine paradise. What then happened to it? The pollen grains and the bones yield a grim answer.

Pollen records show that destruction of Easter's forests was well under way by the year 800, just a few centuries after the start of human settlement. Then charcoal from wood fires came to fill the sediment cores, while pollen of palms and other trees and woody shrubs decreased or disappeared, and pollen of the grasses that replaced the forest became more abundant. Not long after 1400 the palm finally became extinct, not only as a result of being chopped down but also because the now ubiquitous rats prevented its regeneration: of the dozens of preserved palm nuts discovered in caves on Easter, all had been chewed by rats and could no longer germinate. While the hauhau tree did not become extinct in Polynesian times, its numbers declined drastically until there weren't enough left to make ropes from. By the time Heyerdahl visited Easter, only a single, nearly dead toromiro tree remained on the island, and even that lone survivor has now disappeared. (Fortunately, the toromiro still grows in botanical gardens elsewhere.)

The fifteenth century marked the end not only for Easter's palm but for the forest itself. Its doom had been approaching as people cleared land to plant gardens; as they felled trees to build canoes, to transport and erect statues, and to burn; as rats devoured seeds; and probably as the native birds died out that had pollinated the trees' flowers and dispersed their fruit. The overall picture is among the most extreme examples of forest destruction anywhere in the world: the whole forest gone, and most of its tree species extinct.

The destruction of the island's animals was as extreme as that of the forest: without exception, every species of native land bird became extinct. Even shellfish were overexploited, until people had to settle for small sea snails instead of larger cowries. Porpoise bones disappeared abruptly from garbage heaps around 1500; no one could harpoon porpoises anymore, since the trees used for constructing the big seagoing canoes no longer existed. The colonies of more than half of the seabird species breeding on Easter or on its offshore islets were wiped out.

In place of these meat supplies, the Easter Islanders intensified their production of chickens, which had been only an occasional food item. They also turned to the largest remaining meat source available: humans, whose bones became common in late Easter Island garbage heaps. Oral traditions of the islanders are rife with cannibalism; the most inflammatory taunt that could be snarled at an enemy was "The flesh of your mother sticks between my teeth." With no wood available to cook these new goodies, the islanders resorted to sugarcane scraps, grass, and sedges to fuel their fires.

30   All these strands of evidence can be wound into a coherent narrative of a society's decline and fall. The first Polynesian colonists found themselves on an island with fertile soil, abundant food, bountiful building materials, ample *lebensraum* and all the prerequisites for comfortable living. They prospered and multiplied.

After a few centuries, they began erecting stone statues on platforms, like the ones their Polynesian forebears had carved. With passing years, the statues and platforms became larger and larger, and the statues began sporting ten-ton red crowns – probably in an escalating spiral of one-upmanship, as rival clans tried to surpass each other with shows of wealth and power. (In the same way, successive Egyptian pharaohs built ever-larger pyramids. Today Hollywood movie moguls near my home in Los Angeles are displaying their wealth and power by building ever more ostentatious mansions. Tycoon Marvin Davis topped previous moguls with plans for a 50,000-square-foot house, so now Aaron Spelling has topped Davis with a 56,000-square-foot house. All that those buildings lack to make the message explicit are ten-ton red crowns.) On Easter, as in modern America, society was held together by a complex political system to redistribute locally available resources and to integrate the economies of different areas.

Eventually Easter's growing population was cutting the forest more rapidly than the forest was regenerating. The people used the land for gardens and the wood for fuel, canoes, and houses – and, of course, for lugging statues. As forest disappeared, the islanders ran out of timber and rope to transport and erect their statues. Life became more uncomfortable – springs and streams dried up, and wood was no longer available for fires.

People also found it harder to fill their stomachs, as land birds, large sea snails, and many seabirds disappeared. Because timber for building seagoing canoes vanished, fish catches declined and porpoises disappeared from the table. Crop yields also declined, since deforestation allowed the soil to be eroded by rain and wind, dried by the sun, and its nutrients to be leeched from it. Intensified chicken production and cannibalism replaced only part of all those lost foods. Preserved statuettes with sunken cheeks and visible ribs suggest that people were starving.

With the disappearance of food surpluses, Easter Island could no longer feed the chiefs, bureaucrats, and priests who had kept a complex society running. Surviving islanders described to early European visitors how local chaos replaced centralized government and a warrior class took over from the hereditary chiefs. The stone points of spears and daggers, made by the warriors during their heyday in the 1600s and 1700s, still litter the ground of Easter today. By around 1700, the population began to crash toward between one-quarter and one-tenth of its former number. People took to living in caves for protection against their enemies. Around 1770 rival clans started to topple each other's statues, breaking the heads off. By 1864 the last statue had been thrown down and desecrated.

As we try to imagine the decline of Easter's civilization, we ask ourselves, "Why didn't they look around, realize what they were doing, and stop before it was too late? What were they thinking when they cut down the last palm tree?"

I suspect, though, that the disaster happened not with a bang but with a whimper. After all, there are those hundreds of abandoned statues to consider. The forest the islanders depended on for rollers and rope didn't simply disappear one day – it vanished slowly, over decades. Perhaps war interrupted the moving teams; perhaps by the time the carvers had finished their work, the last rope snapped. In the meantime, any islander who tried to warn about the dangers of progressive deforestation would have been overridden by vested interests of carvers, bureaucrats, and chiefs, whose jobs depended on continued deforestation. Our Pacific Northwest loggers are only the latest in a long line of loggers to cry, "Jobs over trees!" The changes in forest cover from year to year would have been hard to detect: yes, this year we cleared those woods over there, but trees are starting to grow back again on this abandoned garden site here. Only older people, recollecting their childhoods decades earlier, could have recognized a difference. Their children could no more have comprehended their parents' tales than my eight-year-old sons today can comprehend my wife's and my tales of what Los Angeles was like 30 years ago.

Gradually trees became fewer, smaller, and less important. By the time the last fruit-bearing adult palm tree was cut, palms had long since ceased to be of economic significance. That left only smaller and smaller palm saplings to clear each year, along with other bushes and treelets. No one would have noticed the felling of the last small palm.

By now the meaning of Easter Island for us should be chillingly obvious. Easter Island is Earth writ small. Today, again, a rising population confronts shrinking resources. We too have no emigration valve, because all human societies are linked by international transport, and we can no more escape into space than the Easter Islanders could flee into the ocean. If we continue to follow our present course, we shall have exhausted the world's major fisheries, tropical rain forests, fossil fuels, and much of our soil by the time my sons reach my current age.

Every day newspapers report details of famished countries – Afghanistan, Liberia, Rwanda, Sierra Leone, Somalia, the former Yugoslavia, Zaire – where soldiers have appropriated the wealth or where central government is yielding to local gangs of thugs. With the risk of nuclear war receding, the threat of our ending with a bang no longer has a chance of galvanizing us to halt our course. Our risk now is of winding down, slowly, in a whimper. Corrective action is blocked by vested interests, by well-intentioned political and business leaders, and by their electorates, all of whom are perfectly correct in not noticing big changes from year to year. Instead, each year there are just somewhat more people, and somewhat fewer resources, on Earth.

40   It would be easy to close our eyes or to give up in despair. If mere thousands of Easter Islanders with only stone tools and their own muscle power sufficed to destroy their society, how can billions of people with metal tools and machine power fail to do worse? But there is one crucial difference. The Easter Islanders had no books and no histories of other doomed societies. Unlike the Easter Islanders, we have histories of the past – information that can save us. My main hope for my sons' generation is that we may now choose to learn from the fates of societies like Easter's.

(1995)

Questions:

1. Who is the article's intended reader? Is it likely to be a Polynesian person? How can you tell?

2. Do you find the article expresses more support for some theories about Easter Island's history than for others? Look particularly at the discussions of the ideas of Jacob Roggeveen, Thor Heyerdahl, and Erich von Däniken, and compare these to the discussions of modern scientific exploration of the island.

3. Did you know who Roggeveen, Heyerdahl, and von Däniken were before reading this article? How does the article integrate background information about these people without losing your interest?

4. How did the Easter Islanders' diet change over the centuries, and how was this related to changes in their environment as revealed by pollen analysis? Support your answer with specific evidence from the article.

5. How does the article demonstrate the Easter Islanders' Polynesian ancestry? Do you find this evidence persuasive? What elements are particularly convincing or unconvincing?

6. Consider the evidence found by modern scientists. What, if any, time-specific scientific evidence matches the observations made by Roggeveen and/or Cook? What, if any, time-specific scientific evidence is contradicted by either explorer's observations?

7. Find a discussion of exploration of the Pacific Islands as carried out by either Cook or Roggeveen. What attitudes, biases, or misinformation did these explorers hold that might have distorted their understanding of the people and environments they encountered?

8. The article does not include formal citations or footnotes. Why do you think this is? How does this absence affect your experience as a reader? What authority-claim strategies does the article use instead of formal references?

9. What does "*lebensraum*" mean? Why do you think the article uses the word without providing an explicit explanation?

10. According to the article, why do the statues on Easter Island suggest a high level of social organization?

11. What methods does the article use to make this material seem pertinent to you as a contemporary reader?

two

Luis W. Alvarez, Walter Alvarez, Frank Asaro, and Helen V. Michel

from Extraterrestrial Cause for the Cretaceous-Tertiary Extinction

This article, originally published in Science, *led to a breakthrough concerning the cause of the great extinction that eliminated the dinosaurs sixty-five million years ago. The article posits the hypothesis, now well known but extremely controversial at the time, that an asteroid impact led to the extinction.*

❧

Summary

Platinum metals are depleted in the earth's crust relative to their cosmic abundance; concentrations of these elements in deep-sea sediments may thus indicate influxes of extraterrestrial material. Deep-sea limestones exposed in Italy, Denmark, and New Zealand show iridium increases of about 30, 160, and 20 times, respectively, above the background level at precisely the time of the Cretaceous-Tertiary extinctions, 65 million years ago. Reasons are given to indicate that this iridium is of extraterrestrial origin, but did not come from a nearby supernova. A hypothesis is suggested which accounts for the extinctions and the iridium observations. Impact of a large earth-crossing asteroid would inject about 60 times the object's mass into the atmosphere as pulverized rock; a fraction of this dust would stay in the stratosphere for several years and be distributed worldwide. The resulting darkness would suppress photosynthesis, and the expected biological consequences match quite closely the extinctions observed in the paleontological record. One prediction of this hypothesis has been verified: the chemical composition of the boundary clay, which is thought to come from the stratospheric dust, is markedly different from that of clay mixed with the Cretaceous and Tertiary limestones, which are chemically similar to each other. Four different independent estimates of the diameter of the asteroid give values that lie in the range 10 ± 4 kilometers.

In the 570-million-year period for which abundant fossil remains are available, there have been five great biological crises, during which many groups of organisms died out. The most recent of the great extinctions is used to define the boundary between the Cretaceous and Tertiary periods, about 65 million years ago. At this time, the marine reptiles, the flying reptiles, and both orders of dinosaurs died out,[1] and extinctions occurred at various taxonomic levels among the marine invertebrates. Dramatic extinctions occurred among the microscopic floating animals and plants; both the calcareous planktonic foraminifera and the calcareous nannoplankton[a] were nearly exterminated, with only a few species surviving the crisis. On the other hand, some groups were little affected, including the land plants, crocodiles, snakes, mammals, and many kinds of invertebrates. Russell[2] concludes that about half of the genera living at that time perished during the extinction event.

Many hypotheses have been proposed to explain the Cretaceous-Tertiary (C-T) extinctions[3,4] and two recent meetings on the topic[5,6] produced no sign of a consensus. Suggested causes include gradual or rapid changes in oceanographic, atmospheric, or climatic conditions[7] due to a random[8] or a cyclical[9] coincidence of causative factors; a magnetic reversal;[10] a nearby supernova;[11] and the flooding of the ocean surface by fresh water from a postulated arctic lake.[12]

A major obstacle to determining the cause of the extinction is that virtually all the available information on events at the time of the crisis deals with biological changes seen in the paleontological record and is therefore inherently indirect. Little physical evidence is available, and it also is indirect. This includes variations in stable oxygen and carbon isotopic ratios[b] across the boundary in pelagic sediments,[c] which may reflect changes in temperature, salinity, oxygenation, and organic productivity of the ocean water, and which are not easy to interpret.[13,14] These isotopic changes are not particularly striking and, taken by themselves, would not suggest a dramatic crisis. Small changes in minor and trace element levels at the C-T boundary have been noted from limestone sections in Denmark and Italy,[15] but these data also present interpretational difficulties. It is noteworthy that in pelagic marine sequences, where nearly continuous deposition is to be expected, the C-T boundary is commonly marked by a hiatus.[3,16]

---

a *calcareous planktonic ... nannoplankton* Drifting water organisms with shells or other body structures made of calcium carbonate.

b *isotopic ratios* Ratios between different atom types of the same chemical element; e.g., an oxygen isotopic ratio is the ratio of oxygen with eight neutrons to that with ten neutrons.

c *pelagic sediments* Accumulations of clay and organic matter on the ocean floor.

In this article we present direct physical evidence for an unusual event at exactly the time of the extinctions in the planktonic realm. None of the current hypotheses adequately accounts for this evidence, but we have developed a hypothesis that appears to offer a satisfactory explanation for nearly all the available paleontological and physical evidence.

IDENTIFICATION OF EXTRATERRESTRIAL PLATINUM METALS IN DEEP-SEA SEDIMENTS

5   This study began with the realization that the platinum group elements (platinum, iridium, osmium, and rhodium) are much less abundant in the earth's crust and upper mantle than they are in chondritic meteorites[a] and average solar system material. Depletion of the platinum-group elements in the earth's crust and upper mantle is probably the result of concentration of these elements in the earth's core.

Pettersson and Rotschi[17] and Goldschmidt[18] suggested that the low concentrations of platinum group elements in sedimentary rocks might come largely from meteoritic dust formed by ablation[b] when meteorites passed through the atmosphere. Barker and Anders[19] showed that there was a correlation between sedimentation rate and iridium concentration, confirming the earlier suggestions. Subsequently, the method was used by Ganapathy, Brownlee, and Hodge[20] to demonstrate an extraterrestrial origin for silicate spherules in deep-sea sediments. Sarna-Wojcicki et al.[21] suggested that meteoric dust accumulation in soil layers might enhance the abundance of iridium sufficiently to permit its use as a dating tool. Recently, Crocket and Kuo[22] reported iridium abundances in deep-sea sediments and summarized other previous work.

Considerations of this type[23] prompted us to measure the iridium concentration in the 1-centimeter-thick clay layer that marks the C-T boundary in some sections in the Umbrian Apennines,[c] in the hope of determining the length of time represented by that layer. Iridium can easily be determined at low levels by neutron activation analysis[d] (NAA).[24][...] The other platinum group elements are more difficult to determine by NAA.

---

a   *chondritic meteorites*   Meteorites composed of chondrules, millimeter-sized objects that were originally free-floating in space.

b   *ablation*   I.e., the vaporization of outer layers caused by high temperatures.

c   *Umbrian Apennines*   A mountain range in the Italian Peninsula.

d   *neutron activation analysis*   A process for determining the composition of elements in a material by bombarding samples with neutrons and measuring the resulting radioactive emissions.

ITALIAN STRATIGRAPHIC SECTIONS

Many aspects of earth history are best recorded in pelagic sedimentary rocks, which gradually accumulate in the relatively quiet waters of the deep sea as individual grains settle to the bottom. In the Umbrian Apennines of northern peninsular Italy there are exposures of pelagic sedimentary rocks representing the time from Early Jurassic to Oligocene, around 185 to 30 million years ago.[25] [...]

In well-exposed, complete sections[a] there is a bed of clay about 1 cm thick between the highest Cretaceous and the lowest Tertiary limestone beds.[28] This bed is free of primary $CaCO_3$,[b] so there is no record of the biological changes during the time interval represented by the clay. The boundary is further marked by a zone in the uppermost Cretaceous in which the normally pink limestone is white in color. This zone is 0.3 to 1.0 meter thick, varying from section to section. Its lower boundary is a gradational color change; its upper boundary is abrupt and coincides with the faunal and floral extinctions. In one section (Contessa) we can see that the lower 5 mm of the boundary clay is gray and the upper 5 mm is red, thus placing the upper boundary of the zone in the middle of the clay layer. [...]

RESULTS FROM THE ITALIAN SECTIONS

Our first experiments involved NAA of nine samples from the Bottaccione section (two limestone samples from immediately above and below the boundary plus seven limestone samples spaced over 325 m of the Cretaceous). This was supplemented by three samples from the nearby Contessa section (two from the boundary clay and one from the basal[c] Tertiary bed). [...]

Figure 5 shows the results of 29 Ir[d] analyses completed on Italian samples. Note that the section is enlarged and that the scale is linear in the vicinity of the C-T boundary, where details are important, but changes to logarithmic to show results from 350 m below to 50 m above the boundary. It is also important to note that analyses from five stratigraphic sections are plotted on the same diagram on the basis of their stratigraphic position above or below the boundary. Because slight differences in sedimentation rate probably exist from one section to the next, the chronologic sequence of samples from different sections may not be exactly correct. Nevertheless, Fig. 5 gives a clear picture of the general trend of iridium concentrations as a function of stratigraphic level.

10

---

a   *sections*   I.e., stratigraphic sections, sequences of layered rocks and sediment.
b   *CaCO₃*   Calcium carbonate, the main component of limestone.
c   *basal*   Lowest.
d   *Ir*   Iridium.

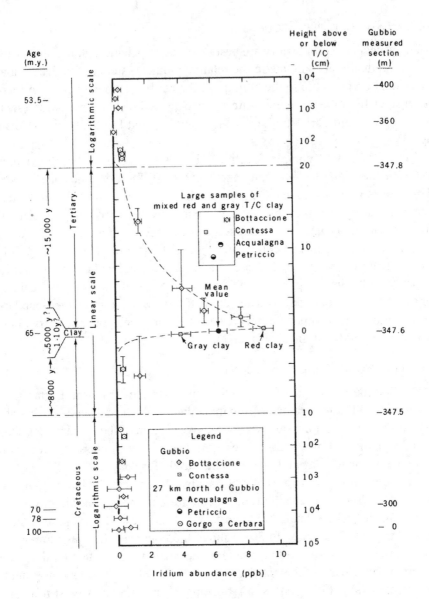

FIGURE 5

Iridium abundances per unit weight of $2N$ HNO$_3$ acid-insoluble residues from Italian limestones near the Tertiary-Cretaceous boundary. Error bars on abundances are the standard deviations in counting radioactivity. Error bars on stratigraphic position indicate the stratigraphic thickness of the sample. The dashed line above the boundary is an "eyeball fit" exponential with a half-height of 4.6 cm. The dashed line below the boundary is a best fit exponential (two points) with a half-height of 0.43 cm. The filled circle and error bar are the mean and standard deviation of Ir abundances in four large samples of boundary clay from different locations.

The pattern, based especially on the samples from the Bottaccione Gorge and Gorgo a Cerbara, shows a steady background level of ~ 0.3 ppb[a] throughout the Upper Cretaceous, continuing into the uppermost bed of the Cretaceous. The background level in the acid-insoluble residues[b] is roughly comparable to the iridium abundance measured by other workers[19,22,32] in deep-sea clay sediments. This level increases abruptly, by a factor of more than 30, to 9.1 ppb, the Ir abundance in the red clay from the Contessa section. Iridium levels are high in clay residues from the first few beds of Tertiary limestone, but fall off to background levels by 1 m above the boundary. For comparison, the upper dashed line in Fig. 5 shows an exponential decay from the boundary clay Ir level with a half-height of 4.6 cm.

To test the possibility that iridium might somehow be concentrated in clay layers, we subsequently analyzed two red clay samples from a short distance below the C-T boundary in the Bottaccione section. One is from a distinctive clay layer 5 to 6 mm thick, 1.73 m below the boundary; the other is from a 1- to 2-mm bedding-plane[c] clay seam 0.85 m below the boundary. The whole-rock analyses[d] of these clays showed no detectable Ir with limits of 0.5 and 0.24 ppb, respectively. Thus neither clay layers from below the C-T boundary nor clay components in the limestone show evidence of Ir above the background level.

THE DANISH SECTION

To test whether the iridium anomaly is a local Italian feature, it was desirable to analyze segments of similar age from another region. The sea cliff of Stevns Klint, about 50 km south of Copenhagen, is a classical area for the C-T boundary and for the Danian or basal stage of the Tertiary. A collection of up-to-date papers on this and nearby areas has recently been published, which includes a full bibliography of earlier works.[6, vol. 1] [...]

A SUDDEN INFLUX OF EXTRATERRESTRIAL MATERIAL

To test whether the anomalous iridium at the C-T boundary in the Gubbio sections is of extraterrestrial origin, we considered the increases in 27 of the 28 elements measured by NAA that would be expected if iridium in excess of the background level came from a source with the average composition of the

15

---

a   *ppb*   Parts per billion.

b   *acid-insoluble residues*   Products of a chemical process used to determine the proportions of mineral components in a rock, such as the quantity of iridium.

c   *bedding-plane*   Division between two rock layers.

d   *whole-rock analyses*   Results of a chemical process that, like acid-insoluble residue analysis, can be used to determine the proportion of elements, such as iridium, in a rock.

earth's crust. The crustal Ir abundance, less than 0.1 ppb,[19,22] is too small to be a worldwide source for material with an Ir abundance of 6.3 ppb, as found near Gubbio. Extraterrestrial sources with Ir levels of hundreds of parts per billion or higher are more likely to have produced the Ir anomaly. Figure 10 shows that if the source had an average earth's crust composition,[46] increases significantly above those observed would be expected in all 27 elements. However, for a source with average carbonaceous chondrite[a] composition,[46] only nickel should show an elemental increase greater than that observed. As shown in Fig. 11, such an increase in nickel was not observed, but the predicted effect is small and, given appropriate conditions, nickel oxide would dissolve in seawater.[47] We conclude that the pattern of elemental abundances in the Gubbio sections is compatible with an extraterrestrial source for the anomalous iridium and incompatible with a crustal source. [...]

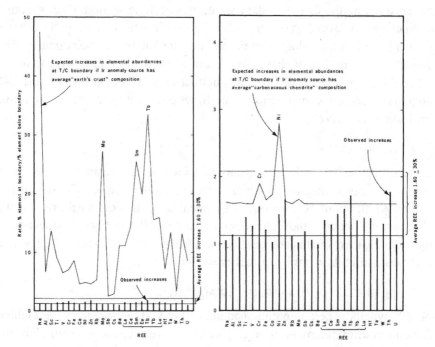

FIGURE 10 (LEFT)
Comparison of observed elemental abundance patterns in the Gubbio section samples with average patterns expected for crustal material (*46*).

FIGURE 11 (RIGHT)
Comparison of observed elemental abundance patterns in the Gubbio section samples with patterns expected for carbonaceous chondrites (*46*).

a *carbonaceous chondrite* Chondritic meteorite containing high proportions of water and organic compounds.

We next consider whether the Ir anomaly is due to an abnormal influx of extraterrestrial material at the time of the extinctions, or whether it was formed by the normal, slow accumulation of meteoric material,[19] followed by concentration in the boundary rocks by some identifiable mechanism.

There is prima facie evidence for an abnormal influx in the observations that the excess iridium occurs exactly at the time of one of the extinctions; that the extinctions were clearly worldwide; and that the iridium anomaly is now known from two different areas in western Europe and in New Zealand. Furthermore, we will show in a later section that impact of a 10-km earth-crossing asteroid, an event that probably occurs with about the same frequency as major extinctions, may have produced the observed physical and biological effects. Nevertheless, one can invent two other scenarios that might lead to concentration of normal background iridium at the boundary. These appear to be much less likely than the sudden-influx model, but we cannot definitely rule out either one at present.

The first scenario requires a physical or chemical change in the ocean waters at the time of the extinctions, leading to extraction of iridium resident in the seawater. This would require iridium concentrations in seawater that are higher than those presently observed. In addition, it suggests that the positive iridium anomaly should be accompanied by a compensating negative anomaly immediately above, but this is not seen.

The second scenario postulates a reduction in the deposition rate of all components of the pelagic sediment except for the meteoritic dust that carries the concentrated iridium. This scenario requires removal of clay but not of iridium-bearing particles, perhaps by currents of exactly the right velocity. These currents must have affected both the Italian and Danish areas at exactly the time of the C-T extinctions, but at none of the other times represented by our samples. We feel that this scenario is too contrived, a conclusion that is justified in more detail elsewhere.[23]

In summary, we conclude that the anomalous iridium concentration at the C-T boundary is best interpreted as indicating an abnormal influx of extraterrestrial material. [...]

20

THE ASTEROID IMPACT HYPOTHESIS

After obtaining negative results in our tests of the supernova hypothesis,[a] we were left with the question of what extraterrestrial source within the solar system could supply the observed iridium and also cause the extinctions. We

a *supernova hypothesis* In the previous section, not included here, Alvarez et al. consider and reject the possibility that a supernova was the source of the iridium.

considered and rejected a number of hypotheses;[23] finally, we found that an extension of the meteorite impact hypothesis[55,56] provided a scenario that explains most or all of the biological and physical evidence. In brief, our hypothesis suggests that an asteroid struck the earth, formed an impact crater, and some of the dust-sized material ejected from the crater reached the stratosphere and was spread around the globe. This dust effectively prevented sunlight from reaching the surface for a period of several years, until the dust settled to earth. Loss of sunlight suppressed photosynthesis, and as a result most food chains collapsed and extinctions resulted. Several lines of evidence support this hypothesis, as discussed in the next few sections. The size of the impacting object can be calculated from four independent sets of observations, with good agreement among the four different diameter estimates.

Earth-Crossing Asteroids and Earth Craters

Two quite different data bases show that for the last billion years the earth has been bombarded by a nearly constant flux of asteroids that cross the earth's orbit. One data base comes from astronomical observations of such asteroids and a tabulation of their orbital parameters and their distribution of diameters.[57] Öpik[58] computed that the mean time to collision with the earth for a given earth-crossing asteroid is about 200 million years. [...] Shoemaker[61] estimates that a mean collision time of 100 million years is consistent with a diameter of 10 km, which is the value we will adopt. A discussion of cratering data, which leads to similar estimates, is given in Grieve and Robertson's review article[62] on the size and age distribution of large impact craters on the earth. Rather than present our lengthy justification[23] for the estimates based on the cratering data, we will simply report the evaluation of Grieve,[63] who wrote: "I can find nothing in our data that is at odds with your premise." Grieve also estimates that the diameter of the crater formed by the impact of a 10-km asteroid would be about 200 km.[63] This section of our article has thus been greatly condensed now that we have heard from experienced students of the two data bases involved.

Krakatoa

The largest well-studied terrestrial explosion in historical times was that of the island volcano, Krakatoa, in the Sunda Strait, between Java and Sumatra.[64] Since this event provides the best available data on injection of dust into the stratosphere, we give here a brief summary of relevant information.

On 26 and 27 August 1883, Krakatoa underwent volcanic eruptions that shot an estimated 18 km^3 of material into the atmosphere, of which about 4 km^3 ended up in the stratosphere, where it stayed for 2 to 2.5 years. Dust from the explosion circled the globe, quickly giving rise to brilliant sunsets seen worldwide. Recent measurements of the ^{14}C[a] injected into the atmosphere by nuclear bomb tests confirm the rapid mixing (about 1 year) between hemi-spheres.[65] If we take the estimated dust mass in the stratosphere (4 km^3 times the assumed low density of 2 g/cm^3) and spread it uniformly over the globe, it amounts to 1.6 x 10^{-3} g/cm^2. This layer did not absorb much of the incident radiation[b] on a "straight-through" basis. However, if it were increased by a factor of about 10^3 (a rough prediction of our theory), it is most probable that the sunlight would be attenuated to a high degree.

Since the time for the colored sunsets to disappear after Krakatoa is frequently given as 2 to 2.5 years, we have assumed that the asteroid impact material in the stratosphere settled in a few years. Thus, 65 million years ago, day could have been turned into night for a period of several years, after which time the atmosphere would return relatively quickly to its normal transparent state.

What happened during the Krakatoa explosions can be expected to happen to a much greater extent during the impact of a large asteroid. An interesting difference is that extreme atmospheric turbulence would follow the impact. The asteroid would enter the atmosphere at roughly 25 km/sec and would "punch a hole" in the atmosphere about 10 km across. The kinetic energy of the asteroid is approximately equivalent to that of 10^8 megatons of TNT.

SIZE OF THE IMPACTING OBJECT

If we are correct in our hypothesis that the C-T extinctions were due to the impact of an earth-crossing asteroid, there are four independent ways to calculate the size of the object. The four ways and the results obtained are outlined below.

1) The postulated size of the incoming asteroid was first computed from the iridium measurements in the Italian sections, the tabulated Ir abundances[66] in type I carbonaceous chondrites (CI), which are considered to be typical solar system material, and the fraction of erupted material estimated to end up in the stratosphere. [...]

2) The second estimate comes from data on earth-crossing asteroids and the craters they have made on the earth's surface. In a sense, the second estimate comes from two quite different data bases – one from geology and the other from astronomy. Calculations of the asteroid diameter can be made

---

a ^{14}C Carbon-14, a radioactive isotope of carbon.
b *incident radiation* I.e., sunlight.

from both data bases, but they will not really be independent since the two data bases are known to be consistent with each other. As shown in an earlier section, the most believable calculation of the mean time between collisions of the earth and asteroids equal to or larger than 10 km in diameter is about 100 million years. The smaller the diameter the more frequent are the collisions, so our desire to fit not only the C-T extinction, but earlier ones as well, sets the mean time between extinctions at about 100 million years and the diameter at about 10 km. [...]

30 3) The third method of estimating the size of the asteroid comes from the possibility that the 1-cm boundary layer at Gubbio and Copenhagen is composed of material that fell out of the stratosphere, and is not related to the clay that is mixed in with the limestone above and below it. This is quite a surprising prediction of the hypothesis, since the most obvious explanation for the origin of the clay is that it had the same source as the clay impurity in the rest of the Cretaceous and Tertiary limestone, and that it is nearly free of primary $CaCo_3$ because the extinction temporarily destroyed the calcite-producing plankton for about 5000 years. But as discussed earlier, the material in the boundary layer is of a different character from the clay above and below it, whereas the latter two clays are very similar. [...]

The first and third methods are independent, even though they both depend on measurements made on the boundary material. This can best be appreciated by noting that if the Ir abundance were about the same in the earth's crust as it is in meteorites, the iridium anomaly seen in Fig. 5 would not exist. Therefore, method 1 would not exist either. The fact that method three could still be used is the indicator of the relative independence of the two methods.

4) The fourth method is not yet able to set close limits on the mass of the incoming asteroid, but it leads to consistent results. This method derives from the need to make the sky much more opaque than it was in the years following the Krakatoa explosion. If it is assumed that the Krakatoa dust cloud attenuated the vertically incident sunlight by about 3 percent, then an explosion involving 33 times as much material would reduce the light intensity to $1/e$. The stratospheric mass due to an explosion of the magnitude calculated in the three earlier methods – about 1000 times that of Krakatoa – would then be expected to reduce the sunlight to $\exp(-30) = 10^{-13}$. This is, of course, much more light attenuation than is needed to stop photosynthesis. But the model used in this simplistic calculation assumes that the dust is a perfect absorber of the incident light. A reasonable albedo[a] coupled with a slight reduction in the

---

a *albedo* I.e., ratio of light striking the earth to light reflected from the earth.

mass of dust can raise the light intensity under the assumed "optical depth" to 10^{-7} or normal sunlight, corresponding to 10 percent of full moonlight.

Although it is impossible to make an accurate estimate of the asteroid's size from the Krakatoa extrapolation, it would have been necessary to abandon the hypothesis had a serious discrepancy been apparent. In the absence of good measurements of the solar constant in the 1880s, it can only be said that the fourth method leads to asteroid sizes that are consistent with the other three. [...] We conclude that the data are consistent with an impacting asteroid with a diameter of about 10 ± 4 km.

Biological Effects

A temporary absence of sunlight would effectively shut off photosynthesis and thus attack food chains at their origins. In a general way the effects to be expected from such an event are what one sees in the paleontological record of the extinction.

The food chain in the open ocean is based on microscopic floating plants, such as the coccolith-producing[a] algae, which show a nearly complete extinction. The animals at successively higher levels of the food chain were also very strongly affected, with nearly total extinction of the foraminifera and complete disappearance of the belemnites, ammonites,[b] and marine reptiles.

A second food chain is based on land plants. Among these plants, existing individuals would die, or at least stop producing new growth, during an interval of darkness, but after light returned they would regenerate from seeds, spores, and existing root systems. However, the large herbivorous and carnivorous animals that were directly or indirectly dependent on this vegetation would become extinct. Russell[2] states that "no terrestrial vertebrate heavier than about 25 kg is known to have survived the extinctions." Many smaller terrestrial vertebrates did survive, including the ancestral mammals, and they may have been able to do this by feeding on insects and decaying vegetation.

The situation among shallow marine bottom-dwelling invertebrates is less clear; some groups became extinct and others survived. A possible base for a temporary food chain in this environment is nutrients originating from the decaying land plants and animals and brought by rivers to the shallow marine waters.

We will not go further into this matter but we will refer the reader to the proceedings of the 1976 Ottawa meeting on the C-T extinction. [...]

a   *coccolith*   Calcium carbonate plate.
b   *belemnites*   Cephalopods similar to cuttlefish; *ammonites*   Cephalopods with coiled shells.

IMPLICATIONS

Among the many implications of the asteroid impact hypothesis, if it is correct, two stand out prominently. First, if the C-T extinctions were caused by an impact event, the same could be true of the earlier major extinctions as well. There have been five such extinctions since the end of the Precambrian, 570 million years ago, which matches well the probable interval of about 100 million years between collisions with 10-km diameter objects. Discussions of these extinction events generally list the organisms affected according to taxonomic groupings; it would be more useful to have this information given in terms of interpreted ecological or food-chain groupings. It will also be important to carry out iridium analyses in complete stratigraphic sections across these other boundaries. However, E. Shoemaker (private communication) predicts that if some of the extinctions were caused by the collision of a "fresh" comet (mostly ice), the Ir anomaly would not be seen even though the extinction mechanism was via the same dust cloud of crustal material, so the absence of a higher Ir concentration at, for example, the Permian-Triassic boundary would not invalidate our hypothesis. According to Shoemaker, cometary collisions in this size range could be twice as frequent as asteroidal collisions.

40   Second, we would like to find the crater produced by the impacting object. Only three craters 100 km or more in diameter are known.[62] Two of these (Sudbury and Vredefort) are of Precambrian age. For the other, Popigay crater in Siberia, a stratigraphic age of Late Cretaceous to Quaternary and a potassium-aragon date of 28.8 million years (no further details given) have been reported.[72,73] Thus, Popigay crater is probably too young, and at 100-km-diameter probably also too small, to be the C-T impact site. There is about a 2/3 probability that the object fell in the ocean. Since the probable diameter of the object, 10 km, is twice the typical oceanic depth, a crater would be produced on the ocean bottom and pulverized rock would be ejected. However, in this event we are unlikely to find the crater, since bathymetric information[a] is not sufficiently detailed and since a substantial portion of the pre-Tertiary ocean has been subducted.[b]

(1980)

---

a   *bathymetric information*   Measurements of the contours of ocean beds.

b   *subducted*   Subduction is the movement of one of the Earth's plates under another plate, forcing the lower plate down into the Earth's mantle.

REFERENCES AND NOTES[a]

1. D.A. Russell, *Geol. Assoc. Can. Spec. Rep. 13* (1975), p. 119.
2. _____, in (*5*), p. 11.
3. M.B. Cita and I. Premoli Silva, *Riv. Ital. Paleontol. Stratigr. Mem. 14* (1974), p. 193.
4. D.A. Russell, *Annu. Rev. Earth Planet. Sci.* 7, 163 (1979).
5. K-TEC group (P. Béland *et al.*), *Cretaceous-Tertiary Extinctions and Possible Terrestrial and Extraterrestrial Causes* (Proceedings of Workshop, National Museum of Natural Sciences, Ottawa, 16 and 17 Nov. 1976).
6. T. Birkelund and R.G. Bromley, Eds., *Cretaceous-Tertiary Boundary Events*, vol. 1, *The Maastrichtian and Danian of Denmark* (Symposium, University of Copenhagen, Copenhagen, 1979); W.K. Christiansen and T. Birkelund, eds., *ibid.*, vol. 2, *Proceedings*.
7. H. Tappan, *Palaeogeogr. Paleoclimatol. Palaeoecol.* 4, 187 (1968): T.R. Worsley, *Nature (London)* 230, 403 (1977); D.M. McLean, *Science* 200, 1060 (1978); *ibid.* 201, 401 (1978); S. Gartner and J. Keany, *Geology* 6, 708 (1978).
8. E.G. Kauffman, in (*6*), vol. 2, p. 29.
9. A.G. Fischer, in (*6*), vol. 2, p. 11; and M.A. Arthur, *Soc. Econ. Paleontol. Mineral. Spec. Publ. 25* (1977), p. 19.
10. J.F. Simpson, *Geol. Soc. Am. Bull.* 77, 197 (1966); J.D. Hays, *ibid.* 82, 2433 (1971); C.G.A. Harrison and J.M. Prospero, *Nature (London)* 250, 563 (1974).
11. O.H. Schindewolf, *Neues Jahrb. Geol. Paleontol. Monatsh.* 1954, 451 (1954); *ibid.* 1958, 270 (1958); A.R. Leoblich, Jr., and H. Tappan, *Geol. Soc. Am. Bull.* 75, 367 (1964); V.I. Krasovski and I.S. Shivlovsky, *Dokl. Akad. Nauk SSSR* 116, 197 (1957); K.D. Terry and W.H. Tucker, *Science* 159, 421 (1968); H. Laster, *ibid.* 160, 1138 (1968); W.H. Tucker and K.D. Terry, *ibid.*, p. 1138; D. Russell and W.H. Tucker, *Nature (London)* 229, 553 (1971); M.A. Ruderman, *Science* 184, 1079 (1974); R.C. Whitten, J. Cuzzi, W.J. Borucki, J.H. Wolfe, *Nature (London)* 263, 398 (1976).
12. S. Gartner and J.P. McGuirk, *Science* 206, 1272 (1979).
13. A. Boersma and N. Shackleton, in (*6*), vol. 2, p. 50; B. Buchardt and N.O. Jorgensen, in (*6*), vol. 2, p. 54.
14. L. Christensen, S. Fregerslev, A. Simonsen, J. Thiede, *Bull. Geog. Soc. Den* 22, 193 (1973).
15. N.O. Jorgensen, in (*6*), vol. 1, p. 33, vol. 2, p. 62; M. Renard, in (*6*), vol. 2, p. 70.
16. H.P. Luterbacher and I. Premoli Silva, *Riv. Ital. Paleontol. Stratigr.* 70, 67 (1964).
17. H. Pettersson and H. Rotschi, *Geochim. Cosmochim. Acta* 2, 81 (1952).
18. V.M. Goldschmidt, *Geochemistry* (Oxford Univ. Press, New York, 1954).
19. J.L. Barker, Jr., and E. Anders, *Geochim. Cosmochim. Acta* 32, 627 (1968).
20. R. Ganapathy, D.E. Brownlee, P.W. Hodge, *Science* 201, 1119 (1978).
21. A.M. Sarna-Wojcicki, H.R. Bowman, D. Marchand, E. Helley, private communication.
22. J.H. Crocket and H.Y. Kuo, *Geochim. Cosmochim. Acta* 43, 831 (1979).
23. These are briefly discussed in L.W. Alvarez, W. Alvarez, F. Asaro, H.V. Michel, *Univ. Calif. Lawrence Berkeley Lab. Rep. LBL-9666* (1979).
24. A description of the NAA techniques is given in Alvarez *et al.* (*23*), appendix II; I. Perlman and F. Asaro, in *Science and Archaeology*, R.H. Brill, Ed. (MIT Press, Cambridge, Mass., 1971), p. 182.
25. These limestones belong to the Umbrian sequence, of Jurassic to Miocene age, which has been described in V. Bortolotti, P. Passerini, M. Sagri, G. Sestini, *Sediment. Geol.* 4, 341 (1970);

---

a With the exception of the authors' acknowledgments, references have been cut to show only those cited in the included material.

A. Jacobacci, E. Centamore, M. Chiocchini, N. Malferrari, G. Martelli, A. Micarelli, *Note Esplicative Carta Geologica d'Italia (1:50,000), Foglio 190: "Cagli"* (Rome, 1974).

28. D.V. Kent, *Geology* 5, 769 (1977); M.A. Arthur, thesis, Princeton University (1979).

32. J.H. Crocket, J.D. McDougall, R.C. Harriss, *Geochim. Cosmochim. Acta* 37, 2547 (1973).

46. *Encyclopaedia Britannica* (Benton, Chicago, ed. 15, 1974), vol. 6, p. 702.

47. K.K. Turekian, *Oceans* (Prentice-Hall, Englewood Cliffs, N.J., 1976), p. 122.

55. H.C. Urey, [*Nature (London)*] 242, 32 (1973).

56. E.J. Öpik, *Ir. Astron. J.* 5 (No. 1), 34 (1958).

57. E.M. Shoemaker, J.G. Williams, E.F. Helin, R.F. Wolfe, in *Asteroids*, T. Gehrels, Ed. (Univ. of Arizona Press, Tucson, 1979), pp. 253-282.

58. E.J. Öpik, *Adv. Astron. Astrophys.* 2, 220 (1963); *ibid.* 4, 302 (1964); *ibid.* 8, 108 (1971). These review articles give references to Öpik's extensive bibliography on meteorites, Apollo objects, and asteroids.

61. E.M. Shoemaker, personal communication.

62. R.A.F. Grieve and P.B. Robertson, *Icarus* 38, 212 (1979).

63. R.A.F. Grieve, personal communication.

64. G.J. Symons, Ed., *The Eruption of Krakatoa and Subsequent Phenomena* (Report of the Krakatoa Committee of the Royal Society, Harrison, London, 1888).

65. I.U. Olson and I. Karlen, *Am. J. Sci. Radio-carbon Suppl.* 7 (1965), p. 331; T.A. Rafter and B.J. O'Brien, *Proc. 8th Int. Conf. Radiocarbon Dating* 1, 241 (1972).

66. U. Krähenbühl, *Geochim. Cosmochim. Acta* 37, 1353 (1973).

72. V.L. Masaytis, M.V. Mikhaylov, T.V. Selivanovskaya, *Sov. Geol. No. 6* (1971), pp. 143-147; translated in *Geol. Rev.* 14, 327 (1972).

73. V.L. Masaytis, *Sov. Geol. No. 11* (1975), pp. 52-64; translated in *Int. Geol. Rev.* 18, 1249 (1976).

74. It will be obvious to anyone reading this article that we have benefited enormously from conversations and correspondence with many friends and colleagues throughout the scientific community. We would particularly like to acknowledge the help we have received from E. Anders, J.R. Arnold, M.A. Arthur, A. Buffington, I.S.E. Carmichael, G. Curtis, P. Eberhard, S. Gartner, R.L. Garwin, R.A.F. Grieve, E.K. Hyde, W. Lowrie, C. McKee, M.C. Michel (who was responsible for the mass spectrometric measurements), J. Neil, B.M. Oliver, C. Orth, B. Pardoe, I. Perlman, D.A. Russell, A.M. Sessler, and E. Shoemaker. One of us (W.A.) thanks the National Science foundation for support, the other three authors thank the Department of Energy for support, and one of us (L.W.A.) thanks the National Aeronautics and Space Administration for support. The x-ray fluorescence measurements of trace elements Fe and Ti by R.D. Giaque and of major elements by S. Flexser and M. Sturz were most appreciated. We appreciate the assistance of D. Jackson and C. Nguyen in the sample preparation procedures. We are grateful to T. Lim and the staff of the Berkeley Research Reactor for many neutron irradiations used in this work. We also appreciate the efforts of G. Pefley and the staff of the Livermore Pool Type Reactor for the irradiations used for the Ir isotopic ratio measurements.

Questions:

1. Summarize the asteroid impact hypothesis as put forward by the article. What biological and geological events does it attempt to explain, and how?

2. Rewrite the summary as a paraphrase. What concepts are most difficult to explain without resorting to the terminology or phrasing used in the article?

3. Why does the study measure concentrations of iridium, as opposed to another substance?

4. What kinds of organisms died during the mass extinction, and what kinds survived? What hypothesis explains the survivals?

5. Alvarez et al. refer to the "many implications of the asteroid impact hypothesis," and they discuss two of these implications. What other implications are contained in the hypothesis?

6. Apart from the asteroid impact hypothesis, Alvarez et al. provide several scenarios that would explain the anomaly in iridium concentrations. Why do you think they provide these other options? What types of justifications support the asteroid impact hypothesis?

7. How persuasive is the information presented in Figure 5? Why? What elements of the graph are most persuasive?

8. Why does the graph in Figure 5 change from linear to logarithmic? Is this a usual shift in this type of graph?

9. What is "prima facie evidence" (paragraph 17)? How persuasive is the use of this term here?

10. Summarize the information presented in Figures 10 and 11. According to Alvarez et al., what is the significance of this information? Do the figures reproduce textual material? How do they function in relation to the text? Would the text make as much sense without these figures?

11. What relevant information do Alvarez et al. suggest can be gleaned from the 1883 eruption of Krakatoa? Name two significant differences between that event and the hypothesized asteroid impact. To what extent is the information about the effects of Krakatoa still useful despite these differences?

12. The article describes four independent methods used to estimate the size of the asteroid. Explain why the first and third methods are independent of each other.

13. Since the publication of this article in 1980, the asteroid impact hypothesis has become widely accepted, while complementary explanations have been advanced that suggest the impact was not the sole cause of the extinction. Find a recent article on the Cretaceous-Tertiary extinction and summarize its findings. Does this article reference Alvarez et al.? If so, what function does the reference serve?

three

ELIZABETH KOLBERT

The Sixth Extinction?

This article, written by an environmentalist journalist for
The New Yorker, *investigates the current worldwide
reduction in natural diversity – and its disturbing
resemblance to the mass extinctions of our planet's past.*

❦

The town of El Valle de Antón, in central Panama, sits in the middle of a volcanic crater formed about a million years ago. The crater is almost four miles across, but when the weather is clear you can see the jagged hills that surround the town, like the walls of a ruined tower. El Valle has one main street, a police station, and an open-air market that offers, in addition to the usual hats and embroidery, what must be the world's largest selection of golden-frog figurines. There are golden frogs sitting on leaves and – more difficult to understand – golden frogs holding cell phones. There are golden frogs wearing frilly skirts, and golden frogs striking dance poses, and ashtrays featuring golden frogs smoking cigarettes through a holder, after the fashion of F.D.R.[a] The golden frog, which is bright yellow with dark brown splotches, is endemic to the area around El Valle. It is considered a lucky symbol in Panama – its image is often printed on lottery tickets – though it could just as easily serve as an emblem of disaster.

In the early nineteen-nineties, an American graduate student named Karen Lips established a research site about two hundred miles west of El Valle, in the Talamanca Mountains, just over the border in Costa Rica. Lips was planning to study the local frogs, some of which, she later discovered, had never been identified. In order to get to the site, she had to drive two hours from the nearest town – the last part of the trip required tire chains – and then hike for an hour in the rain forest.

Lips spent two years living in the mountains. "It was a wonderland," she recalled recently. Once she had collected enough data, she left to work on her

---

a *F.D.R.*   Franklin D. Roosevelt, President of the United States from 1933 to 1945.

dissertation. She returned a few months later, and though nothing seemed to have changed, she could hardly find any frogs. Lips couldn't figure out what was happening. She collected all the dead frogs that she came across – there were only a half dozen or so – and sent their bodies to a veterinary pathologist in the United States. The pathologist was also baffled: the specimens, she told Lips, showed no signs of any known disease.

A few years went by. Lips finished her dissertation and got a teaching job. Since the frogs at her old site had pretty much disappeared, she decided that she needed to find a new location to do research. She picked another isolated spot in the rain forest, this time in western Panama. Initially, the frogs there seemed healthy. But, before long, Lips began to find corpses lying in the streams and moribund animals sitting on the banks. Sometimes she would pick up a frog and it would die in her hands. She sent some specimens to a second pathologist in the US, and, once again, the pathologist had no idea what was wrong.

Whatever was killing Lips's frogs continued to move, like a wave, east 5 across Panama. By 2002, most frogs in the streams around Santa Fé, a town in the province of Veraguas, had been wiped out. By 2004, the frogs in the national park of El Copé, in the province of Coclé, had all but disappeared. At that point, golden frogs were still relatively common around El Valle; a creek not far from the town was nicknamed Thousand Frog Stream. Then, in 2006, the wave hit.

Of the many species that have existed on earth – estimates run as high as fifty billion – more than ninety-nine per cent have disappeared. In the light of this, it is sometimes joked that all of life today amounts to little more than a rounding error.

Records of the missing can be found everywhere in the world, often in forms that are difficult to overlook. And yet extinction has been a much contested concept. Throughout the eighteenth century, even as extraordinary fossils were being unearthed and put on exhibit, the prevailing view was that species were fixed, created by God for all eternity. If the bones of a strange creature were found, it must mean that the creature was out there somewhere.

"Such is the economy of nature," Thomas Jefferson wrote, "that no instance can be produced, of her having permitted any one race of her animals to become extinct; of her having formed any link in her great work so weak as to be broken." When, as President, he dispatched Meriwether Lewis and William Clark to the Northwest, Jefferson hoped that they would come upon live mastodons roaming the region.

The French naturalist Georges Cuvier was more skeptical. In 1812, he published an essay on the "Revolutions of the Surface of the Globe," in which he asked, "How can we believe that the immense mastodons, the gigantic megatheriums, whose bones have been found in the earth in the two Americas, still live on this continent?" Cuvier had conducted studies of the fossils found in gypsum mines in Paris, and was convinced that many organisms once common to the area no longer existed. These he referred to as *espèces perdues*, or lost species. Cuvier had no way of knowing how much time had elapsed in forming the fossil record. But, as the record indicated that Paris had, at various points, been under water, he concluded that the *espèces perdues* had been swept away by sudden cataclysms.

10   "Life on this earth has often been disturbed by dreadful events," he wrote. "Innumerable living creatures have been victims of these catastrophes." Cuvier's essay was translated into English in 1813 and published with an introduction by the Scottish naturalist Robert Jameson, who interpreted it as proof of Noah's flood. It went through five editions in English and six in French before Cuvier's death, in 1832.

Charles Darwin was well acquainted with Cuvier's ideas and the theological spin they had been given. (He had studied natural history with Jameson at the University of Edinburgh.) In his theory of natural selection, Darwin embraced extinction; it was, he realized, essential that some species should die out as new ones were created. But he believed that this happened only slowly. Indeed, he claimed that it took place even more gradually than speciation: "The complete extinction of the species of a group is generally a slower process than their production." In "On the Origin of Species," published in the fall of 1859, Darwin heaped scorn on the catastrophist approach:

> So profound is our ignorance, and so high our presumption, that we marvel when we hear of the extinction of an organic being; and as we do not see the cause, we invoke cataclysms to desolate the world.

By the start of the twentieth century, this view had become dominant, and to be a scientist meant to see extinction as Darwin did. But Darwin, it turns out, was wrong.

Over the past half-billion years, there have been at least twenty mass extinctions, when the diversity of life on earth has suddenly and dramatically contracted. Five of these – the so-called Big Five – were so devastating that they are usually put in their own category. The first took place during the late Ordovician period, nearly four hundred and fifty million years ago, when life was still confined mainly to water. Geological records indicate that more than eighty per cent of marine species died out. The fifth occurred at the end of the

Cretaceous period, sixty-five million years ago. The end-Cretaceous event exterminated not just the dinosaurs but seventy-five per cent of all species on earth.

The significance of mass extinctions goes beyond the sheer number of organisms involved. In contrast to ordinary, or so-called background, extinctions, which claim species that, for one reason or another, have become unfit, mass extinctions strike down the fit and the unfit at once. For example, brachiopods, which look like clams but have an entirely different anatomy, dominated the ocean floor for hundreds of millions of years. In the third of the Big Five extinctions – the end-Permian – the hugely successful brachiopods were nearly wiped out, along with trilobites, blastoids, and eurypterids. (In the end-Permian event, more than ninety per cent of marine species and seventy per cent of terrestrial species vanished; the event is sometimes referred to as "the mother of mass extinctions" or "the great dying.")

Once a mass extinction occurs, it takes millions of years for life to recover, and when it does it generally has a new cast of characters; following the end-Cretaceous event, mammals rose up (or crept out) to replace the departed dinosaurs. In this way, mass extinctions, though missing from the original theory of evolution, have played a determining role in evolution's course; as Richard Leakey has put it, such events "restructure the biosphere" and so "create the pattern of life." It is now generally agreed among biologists that another mass extinction is under way. Though it's difficult to put a precise figure on the losses, it is estimated that, if current trends continue, by the end of this century as many as half of earth's species will be gone.

The El Valle Amphibian Conservation Center, known by the acronym EVACC (pronounced "e-vac"), is a short walk from the market where the golden-frog figurines are sold. It consists of a single building about the size of an average suburban house. The place is filled, floor to ceiling, with tanks. There are tall tanks for species that, like the Rabb's fringe-limbed tree frog, live in the forest canopy, and short tanks for species that, like the big-headed robber frog, live on the forest floor. Tanks of horned marsupial frogs, which carry their eggs in a pouch, sit next to tanks of casque-headed frogs, which carry their eggs on their backs.

The director of EVACC is a herpetologist named Edgardo Griffith. Griffith is tall and broad-shouldered, with a round face and a wide smile. He wears a silver ring in each ear and has a large tattoo of a toad's skeleton on his left shin. Griffith grew up in Panama City, and fell in love with amphibians one day in college when a friend invited him to go frog hunting. He collected most of the frogs at EVACC – there are nearly six hundred – in a rush, just as corpses were beginning to show up around El Valle. At that point, the center was little more than a hole in the ground, and so the frogs had to spend several months in temporary tanks at a local hotel. "We got a very good rate," Griffith assured me. While the amphibians were living in rented rooms, Griffith and his wife, a former Peace Corps volunteer, would go out into a nearby field to catch crickets for their dinner. Now EVACC raises bugs for the frogs in what looks like an oversized rabbit hutch.

EVACC is financed largely by the Houston Zoo, which initially pledged twenty thousand dollars to the project and has ended up spending ten times that amount. The tiny center, though, is not an outpost of the zoo. It might be thought of as a preserve, except that, instead of protecting the amphibians in their natural habitat, the center's aim is to isolate them from it. In this way, EVACC represents an ark built for a modern-day deluge. Its goal is to maintain twenty-five males and twenty-five females of each species – just enough for a breeding population.

20   The first time I visited, Griffith pointed out various tanks containing frogs that have essentially disappeared from the wild. These include the Panamanian golden frog, which, in addition to its extraordinary coloring, is known for its unusual method of communication; the frogs signal to one another using a kind of semaphore. Griffith said that he expected between a third and a half of all Panama's amphibians to be gone within the next five years. Some species, he said, will probably vanish without anyone's realizing it: "Unfortunately, we are losing all these amphibians before we even know that they exist."

Griffith still goes out collecting for EVACC. Since there are hardly any frogs to be found around El Valle, he has to travel farther afield, across the Panama Canal, to the eastern half of the country.

One day this winter, I set out with him on one of his expeditions, along with two American zookeepers who were also visiting EVACC. The four of us spent a night in a town called Cerro Azul and, at dawn the next morning, drove in a truck to the ranger station at the entrance to Chagres National Park. Griffith was hoping to find females of two species that EVACC is short of. He pulled out his collecting permit and presented it to the sleepy officials manning the station. Some underfed dogs came out to sniff around.

Beyond the ranger station, the road turned into a series of craters connected by ruts. Griffith put Jimi Hendrix on the truck's CD player, and we bounced along to the throbbing beat. (When the driving got particularly gruesome, he would turn down the volume.) Frog collecting requires a lot of supplies, so Griffith had hired two men to help with the carrying. At the very last cluster of houses, in the village of Los Ángeles, they materialized out of the mist. We bounced on until the truck couldn't go any farther; then we all got out and started walking.

The trail wound its way through the rain forest in a slather of red mud. Every few hundred yards, the main path was crossed by a narrower one; these paths had been made by leaf-cutter ants, making millions – perhaps billions – of trips to bring bits of greenery back to their colonies. (The colonies, which look like mounds of sawdust, can cover an area the size of a suburban back yard.) One of the Americans, Chris Bednarski, from the Houston Zoo, warned me to avoid the soldier ants, which will leave their jaws in your shin even after they're dead. "Those'll really mess you up," he observed. The other American, John Chastain, from the Toledo Zoo, was carrying a long hook, for use against venomous snakes. "Fortunately, the ones that can really mess you up are pretty rare," Bednarski said. Howler monkeys screamed in the distance. Someone pointed out jaguar prints in the soft ground.

After about five hours, we emerged into a small clearing. While we were   25
setting up camp, a blue morpho butterfly flitted by, its wings the color of the sky.

That evening, after the sun set, we strapped on headlamps and clambered down to a nearby stream. Many amphibians are nocturnal, and the only way to see them is to go looking in the dark, an exercise that's as tricky as it sounds. I kept slipping, and violating Rule No. 1 of rain-forest safety: never grab on to something if you don't know what it is. After one of my falls, Bednarski showed me a tarantula the size of my fist that he had found on a nearby tree.

One technique for finding amphibians at night is to shine a light into the forest and look for the reflecting glow of their eyes. The first amphibian sighted this way was a San José Cochran frog, perched on top of a leaf. San José Cochran frogs are part of a larger family known as "glass frogs," so named because their translucent skin reveals the outline of their internal organs. This particular glass frog was green, with tiny yellow dots. Griffith pulled a pair of surgical gloves out of his pack. He stood entirely still and then, with a heronlike gesture, darted to scoop up the frog. With his free hand, he took what looked like the end of a Q-tip and swabbed the frog's belly. Finally, he put the Q-tip in a little plastic vial, placed the frog back on the leaf, and pulled out his camera. The frog stared into the lens impassively.

We continued to grope through the blackness. Someone spotted a La Loma robber frog, which is an orangey-red, like the forest floor; someone else spotted a Warzewitsch frog, which is bright green and shaped like a leaf. With every frog, Griffith went through the same routine – snatching it up, swabbing its belly, photographing it. Finally, we came upon a pair of Panamanian robber frogs locked in amplexus – the amphibian version of sex. Griffith left those two alone.

One of the frogs that Griffith was hoping to catch, the horned marsupial frog, has a distinctive call that's been likened to the sound of a champagne bottle being uncorked. As we sloshed along, the call seemed to be emanating from several directions at once. Sometimes it sounded as if we were right nearby, but then, as we approached, it would fall silent. Griffith began imitating the call, making a cork-popping sound with his lips. Eventually, he decided that the rest of us were scaring the frogs with our splashing. He waded ahead, while we stood in the middle of the stream, trying not to move. When Griffith gestured us over, we found him standing in front of a large yellow frog with long toes and an owlish face. It was sitting on a tree limb, just above eye level. Griffith grabbed the frog and turned it over. Where a female marsupial frog would have a pouch, this one had none. Griffith swabbed it, photographed it, and put it back in the tree.

30   "You are a beautiful boy," he told the frog.

Amphibians are among the planet's greatest survivors. The ancestors of today's frogs and toads crawled out of the water some four hundred million years ago, and by two hundred and fifty million years ago the earliest representatives of what became the modern amphibian clades – one includes frogs and toads, a second newts and salamanders – had evolved. This means that amphibians have been around not just longer than mammals, say, or birds; they have been around since before there were dinosaurs. Most amphibians – the word comes from the Greek meaning "double life" – are still closely tied to the aquatic realm from which they emerged. (The ancient Egyptians thought that frogs were produced by the coupling of land and water during the annual flooding of the Nile.) Their eggs, which have no shells, must be kept moist in order to develop. There are frogs that lay their eggs in streams, frogs that lay them in temporary pools, frogs that lay them underground, and frogs that lay them in nests that they construct out of foam. In addition to frogs that carry their eggs on their backs and in pouches, there are frogs that carry them in their vocal sacs, and, until recently at least, there were frogs that carried their eggs in their stomachs and gave birth through their mouths. Amphibians emerged at a time when all the land on earth was part of one large mass; they have since adapted to conditions on every continent except Antarctica. Worldwide, more than six thousand species have been identified, and while the greatest number are found in the tropical rain forests, there are amphibians that, like the sandhill frog of Australia, can live in the desert, and also amphibians that, like the wood frog, can live above the Arctic Circle.

Several common North American frogs, including spring peepers, are able to survive the winter frozen solid.

When, about two decades ago, researchers first noticed that something odd was happening to amphibians, the evidence didn't seem to make sense. David Wake is a biologist at the University of California at Berkeley. In the early nineteen-eighties, his students began returning from frog-collecting trips in the Sierra Nevadas empty-handed. Wake remembered from his own student days that frogs in the Sierras had been difficult to avoid. "You'd be walking through meadows, and you'd inadvertently step on them," he told me. "They were just everywhere." Wake assumed that his students were just going to the wrong spots, or that they just didn't know how to look. Then a postdoc with several years of experience told him that he couldn't find any, either. "I said, 'OK, I'll go up with you and we'll go out to some proven places,'" Wake recalled. "And I took him out to this proven place and we found, like, two toads."

Around the same time, other researchers, in other parts of the world, reported similar difficulties. In the late nineteen-eighties, a herpetologist named Marty Crump went to Costa Rica to study golden toads; she was forced to change her project because, from one year to the next, the toad essentially vanished. (The golden toad, now regarded as extinct, was actually orange; it is not to be confused with the Panamanian golden frog, which is technically also a toad.) Probably simultaneously, in central Costa Rica the populations of twenty species of frogs and toads suddenly crashed. In Ecuador, the jambato toad, a familiar visitor to back-yard gardens, disappeared in a matter of years. And in northeastern Australia biologists noticed that more than a dozen amphibian species, including the southern day frog, one of the more common in the region, were experiencing drastic declines.

But, as the number of examples increased, the evidence only seemed to grow more confounding. Though amphibians in some remote and – relatively speaking – pristine spots seemed to be collapsing, those in other, more obviously disturbed habitats seemed to be doing fine. Meanwhile, in many parts of the world there weren't good data on amphibian populations to begin with, so it was hard to determine what represented terminal descent and what might be just a temporary dip.

"It was very controversial to say that amphibians were disappearing," Andrew Blaustein, a zoology professor at Oregon State University, recalls. Blaustein, who was studying the mating behavior of frogs and toads in the Cascade Mountains, had observed that some long-standing populations simply weren't there anymore. "The debate was whether or not there really was an amphibian population problem, because some people were saying it was just natural variation." At the point that Karen Lips went to look for her first research site, she purposefully tried to steer clear of the controversy.

"I didn't want to work on amphibian decline," she told me. "There were endless debates about whether this was a function of randomness or a true pattern. And the last thing you want to do is get involved when you don't know what's going on."

But the debate was not to be avoided. Even amphibians that had never seen a pond or a forest started dying. Blue poison-dart frogs, which are native to Suriname, had been raised at the National Zoo, in Washington, DC, for several generations. Then, suddenly, the zoo's tank-bred frogs were wiped out.

It is difficult to say when, exactly, the current extinction event – sometimes called the sixth extinction – began. What might be thought of as its opening phase appears to have started about fifty thousand years ago. At that time, Australia was home to a fantastic assortment of enormous animals; these included a wombatlike creature the size of a hippo, a land tortoise nearly as big as a VW Beetle, and the giant short-faced kangaroo, which grew to be ten feet tall. Then all of the continent's largest animals disappeared. Every species of marsupial weighing more than two hundred pounds – there were nineteen of them – vanished, as did three species of giant reptiles and a flightless bird with stumpy legs known as *Genyornis newtoni*.

This die-off roughly coincided with the arrival of the first people on the continent, probably from Southeast Asia. Australia is a big place, and there couldn't have been very many early settlers. For a long time, the coincidence was discounted. Yet, thanks to recent work by geologists and paleontologists, a clear global pattern has emerged. About eleven thousand years ago, three-quarters of North America's largest animals – among them mastodons, mammoths, giant beavers, short-faced bears, and sabre-toothed tigers – began to go extinct. This is right around the time the first humans are believed to have wandered across the Bering land bridge. In relatively short order, the first humans settled South America as well. Subsequently, more than thirty species of South American "megamammals," including elephant-size ground sloths and rhino-like creatures known as toxodons, died out.

40   And what goes for Australia and the Americas also goes for many other parts of the world. Humans settled Madagascar around two thousand years ago; the island subsequently lost all mammals weighing more than twenty pounds, including pygmy hippos and giant lemurs. "Substantial losses have occurred throughout near time," Ross MacPhee, a curator at the American Museum of Natural History, in New York, and an expert on extinctions of

the recent geological past, has written. "In the majority of cases, these losses occurred when, and only when, people began to expand across areas that had never before experienced their presence." The Maori arrived in New Zealand around eight hundred years ago. They encountered eleven species of moas – huge ostrichlike creatures without wings. Within a few centuries – and possibly within a single century – all eleven moa species were gone. While these "first contact" extinctions were most pronounced among large animals, they were not confined to them. Humans discovered the Hawaiian Islands around fifteen hundred years ago; soon afterward, ninety per cent of Hawaii's native bird species disappeared.

"We expect extinction when people arrive on an island," David Steadman, the curator of ornithology at the Florida Museum of Natural History, has written. "Survival is the exception."

Why was the first contact with humans so catastrophic? Some of the animals may have been hunted to death; thousands of moa bones have been found at Maori archaeological sites, and man-made artifacts have been uncovered near mammoth and mastodon remains at more than a dozen sites in North America. Hunting, however, seems insufficient to account for so many losses across so many different taxa in so many parts of the globe. A few years ago, researchers analyzed hundreds of bits of emu and *Genyornis newtoni* eggshell, some dating from long before the first people arrived in Australia and some from after. They found that around forty-five thousand years ago, rather abruptly, emus went from eating all sorts of plants to relying mainly on shrubs. The researchers hypothesized that Australia's early settlers periodically set the countryside on fire – perhaps to flush out prey – a practice that would have reduced the variety of plant life. Those animals which, like emus, could cope with a changed landscape survived, while those which, like *Genyornis*, could not died out.

When Australia was first settled, there were maybe half a million people on earth. There are now more than six and a half billion, and it is expected that within the next three years the number will reach seven billion.

Human impacts on the planet have increased proportionately. Farming, logging, and building have transformed between a third and a half of the world's land surface, and even these figures probably understate the effect, since land not being actively exploited may still be fragmented. Most of the world's major waterways have been diverted or dammed or otherwise manipulated – in the United States, only two per cent of rivers run unimpeded – and people now use half the world's readily accessible freshwater runoff. Chemical plants fix more atmospheric nitrogen than all natural terrestrial processes combined, and fisheries remove more than a third of the primary production of the temperate coastal waters of the oceans. Through global trade and international travel, humans have transported countless species into ecosystems that are not prepared for them. We have pumped enough carbon dioxide into the air to alter the climate and to change the chemistry of the oceans.

45   Amphibians are affected by many – perhaps most – of these disruptions. Habitat destruction is a major factor in their decline, and agricultural chemicals seem to be causing a rash of frog deformities. But the main culprit in the wavelike series of crashes, it's now believed, is a fungus. Ironically, this fungus, which belongs to a group known as chytrids (pronounced "kit-rids"), appears to have been spread by doctors.

Chytrid fungi are older even than amphibians – the first species evolved more than six hundred million years ago – and even more widespread. In a manner of speaking, they can be found – they are microscopic – just about everywhere, from the tops of trees to deep underground. Generally, chytrid fungi feed off dead plants; there are also species that live on algae, species that live on roots, and species that live in the guts of cows, where they help break down cellulose. Until two pathologists, Don Nichols and Allan Pessier, identified a weird microorganism growing on dead frogs from the National Zoo, chytrids had never been known to attack vertebrates. Indeed, the new chytrid was so unusual that an entire genus had to be created to accommodate it. It was named *Batrachochytrium dendrobatidis* – *batrachos* is Greek for "frog" – or Bd for short.

Nichols and Pessier sent samples from the infected frogs to a mycologist at the University of Maine, Joyce Longcore, who managed to culture the Bd fungus. Then they exposed healthy blue poison-dart frogs to it. Within three weeks, the animals had sickened and died.

The discovery of Bd explained many of the data that had previously seemed so puzzling. Chytrid fungi generate microscopic spores that disperse in water; these could have been carried along by streams, or in the runoff after a rainstorm, producing what in Central America showed up as an eastward-moving scourge. In the case of zoos, the spores could have been brought in on other frogs or on tracked-in soil. Bd seemed to be able to live on just about any frog or toad, but not all amphibians are as susceptible to it, which would account for why some populations succumbed while others appeared to be unaffected.

Rick Speare is an Australian pathologist who identified Bd right around the same time that the National Zoo team did. From the pattern of decline, Speare suspected that Bd had been spread by an amphibian that had been moved around the globe. One of the few species that met this condition was *Xenopus laevis*, commonly known as the African clawed frog. In the early nineteen-thirties, a British zoologist named Lancelot Hogben discovered that female *Xenopus laevis*, when injected with certain types of human hormones, laid eggs. His discovery became the basis for a new kind of pregnancy test and, starting in the late nineteen-thirties, thousands of African clawed frogs were exported out of Cape Town. In the nineteen-forties and fifties, it was not uncommon for obstetricians to keep tanks full of the frogs in their offices.

To test his hypothesis, Speare began collecting samples from live African 50 clawed frogs and also from specimens preserved in museums. He found that specimens dating back to the nineteen-thirties were indeed already carrying the fungus. He also found that live African clawed frogs were widely infected with Bd, but seemed to suffer no ill effects from it. In 2004, he co-authored an influential paper that argued that the transmission route for the fungus began in southern Africa and ran through clinics and hospitals around the world.

"Let's say people were raising African clawed frogs in aquariums, and they just popped the water out," Speare told me. "In most cases when they did that, no frogs got infected, but then on that hundredth time, one local frog might have been infected. Or people might have said, 'I'm sick of this frog. I'm going to let it go.' And certainly there are populations of African clawed frogs established in a number of countries around the world, to illustrate that that actually did occur."

At this point, Bd appears to be, for all intents and purposes, unstoppable. It can be killed by bleach – Clorox is among the donors to EVACC – but it is impossible to disinfect an entire rain forest. Sometime in the last year or so, the fungus jumped the Panama Canal. (When Edgardo Griffith swabbed the frogs on our trip, he was collecting samples that would eventually be analyzed for it.) It also seems to be heading into Panama from the opposite direction, out of Colombia. It has spread through the highlands of South America, down the eastern coast of Australia, and into New Zealand, and has been detected in Italy, Spain, and France. In the US, it appears to have radiated from several points, not so much in a wavelike pattern as in a series of ripples.

In the fossil record, mass extinctions stand out, so sharply that the very language scientists use to describe the earth's history derives from them. In 1840, the British geologist John Phillips divided life into three chapters: the Paleozoic (from the Greek for "ancient life"), the Mesozoic ("middle life"), and the Cenozoic ("new life"). Phillips fixed as the dividing point between the first and second eras what would now be called the end-Permian extinction, and between the second and the third the end-Cretaceous event. The fossils from these eras were so different that Phillips thought they represented three distinct episodes of creation.

Darwin's resistance to catastrophism meant that he couldn't accept what the fossils seemed to be saying. Drawing on the work of the eminent geologist Charles Lyell, a good friend of his, Darwin maintained that the apparent discontinuities in the history of life were really just gaps in the archive. In "On the Origin of Species," he argued:

55   With respect to the apparently sudden extermination of whole families or orders, as of Trilobites at the close of the palaeozoic period and of Ammonites at the close of the secondary period, we must remember what has been already said on the probable wide intervals of time between our consecutive formations; and in these intervals there may have been much slow extermination.

All the way into the nineteen-sixties, paleontologists continued to give talks with titles like "The Incompleteness of the Fossil Record." And this view might have persisted even longer had it not been for a remarkable, largely inadvertent discovery made in the following decade.

In the mid-nineteen-seventies, Walter Alvarez, a geologist at the Lamont Doherty Earth Observatory, in New York, was studying the earth's polarity. It had recently been learned that the orientation of the planet's magnetic field reverses, so that every so often, in effect, south becomes north and then vice versa. Alvarez and some colleagues had found that a certain formation of pinkish limestone in Italy, known as the *scaglia rossa*, recorded these occasional reversals. The limestone also contained the fossilized remains of millions of tiny sea creatures called foraminifera. In the course of several trips to Italy, Alvarez became interested in a thin layer of clay in the limestone that seemed to have been laid down around the end of the Cretaceous. Below the layer, certain species of foraminifera – or forams, for short – were preserved. In the clay layer there were no forams. Above the layer, the earlier species disappeared and new forams appeared. Having been taught the uniformitarian view, Alvarez wasn't sure what to make of what he was seeing, because the change, he later recalled, certainly "looked very abrupt."

Alvarez decided to try to find out how long it had taken for the clay layer to be deposited. In 1977, he took a post at the University of California at Berkeley, where his father, the Nobel prize-winning physicist Luis Alvarez, was also teaching. The older Alvarez suggested using the element iridium to answer the question.

Iridium is extremely rare on the surface of the earth, but more plentiful in meteorites, which, in the form of microscopic grains of cosmic dust, are constantly raining down on the planet. The Alvarezes reasoned that, if the clay layer had taken a significant amount of time to deposit, it would contain detectable levels of iridium, and if it had been deposited in a short time it wouldn't. They enlisted two other scientists, Frank Asaro and Helen Michel, to run the tests, and gave them samples of the clay. Nine months later, they got a phone call. There was something seriously wrong. Much too much iridium was showing up in the samples. Walter Alvarez flew to Denmark to take samples of another layer of exposed clay from the end of the Cretaceous. When they were tested, these samples, too, were way out of line.

The Alvarez hypothesis, as it became known, was that everything – the clay layer from the *scaglia rossa*, the clay from Denmark, the spike in iridium, the shift in the fossils – could be explained by a single event. In 1980, the Alvarezes and their colleagues proposed that a six-mile-wide asteroid had slammed into the earth, killing off not only the forams but the dinosaurs and all the other organisms that went extinct at the end of the Cretaceous. "I can remember working very hard to make that 1980 paper just as solid as it could possibly be," Walter Alvarez recalled recently. Nevertheless, the idea was greeted with incredulity.

"The arrogance of these people is simply unbelievable," one paleontologist told the *Times*.

"Unseen bodies dropping into an unseen sea are not for me," another declared.

Over the next decade, evidence in favor of an enormous impact kept accumulating. Geologists looking at rocks from the end of the Cretaceous in Montana found tiny mineral grains that seemed to have suffered a violent shock. (Such "shocked quartz" is typically found in the immediate vicinity of meteorite craters.) Other geologists, looking in other parts of the world, found small, glasslike spheres of the sort believed to form when molten-rock droplets splash up into the atmosphere. In 1990, a crater large enough to have been formed by the enormous asteroid that the Alvarezes were proposing was found, buried underneath the Yucatán. In 1991, that crater was dated, and discovered to have been formed at precisely the time the dinosaurs died off.

"Those eleven years seemed long at the time, but looking back they seem very brief," Walter Alvarez told me. "Just think about it for a moment. Here you have a challenge to a uniformitarian viewpoint that basically every geologist and paleontologist had been trained in, as had their professors and their professors' professors, all the way back to Lyell. And what you saw was people looking at the evidence. And they gradually did come to change their minds."

Today, it's generally accepted that the asteroid that plowed into the Yucatán led, in very short order, to a mass extinction, but scientists are still uncertain exactly how the process unfolded. One theory holds that the impact raised a cloud of dust that blocked the sun, preventing photosynthesis and causing widespread starvation. According to another theory, the impact kicked up a plume of vaporized rock traveling with so much force that it broke through the atmosphere. The particles in the plume then recondensed, generating, as they fell back to earth, enough thermal energy to, in effect, broil the surface of the planet.

Whatever the mechanism, the Alvarezes' discovery wreaked havoc with the uniformitarian idea of extinction. The fossil record, it turned out, was marked by discontinuities because the history of life was marked by discontinuities.

In the nineteenth century, and then again during the Second World War, the Adirondacks were a major source of iron ore. As a result, the mountains are now riddled with abandoned mines. On a gray day this winter, I went to visit one of the mines (I was asked not to say which) with a wildlife biologist named Al Hicks. Hicks, who is fifty-four, is tall and outgoing, with a barrel chest and ruddy cheeks. He works at the headquarters of the New York State Department of Environmental Conservation, in Albany, and we met in a parking lot not far from his office. From there, we drove almost due north.

Along the way, Hicks explained how, in early 2007, he started to get a lot of strange calls about bats. Sometimes the call would be about a dead bat that had been brought inside by somebody's dog. Sometimes it was about a live – or half-alive – bat flapping around on the driveway. This was in the middle of winter, when any bat in the Northeast should have been hanging by its feet in a state of torpor. Hicks found the calls bizarre, but, beyond that, he didn't know what to make of them. Then, in March 2007, some colleagues went to do a routine census of hibernating bats in a cave west of Albany. After the survey, they, too, phoned in.

"They said, 'Holy shit, there's dead bats everywhere,'" Hicks recalled. He instructed them to bring some carcasses back to the office, which they did. They also shot photographs of live bats hanging from the cave's ceiling. When Hicks examined the photographs, he saw that the animals looked as if they had been dunked, nose first, in talcum powder. This was something he had never run across before, and he began sending the bat photographs to all the bat specialists he could think of. None of them could explain it, either.

70   "We were thinking, Oh boy, we hope this just goes away," he told me. "It was like the Bush Administration. And, like the Bush Administration, it just wouldn't go away." In the winter of 2008, bats with the white powdery substance were found in thirty-three hibernating spots. Meanwhile, bats kept

dying. In some hibernacula, populations plunged by as much as ninety-seven per cent.

That winter, officials at the National Wildlife Health Center, in Madison, Wisconsin, began to look into the situation. They were able to culture the white substance, which was found to be a never before identified fungus that grows only at cold temperatures. The condition became known as white-nose syndrome, or W.N.S. White nose seemed to be spreading fast; by March, 2008, it had been found on bats in three more states – Vermont, Massachusetts, and Connecticut – and the mortality rate was running above seventy-five per cent. This past winter, white nose was found to have spread to bats in five more states: New Jersey, New Hampshire, Virginia, West Virginia, and Pennsylvania.

In a paper published recently in *Science*, Hicks and several co-authors observed that "parallels can be drawn between the threat posed by W.N.S. and that from chytridiomycosis, a lethal fungal skin infection that has recently caused precipitous global amphibian population declines."

When we arrived at the base of a mountain not far from Lake Champlain, more than a dozen people were standing around in the cold, waiting for us. Most, like Hicks, were from the D.E.C., and had come to help conduct a bat census. In addition, there was a pair of biologists from the US Fish and Wildlife Service and a local novelist who was thinking of incorporating a subplot about white nose into his next book. Everyone put on snowshoes, except for the novelist, who hadn't brought any, and began tromping up the slope toward the mine entrance.

The snow was icy and the going slow, so it took almost half an hour to reach an outlook over the Champlain Valley. While we were waiting for the novelist to catch up – apparently, he was having trouble hiking through the three-foot-deep drifts – the conversation turned to the potential dangers of entering an abandoned mine. These, I was told, included getting crushed by falling rocks, being poisoned by a gas leak, and plunging over a sheer drop of a hundred feet or more.

After another fifteen minutes or so, we reached the mine entrance – essentially, a large hole cut into the hillside. The stones in front of the entrance were white with bird droppings, and the snow was covered with paw prints. Evidently, ravens and coyotes had discovered that the spot was an easy place to pick up dinner.

"Well, shit," Hicks said. Bats were fluttering in and out of the mine, and in some cases crawling on the ground. Hicks went to catch one; it was so lethargic that he grabbed it on the first try. He held it between his thumb and his forefinger, snapped its neck, and placed it in a ziplock bag.

"Short survey today," he announced.

At this point, it's not known exactly how the syndrome kills bats. What is known is that bats with the syndrome often wake up from their torpor and fly around, which leads them to die either of starvation or of the cold or to get picked off by predators.

We unstrapped our snowshoes and put on helmets. Hicks handed out headlamps – we were supposed to carry at least one extra – and packages of batteries; then we filed into the mine, down a long, sloping tunnel. Shattered beams littered the ground, and bats flew up at us through the gloom. Hicks cautioned everyone to stay alert. "There's places that if you take a step you won't be stepping back," he warned. The tunnel twisted along, sometimes opening up into concert-hall-size chambers with side tunnels leading out of them.

80   Over the years, the various sections of the mine had acquired names; when we reached something called the Don Thomas section, we split up into groups to start the survey. The process consisted of photographing as many bats as possible. (Later on, back in Albany, someone would have to count all the bats in the pictures.) I went with Hicks, who was carrying an enormous camera, and one of the biologists from the Fish and Wildlife Service, who had a laser pointer. The biologist would aim the pointer at a cluster of bats hanging from the ceiling. Hicks would then snap a photograph. Most of the bats were little brown bats; these are the most common bats in the US, and the ones you are most likely to see flying around on a summer night. There were also Indiana bats, which are on the federal endangered-species list, and small-footed bats, which, at the rate things are going, are likely to end up there. As we moved along, we kept disturbing the bats, which squeaked and started to rustle around, like half-asleep children.

Since white nose grows only in the cold, it's odd to find it living on mammals, which, except when they're hibernating (or dead), maintain a high body temperature. It has been hypothesized that the fungus normally subsists by breaking down organic matter in a chilly place, and that it was transported to bat hibernacula, where it began to break down bats. When news of white nose began to get around, a spelunker sent Hicks photographs that he had shot in Howe's Cave, in central New York. The photographs, which had been taken in 2006, showed bats with clear signs of white nose and are the earliest known record of the syndrome. Howe's Cave is connected to Howe's Caverns, a popular tourist destination.

"It's kind of interesting that the first record we have of this fungus is photographs from a commercial cave in New York that gets about two hundred thousand visits a year," Hicks told me.

Despite the name, white nose is not confined to bats' noses; as we worked our way along, people kept finding bats with freckles of fungus on their wings and ears. Several of these were dispatched, for study purposes, with a thumb and forefinger. Each dead bat was sexed – males can be identified by their tiny penises – and placed in a ziplock bag.

At about 7 pm, we came to a huge, rusty winch, which, when the mine was operational, had been used to haul ore to the surface. By this point, we were almost down at the bottom of the mountain, except that we were on the inside of it. Below, the path disappeared into a pool of water, like the River Styx.[a] It was impossible to go any further, and we began working our way back up.

Bats, like virtually all other creatures alive today, are masters of adaptation descended from lucky survivors. The earliest bat fossil that has been found dates from fifty-three million years ago, which is to say twelve million years after the impact that ended the Cretaceous. It belongs to an animal that had wings and could fly but had not yet developed the specialized inner ear that, in modern bats, allows for echolocation. Worldwide, there are now more than a thousand bat species, which together make up nearly a fifth of all species of mammals. Most feed on insects; there are also bats that live off fruit, bats that eat fish – they use echolocation to detect minute ripples in the water – and a small but highly celebrated group that consumes blood. Bats are great colonizers – Darwin noted that even New Zealand, which has no other native mammals, has its own bats – and they can be found as far north as Alaska and as far south as Tierra del Fuego.

In the time that bats have evolved and spread, the world has changed a great deal. Fifty-three million years ago, at the start of the Eocene, the planet was very warm, and tropical palms grew at the latitude of London. The climate cooled, the Antarctic ice sheet began to form, and, eventually, about two million years ago, a period of recurring glaciations began. As recently as fifteen thousand years ago, the Adirondacks were buried under ice.

One of the puzzles of mass extinction is why, at certain junctures, the resourcefulness of life seems to falter. Powerful as the Alvarez hypothesis proved to be, it explains only a single mass extinction.

"I think that, after the evidence became pretty strong for the impact at the end of the Cretaceous, those of us who were working on this naïvely expected that we would go out and find evidence of impacts coinciding with the other events," Walter Alvarez told me. "And, of course, it's turned out to

85

---

a *River Styx* In Greek mythology, a river at the edge of the Underworld.

be much more complicated. We're seeing right now that a mass extinction can be caused by human beings. So it's clear that we do not have a general theory of mass extinction."

Andrew Knoll, a paleontologist at Harvard, has spent most of his career studying the evolution of early life. (Among the many samples he keeps in his office are fossils of microorganisms that lived 2.8 billion years ago.) He has also written about more recent events, like the end-Permian extinction, which took place two hundred and fifty million years ago, and the current extinction event.

90 Knoll noted that the world can change a lot without producing huge losses; ice ages, for instance, come and go. "What the geological record tells us is that it's time to worry when the rate of change is fast," he told me. In the case of the end-Permian extinction, Knoll and many other researchers believe that the trigger was a sudden burst of volcanic activity; a plume of hot mantle rock from deep in the earth sent nearly a million cubic miles' worth of flood basalts streaming over what is now Siberia. The eruption released enormous quantities of carbon dioxide, which presumably led – then as now – to global warming, and to significant changes in ocean chemistry.

"CO_2 is a paleontologist's dream," Knoll told me. "It can kill things directly, by physiological effects, of which ocean acidification is the best known, and it can kill things by changing the climate. If it gets warmer faster than you can migrate, then you're in trouble."

In the end, the most deadly aspect of human activity may simply be the pace of it. Just in the past century, CO_2 levels in the atmosphere have changed by as much – a hundred parts per million – as they normally do in a hundred-thousand year glacial cycle. Meanwhile, the drop in ocean pH levels that has occurred over the past fifty years may well exceed anything that happened in the seas during the previous fifty million. In a single afternoon, a pathogen like Bd can move, via United or American Airlines, halfway around the world. Before man entered the picture, such a migration would have required hundreds, if not thousands, of years – if, indeed, it could have been completed at all.

Currently, a third of all amphibian species, nearly a third of reef-building corals, a quarter of all mammals, and an eighth of all birds are classified as "threatened with extinction." These estimates do not include the species that humans have already wiped out or the species for which there are insufficient data. Nor do the figures take into account the projected effects of global warming or ocean acidification. Nor, of course, can they anticipate the kinds of sudden, terrible collapses that are becoming almost routine.

I asked Knoll to compare the current situation with past extinction events. He told me that he didn't want to exaggerate recent losses, or to suggest that an extinction on the order of the end-Cretaceous or end-Permian was imminent. At the same time, he noted, when the asteroid hit the Yucatán, "it was one terrible afternoon." He went on, "But it was a short-term event, and then things started getting better. Today, it's not like you have a stress and the stress is relieved and recovery starts. It gets bad and then it keeps being bad, because the stress doesn't go away. Because the stress is us."

Aeolus Cave, in Dorset, Vermont, is believed to be the largest bat hibernacu- 95
lum in New England; it is estimated that, before white nose hit, more than two hundred thousand bats – some from as far away as Ontario and Rhode Island – came to spend the winter there.

In late February, I went with Hicks to visit Aeolus. In the parking lot of the local general store, we met up with officials from the Vermont Fish and Wildlife Department, who had organized the trip. The entrance to Aeolus is about a mile and a half from the nearest road, up a steep, wooded hillside. This time, we approached by snowmobile. The temperature outside was about twenty-five degrees – far too low for bats to be active – but when we got near the entrance we could, once again, see bats fluttering around. The most senior of the Vermont officials, Scott Darling, announced that we'd have to put on latex gloves and Tyvek suits before proceeding. At first, this seemed to me to be paranoid; soon, however, I came to see the sense of it.

Aeolus is a marble cave that was created by water flow over the course of thousands of years. The entrance is a large, horizontal tunnel at the bottom of a small hollow. To keep people out, the Nature Conservancy, which owns the cave, has blocked off the opening with huge iron slats, so that it looks like the gate of a medieval fortress. With a key, one of the slats can be removed; this creates a narrow gap that can be crawled (or slithered) through. Despite the cold, there was an awful smell emanating from the cave – half game farm, half garbage dump. When it was my turn, I squeezed through the gap and immediately slid on the ice, into a pile of dead bats. The scene, in the dimness, was horrific. There were giant icicles hanging from the ceiling, and from the floor large knobs of ice rose up, like polyps. The ground was covered with dead bats; some of the ice knobs, I noticed, had bats frozen into them. There were torpid bats roosting on the ceiling, and also wide-awake ones, which would take off and fly by or, sometimes, right into us.

Why bat corpses pile up in some places, while in others they get eaten or in some other way disappear, is unclear. Hicks speculated that the weather conditions at Aeolus were so harsh that the bats didn't even make it out of the cave before dropping dead. He and Darling had planned to do a count of the bats in the first chamber of the cave, known as Guano Hall, but this plan was soon abandoned, and it was decided just to collect specimens. Darling explained that the specimens would be going to the American Museum of Natural History, so that there would at least be a record of the bats that had once lived in Aeolus. "This may be one of the last opportunities," he said. In contrast to a mine, which has been around at most for centuries, Aeolus, he pointed out, has existed for millennia. It's likely that bats have been hibernating there, generation after generation, since the end of the last ice age.

"That's what makes this so dramatic – it's breaking the evolutionary chain," Darling said.

100   He and Hicks began picking dead bats off the ground. Those which were too badly decomposed were tossed back; those which were more or less intact were sexed and placed in two-quart plastic bags. I helped out by holding open the bag for females. Soon, it was full and another one was started. It struck me, as I stood there holding a bag filled with several dozen stiff, almost weightless bats, that I was watching mass extinction in action.

Several more bags were collected. When the specimen count hit somewhere around five hundred, Darling decided that it was time to go. Hicks hung back, saying that he wanted to take some pictures. In the hours we had been slipping around the cave, the carnage had grown even more grotesque; many of the dead bats had been crushed and now there was blood oozing out of them. As I made my way up toward the entrance, Hicks called after me: "Don't step on any dead bats." It took me a moment to realize that he was joking.

(2009)

Questions:

1. Identify what material in this article is "evidence," and what material functions through other forms of persuasion.

2. The article is divided into sections using line breaks. Outline the structure of the article in terms of these divisions.

3. This article quotes Hicks, who says of white nose syndrome, "It's kind of interesting that the first record we have of this fungus is photographs from a commercial cave in New York that gets about two hundred thousand visits a year." What does this use of juxtaposition or correlation imply? To what extent does the information in the article support this implication?

4. This article incorporates statistics and scientific terminology. Give some examples of this incorporation. What strategies does the article use to make scientific information accessible to the reader?

5. Discuss the significance of the question mark in this article's title. Does it reflect the tone of the article itself? Why do you think it is used?

6. Why are bats and frogs used as examples in this article? What do they have in common? Why are frogs seen as a "sentinel species"?

7. In the second-last paragraph, Kolbert writes, "It struck me, as I stood there holding a bag filled with several dozen stiff, almost weightless bats, that I was watching mass extinction in action." Consider the emotional and metaphoric impact of this statement. For instance, how does the writer's use of the first person function here?

8. Compare this article to those on the problems facing bees and frogs (see docs. 31 and 32 in this anthology, or find other articles on "colony collapse disorder" and "amphibian die-off"). Do you see any common denominators? How might these problems, and other extinctions, be similar to the mass extinctions of the past? Are there significant differences?

9. This article focuses on frogs and bats as illustrative examples of the current extinctions. Find and research another example, and write about it in the style of Kolbert.

10. Kolbert's article describes the research process behind Alvarez et al.'s article, "Extraterrestrial Cause for the Cretaceous-Tertiary Extinction," also included in this anthology. Read that article and answer the following questions:

    a. How does Kolbert's description of Alvarez et al.'s research process differ from their own description of it in their article? How do the different descriptions function, textually?

    b. Both articles make use of numerical data. In each instance, what kind of data is used? How is it presented? Is one presentation more accessible than the other? More persuasive? More objective?

11. Walter Alvarez is quoted in this article as saying that the asteroid impact hypothesis does not provide "'a general theory of mass extinction.'" What does this mean?

four

The "Hockey Stick" Graph

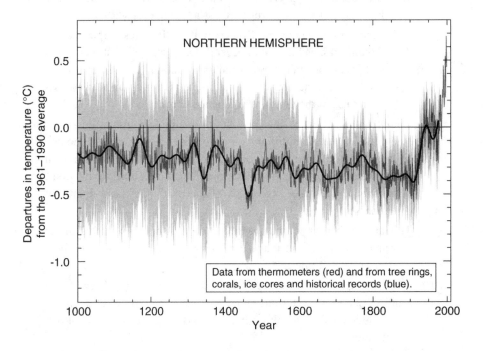

This graph, which shows a reconstruction of the average temperatures in the Northern Hemisphere over the past thousand years, derives its popular name from its shape: the relatively even temperatures up to the twentieth century constitute the "handle," while the sharp rise in temperatures during the twentieth century constitutes the "blade." Thermometer data is available for the 1850s onward, and the graph uses thermometer data from that point, but up to the late 1900s the graph also relies on "proxy" sources of data – sources such as tree rings, ice core samples taken from ice sheets, and bands in corals, all of which can be analyzed to produce estimates of past temperatures.

Climate scientist Michael Mann published the first version of this graph in 1998, and the version reprinted here was included in a 2001 report issued by the Intergovernmental Panel on Climate Change. After the release of that report, the "hockey stick" became a focus of attack from climate change deniers; see the next selections in this anthology for further discussion of the controversy surrounding the graph. Although some details of the graph have been successfully challenged, its fundamental conclusion – that average temperatures in the Northern Hemisphere have been higher in the past century than at any other point in the past thousand years – has held up to scrutiny, and has been confirmed by many independent temperature reconstructions.

Questions:

1. Assess the design elements of this graph on its own merits. For instance, what temperature range does it cover?
2. Does the graph look like a hockey stick to you? Why do you think that term was chosen?
3. The term "hockey stick graph" is now a well-known phrase. How would you go about finding out who originally used it?

five

MICHAEL MANN

Myth vs. Fact Regarding the "Hockey Stick"

This article appeared on the climate science website RealClimate.
Written by one of the creators of the "hockey stick" graph,
it attempts to address various myths surrounding the graph and
its role in the scientific discussion of climate change. A version
of the graph itself is reproduced in this anthology.

☙

Numerous myths regarding the so-called "hockey stick" reconstruction of past temperatures can be found on various non-peer reviewed websites, internet newsgroups and other non-scientific venues. The most widespread of these myths are debunked below:

MYTH #0: Evidence for modern human influence on climate rests entirely upon the "Hockey Stick" Reconstruction of Northern Hemisphere mean temperatures indicating anomalous late 20th century warmth.

This peculiar suggestion is sometimes found in op-ed[a] pieces and other dubious propaganda, despite its transparent absurdity. Paleoclimate[b] evidence is simply one in a number of independent lines of evidence indicating the strong likelihood that human influences on climate play a dominant role in the observed 20th century warming of the earth's surface. Perhaps the strongest piece of evidence in support of this conclusion is the evidence from so-called "Detection and Attribution Studies." Such studies demonstrate that the pattern of 20th century climate change closely matches that predicted by state-of-the-art models of the climate system in response to 20th century anthropogenic

---

a *op-ed* Page in a newspaper that is "opposite" the "editorial" page, usually containing comment and feature articles related to current events.

b *Paleoclimate* Climate of the past, especially when studied on a geologic time scale.

forking[a] (due to the combined influence of anthropogenic greenhouse gas concentrations and industrial aerosol increases).

MYTH #1: The "Hockey Stick" Reconstruction is based solely on two publications by climate scientist Michael Mann and colleagues (Mann et al., 1998; 1999).

This is patently false. Nearly a dozen model-based and proxy-based[b] reconstructions of Northern Hemisphere mean temperature by different groups all suggest that late 20th century warmth is anomalous in a long-term (multi-century to millennial) context.... 5

Some proxy-based reconstructions suggest greater variability than others. This greater variability may be attributable to different emphases in seasonal and spatial emphasis (see Jones and Mann, 2004; Rutherford et al., 2004; Cook et al., 2004). However, even for those reconstructions which suggest a colder "Little Ice Age" and greater variability in general in past centuries, such as that of Esper et al. (2002), late 20th century hemispheric warmth is still found to be anomalous in the context of the reconstruction (see Cook et al., 2004).

MYTH #2: Regional proxy evidence of warm or anomalous (wet or dry) conditions in past centuries contradicts the conclusion that late 20th century hemispheric mean warmth is anomalous in a long-term (multi-century to millennial) context.

Such claims reflect a lack of awareness of the distinction between regional and large-scale climate change. Similar such claims were recently made in two articles by astronomer Willie Soon and co-authors (Soon and Baliunas, 2003; Soon et al., 2003). These claims were subsequently rebutted by a group of more than a dozen leading climate scientists in an article in the journal *Eos* of the American Geophysical Union (Mann et al., *Eos*, 2003). The rebuttal raised, among other points, the following two key points:

(1) In drawing conclusions regarding past regional temperature changes from proxy records, it is essential to assess proxy data for actual sensitivity to past temperature variability. In some cases (Soon and Baliunas, 2003, Soon et al., 2003) a global 'warm anomaly' has been defined for any period

---

a   *anthropogenic forcing*   Influence on climate caused by the products of human activity.

b   *model-based*   Created through mathematical modeling;   *proxy-based*   Created from data gathered from "climate proxies" (information sources, such as tree rings and historical records, that can be used to estimate past temperatures).

during which various regions appear to indicate climate anomalies that can be classified as being *either* 'warm', 'wet', *or* 'dry' relative to '20th century' conditions. Such a criterion could be used to define *any* period of climate as 'warm' or 'cold', and thus cannot meaningfully characterize past large-scale surface temperature changes.

10   (2) It is essential to distinguish (e.g. by compositing or otherwise assimilating different proxy information in a consistent manner – e.g., Jones et al., 1998; Mann et al., 1998, 1999; Briffa et al., 2001) between regional temperature changes and changes in global or hemispheric mean temperature. Specific periods of cold and warmth differ from region to region over the globe (see Jones and Mann, 2004), as changes in atmospheric circulation over time exhibit a wave-like character, ensuring that certain regions tend to warm (due, for example, to a southerly flow in the Northern Hemisphere winter mid-latitudes) when other regions cool (due to the corresponding northerly flow that must occur elsewhere). Truly representative estimates of global or hemispheric average temperature must therefore average temperature changes over a sufficiently large number of distinct regions to average out such offsetting regional changes. The specification of a warm period, therefore, requires that warm anomalies in different regions should be truly synchronous and not merely required to occur within a very broad interval in time, such as 800–1300 CE (as in Soon et al., 2003; Soon and Baliunas, 2003).

MYTH #3: The "Hockey Stick" studies claim that the 20th century on the whole is the warmest period of the past 1000 years.

This is a mis-characterization of the actual scientific conclusions. Numerous studies suggest that hemispheric mean warmth for the late 20th century (that is, the past few decades) appears to exceed the warmth of any comparable length period over the past thousand years or longer, taking into account the uncertainties in the estimates.... On the other hand, in the context of the long-term reconstructions, the early 20th century appears to have been a relatively cold period while the mid 20th century was comparable in warmth, by most estimates, to peak Medieval warmth (i.e., the so-called "Medieval Warm Period"). It is not the average 20th century warmth, but the magnitude of warming during the 20th century, and the level of warmth observed during the past few decades, which appear to be anomalous in a long-term context. Studies such as those of Soon and associates (Soon and Baliunas, 2003; Soon et al., 2003) that consider only '20th century' conditions, or interpret past temperature changes using evidence incapable of resolving trends in recent decades, cannot meaningfully address the question of whether late 20th century warmth is anomalous in a long-term and large-scale context.

MYTH #4: Errors in the "Hockey Stick" undermine the conclusion that late 20th century hemispheric warmth is anomalous.

This statement embraces at least two distinct falsehoods. The first falsehood holds that the "Hockey Stick" is the result of one analysis or the analysis of one group of researchers (i.e., that of Mann et al., 1998 and Mann et al., 1999). However, as discussed in the response to Myth #1 above, the basic conclusions of Mann et al. (1998, 1999) are affirmed in multiple independent studies. Thus, even if there were errors in the Mann et al. (1998) reconstruction, numerous other studies independently support the conclusion of anomalous late 20th century hemispheric-scale warmth.

The second falsehood holds that there are errors in the Mann et al. (1998, 1999) analyses, and that these putative errors compromise the "hockey stick" shape of hemispheric surface temperature reconstructions. Such claims seem to be based in part on the misunderstanding or misrepresentation by some individuals of a corrigendum[a] that was published by Mann and colleagues in *Nature*. This corrigendum simply corrected the descriptions of supplementary information that accompanied the Mann et al. article detailing precisely what data were used. As clearly stated in the corrigendum, these corrections have no influence at all on the actual analysis or any of the results shown in Mann et al. (1998). Claims that the corrigendum reflects any errors at all in the Mann et al. (1998) reconstruction are entirely false.

False claims of the existence of errors in the Mann et al. (1998) reconstruction can also be traced to spurious allegations made by two individuals, McIntyre and McKitrick (McIntyre works in the mining industry, while McKitrick is an economist). The false claims were first made in an article (McIntyre and McKitrick, 2003) published in a non-scientific (social science) journal *Energy and Environment* and later, in a separate "Communications Arising" comment that was rejected by *Nature* based on negative appraisals by reviewers and editor [as a side note, we find it peculiar that the authors have argued elsewhere that their submission was rejected due to 'lack of space'. *Nature* makes their policy on such submissions quite clear: "The Brief Communications editor will decide how to proceed on the basis of whether the central conclusion of the earlier paper is brought into question; of the length of time since the original publication; and of whether a comment

15

---

a *corrigendum* Correction.

or exchange of views is likely to seem of interest to nonspecialist readers. Because Nature receives so many comments, those that do not meet these criteria are referred to the specialist literature." Since *Nature* chose to send the comment out for review in the first place, the "time since the original publication" was clearly not deemed a problematic factor. One is logically left to conclude that the grounds for rejection were the deficiencies in the authors' arguments explicitly noted by the reviewers].[a] The rejected criticism has nonetheless been posted on the internet by the authors, and promoted in certain other non-peer-reviewed venues....

The claims of McIntyre and McKitrick, which hold that the "Hockey-Stick" shape of the MBH98 reconstruction[b] is an artifact of the use of series with infilled data and the convention by which certain networks of proxy data were represented in a Principal Components Analysis[c] ("PCA"), are readily seen to be false, as detailed in a response by Mann and colleagues to their rejected *Nature* criticism demonstrating that (1) the Mann et al. (1998) reconstruction is robust with respect to the elimination of any data that were infilled in the original analysis, (2) the main features of the Mann et al. (1998) reconstruction are entirely insensitive to whether or not proxy data networks are represented by PCA, (3) the putative 'correction' by McIntyre and McKitrick, which argues for anomalous 15th century warmth (in contradiction to all other known reconstructions), is an artifact of the censoring by the authors of key proxy data in the original Mann et al. (1998) dataset, and finally, (4) unlike the original Mann et al. (1998) reconstruction, the so-called 'correction' by McIntyre and McKitrick fails statistical verification exercises, rendering it statistically meaningless and unworthy of discussion in the legitimate scientific literature.

The claims of McIntyre and McKitrick have now been further discredited in the peer-reviewed scientific literature, in a paper to appear in the American Meteorological Society journal, *Journal of Climate* by Rutherford and colleagues (2004).... Rutherford et al. (2004) demonstrate nearly identical results to those of MBH98, using the same proxy dataset as Mann et al. (1998) but addressing the issues of infilled/missing data raised by McIntyre and McKitrick, and using an alternative climate field reconstruction (CFR) methodology that does not represent any proxy data networks by PCA at all.

(2004)

---

a *[as a side ... reviewers]* This material was given in square brackets in the original article,

b *MBH98 reconstruction* The reconstruction put forward by Mann et al. in their 1998 article.

c *Principal Components Analysis* Mathematical process used to condense and find patterns in a set of data.

REFERENCES

Cook, E.R., J. Esper and R.D. D'Arrigo, Extra-tropical Northern Hemisphere land temperature variability over the past 1000 years, *Quat. Sci. Rev.*, 23, 2063-2074, 2004.

Crowley, T.J., and T. Lowery, How Warm Was the Medieval Warm Period?, *Ambio*, 29, 51-54, 2000.

Esper, J., E.R. Cook and F.H. Schweingruber, Low-frequency signals in long tree-line chronologies for reconstructing past temperature variability, *Science*, 295, 2250-2253, 2002.

Jones, P.D., K.R. Briffa, T.P. Barnett and S.F.B. Tett, High-resolution palaeoclimatic records for the last millennium: Integration, interpretation and comparison with General Circulation Model control run temperatures, *Holocene*, 8, 455-471, 1998.

Jones, P.D., Mann, M.E., Climate Over Past Millennia, *Reviews of Geophysics*, 42, RG2002, doi: 10.1029/2003RG000143, 2004.

Mann, M.E., R.S. Bradley, and M.K. Hughes, Global-scale temperature patterns and climate forcing over the past six centuries, *Nature*, 392, 779-787, 1998.

Mann, M.E., R.S. Bradley, and M.K. Hughes, Northern Hemisphere Temperatures During the Past Millennium: Inferences, Uncertainties, and Limitations, *Geophysical Research Letters*, 26, 759-762, 1999.

Mann, M.E., Ammann, C.M., Bradley, R.S., Briffa, K.R., Crowley, T.J., Hughes, M.K., Jones, P.D., Oppenheimer, M., Osborn, T.J., Overpeck, J.T., Rutherford, S., Trenberth, K.E., Wigley, T.M.L., On Past Temperatures and Anomalous Late 20th Century Warmth, *Eos*, 84, 256-258, 2003.

Rutherford, S., Mann, M.E., Osborn, T.J., Bradley, R.S., Briffa, K.R., Hughes, M.K., Jones, P.D., Proxy-based Northern Hemisphere Surface Temperature Reconstructions: Sensitivity to Methodology, Predictor Network, Target Season and Target Domain, *Journal of Climate*, in press, 2004.

Soon, W., and S. Baliunas, Proxy climatic and environmental changes over the past 1000 years, *Climate Research*, 23, 89-110, 2003.

Soon, W., S. Baliunas, C. Idso, S. Idso and D.R. Legates, Reconstructing climatic and environmental changes of the past 1000 years, *Energy and Environment*, 14, 233-296, 2003.

Questions:

1. Reflect on the paragraphs explaining Myth #0. What is the tone of the author's writing? What specific word choices contribute to the development of this tone?

2. According to the article, what role do anthropogenic greenhouse gases play in climate change? What other elements does the article suggest might contribute to climate change? Do you find the breakdown convincing? Why or why not?

3. Consider the kinds of references this article cites. Are they reliable sources? How does the inclusion of these sources affect your interpretation of the article?

4. The article addresses myths found on "various non-peer reviewed websites, internet newsgroups and other non-scientific venues." Is it

important to address myths posted on non-scholarly sites and venues? Why or why not?

5. Read Myth #2. Define and compare the terms "regional climate change" and "large-scale climate change." How is the distinction between those terms important to understanding Myth #2?

6. Is the article's information organized in a logical manner? Outline this organization and rearrange the material in another way.

7. The article frequently uses the term "anomalous." Define this term. Why do you think this term is important in the study of climate change?

8. Myth #3 focuses on the climate of the twentieth century, though climate change data is available for a much longer period of time. List three reasons why the twentieth century is a major focus of the "hockey stick" debate.

9. In Myth #3, the author mentions the "Medieval Warm Period." Research this event and summarize your findings. In what ways might knowing about this event change your understanding of climate fluctuation?

10. Consider the language used in the paragraphs explaining Myth #4. How do terms like "falsehood" and "spurious allegations" add to the tone of this section?

11. What is the article's attitude toward the work done by McIntyre and McKitrick, as discussed in Myth #4? Does this approach reveal a biased or unbiased attitude? If there is a bias, do you think it is justified?

12. Choose one of the five myths and research its background. Compare the information you find with the information in the article. Assess the reliability of the various sources involved. Which do you find most convincing, and why?

six

FRED PEARCE

Battle over Climate Data Turned into War between Scientists and Sceptics

In this article – the first in a series of twelve written for the Guardian – Fred Pearce examines the so-called "climategate scandal." In 2009, hackers accessed and published a series of private emails between scientists studying climate change; the contents of the emails led to allegations that leading climatologists had attempted to hide flaws in their work and to wrongfully prevent climate change deniers from accessing data.

❧

This story is dark; there are no heroes. Environmentalists will be distressed at what happens in the labs; many may think we should not publish for fear of wrecking the already battered cause of fighting climate change. But some of it, according to the British government's Information Commissioner, may have been illegal.

Remember two other things. First, this was war. The scientists were under intense and prolonged attack, they believed, from politically and commercially motivated people who wanted to prevent them from doing their science and trash their work. And they had, as their most vocal protagonist Professor Michael Mann puts it in one email, "dirty laundry one doesn't want to fall into the hands of those who might potentially try to distort things...."

Meanwhile, their attackers came to believe that the scientists were fraudsters. In many ways, what follows is a Shakespearean tragedy of misunderstood motives.

There are two competing analyses of what "climategate" means. One sees it as the mob entering the lab – the story of a malicious attempt to disrupt, cross-question, belittle and trash the work of mainstream scientists. This may or may not have been the motivation for the original hack, but it has certainly been the motive of some who have driven the news agenda since.

The second analysis sees it as democracy in action – the outcome of an 5 entirely laudable effort by amateur scientists and others outside the scientific mainstream, headed by Canadian mathematician Steve McIntyre, to gain access to the complex data sets behind some of the climate scientists' conclusions, and to subject them to their own analysis.

The interweaving of these two narratives has created the tragedy of climategate. The bunker mentality of climate scientists such as the key email correspondents – headed by the director of the Climatic Research Unit (CRU) at the University of East Anglia, Phil Jones – is exposed in the emails. But so too is the chaos caused in the labs by the efforts of outsiders to question what was going on, without using the established rules of science, like working through publication in peer-reviewed literature. The clash of cultures between the blogosphere and the pages of august journals such as *Nature* could not be greater.

All this happened against the backdrop of a long-term assault by politically motivated, and commercially funded, climate-change deniers against the activities of many of the key scientists featuring in the emails. Indeed it is striking that people with a limited scientific involvement with CRU who have been victims of past attacks – such as Kevin Trenberth of the US government's National Center for Atmospheric Research (NCAR) and Ben Santer of the Lawrence Livermore National Laboratory – became regular email correspondents with Jones and his colleagues. They were huddling together in the storm.

Through the emails we also see that some insiders were always demanding more openness from their colleagues and providing candid criticism of shoddy or mistaken work. One person stands out in this: Tom Wigley. He was Jones's former boss, having preceded him as head of CRU. Now based at the University Corporation for Atmospheric Research, in Boulder, Colorado, Wigley kept up a vigil for honesty and integrity in emails over many years. If there is a hero in this sorry tale, perhaps it is Wigley.

The science discussed in the emails is mostly from one small area of climate research – the taking of raw temperature data from thermometers, satellites and proxy measures of historical temperatures such as tree rings and turning it into useable information on temperature trends. The result being iconic graphs like the famous "hockey stick," first published 12 years ago and one of climate science's most famous and controversial products. It shows a long period of natural stable temperatures followed by a sharp, exceptional warming in the late 20th century.

10   In this area of work, CRU has been crucial. Under Jones's management, it has assembled the most comprehensive thermometer data record in the world, much of it under contract to the US Department of Energy. It is also home to some leading tree-ring researchers like the deputy head of the CRU, Dr Keith Briffa. The acerbic correspondence of Jones and Briffa with Michael Mann of

Penn State University, the chief creator of the hockey stick graph, is a central feature of the emails.

CRU's work is the prime (though not the only) basis for the claim that man-made global warming is happening now and is exceptional in history. But as it comes under assault, it is worth remembering that it does not directly touch on other key issues like the physics of climate change, forecasts of future climate change and so on. Even if all the work of CRU were revealed as entirely phony, which is far from being true, it would not demonstrate climate change was a hoax, or even much alter predictions of future climate.

The emails reveal that Jones, Briffa, Mann and other emailers were the gatekeepers of the science on which they worked. These men (there are virtually no women in the emails) reviewed papers by colleagues and rivals. They held key writing positions with the Intergovernmental Panel on Climate Change (IPCC) in its assessments of the science of climate change. So if they are damaged, then so is the IPCC.

Their correspondence reveals that there is some basis to the charge, made in October 2009 by climate contrarian Ross McKitrick, an environmental economist at the University of Guelph in Canada, that "the IPCC review process is nothing at all like what the public has been told. Conflicts of interest are endemic, critical evidence is systematically ignored and there are no effective checks and balances against bias or distortion." There are more than a thousand leaked files of emails to and from scientists and CRU. The emails are clearly a small subset of all the emails that would have been sent and received by CRU scientists since the first one in 1996. Nobody is yet clear why this set made it into the public domain, but they are overwhelmingly between CRU scientists and foreign compatriots. They include technical discussions about tree ring chronologies and data analysis, scheming about how to repel Freedom of Information (FoI) requests, and bitching about their enemies among the sceptics – the group the scientists referred to as "the contrarians."

Our analysis finds previously undisclosed evidence of slipshod use of data and apparent efforts to cover that up. It also finds persistent efforts to censor work by climatic sceptics regarded as hostile – especially those outside the scientific priesthood of peer review – or those able to generate headlines in media outlets thought unfriendly, like Fox News.

We would agree with Judy Curry of the Georgia Institute of Technology, [15] a leading climate scientist who maintains contacts with both camps, who says: "There are two broad issues raised by these emails ... lack of transparency in climate data, and 'tribalism' in some segments of the climate research community."

McIntyre's War

Climategate would not have happened without one man: a Canadian squash-playing blogger and data obsessive in his 60s called Steve McIntyre. Hero or villain, his data wars with Mann, Jones, Briffa and Santer largely created the siege mentality among the scientists, set them on a path of opposition to freedom of information, and by drawing in scores of data liberationists inside and outside the science community, almost certainly inspired whoever stole and released the emails.

McIntyre, a trained mathematician, had a successful career heading small Canadian minerals companies, often using his statistical prowess to analyse mineral prospecting data and out-bet his rivals. In 2002, he took up a new hobby – investigating climate change science. It started with an email from his home in Toronto to Jones at CRU asking for some weather station data. Initially the exchanges, as revealed on McIntyre's website Climate Audit, were civilised. But as the years passed, and his data demands grew greater, relations soured.

From the start, McIntyre deconstructed studies that claim to show evidence of large-scale warming of the planet and of the human fingerprint in that warming. He pioneered the use of freedom of information legislation in the US and UK to demand the raw data behind the studies. It was not normal practice for scientists to publish this full data, nor the computer programmes they devised to analyse it.

McIntyre clearly doubted the statistical techniques being employed by the climatologists, and felt that, as a trained mathematician, he could do better despite his ignorance of climate science. And, as he grew more suspicious, he suspected them of cherry-picking data. He wondered exactly how Mann turned dozens of studies on the past climate, including a series of tree rings studies managed by Briffa at CRU, into his neat hockey stick graph. And he questioned the reliability of the thermometer data used by Jones to produce his graphs of warming over the past 160 years.

20　　He found that no independent researchers had seriously tried to replicate the findings – a cornerstone of scientific inquiry. "Nobody's ever checked this stuff with any sort of due diligence," he said recently. He says too much is taken on trust in the cozy, collegiate world of science.

The climate scientists came to regard him as a meddling, time-wasting and probably politically motivated wrecker, who rarely published his own papers and devoted his retirement to trashing theirs. So when he tried to access their raw data and computer programmes, they resisted. The emails reveal that the researchers shared tactics, encouraged each other and competed for the rudest invective against McIntyre. And they grew even angrier as other wannabe investigators joined the data hunt. Men such as Doug Keenan, a former financial trader on Wall Street and the City of London, and a retired electrical engineer from Northampton called David Holland.

Many have accused McIntyre, Keenan and others of being hired hands of corporations out to fight climate change legislation. *The Guardian* has found no evidence of that. Instead, they appear to be an unanticipated outpost of the rise of "grey power," retired numerate professionals with time on their hands, an obsessive streak in their heads and a cause to pursue. The story of the battles of McIntyre and his acolytes[a] to access the raw data, and the protracted and generally failed attempts by the scientists to repel him, is the central story of the leaked emails from 2003 onwards.

At first McIntyre published regular peer-reviewed scientific papers, co-authoring a couple with Ross McKitrick. The mainstream climate scientists responded angrily to them. They often used their influence to exclude what they regarded as substandard papers from major journals. So McIntyre, McKitrick and other sceptical authors, like Patrick Michaels of the University of Virginia and the Cato Institute and later Keenan, increasingly used *Climate Research* and *Energy and Environment* – two peer-reviewed journals widely disliked by mainstream climate scientists.

Tensions were strained further when McIntyre published more of his deconstructions of published papers on his website, but without scientific peer review.

Strident though his website often is, McIntyre has usually avoided outright personal abuse. The abuse was usually only a link away on other sites, however. And few of McIntyre's targets distinguished him from more politically motivated foes. Santer, for instance, concluded in one email in 2008 that McIntyre "has no interest in rational scientific discourse. He deals in the currency of threats and intimidation." He believes McIntyre saw himself as the "self-appointed Joe McCarthy[b] of climate science."

Last September, RealClimate, a website run by Mann and other climate scientists, summed up how mainstream scientists felt about this kind of scientific discourse. "The timeline for these mini-blogstorms is always similar. An unverified accusation of malfeasance is based on nothing, and it is instantly telegraphed across the denial-o-sphere while being embellished along the way to apply to anything hockey-stick shaped and any and all scientists. The

25

---

a   *acolytes*   Assistants or followers.

b   *Joe McCarthy*   Joseph McCarthy (1908–57), an American senator best known for leading a witch hunt for gay government employees and supposed communist spies during the Cold War.

usual suspects become hysterical with glee that finally the 'hoax' has been revealed.... After a while it is clear that no scientific edifice has collapsed and the search goes on.... Net effect on lay people? Confusion. Net effect on science. Zip."

McIntyre, they complained, kept his hands relatively clean. He never talked about a hoax being exposed, and rarely questioned the "edifice" of climate science. He just picked away, providing fodder for his more excitable and less fastidious fans. As the RealClimate post went on: "Science is made up of people challenging assumptions and other people's results.... What *is* objectionable is the conflation of technical criticism with unsupported, unjustified and unverified accusations of scientific mal-conduct." McIntyre rarely makes such charges personally but, they complained, he "continues to take absolutely no responsibility for the ridiculous fantasies and exaggerations that his supporters broadcast."

There was a clash of cultures, too, between the ways of Canadian mining prospectors and those of academia. As one academic put it to me: "I think McIntyre confuses the more aggressive and confrontational style of business he used as a geophysical consultant with the more even responses in scholarship exchanges." On the other hand, the CRU emails hardly suggest that the scientists are shrinking violets. When Australian climate sceptic John Daly died, Jones commented, "In an odd way this is cheering news."

In the final months before climategate, the battle was not a cultural one, or even really about climate change. It was about data pure and simple. McIntyre wanted the scientists' data. In one week in the summer of 2009, he showered CRU with 58 freedom of information requests. He often made it clear that he did not have any particular reason for requiring the data. He just wanted to liberate it. It was a battle to break down the walls of the ivory towers, to blow apart the cosy world of peer review. It was a battle for the heart and soul of science, and for its lifeblood: data.

30   Then came the stolen emails. Whether hacked from outside or leaked from inside, the emails lit a fuse, but the fuel of mistrust had been piling up for years. As a result, the bonfire has been spectacular.

SCIENTISTS IN THE FIRING LINE

Many of the researchers caught up in the "climategate" saga have spent years in the firing line of sceptics. And they have felt the heat.

In late 2006, I interviewed a number of them for an article in *New Scientist* magazine, which focused on how the propaganda war was shaping up prior to the publication of the next Intergovernmental Panel on Climate Change (IPCC) assessment the following year.

Kevin Trenberth had suffered abuse for publicly linking global warming to the exceptional 2005 Atlantic hurricane season, which culminated in hurricane Katrina. He told me: "The attacks on me are clearly designed to get me fired or to resign."

Ben Santer of the Lawrence Livermore laboratory in California, and formerly of the Climatic Research Unit at the University of East Anglia, was attacked for his role in writing the 1995 IPCC report, which claimed to see the hand of man in climate change. He said: "There is a strategy to single out individuals, tarnish them and try to bring the whole of science into disrepute."

Prof Mike Mann of Pennsylvania State University, fresh from his battle over the hockey stick in 2001,[a] said: "There is an orchestrated campaign against the IPCC." 35

Funding trails to some of the more prominent sceptics also emerged at that time. Steve McIntyre, who runs the influential sceptic blog Climate Audit, was free of financial conflicts of interest, but it emerged that prominent sceptic Patrick Michaels received hundreds of thousands of dollars in "consultancy" fees from the Intermountain Rural Electric Association, a coal-burning electric company based in Colorado. A leaked letter from the company's general manager, Stanley Lewandowski, said: "We believe it is necessary to support the scientific community that is willing to stand up against the alarmists."

The funding of climate sceptics has a long and probably ongoing history. In 1998, I revealed in the *Guardian* leaked documents showing that the powerful American Petroleum Institute (API) was planning to recruit a team of "independent scientists" to do battle against climatologists on global warming. The aim was to bolster a campaign to prevent the US government ratifying the Kyoto protocol.

The API's eight-page Global Climate Science Communications Plan said it aimed to change the US political climate so that "those promoting the Kyoto treaty on the basis of extant science appear to be out of touch with reality."

The leaked document said: "If we can show that science does not support the Kyoto treaty ... this puts the US in a stronger moral position and frees its negotiators from the need to make concessions as a defence against perceived selfish economic concerns."

Its first task was to "identify, recruit and train a team of five independent scientists to participate in media outreach." It is not clear if the plan went ahead, but the policy objective was achieved. 40

(2010)

---

a *battle over ... in 2001* After it was included in a 2001 IPCC report, Mann's "hockey stick" graph became a focus of attack among climate-change deniers.

Questions:

1. Describe the language used in the first two paragraphs of the article. What is the tone, and what comparison is being made? Why might the article use this language? Rewrite the paragraphs using a more neutral style.

2. Outline the events of "climategate." To what extent (if at all) do these events affect your opinion of the scientific study of climate change?

3. Compare the representation of McIntyre in this article with the representation of him in "Myth vs. Fact Regarding the 'Hockey Stick'" (in this anthology). Which representation is more believable? Why?

4. According to the article, why is peer review important to scientific research?

5. Look up either the "Climatic Research Unit" or the "Intergovernmental Panel on Climate Change." Note the major objectives of this group. Is the group accurately represented in the article? Why or why not?

6. If you are a student at a post-secondary institution, think about one project you have completed or intend to complete. Should all your research data and the associated emails be available to anyone who submits a Freedom of Information request? Why or why not?

7. In the quotation in paragraph 13, what concerns about the IPCC review process does McKitrick express? Based on your reading of this article, are these concerns reasonable? Why or why not?

8. The article suggests that climate scientists came to regard McIntyre as a "politically motivated wrecker." Discuss some ways in which "politics" might be driving many elements of this debate.

9. Find two examples of sarcasm used in the article. What is the effect of this sarcasm? Do you think the use of sarcasm makes this article seem any more or less reliable? Why?

10. Research the Kyoto Protocol. List the key objectives of this treaty. Why might the American Petroleum Institute not have wanted the US government to ratify the Kyoto Protocol? What was Canada's ostensible reason for withdrawing from the Protocol in 2011, and do you find it convincing? Why or why not?

11. What roles do various elements of the media play in influencing the direction of modern scientific research? Support your position with reference to "climategate" or to another scandal.

seven

Fred Pearce

Climate Change Debate Overheated after Sceptic Grasped "Hockey Stick"

*This article is the fourth in a series of twelve on climate change
controversy written by science journalist Fred Pearce for the* Guardian*;
the first in the series is included above in this anthology.*

❧

After the publication of the IPCC[a] report in 2001, the controversy about the hockey stick[b] spread beyond the science community. Political opponents of climate scientists cried foul, and they have stayed on Michael Mann's trail for years.

Republican senator James Inhofe of Oklahoma, who calls global warming a "hoax," repeatedly attacked the Penn State University professor's hockey stick graph. In 2005, Congressman Joe Barton of Texas ordered Mann to provide the House Committee on Energy and Commerce, which he chaired, with extensive details of his working procedures, computer programs and past funding. "There are people who believe that if they bring down Mike Mann, they can bring down the IPCC," Ben Santer of the Lawrence Livermore Laboratory in California told me at the time.

Mann's voluble, self-confident style did not help matters. "The goddam guy is a slick talker and super-confident. He won't listen to anyone else," one of climate science's most senior figures, Wally Broecker of the Lamont-Doherty Earth Observatory at Columbia University in New York, told me. "I don't trust people like that. A lot of the data sets he uses are shitty, you know. They are just not up to what he is trying to do.... If anyone deserves to get hit it is goddam Mann."

---

a   *IPCC*   Intergovernmental Panel on Climate Change.
b   *hockey stick*   Graph showing temperature changes over many hundreds of years; see the version reprinted in this anthology.

It should be said that Broecker has a reputation among some scientists for bad-mouthing young researchers.[a]

5   The temperature of the debate soared in 2003 with the intervention of Canadian sceptic Steve McIntyre and his economist co-author Ross McKitrick of the University of Guelph. In a paper published in what was becoming the house journal of the sceptics, *Energy and Environment*, McIntyre and McKitrick widened the attack on the hockey stick by calling into question the statistical methods employed by Mann to amalgamate his different data sets. They even suggested that the hockey stick was entirely an artefact of those methods.

Mann replied in kind. The emails reveal that he heard about the "M&M" paper for the first time the day before it was published. He was angry that the journal had not asked him to review the paper, or at least comment on it, before publication. He put his friends on attack alert. "My suggested response is to dismiss this as a stunt appearing in a 'journal' already known to have defied standard practices of peer-review. It is clear, for example, that nobody we know has been asked to 'review' this so-called paper ... the claim is nonsense."

He went on: "Who knows what sleight of hand the authors have pulled. Of course the usual suspects are going to try to peddle this crap. The important thing is to deny that this has any intellectual credibility whatsoever."

In an ironic twist, he appended the anonymous note that had alerted him to the paper, apparently after being distributed among several scientists. It said that, far from being nonsense, the M&M paper reveals what "was known by most people who understand Mann's methodology [that] it can be quite sensitive to the input data in the early centuries." It went on: "There's going to be a lot of noise about this one, and knowing Mann's very thin skin, I am afraid he will react strongly, unless he has learned (as I hope he has) from the past...."

M&M's statistical complaint was that the analysis Mann pioneered, in which different proxy records are merged, involved sorting and aggregating these signals and smoothing the result. It had the effect of flattening the hockey stick shaft. Any graph of real temperatures would have been much less smooth. That was reasonable when all the data used along the graph had been subjected to the same smoothing. But, they complained, if you then added a graph of real temperatures onto the end, to cover the final decades,

---

a   *It should ... young researchers*   The *Guardian* appended the following statement to this article: Prof Wally Broecker said that he does not recognise this characterisation. He said, "if you check around, you will find that I am extremely supportive of young researchers. The reality is that I have absolutely no patience with bad science whether it be by junior or senior people."

it gave a misleading impression. Because there was no smoothing in this real data. Their point was that the shaft had been smoothed, but the blade had not. If a few decades of unusually warm temperature had showed up in, say, the 11th century they might have been smoothed away to nothing.

Mann didn't try to hide this in his papers. He put in error bars above and below the main line on his graph, showing how much temperature change the smoothing might have removed. He was among the first paleoclimatologists to do this. What is noticeable is that the error bars are huge. Most of the "blade" of 20th century warming would have fitted within the errors. It wasn't his fault that in future renditions, those very wide error bars sometimes disappeared.

Another criticism was that Mann analysed temperatures in terms of their divergence from the 20th-century mean. Mann agrees this would have highlighted differences from that period and accentuated any hockey stick shape. When M&M repeated Mann's analysis using different statistical methods they said they found a big rise in temperatures in the middle ages.

Finally, and perhaps most troublingly, M&M raised questions about the reliability of tree rings as a measure of temperature at all. Tree ring analysts are pretty sure that from the mid-19th century, when we have useable thermometer data, through to the mid-20th century, the width of rings faithfully represents real temperatures. Some detail is lost but the overall measure is good. But since around 1960, a "divergence" problem has emerged. Most tree ring data sets do not reflect the warming seen in thermometer readings (and indeed in nature, as glaciers melt, sea ice disappears, springs come earlier and so on).

Most scientists believe this divergence is a result of some other human-caused factor, but nobody is sure what. And until that is clear, there must be a question mark over the reliability of tree ring data for eras before we have thermometers. In fact this criticism ought to make Mann's hockey stick, which uses a range of different proxies, more reliable than temperature reconstructions based solely on tree rings. And, while the emphasis has mostly been on the probity of Mann's hockey stick, most researchers I have spoken to regard the M&M study as far more deeply flawed. They say it also includes subjective decisions about choice of data sets that seem hard to explain.

There are two take-home questions from this complex saga. Was Mann wrong to do as he did? And did it make any difference to his findings? In the aftermath of the M&M attack on Mann, a number of groups of researchers scrutinised the competing claims.

Hans von Storch of the GKSS Research Centre in Geesthacht, Germany, concluded that M&M were right to say that temperatures should be analysed relative to the 1,000-year mean, not the 20th-century mean. But he also found that even when this was done, it did not have much effect on the result. This didn't stop Mann bad-mouthing von Storch's work in a succession of emails through 2005.

Meanwhile, two people closer to Mann – Caspar Ammann of the National Center for Atmospheric Research in Boulder, Colorado and Eugene Wahl of Alfred University, New York – claimed that most of the difference between the findings of Mann and M&M had nothing to do with statistical methods. M&M had not "repeated" Mann's study as they claimed. In fact they had done a different study, leaving out some of the sets of tree-ring data that Mann included. In particular, they had excluded tree-ring studies based on ancient bristlecone pines in the south-west of the US. "Basically, the M&M case boiled down to whether selected North American tree rings should have been included, and not that there was a mathematical flaw in Mann's analysis," Ammann told me in 2006.

Interestingly, McKitrick now says he partially agrees. In a newspaper article in the Canadian *Financial Post* in October 2009, while still complaining that Mann's statistical methods skewed the data, he said of the hockey stick "its shape was determined by suspect bristlecone tree ring data."

Mann has always accepted that his graph was a work in progress, and most researchers in the field accept that he is honest if hot-headed. "I'm not slamming what he did overall. It was a great effort, a great step," Jacoby[a] told me in 2005. "But he got into hot water by defending it too hard in places where he shouldn't." But there is a troublingly arbitrary nature about temperature reconstructions when the choices made about which data to include and which not seem often to be based on researchers' hunches. However honest, they are open to the charge of cherry-picking their data. That applies as much to M&M as to Mann.

What counts in science, however, is not a single study. It is whether its finding can be replicated by others. Here Mann has been on a winning streak. Upwards of a dozen studies, using different statistical techniques or different combinations of proxy records, have produced reconstructions broadly similar to the original hockey stick. These reconstructions all have a hockey stick shaft and blade. While the shaft is not always as flat as Mann's version, it is present. Almost all support the main claim in the IPCC summary: that the 1990s was then probably the warmest decade for 1000 years.

a   *Jacoby*   Gordon Jacoby, a research scientist at Columbia University's Tree Ring Laboratory. In an earlier article in this series, Pearce cites Jacoby's concerns regarding some of the tree-ring data Mann used.

A decade on, Mann's original work emerges remarkably unscathed. 20
Briffa's[a] more recent reconstructions are closer to Mann's than those he had
in the late 1990s. Folland[b] says: "The Mann work still stands."

McIntyre remains unimpressed. "There is a distinct possibility that researchers have either purposefully or subconsciously selected series with the
hockey stick shape," he says.

McKitrick similarly insists that there is a cabal of paleoclimatologists
who have their favourite data sets that produce the required shape. In the
Financial Post he singled out dodgy data from the US bristlecone pines and
another set of tree rings from the remote Yamal peninsula in Siberia. He said
they occurred in so many studies that they skewed the lot.

This is not so. The Yamal tree rings were not in the famous hockey sticks
of the late 1990s. They were not even published then. According to Jones, of
the 12 reconstructions of temperature over the past thousand years used in the
last IPCC assessment, only three contained Yamal data.

In 2006, the US National Academy of Sciences published the results of
a long inquiry into Mann's findings, triggered by a request from Congress.
It upheld most of Mann's findings, albeit with some caveats. "There is sufficient evidence ... of past surface temperatures to say with a high level of
confidence that the last few decades of the 20th century were warmer than
any comparable period in the last 400 years. Less confidence can be placed
in proxy-based reconstructions of surface temperatures for AD 900 to 1600,
although the available proxy evidence does indicate that many locations were
warmer during the past 25 years than during any other 25-year period since
900."

It agreed that there were statistical failings of the kind highlighted by 25
M&M, but like von Storch it found that they had little effect on the overall
result. One panel member, Kurt Cuffey of the University of California at
Berkeley, reserved his criticism for the way the graph had been used by the
IPCC. "I think that sent a very misleading message about how resolved this
part of the scientific research was," he said. In retrospect, Mann rather agrees.
"Given its place in the IPCC summary with the uncertainties not even shown,
we were a target from the beginning," he admitted to me later.

The hockey stick, a pioneering piece of work in progress, became victim
of the notoriety it gained from being included in the IPCC summary. And of
course its catchy title.

---

a *Briffa* Keith Briffa, deputy director of the University of East Anglia's Climatic Research Unit.
b *Folland* Chris Folland, a leading climatologist.

"The label was always a caricature and it became a stick to beat us with," Mann said later. Was it flawed research? Yes. Was it hyped by the IPCC? Yes. Has it been disproved? Despite all the efforts, no. So far, it has survived the ultimate scientific test of repeated replication.

(2010)

Questions:

1. Describe Wally Broecker's opinion of Mann and his work. Does this language focus mostly on Mann or mostly on his work? Which of the two researchers do the comments make you most sympathetic with? Why?

2. According to the information given in the article, do the personal interactions of climatologists have a positive or negative effect on scientific progress? Use evidence from the article to support your answer.

3. Discuss the role of intellectual credibility and ethics in the process of journal publication. How does peer review affect the integrity of published scientific articles? Why is it important?

4. The journal *Energy & Environment*'s mission statement includes the following information:

   > E&E seeks to encourage communications between the many branches of the policy-making and research worlds that deal with 'energy' and 'environment' issues and encourage excursions into theory and futuristic speculation. E&E has consistently striven to publish many 'voices' and to challenge conventional wisdoms. Perhaps more so than other European energy journal [sic], the editor has made E&E a forum for more sceptical analyses of 'climate change' and the advocated solutions. We look for contributions that make energy technology a contributor to improving social and environmental conditions where this is most needed.

   The editorial board includes policy makers and "experts from the carbon fuels industries." In what ways is this focus different from that of most academic journals? Look at the rest of the mission statement and assess the journal's focus and objectivity: http://www.multi-science. co.uk/ee-mission.htm.

5. Paragraph 8 describes an "ironic twist." What is the twist and why is it ironic?

6. Why does McIntyre and McKitrick's work suggest that combining proxy records and real temperature data is problematic? What is the

resulting effect on the "hockey stick" graph? Did the original piece by Mann take this problem into account?

7. According to the article, why are tree rings sometimes a reliable source of data, and sometimes not?

8. How is this article organized? Outline the main ideas. Would additional information help a general reader's understanding of the article content? If so, what should be included?

9. McIntyre and McKitrick have raised a number of concerns regarding Mann's graph; list three of them. According to this article, how valid are these objections to Mann's data, formulae, and results? Select one of the objections and evaluate it yourself.

10. In terms of the scientific process, why is it important that Mann's data can be replicated by other scientists? What elements of the data are easy to replicate? What elements are more debatable? Was the debatability of those elements identified in the original article?

11. Did Mann's errors ultimately discredit his overall findings? Why or why not? Use evidence from the article (or from your own research) to support your answer.

eight

What's Up? South! Map

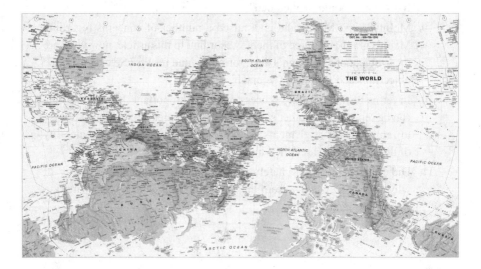

There is, of course, no logical reason for maps to show the north at the top, but it has become the convention for them to do so – a convention that maps such as the "What's Up? South! Map" aim to challenge.

In projecting the spherical surface of the Earth onto a two-dimensional surface, all maps must distort either the shape or the size of the Earth's features. This map employs a Van der Grinten projection, which attempts to compromise by distorting both shape and size, each to a lesser extent than maps that only distort one.

Questions:

1. What historical circumstances can you think of that would have contributed to the northern hemisphere being considered Earth's "up"?

2. Think of a globe. Where is the center? Where do you consider the Earth's geographical center to be? Where do you consider its political center to be? Has that center shifted during your lifetime? What about in the last 200 years?

3. Look again at this map. Almost all flat maps have to distort the size of the map's elements to fit the two-dimensional structure. Which countries are smaller in this version than you expected? Which are larger?

4. What particular distortions show up on this map? Pick one distortion and explore a consequence of looking at this particular element this way (for instance, is Greenland really that large? If so, should it not have more say over what goes on in the North to do with shipping?).

5. This map gives the Pacific's various islands more prominence than most other maps do. Why do you think that might be?

6. Look at one of the other redesigned maps on the website odtmaps.com. Which one comes closest to showing how you imagine the world to look? Why?

nine

ALEXIS DE GREIFF AND MAURICIO NIETO

from What We Still Do Not Know about South-North Technoscientific Exchange: North-Centrism, Scientific Diffusion, and the Social Studies of Science

❧

THE RACE FOR THIRD WORLD HEARTS AND MINDS: THE SEDUCTION OF DEVELOPMENT

During the 1950s and 1960s, large portions of the world's population strove to build a national identity within the context of international tension, national class struggles, and ideological debate. Although the state of revolt is a better-known face of the Third World, it is just one aspect of these nations' history. As Arturo Escobar and others point out, this period was marked by the construction of the "development discourse" as a new form of domination over the new nations and, more generally, over the so called "developing countries."[21] International institutions, notably the World Bank and the United Nations technical agencies, played a central role in nurturing the discourse and practice of development programs. The effect of these programs has meant the creation of ever-greater gaps between rich and poor countries, the widening of the internal social, cultural, and economic contrasts, the degradation of the environment, and other social ills. The workings of public rhetoric and practice of development are still largely unexplored in current historiography,[a] although some efforts have been made to investigate the phenomenon in studies of discrimination through literary analysis, as well as in anthropological works on modernization and resistance, especially in

---

a *historiography* Study of the writing of history.

Asia and Africa, and Latin America.[22] Curiously, even these works fail to tackle the issue of science and development to the extent that some historians have done in studying the close link between colonial domination, science, and technology. An important lesson to learn from these works is the necessity of shifting the object of study "from the people to be 'developed' to the institutional apparatus that is doing the 'developing.'"[23] In other words, we need to abandon the idea that development and modernity are "unfinished" projects everywhere but in Western Europe and North America due to the cultural and/or structural obstacles of the peoples to be "developed." We need to start looking at the institutions for development as instruments of control and domination and realize that scientific programs are political programs.

Escobar focuses on development as a practice – that is, the establishment and operation of institutions concerned with the implementation of programs, mobilization of resources, and creation of new spaces of representation based on the idea of "development." By contrast, Gilbert Rist, from a slightly different perspective, analyzes the history of the concept and how it shaped the views of world history during the twentieth century. He points out that "development" is a central element of the religion of modernity. It is therefore a set of beliefs deeply rooted in our conception of social, political, and economic relations in both national and international arenas. Development is a collective certainty, a dogmatic truth that is not debatable and, therefore, becomes a coercive force. "The action determined by the belief is obligatory," Rist notes, "and does not rest upon any choice."[24] In the name of development, mistakes are made and people are aware of them. In this sense the parallel with religion helps to understand the phenomenon. Religious believers realize the deep contradictions between the doctrine prescribed in the holy books and the practice of ecclesiastical institutions. However, they are tolerant. Thus, Rist defines "development" as "a belief and a series of practices which form a single whole in spite of contradictions between them."[25]

What is the role of science and technology in development theories? As John Agnew points out, all theories of development and social change contain within them positions on the role and impact of science and technology on development.[26] Surprisingly, the dialogue between science and technology studies (including the history of contemporary science, technology, and medicine) and critical analyses of development is poor. For instance, Escobar is concerned with the negative impact caused by producing technology in one place to be "applied" in another. He thus advocates for "a policy of technological research and development in support of autonomous peasant production systems."[27] According to Escobar, the only way to tackle effectively the problem of poverty through useful knowledge is with peasants' self-understanding – then proceed to build a system of communication involving peasants, institutions, and researchers.

This is an all too familiar prescription to historians and sociologists of science and technology, who have insisted that science and technology are essentially local practices. While historians of recent science and technology have been virtually oblivious to South-North exchange for development, leading scholars of development continue to treat science and technology as black boxes. Rist, for instance, explains, "Belief is so made to tolerate contradictions – especially as, unlike scientific theories, it cannot be refuted. This is why science changes faster than belief, which has immunity against anything that might place it in question."[28] This pre-Kuhnian[a] view of science contrasts with empirical studies that show scientists as conservative professionals, committed to their local research traditions. One should ask about the articulation between faith in development and faith in certain scientific theories and technical innovations. Indeed, Escobar has observed that science and technology act not just as "promises" (a word reminiscent of Kuhn's paradigm) but "makers of civilization." But in what way? Michael Adas' work – on the way science and technology as ideological instruments to establish colonial power – ends in the Great War.[29] However, the role of science and technology in international relations has intensified since then. After the Second World War international development programs translated into technical assistance and scientific manpower building. How new forms of domination (development discourse) worked back-to-back with science and technology is a subject that needs further exploration. While the workings of social scientists in the diagnoses and construction of social representations of the "developing" world are studied in detail, natural scientists are missing actors in most works. Several questions require study. For instance, what roles did technical personnel, engineers, scientific administrators, and scientists play in scientific and technological projects for development? What can we learn about their local negotiations and the use of local cultural resources to gain epistemological supremacy and, thus, access to resources? What image of science and technology did these agents try to establish? What kind of hybrid image resulted from that effort once joined with local knowledge?

---

a   *Kuhnian*   Reference to Thomas Kuhn (1922–96), a historian and philosopher of science best known for his argument that, at any given point in history, science operates according to a particular "paradigm," or a set of beliefs, concepts, and methods that are shared by the scientific community. Most scientific work, he argues, is not about "finding the truth" but about reinforcing the current paradigm; revolutions in science occur when there is a "paradigm shift" and the old paradigm is replaced with a new one.

There is an interesting asymmetry in the literature on technoscientific [5] international relations. Whereas the works concerned with international relations between industrialized countries talk of "scientific exchange," the literature on South-North exchange is located in studies of "scientific and technical cooperation," namely analyses of assistance programs for development. It is as though scientific practices, not explicitly tied to development projects in the South, were marginal to political and scientific international relations. Such distinction between exchange and co-operation must be understood as a historical product in itself. Very little has been studied about scientific excellence in the South, to use Marcos Cueto's phrase, and horizontal exchanges with the North.[30] The available resources, professional practices, instruments, and impact are radically different. However, as in development theories, these differences are often perceived as defects, manifestations that southern nations are a step behind the "developed." Indeed, a sort of "sociology of obstacles" is still common in works on science, technology, and economic development. In his seminal book *The Social Function of Science*, J.D. Bernal stressed the constraints imposed by local cultures in Latin America, India, and the Islamic World.[31] Similarly, in *The Two Cultures* (1959), C.P. Snow called for sending an army of scientists and linguists to prevent the Third World fall into communist hands,[32] whereas Ziman and Moravcsik assumed and concluded that Paradisia, an imaginary country in the South, will never progress unless the Western model of science, technology, and its institutions are effectively transferred.[33]

There are, nonetheless, pockets of critical reflection that deserve attention. Critical voices, especially postcolonial studies, are heirs of the *dependency* theory. The process of institutionalization and professionalization of science in Latin America preceded and was instrumental in the establishment of national scientific policies in the 1960s (and 1950s in the case of Argentina and Brazil).[34] Sociological and economic studies stimulated by the dependency theorists provided thoughtful analyses of the problems of science and technology in peripheral countries.[35] Geologist Amílcar Herrera, in his influential book *Ciencia y Política en América Latina*, developed a sociohistorical analysis of scientific research in Latin America. He criticized the contradiction between what he called the "explicit" and the "implicit" science policies operating in the South. The former, the rhetoric of science for development, was the façade covering the local elite's lack of commitment to national development, a disinterest that characterized the implicit policy.[36] Thus dependency theory, which can be placed among the "dissident voices" of development, led to a debate that was particularly fruitful in Latin America.[37] The central thesis was that "underdevelopment" in the periphery was inseparable from capitalist development in the metropolis – here, the metropolis referred to the Western imperialist (and, after decolonization, neoimperialist) powers and the periphery to the colonies (and the Third World). In their view, local elites acted as agents of neocolonialism and underdevelopment. Third World economic and cultural dependency and social crisis were due to collusion between external actors, the colonial powers, and internal ones, the local elites. By extension, scientific and technological development in peripheral economies was severely limited by external interests defended by the local elite.[38]

While these studies have been praised for stimulating a critical discussion of capitalist development at the periphery, they have equally been criticized for their "ideological" bias and lack of empirical evidence.[39] Dependency theorists have also been criticized for the lack of empirical studies that show in detail the dependency character of global knowledge. Sociologists of science and technology who became important science administrators, such as Francisco Sagasti, were enthusiastic supporters of the North American "systems approach" but stayed close to dependency theory. Even his works "remained a largely formal, abstract, reductionist analysis of science and technology development which was difficult to translate into action."[40] Reacting to these criticisms, the heirs of the dependency theories who became engaged in science studies have been investigating case studies, for instance in relation to the Green Revolution.

Dependency theory was perhaps the most original contribution to science studies by and about Third World scientists. Several intellectuals in this region rebelled against the assertiveness of the North and demonstrated the possibility of offering alternative solutions to the problem of the role of science in the Third World.[41] These studies marked a turning point in the history of ideas in Latin America in particular and the Third World in general.

Dependency theory also produced a lively debate within the scientific communities, particularly on the question of the social use of "pure" science in *developing* contexts. During the 1970s, politicians and administrators, especially in countries under military regimes such as Argentina, eager to cut funds for research in universities and to close or reduce other research institutions, strategically invoked the irrelevance of pure science for Third World development. In several Third World countries, resources for research in pure science, already scarce, were radically reduced. The research that did continue was redirected towards projects with "a social utility." The industrialized countries shared this view, discouraging scientific research in the South as well as international cooperation in subjects "not directly related to development." Since the 1950s, science has been considered a luxury that Third World nations cannot afford. For instance, negotiations to create an international center for the promotion of theoretical physics in the Third World met open hostility from delegations of virtually all industrialized countries.[42] Nonetheless, it is worth emphasizing that, in spite of such attitudes – and the concomitant difficulties due to the lack of resources – non-applied research is carried out in Third World countries. Yet, the bulk of the literature is concerned with technology transfer, while science is seldom mentioned.[43]

In our discussion thus far, we referred to scientific institutions. National research centers in the Third World developed interesting and complex intellectual, political, technical, and economic links with institutions in the North. Although the number of works is growing, we are far from having a good map of these institutions and their mutual relations.[44] We need to learn more about the role of academies and scientific societies in the South in the consolidation of local elites who used the science for development discourse, becoming local agents of aid programs offered by the industrialized countries. Political and scientific elites in the Third World often received training in Europe and the United States. However, detailed investigations of the globalization of knowledge after the Second World War are scarce. Although it is obvious for some, we must recall that institutional histories cannot be studied independently from cognitive aspects, for they provide the resources necessary for understanding research and pedagogical practices. In particular, by scrutinizing centers that promote South-North co-operation, we are able to increase the understanding of the role of scientific institutions in the construction of development programs and, concomitantly, learn about the global distribution of knowledge. Science, technology, and training programs were enthusiastically supported by philanthropic foundations. Nevertheless, the existing studies show also the enormous diversity of motivations, mechanisms, and strategies deployed by applicants and foundations alike. Indeed, if we really want to learn about the patterns of funding by American philanthropic foundations, and the kind of knowledge they eagerly promoted, we must focus on their activities in the Third World, where these bodies invested more than twice their total budget for institutions in Europe.[45]

At a different level, we have scientific disciplines. We need to remind ourselves that development discourses and practices produced varying images of science. For many years, the image of modern science and progress was represented by the theoretical physicist rather than by the agronomist. The diffusion of such representations and certain practices associated to them were closely related to the models of development and the role ascribed to technoscience. Of course, local cultures and traditions influenced such images.[46] Thus we can ask: what is the relation between such ideas about science and technology and modernity projects in different cultural settings in the South? Financial support was invariably conditioned to demonstrate that the projects contributed to development. Hence, some areas of research became more "pertinent" than others. Why did governments support certain scientific projects, such as theoretical physics or corrosion, and what was expected from them? What did scientists do to fulfill those expectations or at least give that impression? Analyses of the different "discursive strategies" employed by scientists in their countries and abroad would shed light upon the problem of establishing scientific disciplines in specific cultural environments, showing that development was – like technoscience – essentially a cultural phenomenon. Finally, one must ask whether and how development requirements shaped research. Some areas of research, in both scientific and industrial settings, that came from the North had to be adapted to the South, as cultural traditions, infrastructural facilities, human and natural resources, and so forth were different there. These processes of adaptation are in effect "new uses" of material and conceptual artifacts. The study of such uses may open a very different picture of innovation in the South.[47]

THE OLD AND NEW GREEN "REVOLUTIONS"

Food, poverty, and – since the 1970s – the environment have been the central issues of most development programs. The Green Revolution is perhaps one of the most discussed cases in the literature on science and development. As one environmental historian has explained, it was "a technical and managerial package exported from the First World to the Third beginning in the 1940s but making its major impact in the 1960s and 1970s."[48] In 1970, the American botanist Norman Borlaug, Director of the Division for Wheat Cultivation at the Centro Internacional del Mejoramiento de Maíz y Trigo in Mexico, was awarded the Nobel Peace Prize. He was the main promoter of a worldwide agriculture development program based on the genetic manipulation of seeds to improve production – the Green Revolution. The program was introduced in several Asian countries in 1965. Five years later, the program covered 10 million hectares of cultivated area. The program was promoted and supported by· several institutions from the United States, France, Canada, Germany, Brazil, India, Nigeria and others that constituted the Consultative Group of International Agricultural Research. Philanthropic foundations, such as the Rockefeller and the Ford, participated decisively in the program.

The actual impact of the Green Revolution has stimulated major debates. On the one hand, its effects on national production of wheat and rice became apparent. A number of countries in South America and Asia achieved record harvests. By the end of the 1970s, India became self-sufficient in wheat and rice, tripling its wheat production between 1961 and 1980. This is the positive side of the Green Revolution, according to its apologists.[49] On the other hand, since the 1970s, the Green Revolution has been subjected to severe criticisms. The main one was that for the program to be profitable, it was necessary to employ rich soils, optimal irrigation, intensive use of fertilizers, and chemical pesticides. In addition, by the early 1980s, environmentalists found that intensive fertilization stimulated by the Green Revolution led to euthrophication[a] of rivers and lakes. Although some countries increased their agricultural productions, other regions with little water and lack of credit markets – such as sub-Saharan Africa – suffered. Even in those countries where it was successful, some authors found flaws. J.K. Bajaj argues that rather than improving the agricultural system, it devastated its productivity and increased hunger. Economic dependence increased, for reduction in imported cereals was offset by imports of fertilizers and knowledge dependence on "experts." Thus Bajaj questions the assertion that the Green Revolution made India self-reliant in agricultural production.[50] From the environmental point of view, the speed and scale of dissemination of new breeds made the Green Revolution the largest crop transfer in world history, reducing dramatically biodiversity.[51] Socially, the Green Revolution widened the gap, favoring large landholders who had access to Western education. The drop in the prices of wheat displaced small farmers, which led to the development of urban slums.

These studies of the political dimension of the Green Revolution reveal important features of the twentieth-century tensions. It created a promised land of efficient export-oriented agriculture, which in turn would lead to rapid industrialization – the key to development, according to the economic theories of the day. The Green Revolution was the epitome of a technoscientific solution, an alternative to social revolution. In regions close to the communist border, such as Turkey and Korea, its introduction was a result of the American fear of the spread of Chinese communism.[52] However, the genius of this revolution was presenting itself as apolitical. Analysts such as Edmund Oasa, who was commissioned by the Consultative Group to make an evaluation of the program, concluded that class lines and conflicts worsened as a result of the "inherent contradictions in CG [Consultative Group] policies and the *politically neutral stance that the Group has adopted, at least superficially.*"[53] Promoters of the Green Revolution assumed that a technical solution could solve deep social problems, such as land distribution and the exploitation of

---

a   *euthrophication*   Overabundance of nutrients in a body of water. It causes too much plant growth; when the plants decompose, this reduces the amount of oxygen in the water, and the aquatic animals die.

the work force.[54] The Group isolated itself from political debates, instead of incorporating them as a crucial element of the problem. Vandana Shiva's argument is even more radical. Punjab, on the border between Pakistan and India, was supposedly the Green Revolution's major success. However, the socioeconomic conditions of this region are deplorable, and violence continues to be endemic. This tragedy is presented as an endogenous[a] situation, caused by ethnic conflict between religious groups, and therefore independent of the Green Revolution. Shiva offers an alternative interpretation: "it traces aspects of the conflicts and violence in contemporary Punjab to the ecological and political demands of the Green Revolution as a scientific experiment in development and agricultural transformation." Moreover, Shiva brilliantly demonstrates how science "was offered as a 'miracle' recipe for prosperity. But when discontent and new scarcities emerged, science was delinked from economic processes."[55] Such power of science to vanish from the political scene when things go wrong cements the faith in technoscience as the engine of progress. It erases the contradictions between theory and practice of development. More case studies on South-North exchange programs will be helpful to understand issues on science and democracy today.

15   Genetic engineering (GE) and its products, Genetic Modified Objects (GMOs), is widely regarded as yet another new technological promise to alleviate hunger in the Third World. Furthermore, this technology involves not only transfer of plants, knowledge, techniques and processes from North to South, but it also looks for genes to manipulate and "improve." Thus, the GE firms require germplasms[b] from regions with vast genetic resources, such as the Amazon forest. In other words, the relation is bidirectional: GMOs are moved from the North to the South, while genes are drained in the opposite direction. As far as the technology transfer to the South is concerned, some have argued that the Green Revolution has served as a point of reference to identify the issues at stake.[56] Indeed, GE cannot be understood without a deep analysis of the Green Revolution. Hitherto, the Green Revolution has been studied mainly in the Indian case. But GE companies have interests in other Third World countries. Hence, we need to learn more about the process of inception and the impact of the Green Revolution in other parts of the world. In the Amazon region, for instance, the cases of Colombia, Peru and Ecuador are virtually unexplored. What lesson did GE firms extract from the Green Revolution? This is an important question. The conclusions come from the critics. We need to know more about those who consider it a success and,

---

a *endogenous* Caused by internal factors, as opposed to outside influence.
b *germplasms* Genetic materials that can be used to grow new plants.

therefore, justify GE as an improved version of that first experiment. In terms of the exploitation of Third World genetic resources, there are urgent questions to tackle. As we mentioned earlier,[a] the appropriation of natural resources was a central element of imperialist policies in the late eighteenth century. What kind of technoscientific practices may or may not lead to domination relations? For instance, much analysis, discussion, and debate are needed regarding access to intellectual property and patent regulations. Compared to the development years, the center of power has shifted to the private sector. What are the implications of the leading role of corporate powers, especially in those regions where the State has been endemically weak? International institutes of research that participated in the Green Revolution are engaged in genetic research in associations with partners in the North. For instance, Lawrence Surendra argues that the International Rice Research Institute in the Philippines has been instrumental in the "gene drain" from the South to the North. This trend requires serious attention, analysis and action....

(2006)

REFERENCES

21   Arturo Escobar, *Encountering Development: The Making and Unmaking of the Third World* (Princeton, NJ: Princeton University Press, 1995).

22   V.Y. Mudimbe, *The Invention of Africa: Gnosis, Philosophy, and the Order of Knowledge* (Bloomington, IN: Indiana University Press, 1988); Chandra Mohanty, Ann Russo, and Lourdes Torres, eds, *Third World Women and the Politics of Feminism* (Bloomington: Indiana University Press, 1991); Homi K. Bhabha, *The Location of Culture* (London and New York: Routledge, 1994).

23   Escobar, *Encountering Development*, 107.

24   Gilbert Rist, *The History of Development: From Western Origins to Global Faith*, trans. Patrick Camiller (London and New York: ZED Books, 1999), 22.

25   Ibid., 24.

26   John A. Agnew, "Technology Transfer and Theories of Development," *Journal of Asian and African Studies* 17 (1982): 16–31.

27   Escobar, *Encountering Development*, 151.

28   Rist, *History of Development*, 23.

29   Michael Adas, *Machines as the Measure of Men: Science, Technology, and Ideologies of Western Dominance* (Ithaca, NY and London: Cornell University Press, 1989).

30   Marcos Cueto, *Excelencia Científica en la Periferia: Actividades Científicas e Investigación Biomédica en el Perú, 1890–1950* (Lima, OH: Grade-Concytec, 1989).

31   J.D. Bernal, *The Social Function of Science* (Cambridge: The MIT Press, 1964 [1939]).

32   C.P. Snow, *The Two Cultures* (Cambridge: Cambridge University Press, 1959), 48.

33   Michael Moravcsik and J.M. Ziman, "Paradisia and Dominatia: Science and the Developing World," *Minerva* 53 no. 4 (July 1975): 699–724.

a   *As we mentioned earlier*   This discussion occurred in a part of the paper not included here.

34 Thomas F. Glick, "Science in Twentieth-Century Latin America," in *Ideas and Ideologies in Twentieth Century Latin America since 1870*, ed. Leslie Bethell (Cambridge: Cambridge University Press, 1996), 287–359, on 348–349.

35 Celso Furtado, *La Economía Latinoamericana* (Mexico City: Siglo XXI, 1993).

36 Amílcar Oscar Herrera, *Ciencia y Política en América Latina* (Mexico City: Siglo XXI, 1971).

37 Fernando H. Cardoso and Enzo Faletto, *Dependency and Development in Latin America*, trans. M. Urquidi (Berkeley, CA: University of California Press, 1979). The school has roots in the United States (Paul Baran, Paul Sweezy), in Chile (Oswaldo Sunkel), in Brazil (Cardoso, Faletto, and Celso Furtado), in Colombia (Orlando Fals Borda), and in Mexico (Rodolfo Stavenhagen).

38 Hebe Vessuri, "The Social Study of Science in Latin America," *Social Studies of Science* 17 (1987): 519–554; Glick, "Science," 347–355.

39 See Dudley Seers, ed., *Dependency Theory: A Critical Reassessment* (London: Pinter, 1981).

40 Vessuri, "Social."

41 Terry Shinn, Jack Spaapen, and Venni Krishna, "Science, Technology and Society Studies and Development Perspectives in South-North Transactions," in *Science and Technology in a Developing World*, ed. Terry Shinn, Jack Spaapen, and Venni Krishna (Dordrecht: Kluwer, 1997), 1–34, on 11.

42 Alexis De Greiff, "The Tale of Two Peripheries: The Creation of the International Centre for Theoretical Physics in Trieste," *Historical Studies in the Physical and Biological Sciences* 33, Part 1 (2002): 33–60.

43 See Wesley Shrum, Carl L. Bankston III, and D. Stephen Voss, *Science, Technology, and Society in the Third World: An Annotated Bibliography* (Metchuen, NJ and London: The Scarecrow Press Inc., 1995).

44 Dong-Won Kim, "The Conflict between Image and Role of Physics in South Korea," *Historical Studies in the Physical and Biological Sciences* 33, Part 1 (2002): 107–130; Ana Maria Ribeiro de Andrade, *Físcos, Mésons e Política: A Pinamicia da Ciencia na Sociedade* (Sao Paulo and Rio de Janeiro: Editora HUCITEC; Museu de Astronomia e Ciencias Afins, 1999).

45 Alexis De Greiff, "Supporting Theoretical Physics for the Third World Development: The Ford Foundation and the International Centre for Theoretical Physics in Trieste (1966–1973)," in *American Foundations and Large-Scale Research: Construction and Transfer of Knowledge*, ed. Giuliana Gemelli (Bologna: CLUEB, 2001), 25–50.

46 Kim, "Conflict."

47 We are following Jorge Katz's *De a Importación de Tecnología al Desarollo Local* (Mexico: Fondo de Cultura Económica, 1976); and David Edgerton, ["From Innovation to Use: Ten Eclectic Theses on the Historiography of Technology," *History and Technology* 16, no. 2 (1999): 111–136].

48 John McNeill, *Something New under the Sun: An Environmental History of the Twentieth Century* (London: The Penguin Press, 2000), 219. On the Green Revolution see also: John H. Perkins, *Geopolitics and the Green Revolution: Wheat, Genes, and the Cold War* (New York: Oxford University Press, 1997); N. Cullather, "Miracles of Modernization: The Green Revolution and the Apotheosis of Technology," *Diplomatic History* 28, no. 2 (March 2005): 227–254.

49 McNeill, *Something New*, 219–227; Bernhard Glaeser, ed., *The Green Revolution Revisited: Critique and Alternatives* (London: Allen & Unwin, 1987), 1–9.

50 J.K. Bajaj, "Science and Hunger: A Historical Perspective on the Green Revolution," in *The Revenge of Athena: Science, Exploitation, and the Third World*, Ziauddin Sardar ed., (London: Mansell, 1998): 131–156.

51 McNeill, *Something New*, 224. See also Vandana Shiva, *The Violence of the Green Revolution: Third World Agriculture, Ecology, and Politics* (London and New York: Zed Books Ltd., 1991), Chapter 2.

52 McNeill, *Something New*, 222.

53 Glaeser, *Green Revolution Revisited*, 3.

54 James C. Scott, *Seeing Like a State: How Certain Schemes to Improve the Human Condition Have Failed* (New Haven, CT: Yale University Press, 1999).

55 Shiva, *Violence*, 20.

56 B. Sorj and J., Wilkinson, "Biotechnologies, Multinationals and the Agrofood Systems of Developing Countries," in *From Columbus to ConAgra: The Globalization of Agriculture and Food*, ed. Alessandro Bonanno, Lawrence Busch, William H. Friedland, Lourdes Gouveia, and Enzo Mingione (Lawrence, KS: University Press of Kansas, 1994), 85–104.

Questions:

On "The Race for Third World Hearts and Minds"

1. How easy is it to understand "The Race for Third World Hearts and Minds"? What is its level of jargon use? What do these things tell you about the article's intended audience?

2. Identify three to five jargon or technical terms in this article. Look up the definition of one of these terms and write a sentence or two for inclusion in the article to give enough background that an undergraduate student could understand the term in a basic way.

3. What is historiography? Why might historiographies of science be important?

4. Where does the term "Third World" come from? Why is it outdated? Why do you think this article uses it?

5. Look up Escobar, Rist, or Michael Adas and provide a three-sentence explanation of the importance of the theorist. Do you think the addition of such explanations might be helpful for giving the essay a broader audience? Why or why not?

6. What "scientific programs" can you think of that are also "political programs"? You can give current or historical examples.

7. This article discusses the "religion of modernity" and uses religion as a negative metaphor. Do you think the use might alienate certain readers? If so, rephrase the metaphor to make it more inclusive.

8. The article makes quite a lot of use of Marxist theory (which is why it uses the "religion of modernity" metaphor). Do you think it assumes that the reader is very familiar with Marxism? Why do you think it might use this writing approach?

9. What does the term "epistemological supremacy" mean? What sort of discussion would you expect it to appear in?

10. Rewrite one of the article's paragraphs using the clearest possible language. How difficult did you find this rewriting? To whom does your paragraph now speak?

11. Find a plain-language article critiquing development supremacy in a "Third World" (generally now called a "developing" or "less-developed") country. Summarize the article. What are its strengths?

12. What is pure science? How much of it is encouraged at your particular institution? Does your institution have a policy of applied research as much as possible? What is the difference between science and technology?

On "The Old and New Green 'Revolutions'"

1. This section provides a concrete example to ground the theorizing of the previous piece. Do you find it easier to understand? Why or why not? How does the writing change?

2. Look up Monsanto, bees, and neonicotinoids. How do you think this article would speak about Monsanto's commercial promotion of neonicotinoids?

3. What sort of "Green Revolution" problems can you find in "Third World" countries by surfing today's news? Discuss one of these problems in terms of the benefits it offers to the global North.

4. What is the tone of this article? Does it attempt to convince anyone of its argument? Does it attempt to speak to "apologists" for the Green Revolution? Why do you think it uses this particular tone?

5. Why do you think the article places such emphasis on research as a strategy for dealing with the problems it argues are very prominent?

II

Psychology: Experimenting with Authority – The Milgram Controversy

ten

STANLEY MILGRAM

Behavioral Study of Obedience[1]

*This famous essay by social psychologist Stanley Milgram reports
the surprising results of an experiment on obedience conducted at
Yale University. The selections following this article reflect on
Milgram's influential and controversial work.*

℘

ABSTRACT

This article describes a procedure for the study of destructive obedience in
the laboratory. It consists of ordering a naive S^a to administer increasingly
more severe punishment to a victim in the context of a learning experiment.
Punishment is administered by means of a shock generator with 30 graded
switches ranging from Slight Shock to Danger: Severe Shock. The victim
is a confederate of the $E.^b$ The primary dependent variable is the maximum
shock that S is willing to administer before he refuses to continue further. 26
Ss obeyed the experimental commands fully, and administered the highest
shock on the generator. 14 Ss broke off the experiment at some point after
the victim protested and refused to provide further answers. The procedure
created extreme levels of nervous tension in some Ss. Profuse sweating,
trembling, and stuttering were typical expressions of this emotional disturbance. One unexpected sign of tension – yet to be explained – was the regular
occurrence of nervous laughter, which in some Ss developed into uncontrollable seizures. The variety of interesting behavioral dynamics observed in
the experiment, the reality of the situation for the S, and the possibility of
parametric variation within the framework of the procedure, point to the
fruitfulness of further study.

---

a *S* Subject.
b *confederate* Person who is aware of the experimental conditions and may pretend to be a participant in the experiment, while in fact working with or for the researchers; *E* Experimenter.

Obedience is as basic an element in the structure of social life as one can point to. Some system of authority is a requirement of all communal living, and it is only the man dwelling in isolation who is not forced to respond, through defiance or submission, to the commands of others. Obedience, as a determinant of behavior, is of particular relevance to our time. It has been reliably established that from 1933–45 millions of innocent persons were systematically slaughtered on command. Gas chambers were built, death camps were guarded, daily quotas of corpses were produced with the same efficiency as the manufacture of appliances. These inhumane policies may have originated in the mind of a single person, but they could only be carried out on a massive scale if a very large number of persons obeyed orders.

Obedience is the psychological mechanism that links individual action to political purpose. It is the dispositional cement that binds men to systems of authority. Facts of recent history and observation in daily life suggest that for many persons obedience may be a deeply ingrained behavior tendency, indeed, a prepotent[a] impulse overriding training in ethics, sympathy, and moral conduct. C.P. Snow (1961) points to its importance when he writes:

> When you think of the long and gloomy history of man, you will find more hideous crimes have been committed in the name of obedience than have ever been committed in the name of rebellion. If you doubt that, read William Shirer's "Rise and Fall of the Third Reich." The German Officer Corps were brought up in the most rigorous code of obedience ... in the name of obedience they were party to, and assisted in, the most wicked large scale actions in the history of the world [p. 24].

While the particular form of obedience dealt with in the present study has its antecedents in these episodes, it must not be thought all obedience entails acts of aggression against others. Obedience serves numerous productive functions. Indeed, the very life of society is predicated on its existence. Obedience may be ennobling and educative and refer to acts of charity and kindness, as well as to destruction.

GENERAL PROCEDURE

5 A procedure was devised which seems useful as a tool for studying obedience (Milgram, 1961). It consists of ordering a naive subject to administer electric shock to a victim. A simulated shock generator is used, with 30 clearly marked voltage levels that range from 15 to 450 volts. The instrument bears verbal designations that range from Slight Shock to Danger: Severe Shock. The responses of the victim, who is a trained confederate of the experimenter, are

---

a *prepotent* Most powerful.

standardized. The orders to administer shocks are given to the naive subject in the context of a "learning experiment" ostensibly set up to study the effects of punishment on memory. As the experiment proceeds the naive subject is commanded to administer increasingly more intense shocks to the victim, even to a point of reaching the level marked Danger: Severe Shock. Internal resistances become stronger, and at a certain point the subject refuses to go on with the experiment. Behavior prior to this rupture is considered "obedience," in that the subject complies with the commands of the experimenter. The point of rupture is the act of disobedience. A quantitative value is assigned to the subject's performance based on the maximum intensity shock he is willing to administer before he refuses to participate further. Thus for any particular subject and for any particular experimental condition the degree of obedience may be specified with a numerical value. The crux of the study is to systematically vary the factors believed to alter the degree of obedience to the experimental commands.

The technique allows important variables to be manipulated at several points in the experiment. One may vary aspects of the source of command, content and form of command, instrumentalities for its execution, target object, general social setting, etc. The problem, therefore, is not one of designing increasingly more numerous experimental conditions, but of selecting those that best illuminate the process of obedience from the sociopsychological standpoint.

RELATED STUDIES

The inquiry bears an important relation to philosophic analyses of obedience and authority (Arendt, 1958; Friedrich, 1958; Weber, 1947), an early experimental study of obedience by Frank (1944), studies in "authoritarianism" (Adorno, Frenkel-Brunswik, Levinson, & Sanford, 1950; Rokeach, 1961), and a recent series of analytic and empirical studies in social power (Cartwright, 1959). It owes much to the long concern with *suggestion* in social psychology, both in its normal forms (e.g. Binet, 1900) and in its clinical manifestations (Charcot, 1881). But it derives, in the first instance, from direct observation of a social fact; the individual who is commanded by a legitimate authority ordinarily obeys. Obedience comes easily and often. It is a ubiquitous and indispensable feature of social life.

METHOD

Subjects

The subjects were 40 males between the ages of 20 and 50, drawn from New Haven and surrounding communities. Subjects were obtained by a newspaper advertisement and direct mail solicitations. Those who responded to the appeal believed they were to participate in a study of memory and learning at Yale University. A wide range of occupations is represented in the sample. Typical subjects were postal clerks, high school teachers, salesmen, engineers, and laborers. Subjects ranged in educational level from one who had not finished elementary school, to those who had doctorate and other professional degrees. They were paid $4.50 for their participation in the experiment. However, subjects were told that payment was simply for coming to the laboratory, and that the money was theirs no matter what happened after they arrived. Table 1 shows the proportion of age and occupational types assigned to the experimental condition.

TABLE 1
Distribution of Age and Occupational Types in the Experiment

Occupations	20-29 years	30-39 years	40-50 years	Percentage of total (Occupations)
Workers, skilled and unskilled	4	5	6	37.5
Sales, business, and white collar	3	6	7	40.0
Professional	1	5	3	22.5
Percentage of total (Age)	20	40	40	

Note. Total $N = 40$.

Personnel and Locale

The experiment was conducted on the grounds of Yale University in the elegant interaction laboratory. (This detail is relevant to the perceived legitimacy of the experiment. In further variations, the experiment was dissociated from the university, with consequences for performance.) The role of experimenter was played by a 31-year-old high school teacher of biology. His manner was impassive, and his appearance somewhat stern throughout the experiment. He was dressed in a gray technician's coat. The victim was played by a 47-year-old accountant, trained for the role; he was of Irish-American stock, whom most observers found mild-mannered and likeable.

Procedure

10 One naive subject and one victim (an accomplice) performed in each experiment. A pretext had to be devised that would justify the administration of electric shock by the naive subject. This was effectively accomplished by the cover story. After a general introduction on the presumed relation between punishment and learning, subjects were told:

But actually, we know *very little* about the effect of punishment on learning, because almost no truly scientific studies have been made of it in human beings.

For instance, we don't know how *much* punishment is best for learning – and we don't know how much difference it makes as to who is giving the punishment, whether an adult learns best from a younger or an older person than himself – or many things of that sort.

So in this study we are bringing together a number of adults of different occupations and ages. And we're asking some of them to be teachers and some of them to be learners.

We want to find out just what effect different people have on each other as teachers and learners, and also what effect *punishment* will have on learning in this situation.

Therefore, I'm going to ask one of you to be the teacher here tonight   15
and the other one to be the learner.

Does either of you have a preference?

Subjects then drew slips of paper from a hat to determine who would be the teacher and who would be the learner in the experiment. The drawing was rigged so that the naive subject was always the teacher and the accomplice always the learner. (Both slips contained the word "Teacher.") Immediately after the drawing, the teacher and learner were taken to an adjacent room and the learner was strapped into an "electric chair" apparatus.

The experimenter explained that the straps were to prevent excessive movement while the learner was being shocked. The effect was to make it impossible for him to escape from the situation. An electrode was attached to the learner's wrist, and electrode paste was applied "to avoid blisters and burns." Subjects were told that the electrode was attached to the shock generator in the adjoining room.

In order to improve credibility the experimenter declared, in response to a question by the learner: "Although shocks can be extremely painful, they cause no permanent tissue damage."

Learning task. The lesson administered by the subject was a paired-associate   20
learning task. The subject read a series of word pairs to the learner, and then read the first word of the pair along with four terms. The learner was to indicate which of the four terms had originally been paired with the first word. He communicated his answer by pressing one of four switches in front of him, which in turn lit up one of four numbered quadrants in an answer-box located atop the shock generator.

Shock generator. The instrument panel consists of 30 lever switches set in a horizontal line. Each switch is clearly labeled with a voltage designation that ranges from 15 to 450 volts. There is a 15-volt increment from one switch to the next going from left to right. In addition, the following verbal designations are clearly indicated for groups of four switches going from left to right: Slight Shock, Moderate Shock, Strong Shock, Intense Shock, Extremely Intense Shock, Danger: Severe Shock. (Two switches after this last designation are simply marked XXX.)

Upon depressing a switch: a pilot light corresponding to each switch is illuminated in bright red; an electric buzzing is heard; an electric blue light, labeled "voltage energizer," flashes; the dial on the voltage meter swings to the right; various relay clicks are sounded.

The upper left-hand corner of the generator is labeled Shock generator, Type ZLB, Dyson Instrument Company, Waltham, Mass. Output 15 volts-450 volts.

Details of the instrument were carefully handled to insure an appearance of authenticity. The panel was engraved by precision industrial engravers, and all components were of high quality. No subject in the experiment suspected that the instrument was merely a simulated shock generator.

25 *Sample shock.* Each naive subject is given a sample shock on the shock generator, prior to beginning his run as teacher. This shock is always 45 volts, and is applied by pressing the third switch of the generator. The shock is applied to the wrist of the naive subject, and has its source in a 45-volt battery wired to the generator. This further convinces the subject of the authenticity of the generator.

Shock instructions. The subject is told to administer a shock to the learner each time he gives a wrong response. Moreover – and this is the key command – the subject is instructed to "move one level higher on the shock generator each time the learner flashes a wrong answer." He is also instructed to announce the voltage level before administering a shock. This serves to continually remind subjects of the increasing intensity of shocks administered to the learner.

Preliminary and regular run. Pretests revealed that the procedure of reading words and administering shocks requires some practice before it can be handled smoothly. Therefore, immediately preceding the regular run, the teacher is given a preliminary series of 10 words to read to the learner. There are 3 neutral words in the practice series (i.e., words that the learner answers correctly), so that the shocks are administered for 7 of the words, with the maximum shock of 105 volts (moderate shock). Almost all subjects master the procedure by the time the preliminary run is over.

Subjects are then presented with a second list, and are told that the procedure is the same as for the first list; the experimenter adds however:

> When you get to the bottom of the list, repeat it over again, and continue giving shocks, until the learner has learned all the pairs correctly.

The experimenter instructs the subject to   30

> start from 15 volts and increase the shock level one step each time the learner gives a wrong answer.

Feedback from the victim. In all conditions the learner gives a predetermined set of responses to the word pair test, based on a schedule of approximately three wrong answers to one correct answer. In the present experimental condition no vocal response or other sign of protest is heard from the learner until Shock Level 300 is reached. When the 300-volt shock is administered, the learner pounds on the wall of the room in which he is bound to the electric chair. The pounding can be heard by the subject. From this point on, the learner's answers no longer appear on the four-way panel.

At this juncture, subjects ordinarily turn to the experimenter for guidance. The experimenter instructs the subject to treat absence of a response as a wrong answer, and to shock the subject according to the usual schedule. He advises the subject to allow 5-10 seconds before considering no response as a wrong answer, and to increase the shock level one step each time the learner fails to respond correctly. The learner's pounding is repeated after the 315-volt shock is administered; afterwards he is not heard from, nor do his answers reappear on the four-way signal box.

Experimenter feedback. At various points in the experiment the subject may turn to the experimenter for advice on whether he should continue to administer shocks. Or he may indicate that he does not wish to go on. It was important to standardize the experimenter's responses to these inquiries and reactions.

If the subject indicated his unwillingness to go on, the experimenter responded with a sequence of "prods," using as many as necessary to bring the subject into line.   35

> Prod 1: Please continue. *or* Please go on.
> Prod 2: The experiment requires that you continue.
> Prod 3: It is absolutely essential that you continue.
> Prod 4: You have no other choice, you *must* go on.

40   The prods were always made in sequence: Only if Prod 1 had been unsuccessful, could Prod 2 be used. If the subject still refused to obey the experimenter after Prod 4, the experiment was terminated. The experimenter's tone of voice was at all times firm, but not impolite. The sequence was begun anew on each occasion that the subject balked or showed reluctance to follow orders.

Special prods. If the subject asked if the learner was liable to suffer permanent physical injury, the experimenter said:

> Although the shocks may be painful, there is no permanent tissue damage, so please go on. [Followed by Prods 2, 3, and 4 if necessary.]

If the subject said that the learner did not want to go on, the experimenter replied:

> Whether the learner likes it or not, you must go on until he has learned all the word pairs correctly. So please go on. [Followed by Prods 2, 3, and 4 if necessary.]

Dependent Measures

45 The primary dependent measure for any subject is the maximum shock he administers before he refuses to go any further. In principle this may vary from 0 (for a subject who refuses to administer even the first shock) to 30 (for a subject who administers the highest shock on the generator). A subject who breaks off the experiment at any point prior to administering the thirtieth shock level is termed a *defiant subject*. One who complies with experimental command fully, and proceeds to administer all shock levels commanded, is termed an *obedient* subject.

Further records. With few exceptions, experimental sessions were recorded on magnetic tape. Occasional photographs were taken through one-way mirrors. Notes were kept on any unusual behavior occurring during the course of the experiments. On occasion, additional observers were directed to write objective descriptions of the subjects' behavior. The latency and duration of shocks were measured by accurate timing devices.

Interview and dehoax. Following the experiment, subjects were interviewed; open-ended questions, projective measures, and attitude scales were employed. After the interview, procedures were undertaken to assure that the subject would leave the laboratory in a state of well being. A friendly reconciliation was arranged between the subject and the victim, and an effort was made to reduce any tensions that arose as a result of the experiment.

RESULTS

Preliminary Notions

Before giving an account of the experimental results, it is instructive to consider how persons predict others will perform when commanded to administer powerful shocks to another person. Fourteen Yale seniors, all psychology majors, were provided with a detailed description of the experimental situation. They were asked to reflect carefully on it, and to predict the behavior of 100 hypothetical subjects. More specifically, they were instructed to plot the distribution of obedience of "100 Americans of diverse occupations, and ranging in age from 20 to 50 years," who were placed in the experimental situation.

There was considerable agreement among the respondents on the expected behavior of hypothetical subjects. All respondents predicted that only an insignificant minority would go through to the end of the shock series. (The estimates ranged from 0 to 3%; i.e., the most "pessimistic" member of the class predicted that of 100 persons, 3 would continue through to the most potent shock available on the shock generator – 450 volts.) The class mean was 1.2%. The question was also posed informally to colleagues of the author, and the most general feeling was that few if any subjects would go beyond the designation Very Strong Shock.

Result of Experiment 1

Subjects accept situation. With few exceptions subjects were convinced of 50 the reality of the experimental situation, that they were administering shocks to another person, and the most potent shocks were extremely painful. In the post-experimental interview subjects were asked: "How painful to the learner were the last few shocks that you administered to him?" Subjects were instructed to indicate their answers on a printed 14-point scale ranging from "Not at all painful" to "Extremely painful." The modal response was 14 (Extremely painful) with the mean at 13.42.

Signs of extreme tension. Many subjects showed signs of nervousness in the experimental situation, and especially upon administering the more powerful shocks. In a large number of cases the degree of tension reached extremes that are rarely seen in sociopsychological laboratory studies. Subjects were observed to sweat, tremble, stutter, bite their lips, groan, and dig their fingernails into their flesh. These were characteristic rather than exceptional responses to the experiment.

One sign of tension was the regular occurrence of nervous laughing fits. Fourteen of the 40 subjects showed definite signs of nervous laughter and smiling. The laughter seemed entirely out of place, even bizarre. Full-blown, uncontrollable seizures were observed for 3 subjects. On one occasion we observed a seizure so violently convulsive that it was necessary to call a halt to the experiment. The subject, a 46-year-old encyclopedia salesman, was seriously embarrassed by his untoward and uncontrollable behavior. In the post-experimental interviews subjects took such pains to point out that they were not sadistic types, and that the laughter did not mean they enjoyed shocking the victim.

Distribution of scores. It had been conjectured that persons would not, in general, agree to administer shocks beyond the level designated Very Strong Shock. The distribution of maximum shocks for this experiment is shown in Table 2. The obtained distribution of scores deviates radically from the prediction.

TABLE 2
Distribution of Breakoff Points

Verbal designation and voltage indication	Number of subjects for whom this was maximum shock	Verbal designation and voltage indication	Number of subjects for whom this was maximum shock
Slight Shock		Very Strong Shock	
15	0	195	0
30	0	210	0
45	0	225	0
60	0	240	0
Moderate Shock		Intense Shock	
75	0	255	0
90	0	270	0
105	0	285	0
120	0	300	5
Strong Shock		Extreme Intensity Shock	
135	0	315	4
150	0	330	2
165	0	345	1
180	0	360	1

Verbal designation and voltage indication	Number of subjects for whom this was maximum shock	Verbal designation and voltage indication	Number of subjects for whom this was maximum shock
Danger: Severe Shock		XXX	
375	1	435	0
390	0	450	26
405	0		
420	0		

Upon command of the experimenter, each of the 40 subjects went beyond the expected breakoff point. No subject stopped prior to administering Shock Level 20. (At this level – 300 volts – the victim kicks on the wall and no longer provides answers to the teacher's multiple-choice questions.)

Of the 40 subjects, 5 refused to obey the experimental commands beyond the 330-volt level. Four more subjects administered one further shock, and then refused to go on. Two broke off at the 330-volt level, and 1 each at 345, 360, and 375 volts. Thus a total of 14 subjects defied the experimenter. It is instructive to reprint their remarks at the point of defiance, as transcribed from the tape recordings:

> [0124] I think he's trying to communicate, he's knocking.... Well it's not fair to shock the guy ... these are terrific volts. I don't think this is very humane.... Oh, I can't go on with this; no this isn't right. It's a hell of an experiment. The guy is suffering in there. No, I don't want to go on. This is crazy. [Subject refused to administer more shocks.]

> [0123] He's banging in there. I'm gonna chicken out. I'd like to continue, but I can't do that to a man.... I'm sorry I can't do that to a man. I'll hurt his heart. You take your check.... No really, I couldn't do it.

These subjects were frequently in a highly agitated and even angered state. Sometimes, verbal protest was at a minimum, and the subject simply got up from his chair in front of the shock generator, and indicated that he wished to leave the laboratory.

Of the 40 subjects, 26 obeyed the orders of the experimenter to the end, proceeding to punish the victim until they reached the most potent shock available on the shock generator. At that point, the experimenter called a halt to the sessions. (The maximum shock is labeled 450 volts, and is two steps beyond the designation: Danger: Severe Shock.) Although obedient subjects continued to administer shocks, they often did so under extreme stress. Some expressed reluctance to administer shocks beyond the 300-volt level, and displayed fears similar to those who defied the experimenter; yet they obeyed.

60   After the maximum shocks had been delivered, and the experimenter called a halt to the proceedings, many obedient subjects heaved sighs of relief, mopped their brows, rubbed their fingers over their eyes, or nervously fumbled cigarettes. Some shook their heads, apparently in regret. Some subjects had remained calm throughout the experiment, and displayed only minimal signs of tension from beginning to end.

DISCUSSION

The experiment yielded two findings that were surprising. The first finding concerns the sheer strength of obedient tendencies manifested in this situation. Subjects have learned from childhood that it is a fundamental breach of moral conduct to hurt another person against his will. Yet, 26 subjects abandon this tenet in following the instructions of an authority who has no special powers to enforce his commands. To disobey would bring no material loss to the subject; no punishment would ensue. It is clear from the remarks and outward behavior of many participants that in punishing the victim they are often acting against their own values. Subjects often expressed deep disapproval of shocking a man in the face of his objections, and others denounced it as stupid and senseless. Yet the majority complied with the experimental commands. This outcome was surprising from two perspectives: first, from the standpoint of predictions made in the questionnaire described earlier. (Here, however, it is possible that the remoteness of the respondents from the actual situation, and the difficulty of conveying to them the concrete details of the experiment, could account for the serious underestimation of obedience.)

    But the results were also unexpected to persons who observed the experiment in progress, through one-way mirrors. Observers often uttered expressions of disbelief upon seeing a subject administrate more powerful shocks to the victim. These persons had a full acquaintance with the details of the situation, and yet systematically underestimated the amount of obedience that subjects would display.

    The second unanticipated effect was the extraordinary tension generated by the procedures. One might suppose that a subject would simply break off or continue as his conscience dictated. Yet, this is very far from what happened. There were striking reactions of tension and emotional strain.

    One observer related:

65   I observed a mature and initially poised businessman enter the laboratory smiling and confident. Within 20 minutes he was reduced to a twitching,

stuttering wreck, who was rapidly approaching a point of nervous collapse. He constantly pulled on his earlobe, and twisted his hands. At one point he pushed his fist into his forehead and muttered: "Oh God, let's stop it." And yet he continued to respond to every word of the experimenter, and obeyed to the end.

Any understanding of the phenomenon of obedience must rest on an analysis of the particular conditions in which it occurs. The following features of the experiment go some distance in explaining the high amount of obedience observed in the situation.

1. The experiment is sponsored by and takes place on the grounds of an institution of unimpeachable reputation, Yale University. It may be reasonably presumed that the personnel are competent and reputable. The importance of this background authority is now being studied by conducting a series of experiments outside of New Haven, and without any visible ties to the university.

2. The experiment is, on the face of it, designed to attain a worthy purpose – advancement of knowledge about learning and memory. Obedience occurs not as an end in itself, but as an instrumental element in a situation that the subject construes as significant, and meaningful. He may not be able to see its full significance, but he may properly assume that the experimenter does.

3. The subject perceives that the victim has voluntarily submitted to the authority system of the experimenter. He is not (at first) an unwilling captive impressed for involuntary service. He has taken the trouble to come to the laboratory presumably to aid the experimental research. That he later becomes an involuntary subject does not alter the fact that, initially, he consented to participate without qualification. Thus he has in some degree incurred an obligation toward the experimenter.

4. The subject, too, has entered the experiment voluntarily, and perceives himself under obligation to aid the experimenter. He has made a commitment, and to disrupt the experiment is a repudiation of his initial promise of aid. 70

5. Certain features of the procedure strengthen the subject's sense of obligation to the experimenter. For one, he has been paid for coming to the laboratory. In part this is canceled out by the experimenter's statement that:

> Of course, as in all experiments, the money is yours simply for coming to the laboratory. From this point on, no matter what happens, the money is yours.[4]

6. From the subject's standpoint, the fact that he is the teacher and the other man the learner is purely a chance consequence (it is determined by drawing lots) and he, the subject, ran the same risk as the other man in being

assigned the role of learner. Since the assignment of positions in the experiment was achieved by fair means, the learner is deprived of any basis of complaint on this count. (A similar situation obtains in Army units, in which – in the absence of volunteers – a particularly dangerous mission may be assigned by drawing lots, and the unlucky soldier is expected to bear his misfortune with sportsmanship.)

7. There is, at best, ambiguity with regard to the prerogatives of a psychologist and the corresponding rights of his subject. There is a vagueness of expectation concerning what a psychologist may require of his subject, and when he is overstepping acceptable limits. Moreover, the experiment occurs in a closed setting, and thus provides no opportunity for the subject to remove these ambiguities by discussion with others. There are few standards that seem directly applicable to the situation, which is a novel one for most subjects.

75   8. The subjects are assured that the shocks administered to the subject are "painful but not dangerous." Thus they assume that the discomfort caused the victim is momentary, while the scientific gains resulting from the experiment are enduring.

9. Through Shock Level 20 the victim continues to provide answers on the signal box. The subject may construe this as a sign that the victim is still willing to "play the game." It is only after Shock Level 20 that the victim repudiates the rules completely, refusing to answer further.

These features help to explain the high amount of obedience obtained in this experiment. Many of the arguments raised need not remain matters of speculation, but can be reduced to testable propositions to be confirmed or disproved by further experiments.[5]

The following features of the experiment concern the nature of the conflict which the subject faces.

10. The subject is placed in a position in which he must respond to the competing demands of two persons: the experimenter and the victim. The conflict must be resolved by meeting the demands of one or the other; satisfaction of the victim and the experimenter are mutually exclusive. Moreover, the resolution must take the form of a highly visible action, that of continuing to shock the victim or breaking off the experiment. Thus the subject is forced into a public conflict that does not permit any completely satisfactory solution.

80   11. While the demands of the experimenter carry the weight of scientific authority, the demands of the victim spring from his personal experience of pain and suffering. The two claims need not be regarded as equally pressing and legitimate. The experimenter seeks an abstract scientific datum; the victim cries out for relief from physical suffering caused by the subject's actions.

12. The experiment gives the subject little time for reflection. The conflict comes on rapidly. It is only minutes after the subject has been seated before the shock generator that the victim begins his protests. Moreover, the subject perceives that he has gone through but two-thirds of the shock levels at the time the subject's first protests are heard. Thus he understands that the conflict will have a persistent aspect to it, and may well become more intense as increasingly more powerful shocks are required. The rapidity with which the conflict descends on the subject, and his realization that it is predictably recurrent may well be sources of tension to him.

13. At a more general level, the conflict stems from the opposition of two deeply ingrained behavior dispositions: first, the disposition not to harm other people, and second, the tendency to obey those whom we perceive to be legitimate authorities.

(1961)

SELECTED AUTHOR'S NOTES

1. This research was supported by a grant (NSF G-17916) from the National Science Foundation. Exploratory studies conducted in 1960 were supported by a grant from the Higgins Fund at Yale University. The research assistance of Alan C. Elms and Jon Wayland is gratefully acknowledged.

4. Forty-three subjects, undergraduates at Yale University, were run in the experiment without payment. The results are very similar to those obtained with paid subjects.

5. A series of recently completed experiments employing the obedience paradigm is reported in Milgram (1964).

REFERENCES

Adorno, T., Frenkel-Brunswik, Else, Levinson, D.J., & Sanford, R.N. *The authoritarian personality*. New York: Harper, 1950.

Arendt, H. What was authority? In C.J. Friedrich (Ed.), *Authority*. Cambridge: Harvard Univer. Press, 1958. Pp. 81–112.

Binet, A. *La suggestibilité*. Paris: Schleicher, 1900.

Buss, A.H. *The psychology of aggression*. New York: Wiley, 1961.

Cartwright, S. (Ed.) *Studies in social power*. Ann Arbor: University of Michigan Institute for Social Research, 1959.

Charcot, J.M. *Oeuvres completes*. Paris: Bureax du Progrès Médical, 1881.

Frank, J.D. Experimental studies of personal pressure and resistance. *J. gen. Psychol.*, 1944, 30, 23–64.

Friedrich, C.J. (Ed.) *Authority*. Cambridge: Harvard Univer. Press, 1958.

Milgram, S. Dynamics of obedience. Washington: National Science Foundation, 25 January 1961. (Mimeo)

Milgram, S. Some conditions of obedience and disobedience to authority. *Hum. Relat.*, 1964, in press.

Rokeach, M. Authority, authoritarianism, and conformity. In I.A. Berg & B.M. Bass (Eds.), *Conformity and deviation*. New York: Harper, 1961. Pp. 230–257.

Snow, C.P. Either-or. *Progressive*, 1961 (Feb.), 24.

Weber, M. *The theory of social and economic organization*. Oxford: Oxford Univer. Press, 1947.

Questions:

1. What do you think is the importance of Milgram's study? What possible understandings does it offer in terms of the understanding of human psychology?

2. What is your institution's policy on Human Research? Would Milgram's study have received ethics approval from your Ethics Review Board (or the equivalent)? If not, what guidelines does the study not meet, and in what ways would Milgram's research need to be modified in order to meet those standards?

3. What writing strategies used in the article create an authoritative tone?

4. Review the article's citations. How many are laboratory or "scientific" investigations on a similar topic? How many are texts on the philosophy of human violence? How many citations are to Milgram's own works? Look up the C.P. Snow reference. Why do you think the article uses this reference?

5. "A procedure was devised which seems useful as a tool for studying obedience." Does the article offer an explanation for how and why the procedure was devised?

6. Would it change the article if the research collaborator was referred to as "Experimenter 2," rather than "the victim"?

7. The study enrolled 40 men from a "wide range of occupations." Are the categories described in Table 1 indicative of a "wide range"? What further information could be included in this table to make it more informative?

8. Do Milgram's statistics add anything to his article? If so, what do they add?

9. Consider ways of obtaining unbiased results in surveys. What do you need to know in order to assess the probable bias or absence of bias in the post-experiment survey sent to the participants?

10. Does the article's description of the results actually reflect the process of the experiment and its results accurately? Consider phrases such as the following: "the victim cries out for relief from physical suffering caused by the subject's actions."

eleven

Diana Baumrind

Some Thoughts on Ethics of Research: After Reading Milgram's "Behavioral Study of Obedience"

In this article, originally published in the journal American
Psychology, *a psychologist responds to Stanley Milgram's
"Behavioral Study of Obedience" (included in this anthology),
questioning its ethical and scientific legitimacy.*

❧

Certain problems in psychological research require the experimenter to balance his career and scientific interests against the interests of his prospective subjects. When such occasions arise the experimenter's stated objective frequently is to do the best possible job with the least possible harm to his subjects. The experimenter seldom perceives in more positive terms an indebtedness to the subject for his services, perhaps because the detachment which his functions require prevents appreciation of the subject as an individual.

Yet a debt does exist, even when the subject's reason for volunteering includes course credit or monetary gain. Often a subject participates unwillingly in order to satisfy a course requirement. These requirements are of questionable merit ethically, and do not alter the experimenter's responsibility to the subject.

Most experimental conditions do not cause the subjects pain or indignity, and are sufficiently interesting or challenging to present no problem of an ethical nature to the experimenter. But where the experimental conditions expose the subject to loss of dignity, or offer him nothing of value, then the experimenter is obliged to consider the reasons why the subject volunteered and to reward him accordingly.

The subject's public motives for volunteering include having an enjoyable or stimulating experience, acquiring knowledge, doing the experimenter a favor which may some day be reciprocated, and making a contribution to science. These motives can be taken into account rather easily by the experimenter who is willing to spend a few minutes with the subject afterwards to thank him for his participation, answer his questions, reassure him that he did well, and chat with him a bit. Most volunteers also have less manifest, but equally legitimate, motives. A subject may be seeking an opportunity to have contact with, be noticed by, and perhaps confide in a person with psychological training. The dependent attitude of most subjects toward the experimenter is an artifact of the experimental situation as well as an expression of some subjects' personal need systems at the time they volunteer.

5   The dependent, obedient attitude assumed by most subjects in the experimental setting is appropriate to that situation. The "game" is defined by the experimenter and he makes the rules. By volunteering, the subject agrees implicitly to assume a posture of trust and obedience. While the experimental conditions leave him exposed, the subject has the right to assume that his security and self-esteem will be protected.

There are other professional situations in which one member – the patient or client – expects help and protection from the other – the physician or psychologist. But the interpersonal relationship between experimenter and subject additionally has unique features which are likely to provoke initial anxiety in the subject. The laboratory is unfamiliar as a setting and the rules of behavior ambiguous compared to a clinician's office. Because of the anxiety and passivity generated by the setting, the subject is more prone to behave in an obedient, suggestible manner in the laboratory than elsewhere. Therefore, the laboratory is not the place to study degree of obedience or suggestibility, as a function of a particular experimental condition,[a] since the base line for these phenomena as found in the laboratory is probably much higher than in most other settings. Thus experiments in which the relationship to the experimenter as an authority is used as an independent condition[b] are imperfectly designed for the same reason that they are prone to injure the subjects involved. They disregard the special quality of trust and obedience with which the subject appropriately regards the experimenter.

Other phenomena which present ethical decisions, unlike those mentioned above, *can* be reproduced successfully in the laboratory. Failure experience, conformity to peer judgment, and isolation are among such phenomena. In

---

a   *experimental condition*   Set of circumstances exposing the subjects of an experiment to the variable being studied; each experimental condition exposes subjects to the variable to a different degree. In contrast, the "control condition" should resemble the experimental condition in every respect except that it should not expose the subjects to the variable being studied.

b   *independent condition*   Condition that is varied within an experiment so that its effects can be studied.

these cases we can expect the experimenter to take whatever measures are necessary to prevent the subject from leaving the laboratory more humiliated, insecure, alienated, or hostile than when he arrived. To guarantee that an especially sensitive subject leaves a stressful experimental experience in the proper state sometimes requires special clinical training. But usually an attitude of compassion, respect, gratitude, and common sense will suffice, and no amount of clinical training will substitute. The subject has the right to expect that the psychologist with whom he is interacting has some concern for his welfare, and the personal attributes and professional skill to express his good will effectively.

Unfortunately, the subject is not always treated with the respect he deserves. It has become more commonplace in sociopsychological laboratory studies to manipulate, embarrass, and discomfort subjects. At times the insult to the subject's sensibilities extends to the journal reader when the results are reported. Milgram's (1963) study is a case in point. The following is Milgram's abstract of his experiment:

This article describes a procedure for the study of destructive obedience in the laboratory. It consists of ordering a naive S to administer increasingly more severe punishment to a victim in the context of a learning experiment. Punishment is administered by means of a shock generator with 30 graded switches ranging from Slight Shock to Danger: Severe Shock. The victim is a confederate[a] of E. The primary dependent variable is the maximum shock the S is willing to administer before he refuses to continue further. 26 Ss obeyed the experimental commands fully, and administered the highest shock on the generator. 14 Ss broke off the experiment at some point after the victim protested and refused to provide further answers. The procedure created extreme levels of nervous tension in some Ss. Profuse sweating, trembling, and stuttering were typical expressions of this emotional disturbance. One unexpected sign of tension – yet to be explained – was the regular occurrence of nervous laughter, which in some Ss developed into uncontrollable seizures. The variety of interesting behavioral dynamics observed in the experiment, the reality of the situation for the S, and the possibility of parametric variation within the framework of the procedure, point to the fruitfulness of further study [p. 371].

The detached, objective manner in which Milgram reports the emotional disturbance suffered by his subject contrasts sharply with his graphic account

---

a *confederate* Person who is aware of the experimental conditions and may pretend to be a participant in the experiment, while in fact working with or for the researchers.

of that disturbance. Following are two other quotes describing the effects on his subjects of the experimental conditions:

> I observed a mature and initially poised businessman enter the laboratory smiling and confident. Within 20 minutes he was reduced to a twitching, stuttering wreck, who was rapidly approaching a point of nervous collapse. He constantly pulled on his earlobe, and twisted his hands. At one point he pushed his fist into his forehead and muttered: "Oh God, let's stop it." And yet he continued to respond to every word of the experimenter, and obeyed to the end [p. 377].

> In a large number of cases the degree of tension reached extremes that are rarely seen in sociopsychological laboratory studies. Subjects were observed to sweat, tremble, stutter, bite their lips, groan, and dig their fingernails into their flesh. These were characteristic rather than exceptional responses to the experiment.
>
> One sign of tension was the regular occurrence of nervous laughing fits. Fourteen of the 40 subjects showed definite signs of nervous laughter and smiling. The laughter seemed entirely out of place, even bizarre. Full-blown, uncontrollable seizures were observed for 3 subjects. On one occasion we observed a seizure so violently convulsive that it was necessary to call a halt to the experiment ... [p. 375].

Milgram does state that,

15

> After the interview, procedures were undertaken to assure that the subject would leave the laboratory in a state of well being. A friendly reconciliation was arranged between the subject and the victim, and an effort was made to reduce any tensions that arose as a result of the experiment [p. 374].

It would be interesting to know what sort of procedures could dissipate the type of emotional disturbance just described. In view of the effects on subjects, traumatic to a degree which Milgram himself considers nearly unprecedented in sociopsychological experiments, his casual assurance that these tensions were dissipated before the subject left the laboratory is unconvincing.

What could be the rational basis for such a posture of indifference? Perhaps Milgram supplies the answer himself when he partially explains the subject's destructive obedience as follows, "Thus they assume that the discomfort caused the victim is momentary, while the scientific gains resulting from the experiment are enduring [p. 378]." Indeed such a rationale might suffice to justify the means used to achieve his end if that end were of inestimable value to humanity or were not itself transformed by the means by which it was attained.

The behavioral psychologist is not in as good a position to objectify his faith in the significance of his work as medical colleagues at points of break-through. His experimental situations are not sufficiently accurate models of real-life experience; his sampling techniques are seldom of a scope which would justify the meaning with which he would like to endow his results; and these results are hard to reproduce by colleagues with opposing theoretical views. Unlike the Sabin vaccine,[a] for example, the concrete benefit to human-ity of his particular piece of work, no matter how competently handled, cannot justify the risk that real harm will be done to the subject. I am not speaking of physical discomfort, inconvenience, or experimental deception per se, but of permanent harm, however slight. I do regard the emotional disturbance described by Milgram as potentially harmful because it could easily effect an alteration in the subject's self-image or ability to trust adult authorities in the future. It is potentially harmful to a subject to commit, in the course of an experiment, acts which he himself considers unworthy, particularly when he has been entrapped into committing such acts by an individual he has reason to trust. The subject's personal responsibility for his actions is not erased because the experimenter reveals to him the means which he used to stimulate these actions. The subject realizes that he would have hurt the vic-tim if the current were on. The realization that he also made a fool of himself by accepting the experimental set results in additional loss of self-esteem. Moreover, the subject finds it difficult to express his anger outwardly after the experimenter in a self-acceptant but friendly manner reveals the hoax.

A fairly intense corrective interpersonal experience is indicated wherein the subject admits and accepts his responsibility for his own actions, and at the same time gives vent to his hurt and anger at being fooled. Perhaps an experience as distressing as the one described by Milgram can be integrated by the subject, provided that careful thought is given to the matter. The pro-priety of such experimentation is still in question even if such a reparational experience were forthcoming. Without it I would expect a naive, sensitive subject to remain deeply hurt and anxious for some time, and a sophisticated, cynical subject to become even more alienated and distrustful.

In addition the experimental procedure used by Milgram does not appear 20 suited to the objectives of the study because it does not take into account the special quality of the set which the subject has in the experimental situation.

---

a *Sabin vaccine* Oral vaccine against polio developed by Polish-American Albert Bruce Sabin (1906–93).

Milgram is concerned with a very important problem, namely, the social consequences of destructive obedience. He says,

> Gas chambers were built, death camps were guarded, daily quotas of corpses were produced with the same efficiency as the manufacture of appliances. These inhumane policies may have originated in the mind of a single person, but they could only be carried out on a massive scale if a very large number of persons obeyed orders [p. 371].

But the parallel between authority-subordinate relationships in Hitler's Germany and in Milgram's laboratory is unclear. In the former situation the SS man[a] or member of the German Officer Corps, when obeying orders to slaughter, had no reason to think of his superior officer as benignly disposed towards himself or their victims. The victims were perceived as subhuman and not worthy of consideration. The subordinate officer was an agent in a great cause. He did not need to feel guilt or conflict because within his frame of reference he was acting rightly.

It is obvious from Milgram's own descriptions that most of his subjects were concerned about their victims and did trust the experimenter, and that their distressful conflict was generated in part by the consequences of these two disparate but appropriate attitudes. Their distress may have resulted from shock at what the experimenter was doing to them as well as from what they thought they were doing to their victims. In any case there is not a convincing parallel between the phenomena studied by Milgram and destructive obedience as that concept would apply to the subordinate-authority relationship demonstrated in Hitler's Germany. If the experiments were conducted "outside of New Haven and without any visible ties to the university," I would still question their validity on similar although not identical grounds. In addition, I would question the representativeness of a sample of subjects who would voluntarily participate within a noninstitutional setting.

In summary, the experimental objectives of the psychologist are seldom incompatible with the subject's ongoing state of well being, provided that the experimenter is willing to take the subject's motives and interests into consideration when planning his methods and correctives. Section 4b in *Ethical Standards of Psychologists* (APA, undated) reads in part:

25   Only when a problem is significant and can be investigated in no other way, is the psychologist justified in exposing human subjects to emotional stress or other possible harm. In conducting such research, the psychologist must

---

a   *SS man*   Member of the *Shutzstaffel* (SS), a Nazi organization. During the Holocaust, concentration camps and death camps were managed by a branch of the SS.

seriously consider the possibility of harmful aftereffects, and should be pre-pared to remove them as soon as permitted by the design of the experiment. Where the danger of serious aftereffects exists, research should be conducted only when the subjects or their responsible agents are fully informed of this possibility and volunteer nevertheless [p. 12].

From the subject's point of view procedures which involve loss of dignity, self-esteem, and trust in rational authority are probably most harmful in the long run and require the most thoughtfully planned reparations, if engaged in at all. The public image of psychology as a profession is highly related to our own actions, and some of these actions are changeworthy. It is important that as research psychologists we protect our ethical sensibilities rather than adapt our personal standards to include as appropriate the kind of indignities to which Milgram's subjects were exposed. I would not like to see experiments such as Milgram's proceed unless the subjects were fully informed of the dangers of serious aftereffects and his correctives were clearly shown to be effective in restoring their state of well being.

(1964)

REFERENCES

American Psychological Association. Ethical Standards of Psychologists: A summary of ethical principles. Washington, D.C.: APA, undated.

Milgram, S. Behavioral study of obedience. *J. abnorm. soc. Psychol.,* 1963, 67, 371-378.

Questions:

1. Why does the article postpone introducing Milgram's experiment until the eighth paragraph? What do you think the effect of the framing material is?

2. How does the article depict "subjects"? Given that most subjects in psychology experiments are first-year university students, do you think the representation is accurate?

3. Why does the article suggest that the requirement that first-year students participate in experiments (to receive course credit) is "of questionable merit ethically"? In what ways is the use of such participants likely to skew a study's data?

4. In what ways does an experimenter need to "balance his [sic] career and scientific interests against the interests of his prospective subject"? Does it make a difference to you to know the obedience research was conducted near the beginning of Milgram's career, when he was not yet a tenured professor?

5. What is the article's tone like? Does it shift as the article progresses? What seemingly objective statements do you think might actually be making strong comments about the way Milgram conducted his experiment?

6. Discuss the "special quality of trust and obedience with which the subject appropriately regards the experimenter." If such a quality of trust exists, on what is it based?

7. The article states that Milgram's experiment was one of those designed to "manipulate, embarrass, and discomfort subjects." Do the excerpts taken from "Behavioral Study of Obedience" confirm this assertion?

8. Explain what the article means by the following statement: "Indeed such a rationale might suffice to justify the means used to achieve his end if that end were of inestimable value to humanity or were not itself transformed by the means by which it was obtained."

9. The experimental scientist in this article is consistently referred to as "he." Rephrase one of these sentences so that it is gender neutral.

10. Does the article ever attack Milgram personally? If so, describe where and how.

11. List three reasons this article gives to support its argument that the scientific and historical justifications proposed in "Behavioral Study of Obedience" are false.

12. Why does the article only make reference to "Behavioral Study of Obedience" and *Ethical Standards of Psychologists*? What type of article is it?

twelve

STANLEY MILGRAM

Issues in the Study of Obedience: A Reply to Baumrind

In this response to Diana Baumrind's critique of his "Behavioral Study of Obedience," Stanley Milgram addresses objections to the method and premise of his famous experiment.

℮

Obedience serves numerous productive functions in society. It may be ennobling and educative and entail acts of charity and kindness. Yet the problem of destructive obedience, because it is the most disturbing expression of obedience in our time, and because it is the most perplexing, merits intensive study.

In its most general terms, the problem of destructive obedience may be defined thus: If X tells Y to hurt Z, under what conditions will Y carry out the command of X, and under what conditions will he refuse? In the concrete setting of a laboratory, the question may assume this form: If an experimenter tells a subject to act against another person, under what conditions will the subject go along with the instruction, and under what conditions will he refuse to obey?

A simple procedure was devised for studying obedience (Milgram, 1963). A person comes to the laboratory, and in the context of a learning experiment, he is told to give increasingly severe electric shocks to another person. (The other person is an actor, who does not really receive any shocks.) The experimenter tells the subject to continue stepping up the shock level, even to the point of reaching the level marked "Danger: Severe Shock." The purpose of the experiment is to see how far the naive subject will proceed before he refuses to comply with the experimenter's instructions. Behavior prior to this rupture is considered "obedience" in that the subject does what the experimenter tells him to do. The point of rupture is the act of disobedience. Once the basic procedure is established, it becomes possible to vary conditions of the experiment, to learn under what circumstances obedience to authority is most probable, and under what conditions defiance is brought to the fore (Milgram, in press).

The results of the experiment (Milgram, 1963) showed, first, that it is more difficult for many people to defy the experimenter's authority than was generally supposed. A substantial number of subjects go through to the end of the shock board. The second finding is that the situation often places a person in considerable conflict. In the course of the experiment, subjects fidget, sweat, and sometimes break out into nervous fits of laughter. On the one hand, subjects want to aid the experimenter; and on the other hand, they do not want to shock the learner. The conflict is expressed in nervous reactions.

5   In a recent issue of *American Psychologist*, Diana Baumrind (1964) raised a number of questions concerning the obedience report. Baumrind expressed concern for the welfare of subjects who served in the experiment, and wondered whether adequate measures were taken to protect the participants. She also questioned the adequacy of the experimental design.

Patently, "Behavioral Study of Obedience" did not contain all the information needed for an assessment of the experiment. But it is clearly indicated in the references and footnotes (pp. 373, 378) that this was only one of a series of reports on the experimental program, and Baumrind's article was deficient in information that could have been obtained easily. I thank the editor for allotting space in this journal to review this information, to amplify it, and to discuss some of the issues touched on by Baumrind.

At the outset, Baumrind confuses the unanticipated outcome of an experiment with its basic procedure. She writes, for example, as if the production of stress in our subjects was an intended and deliberate effect of the experimental manipulation. There are many laboratory procedures specifically designed to create stress (Lazarus, 1964), but the obedience paradigm was not one of them. The extreme tension induced in some subjects was unexpected. Before conducting the experiment, the procedures were discussed with many colleagues, and none anticipated the reactions that subsequently took place. Foreknowledge of results can never be the invariable accompaniment of an experimental probe. Understanding grows because we examine situations in which the end is unknown. An investigator unwilling to accept this degree of risk must give up the idea of scientific inquiry.

Moreover, there was every reason to expect, prior to actual experimentation, that subjects would refuse to follow the experimenter's instructions beyond the point where the victim protested; many colleagues and psychiatrists were questioned on this point, and they virtually all felt this would be the case. Indeed, to initiate an experiment in which the critical measure hangs on disobedience, one must start with a belief in certain spontaneous resources in men that enable them to overcome pressure from authority.

It is true that after a reasonable number of subjects had been exposed to the procedures, it became evident that some would go to the end of the shock board, and some would experience stress. That point, it seems to me, is the first legitimate juncture at which one could even start to wonder whether or not to abandon the study. But momentary excitement is not the same as harm. As the experiment progressed there was no indication of injurious effects in the subjects; and as the subjects themselves strongly endorsed the experiment, the judgment I made was to continue the investigation.

Is not Baumrind's criticism based as much on the unanticipated findings as on the method? The findings were that some subjects performed in what appeared to be a shockingly immoral way. If, instead, every one of the subjects had broken off at "slight shock," or at the first sign of the learner's discomfort, the results would have been pleasant, and reassuring, and who would protest?

PROCEDURES AND BENEFITS

A most important aspect of the procedure occurred at the end of the experimental session. A careful post-experimental treatment was administered to all subjects. The exact content of the dehoax varied from condition to condition and with increasing experience on our part. At the very least all subjects were told that the victim had not received dangerous electric shocks. Each subject had a friendly reconciliation with the unharmed victim, and an extended discussion with the experimenter. The experiment was explained to the defiant subjects in a way that supported their decision to disobey the experimenter. Obedient subjects were assured of the fact that their behavior was entirely normal and that their feelings of conflict or tension were shared by other participants. Subjects were told that they would receive a comprehensive report at the conclusion of the experimental series. In some instances, additional detailed and lengthy discussions of the experiments were also carried out with individual subjects.

When the experimental series was complete, subjects received a written report which presented details of the experimental procedure and results. Again their own part in the experiments was treated in a dignified way and their behavior in the experiment respected. All subjects received a follow-up questionnaire regarding their participation in the research, which again allowed expression of thoughts and feelings about their behavior.

The replies to the questionnaire confirmed my impression that participants felt positively toward the experiment. In its quantitative aspect (see Table 1), 84% of the subjects stated they were glad to have been in the experiment; 15% indicated neutral feelings, and 1.3% indicated negative feelings.

To be sure, such findings are to be interpreted cautiously, but they cannot be disregarded.

TABLE 1

Excerpt from Questionnaire Used in a Follow-up Study of the Obedience Research

Now that I have read the report, and all things considered ...	Defiant	Obedient	All
1. I am very glad to have been in the experiment	40.0%	47.8%	43.5%
2. I am glad to have been in the experiment	43.8%	35.7%	40.2%
3. I am neither sorry nor glad to have been in the experiment	15.3%	14.8%	15.1%
4. I am sorry to have been in the experiment	0.8%	0.7%	0.8%
5. I am very sorry to have been in the experiment	0.0%	1.0%	0.5%

Note. Ninety-two percent of the subjects returned the questionnaire. The characteristics of the nonrespondents were checked against the respondents. They differed from the respondents only with regard to age; younger people were overrepresented in the nonresponding group.

Further, four-fifths of the subjects felt that more experiments of this sort should be carried out, and 74% indicated that they had learned something of personal importance as a result of being in the study. The results of the interviews, questionnaire responses, and actual transcripts of the debriefing procedures will be presented more fully in a forthcoming monograph.[a]

15   The debriefing and assessment procedures were carried out as a matter of course, and were not stimulated by any observation of special risk in the experimental procedure. In my judgment, at no point were subjects exposed to danger and at no point did they run the risk of injurious effects resulting from participation. If it had been otherwise, the experiment would have been terminated at once.

Baumrind states that, after he has performed in the experiment, the subject cannot justify his behavior and must bear the full brunt of his actions. By and large it does not work this way. The same mechanisms that allow the subject to perform the act, to obey rather than to defy the experimenter, transcend the moment of performance and continue to justify his behavior for him. The same viewpoint the subject takes while performing the actions is the viewpoint from which he later sees his behavior, that is, the perspective of "carrying out the task assigned by the person in authority."

Because the idea of shocking the victim is repugnant, there is a tendency among those who hear of the design to say "people will not do it." When the results are made known, this attitude is expressed as "if they do it they will

---

a *monograph* Academic essay or book on a single subject, intended for a specialist audience.

not be able to live with themselves afterward." These two forms of denying the experimental findings are equally inappropriate misreadings of the facts of human social behavior. Many subjects do, indeed, obey to the end, and there is no indication of injurious effects.

The absence of injury is a minimal condition of experimentation; there can be, however, an important positive side to participation. Baumrind suggests that subjects derived no benefit from being in the obedience study, but this is false. By their statements and actions, subjects indicated that they had learned a good deal, and many felt gratified to have taken part in scientific research they considered to be of significance. A year after his participation one subject wrote:

> This experiment has strengthened my belief that man should avoid harm to his fellow man even at the risk of violating authority.

Another stated: 20

> To me, the experiment pointed up ... each individual should have or discover firm ground on which to base his decisions, no matter how trivial they appear to be. I think people should think more deeply about themselves and their relation to their world and to other people. If this experiment serves to jar people out of complacency, it will have served its end.

These statements are illustrative of a broad array of appreciative and insightful comments by those who participated.

The 5-page report sent to each subject on the completion of the experimental series was specifically designed to enhance the value of his experience. It laid out the broad conception of the experimental program as well as the logic of its design. It described the results of a dozen of the experiments, discussed the causes of tension, and attempted to indicate the possible significance of the experiment. Subjects responded enthusiastically; many indicated a desire to be in further experimental research. This report was sent to all subjects several years ago. The care with which it was prepared does not support Baumrind's assertion that the experimenter was indifferent to the value subjects derived from their participation.

Baumrind's fear is that participants will be alienated from psychological experiments because of the intensity of experience associated with laboratory procedures. My own observation is that subjects more commonly respond with distaste to the "empty" laboratory hour, in which cardboard procedures are employed, and the only possible feeling upon emerging from the laboratory is that one has wasted time in a patently trivial and useless exercise.

25   The subjects in the obedience experiment, on the whole, felt quite differently about their participation. They viewed the experience as an opportunity to learn something of importance about themselves, and more generally, about the conditions of human action.

A year after the experimental program was completed, I initiated an additional follow-up study. In this connection an impartial medical examiner, experienced in outpatient treatment, interviewed 40 experimental subjects. The examining psychiatrist focused on those subjects he felt would be most likely to have suffered consequences from participation. His aim was to identify possible injurious effects resulting from the experiment. He concluded that, although extreme stress had been experienced by several subjects,

> none was found by this interviewer to show signs of having been harmed by his experience.... Each subject seemed to handle his task [in the experiment] in a manner consistent with well established patterns of behavior. No evidence was found of any traumatic reactions.

Such evidence ought to be weighed before judging the experiment.

OTHER ISSUES

Baumrind's discussion is not limited to the treatment of subjects, but diffuses to a generalized rejection of the work.

30   Baumrind feels that obedience cannot be meaningfully studied in a laboratory setting: The reason she offers is that "The dependent, obedient attitude assumed by most subjects in the experimental setting is appropriate to that situation [p. 421]." Here, Baumrind has cited the very best reason for examining obedience in this setting, namely that it possesses "ecological validity."[a] Here is one social context in which compliance occurs regularly. Military and job situations are also particularly meaningful settings for the study of obedience precisely because obedience is natural and appropriate to these contexts. I reject Baumrind's argument that the observed obedience does not count because it occurred where it is appropriate. That is precisely why it *does* count. A soldier's obedience is no less meaningful because it occurs in a pertinent military context. A subject's obedience is no less problematical because it occurs within a social institution called the psychological experiment.

Baumrind writes: "The game is defined by the experimenter and he makes the rules [p. 421]." It is true that for disobedience to occur the framework of

---

a   *ecological validity*   An experiment is "ecologically valid" when the circumstances created in the experiment closely match a set of real-world circumstances.

the experiment must be shattered. That, indeed, is the point of the design. That is why obedience and disobedience are genuine issues for the subject. *He must really assert himself as a person against a legitimate authority.*

Further, Baumrind wants us to believe that outside the laboratory we could not find a comparably high expression of obedience. Yet, the fact that ordinary citizens are recruited to military service and, on command, perform far harsher acts against people is beyond dispute. Few of them know or are concerned with the complex policy issues underlying martial action; fewer still become conscientious objectors. Good soldiers do as they are told, and on both sides of the battle line. However, a debate on whether a higher level of obedience is represented by (*a*) killing men in the service of one's country, or (*b*) merely shocking them in the service of Yale science, is largely unprofitable. The real question is: What are the forces underlying obedient action?

Another question raised by Baumrind concerns the degree of parallel between obedience in the laboratory and in Nazi Germany. Obviously, there are enormous differences: Consider the disparity in time scale. The laboratory experiment takes an hour; the Nazi calamity unfolded in the space of a decade. There is a great deal that needs to be said on this issue, and only a few points can be touched on here.

1. In arguing this matter, Baumrind mistakes the background metaphor for the precise subject matter of investigation. The German event was cited to point up a serious problem in the human situation: the potentially destructive effect of obedience. But the best way to tackle the problem of obedience, from a scientific standpoint, is in no way restricted by "what happened exactly" in Germany. What happened exactly can *never* be duplicated in the laboratory or anywhere else. The real task is to learn more about the general problem of destructive obedience using a workable approach. Hopefully, such inquiry will stimulate insights and yield general propositions that can be applied to a wide variety of situations.

2. One may ask in a general way: How does a man behave when he is told by a legitimate authority to act against a third individual? In trying to find an answer to this question, the laboratory situation is one useful starting point – and for the very reason stated by Baumrind – namely, the experimenter does constitute a genuine authority for the subject. The fact that trust and dependence on the experimenter are maintained, despite the extraordinary harshness he displays toward the victim, is itself a remarkable phenomenon.

3. In the laboratory, through a set of rather simple manipulations, ordinary persons no longer perceived themselves as a responsible part of the causal chain leading to action against a person. The means through which responsibility is cast off, and individuals become thoughtless agents of action, is of

general import. Other processes were revealed that indicate that the experiments will help us to understand why men obey. That understanding will come, of course, by examining the full account of experimental work and not alone the brief report in which the procedure and demonstrational results were exposed.

At root, Baumrind senses that it is not proper to test obedience in this situation, because she construes it as one in which there is no reasonable alternative to obedience. In adopting this view, she has lost sight of this fact: A substantial proportion of subjects do disobey. By their example, disobedience is shown to be a genuine possibility, one that is in no sense ruled out by the general structure of the experimental situation.

Baumrind is uncomfortable with the high level of obedience obtained in the first experiment. In the condition she focused on, 65% of the subjects obeyed to the end. However, her sentiment does not take into account that within the general framework of the psychological experiment obedience varied enormously from one condition to the next. In some variations, 90% of the subjects *dis*obeyed. It seems to be *not* only the fact of an experiment, but the particular structure of elements within the experimental situation that accounts for rates of obedience and disobedience. And these elements were varied systematically in the program of research.

A concern with human dignity is based on a respect for a man's potential to act morally. Baumrind feels that the experimenter *made* the subject shock the victim. This conception is alien to my view. The experimenter tells the subject to do something. But between the command and the outcome there is a paramount force, the acting person who may obey or disobey. I started with the belief that every person who came to the laboratory was free to accept or to reject the dictates of authority. This view sustains a conception of human dignity insofar as it sees in each man a capacity for *choosing* his own behavior. And as it turned out, many subjects did, indeed, choose to reject the experimenter's commands, providing a powerful affirmation of human ideals.

40   Baumrind also criticizes the experiment on the grounds that "it could easily effect an alteration in the subject's ... ability to trust adult authorities in the future [p. 422]." But I do not think she can have it both ways. On the one hand, she argues the experimental situation is so special that it has no generality; on the other hand, she states it has such generalizing potential that it will cause subjects to distrust all authority. But the experimenter is not just any authority: He is an authority who tells the subject to act harshly and inhumanely against another man. I would consider it of the highest value if participation in the experiment could, indeed, inculcate a skepticism of this kind of authority. Here, perhaps, a difference in philosophy emerges most

clearly. Baumrind sees the subject as a passive creature, completely controlled by the experimenter. I started from a different viewpoint. A person who comes to the laboratory is an active, choosing adult, capable of accepting or rejecting the prescriptions for action addressed to him. Baumrind sees the effect of the experiment as undermining the subject's trust of authority. I see it as a potentially valuable experience insofar as it makes people aware of the problem of indiscriminate submission to authority.

Conclusion

My feeling is that, viewed in the total context of values served by the experiment, approximately the right course was followed. In review, the facts are these: (a) At the outset, there was the problem of studying obedience by means of a simple experimental procedure. The results could not be foreseen before the experiment was carried out. (b) Although the experiment generated momentary stress in some subjects, this stress dissipated quickly and was not injurious. (c) Dehoax and follow-up procedures were carried out to insure the subjects' well-being. (d) These procedures were assessed through questionnaire and psychiatric studies and were found to be effective. (e) Additional steps were taken to enhance the value of the laboratory experience for participants, for example, submitting to each subject a careful report on the experimental program. (f) The subjects themselves strongly endorse the experiment, and indicate satisfaction at having participated.

If there is a moral to be learned from the obedience study, it is that every man must be responsible for his own actions. This author accepts full responsibility for the design and execution of the study. Some people may feel it should not have been done. I disagree and accept the burden of their judgment.

Baumrind's judgment, someone has said, not only represents a personal conviction, but also reflects a cleavage in American psychology between those whose primary concern is with *helping* people and those who are interested mainly in *learning* about people. I see little value in perpetuating divisive forces in psychology when there is so much to learn from every side. A schism may exist, but it does not correspond to the true ideals of the discipline. The psychologist intent on healing knows that his power to help rests on knowledge; he is aware that a scientific grasp of all aspects of life is essential for his work, and is in itself a worthy human aspiration. At the same time, the laboratory psychologist senses his work will lead to human betterment, not only because enlightenment is more dignified than ignorance, but because new knowledge is pregnant with humane consequences.

(1964)

REFERENCES

Baumrind, D. Some thoughts on ethics of research: After reading Milgram's "Behavioral study of obedience." *Amer. Psychologist*, 1964, 19, 421-423.

Lazarus, R. A laboratory approach to the dynamics of psychological stress. *Amer. Psychologist*, 1964, 19, 400-411.

Milgram, S. Behavioral study of obedience. *J. abnorm. soc. Psychol*, 1963, 67, 371-378.

Milgram, S. Some conditions of obedience and disobedience to authority. *Hum. Relat.*, in press.

Questions:

1. This article begins with an objective description of obedience. What does this opening accomplish? For whom might this hook be intended?

2. Paragraph 3 states that "A simple procedure was devised for studying obedience." Is the rationale for this procedure explained here (or in the original article)? What explanation can you give for the "simple procedure" design? Do you think the description of the study design as "a set of rather simple manipulations" is correct?

3. Why do you think a subject's verbal objection to the proceedings was not counted as disobedience? Should it have been?

4. Does Baumrind's "Some Thoughts" read as though "the production of stress was an intended and deliberate effect" in Milgram's experiment? Why do you think Milgram's response makes this distinction between "intended and deliberate effects" and accidental ones?

5. Milgram describes the subjects' responses as examples of "momentary excitement." Where in this article is that assertion contradicted?

6. Did the original subjects behave in a "shockingly immoral way"? What do you think the effect might be on a participant who read this statement?

7. Is the assertion that obedient participants "justify" their behaviors and do not "bear the full brunt" of self-judgment about their actions consistent with the participants' comments quoted in this article?

8. Was the questionnaire the experimenters sent out anonymous in nature? How do you know this? What problems might arise from a non-anonymous approach, especially given susceptible participants?

9. The article suggests that Milgram's original piece "did not contain all the information needed for an assessment of the experiment," but that "Baumrind's article was deficient in information that could have been obtained easily." It later suggests (after a section of new material), that Baumrind should have investigated further before writing her article: "Such evidence ought to be weighed before judging the experiment."

Is it a normal expectation that readers will research other, possibly unpublished, information before responding to a published article? What argumentative function does Milgram's statement seem to serve?

10. List the proof provided that the participants felt the experiment was valuable. Do experimental scientists normally suggest that a participant is the best judge of how an experiment is conducted? Why or why not? Is the participants' evaluation of the experiment relevant to Milgram's defense of this particular study?

11. Research the concept of "ecological validity." Does the obedience lab procedure seem to you to have ecological validity as applied to the military experience? Do you think Milgram's original article "help[s] us to understand why men obey"? If so, in what ways? If not, why not?

12. Argue for or against the assertions made in Milgram's response that an experimenter does not "make" his or her subjects do anything and that an experimental subject is "free to accept or to reject the dictates of authority."

thirteen

Ian Nicholson

from "Torture at Yale": Experimental Subjects, Laboratory Torment and the "Rehabilitation" of Milgram's "Obedience to Authority"

In this article from the journal Theory & Psychology,
*psychologist Ian Nicholson delves into the historical archive
surrounding Stanley Milgram's "Obedience to Authority"
experiments in order to question the ethics of Milgram's work.*

☙

ABSTRACT

Stanley Milgram's experiments on "Obedience to Authority" are among the most criticized in all of psychology. However, over the past 20 years, there has been a gradual rehabilitation of Milgram's work and reputation, a reconsideration that is in turn closely linked to a contemporary "revival" of his Obedience experiments. This paper provides a critical counterpoint to this "Milgram revival" by drawing on archival material from participants in the Obedience study and Milgram himself. This material indicates that Milgram misrepresented (a) the extent of his debriefing procedures, (b) the risk posed by the experiment, and (c) the harm done to his participants. The archival record also indicates that Milgram had doubts about the scientific value of the experiment, thereby compromising his principal ethical justification for employing such extreme methods. The article ends with a consideration of the implications of these historical revelations for contemporary efforts to revive the Milgram paradigm.

"We couldn't possibly conceive that anybody would allow any torture to go through Yale University."

Unnamed participant in the Obedience to
Authority Experiment (Errera, 1963c, p. 10). […]

Since his untimely demise in 1985, Milgram's reputation along with that of the obedience research has undergone a remarkable transformation. In the 1960s, 70s, and early 80s, Milgram was an often embattled figure subject to extensive and forceful ethical and methodological critiques by Baumrind (1964, 1985), Helm and Morelli (1985), Mixon (1972), Patten (1977a, 1977b), and others. Commenting on this critical literature in 1982, Milgram enthusiast and onetime research assistant Alan Elms captured the prevailing view when he described the obedience study as "the most notorious of all ... psychologically stressful research studies" (1982, p. 237). The criticism and subsequent notoriety had an effect: Milgram's obedience research was effectively banned in American psychology and according to Blass (2000), no study using obedience-style procedures was undertaken in the United States for over 30 years. Popular presentations of Milgram's work during this period also displayed a clear sense of discomfort with the obedience research, with the behavior of Milgram himself frequently being the object of concern rather than the allegedly "reprehensible actions" of the experimental participants (Blass, 2004, p. 128). For example, in the 1975 television dramatization of the obedience experiments, "The Tenth Level" (Bellak & Dubin, 1975), Milgram was flamboyantly portrayed by William Shatner as an obsessive scientist who recklessly endangers the lives of his students for the sake of his own career (see also Abse, 1973).

History, however, appears to have been kind to Milgram. Critical media portrayals and forceful, Baumrind-style (1964) critiques have been gradually eclipsed by largely admiring essays and popular presentations which pay tribute to Milgram's imagination and daring. In his comprehensive but generally sympathetic analysis of the Obedience literature, Miller (1986) concluded that Milgram's research had been of "inestimable value, and that the demonstrable costs were relatively negligible" (p. 260). Ten years later, in a special issue of the *Journal of Social Issues* (Miller, Collins, & Brief, 1995) largely uncluttered by dissenting perspectives, Milgram's work was again celebrated for its "groundbreaking" value and "objective, scientific manner" (Elms, 1995, pp. 22, 30). This laudatory tone was maintained in a volume of essays edited by Blass (2000) who praised the Obedience work for its "revelatory power" and "large and wide ... domain of relevance" (p. ix). There was some constructive criticism of Milgram's paradigm in the 1990s (e.g., Darley, 1992), but as Mandel (1998) noted, "the field of social psychology, in general, has demonstrated a rather uncritical acceptance of obedience as the basis for socially organized acts of evil" (p. 90).

Since 9/11 and the emergence of torture and "refined interrogation techniques" as matters of public interest (Henley, 2007), the enthusiastic and largely uncritical discussion of Milgram's work has continued apace and possibly accelerated. Thomas Blass' (2004) biography of Milgram *The Man Who Shocked the World* made light of the voluminous ethical and methodological criticisms of the Obedience ethics while lionizing the study's intellectual import. More recently, Fiske and Harris (2004) and Zimbardo (2007) have mobilized Milgram's work to explain prisoner abuse at Abu Ghraib[a] while Horgan (2010) has cited the obedience research to explain the murder of civilians by U.S. soldiers in Afghanistan. In a similar vein, Sheppard & Young (2007) claimed that the study could contribute to the moral edification of business students in the aftermath of the Enron scandal.[b] In 2009, *American Psychologist* published a special issue on Milgram's Obedience research which, like its predecessors, was largely devoid of any sustained critical content and full of praise for work which "reminds us of the importance of torture in the modern world" (Miller, 2009, p. 22). Most recently, Russell (2010) undertook a thorough but uncritical examination of how Milgram came to develop his experimental procedure in order to "better arm theorists interested in ... the question of why so many participants inflicted every shock" (p. 22). In the face of so much laudatory commentary, it is difficult to resist the assessment of an anonymous *American Psychologist* reviewer who claimed that "time has been 'on the side' of the Milgram studies ... they are less controversial now than ever" (personal communication, June 18, 2008).

Social psychology textbooks provide a further measure of the degree to which Milgram's work has been transformed from a hotly contested study into a "normal," praiseworthy finding. The obedience research has been a staple of social psychology textbooks since 1965 and as Stam, Lubek, and Radtke (1998) have noted, in textbooks from 1965 to 1995 ethical concerns were a standard feature of the discussion. However, over that same period Stam et al. (1998) also noted that textbook discussions of the obedience study "have tended, in recent years, to be more, not less, supportive of experimental social psychological research" (p. 167). This "Milgram-friendly" trend identified by Stam et al. is even more pronounced in contemporary social psychology texts. For example, the well-known text *Exploring Social Psychology* (Myers & Smith, 2009) assigned eight pages to the obedience experiments, but made

---

a   *Abu Ghraib*   Iraq prison where members of the American military subjected prisoners to torture and extreme abuse in 2003–04.

b   *Enron scandal*   2001 scandal in which the energy corporation Enron, then one of the largest companies in the United States, was discovered to have engaged in dishonest accounting practices; the company went bankrupt and a number of executives were sentenced to prison.

light of the furious ethical and epistemological[a] controversy that ensued. The only indication of the contentiousness of the research was a brief paragraph where the authors stated that "the procedures [Milgram] used disturbed many social psychologists" (p. 93). Still, even this limited and heavily bowdlerized[b] Milgram coverage was more than what appeared in several other texts. At the start of the five page section on Milgram in *Mastering Social Psychology* (Baron, Byrne, Branscombe, & Fritzley, 2011), the authors stated that the obedience experiments were "controversial," but they provided no discussion at all of any ethical or epistemological problems with the research. Gilovich, Keltner, and Nisbet (2006) took the sanitizing of the obedience research a step further by omitting any reference to its ethical and epistemological contentiousness. The obedience study was presented as a "now-classic experiment" with no discussion or even acknowledgement of the voluminous disciplinary and cultural criticism that the work attracted (p. 10). So comprehensive is the rehabilitation of Milgram that the word "ethics" is not even mentioned in connection with the obedience study in any of these texts. Thus, it appears that in contemporary social psychology texts the obedience research does not even require a defense; 25 years after the study had been described as the "most notorious of all" (Elms, 1982, p. 237), the morality of Milgram's methods and the validity of his conclusions is a "settled" matter and beyond criticism.

Reviewing recent "Obedience" literature, it is apparent that the revival 5 of Milgram's reputation and the renewed interest in his work among psychologists is not strictly a question of historiography or intellectual curiosity. Animating this work is a desire, and in some cases a consciously stated wish to recover that which was perceived to have been lost. In some cases, this involves an explicit resuscitation of key features of Milgram's obedience work with various compromises made to accommodate contemporary ethical standards (Navarick, 2009). Burger (2009) had a similar "revivalist" ambition insisting that the psychological significance of the Obedience research depended on its viability as a topic for empirical examination: "How long can discussions about the [Obedience] research be sustained without additional data?" (Burger, 2002, p. 666). Convinced of the need for more "data," Burger (2009) undertook an even more faithful recreation than Navarick (2009), modifying the experiment only to "avoid exposing them to the intense stress Milgram's participants often experienced in the subsequent parts of the procedure" (Burger, 2009, p. 2).

---

a *epistemological* Concerned with the theory of knowledge; here, concerned with the validity of the experiment and its findings.

b *bowdlerized* Censored for controversial content.

According to Navarick and Burger, the inspiration for this work is partly empirical, a desire to know "would people still obey today" (Burger, 2009, p. 4). However, the rehabilitation of Milgram and the revival of his work is not limited to such narrow empirical questions. Inspiring much of the Milgram commentary is a sense of frustration with "increasingly stringent IRB[a] rules and regulations" and a feeling that the "pendulum had swung too far" toward the goal of protecting research participants (Benjamin & Simpson, 2009). According to Benjamin & Simpson, these overly "stringent" IRB rules had undermined social and personality psychology's capacity to undertake "high impact studies – those that had exceptionally high experimental realism" (pp. 17–18). This view was echoed by long-time Milgram enthusiast Arthur Miller (2009) who linked the demise of Obedience research with the rise of ethical safeguards. For these scholars, burnishing Milgram's reputation and reprising the "obedience" experiment is a way of pushing the "pendulum" back and restoring to social psychology a spirit of daring, imagination, and excitement lost during the "crisis" of social psychology in the 1970s (Pancer, 1997; Silverman, 1977; Smith, 1972). The professional rewards for this undertaking are clearly apparent in the case of Burger, who by "replicating Milgram" earned himself the aforementioned special issue of *American Psychologist* built around his study, a special feature on ABC television's *Primetime*[3] and flattering coverage in the *New York Times* (Cohen, 2008).

Encouraging more dynamic research is a commendable goal in any discipline and disciplinary "heroes" can be a valuable source of inspiration (Samelson, 1974). However, as journalist Gerald Johnson observed "heroes are created by popular demand, sometimes out of the scantiest materials" (1943, p. 11). In the case of Milgram, I would argue that the material is very scanty indeed and that disciplinary ambitiousness for more imaginative research and wider social relevance has gotten ahead of careful historical scrutiny. Although much valuable work on the history of the Obedience experiments has been undertaken in recent years, several important elements of the Obedience experiments have been omitted or minimized, most crucially the experience of the participants themselves. This paper aims to bring the experiences of the participants back into play and in so doing provide a critical reexamination of both the ethics and meaning of the obedience to authority experiments.

Most scholarly discussions of the Obedience research take Milgram at his word on these matters, essentially paraphrasing what he said in his own defense when called to task for mistreating his participants. In his 1964 reply

---

a *IRB* Institutional Review Board.

to Baumrind and again in his 1974 book, Milgram highlighted his extensive debriefing or "dehoaxing" procedures – unusual for the time – while citing the results of a quantitative follow-up study he undertook which indicated that 84% of participants were glad that they participated and that 4 out of 5 participants felt that more experiments of this sort should be carried out (Milgram, 1964). These are compelling numbers although as Benjamin & Simpson (2009) have noted, the numbers could well reflect a degree of cognitive dissonance on the part of the participants. Milgram did not consider this possibility and he maintained that the retrospective consent of participants justified the obedience procedure. "The participant" Milgram insisted, "rather than the critic must be the ultimate source of judgment" (1974, p. 199) a verdict that contemporary Milgram enthusiasts seem happy to accept.

Milgram's apparent devotion to the importance of his participants' views is commendable, but scholarly consideration of the Obedience participants should not be restricted to a small number of quantitative survey items. What is largely absent from Milgram's own work and from the secondary literature is a consideration of the extensive participant feedback that Milgram gathered but chose to omit from his publications for reasons which in hindsight are clear: they cast doubt on the ethics of the study and the validity of its results. Milgram always publically insisted that his research was ethical insofar as it involved a relatively minimal impact on his participants. Participants were stressed, yes, but it wasn't a matter of any consequence. Milgram also insisted that he took steps to minimize the physical and psychological impact of the extreme stress his experiment engendered. He repeatedly insisted in his published work that all participants were given a thorough and effective debriefing, or to use his term, "dehoaxing" immediately after the experiment (Milgram, 1964, p. 849). Despite this forceful and often-repeated insistence on the innocuous nature of the Obedience experiments, the qualitative records indicate that the debriefing procedures were nowhere near as thorough and effective as Milgram claimed.

The second insight to be gleaned from a consideration of participant narratives involves the very meaning of the study itself. Milgram did concede that his study was right on the line of what was ethically permissible, but he always insisted that the ethical riskiness of his work was more than offset by the extraordinary gains that were accrued, namely the revelation of something "dangerous": the tendency for people to harm others when ordered by an authority figure (Milgram, 1974, p. 188). The important question that arises is one that has, rather curiously, almost completely vanished from contemporary discussion: does the experiment actually reveal a "dangerous" propensity to obey destructive authority?

Milgram obscured this question by blending two definitions of obedience together. He defined "obedience" in operational terms as the "subject who complies with the entire series of experimental commands" (1965, p. 59). However, as Patten (1977a, 1977b) has noted, the cultural power of Milgram's experiment depended heavily on his extensive and largely unacknowledged use of a more comprehensive meaning of "obedience," one that relied on the participant's understanding of the situation. In this second definition of obedience, "one is obedient when one consistently responds to commands by performing acts which one has every reason to believe are immoral, merely because one is ordered to do so" (Patten, 1977b, p. 427). Milgram (1974) did not make a distinction between these two senses of "obedience," assuming that the act of going to the end of the board was something "obviously" and intrinsically unreasonable and "shockingly immoral" (p. 194). Feedback from participants raises doubts about Milgram's claim that it was "shockingly immoral" for his participants to have carried out the instructions of the experimenter.

Before discussing the participant feedback in detail, a note on sources. All of the participant feedback presented in this paper comes from Milgram's own papers at the Yale University Archives. There are two places within the Milgram papers that are relevant. The first series of documents contains brief written comments from participants that were obtained as part of a questionnaire that was sent out to all participants and that a remarkable 95% of participants returned (Reaction of subjects, 1962). The second source of information comes from transcripts of 40 lengthy one-on-one and small-group interviews conducted by Paul Errera, an assistant professor of psychiatry at Yale. Unfortunately, most of these interviews are not yet available to researchers. The Yale Archives is in the process of "sanitizing" these documents – removing names and personal information. I have analyzed four of the interviews that have been sanitized (Errera, 1963a; 1963b; 1963c; 1963d). It is important to note therefore that the participant feedback that I am presenting is limited, however even this partial sample gives plenty of ways and reasons to rethink the ethics and meaning of the Obedience research.[4]

ETHICAL TREATMENT OF PARTICIPANTS

[…] The remarks of several participants make it clear that many were not debriefed after [the experiment] and that some were simply sent home to contemplate what had transpired. "I became somewhat irked when I received your first copy of the results, for after reading it, I felt that I was made a fool of, for I had no idea that this was preplanned." Another participant remarked that "I was pretty well shook up for a few days after the experiment. It would have helped if I had been told the facts shortly after." The psychological impact of being kept in the dark about the experiment is apparent in the remarks of other participants who were not debriefed: "I felt so bad afterward" one of these participants remarked. "I wanted to call and actually apologize ... and I went to the telephone book and looked up the name – he used the name Richardson or something and unfortunately there were three of them with exactly the same first and last name [so] I didn't make the call" (Subjects' conversation, 1963). It was not until weeks later when the participant received his participant report from Milgram that he learned that he had not physically harmed anyone. "I felt wonderful when I received that letter and know – realize that this fellow hadn't received any jolts." A similar combination of extreme discomfort and relief is evident in the remarks of another participant who stated that "I actually checked the death notices in the *New Haven Register* for at least two weeks after the experiment to see if I had been involved and a contributing factor in the death of the so called 'learner' – I was very relieved that his name did not appear in such a column" (Reaction of subjects, 1962).

In a relatively small city like New Haven, CT, word of the experiment soon spread and many participants criticized Milgram for not informing people about the "true" nature of the study immediately afterward: "From what I've learned from others who've taken part, it would seem you have been somewhat irresponsible in permitting disturbed subjects to leave without informing them that they didn't half kill the shockee" (Reaction of subjects, 1962). Milgram's ethical shortcomings in this matter were ably summarized by another participant:

> Upon reflection I seriously question the wisdom and ethics of not completely dehoaxing each subject immediately after the session. The standard decompression treatment I received was not successful in reducing my anger and concern below the boiling point. Probably not many subjects were in a position to take the matter directly to the principal investigator as I did so the question arises as to how many other subjects were more or less seriously disturbed by tensions that could have been resolved by full disclosure at the earliest possible time. Allowing subjects to remain deceived is not justified in my opinion even if such continued deception was thought necessary to avoid contamination. (Reaction of subjects, 1962)

It is clear from this feedback that Milgram deliberately misrepresented his post-experimental procedures in his published work. The reason for this "unsanctioned" deception may be found in the early scholarly reaction to the obedience experiments. Professional criticism began to mount even before the first publication of his results in October, 1963. In 1962 a member of the Yale Psychology Department filed a complaint about Milgram's obedience research with the American Psychological Association. Milgram was subsequently informed that his application for APA membership would be delayed pending an "informal" review by the Secretary of the Committee on Scientific and Professional Ethics and Conduct (Blass, 2004). The following year, Milgram (1963b) informed psychologist Dorwin Cartwright that he had been "clobbered" by professional criticism. If it emerged, in this context, that even a handful of participants had been sent home after so traumatic an experience without any debriefing Milgram's credibility as a responsible researcher would have been undermined. With the ethical integrity and possible future of the obedience study on the line, Milgram evidently decided to lie his way through the criticism. When later pressed by Baumrind (1964) on the extent and efficacy of his dehoaxing procedure, he admitted that the debriefing changed over time and that the "exact content of the dehoax varied from condition to condition and with increasing experience on our part" (Milgram, 1964, p. 849). However, Milgram always insisted – dishonestly – that "all subjects" were told that the shocks were not real and that "procedures were undertaken to assure that the subject would leave the laboratory in a state of well being" (Milgram, 1963a, p. 374).

Serious though this misrepresentation of the extent and efficacy of the debriefing may be, it takes on an added significance in the context of the most serious ethical charge, namely that the obedience research harmed participants. Milgram (1964) was again quite adamant about this, insisting that "at no point did they run the *risk* [emphasis added] of injurious effects" and that critics ought not to confuse "momentary excitement" with "harm" (p. 849). In the context of quantitative survey data, the characterization of participants' experiences as "momentary excitement" seems plausible enough, but a consideration of participant narratives puts a different complexion on the matter. Several participants commented on the toxicity of the experience. "My comment to my wife on arriving home was that this had been the most unpleasant night of my life" one participant remarked, a view echoed by several other participants. "I couldn't remember ever being quite as upset as I was during the experiment" (Reaction of subjects, 1962) one subject remarked while another stated that "I felt real remorse and when I came out – when the experiment was all over, I got home and told my family I had just gone through the most trying thing that I had ever subjected myself to" (Subjects' conversation, 1963). Others indicated that their experience was a matter of some consequence and had effects that extended well beyond the immediate laboratory moment. Some participants were full of strong feelings of self-recrimination after the study was over. "After completing the experiment I really was ashamed of myself. I kept thinking why didn't I refuse to give pain to my fellow man instead of going through as directed to the end [discussed with a friend who also participated and stopped exp.].[a] Thus I hated myself all the more for not doing the same." Another participant commented on his feelings of shame and how difficult it was to shake off a sense of personal diminishment: "Even now I'm ashamed of telling my friends that I took part in this experiment. I just want to forget it" (Reaction of subjects, 1962).

It is important to emphasize that most of Milgram's participants did not appear to take the experience to heart, and a majority indicated that they actually learned something about themselves in the experience. That said, several emotionally sensitive participants took the experience very hard indeed. For example, one participant remarked that his experience had badly jolted his self-image of being a caring, thoughtful, and morally superior person:

> It has bothered me that I went all the way ... I am a person who has had the opportunity to do a great deal of reading and reflecting ... I form my own opinions from the newspapers and mass periodicals ... I also read the smaller circulation publications such as *The Reporter*, *The Nation*, *The New Republic*, and similar publications. I consider myself better informed and I hope more cultured than the average non-college student. In spite of all this I gave the same performance that the average slob, taken off the street, would probably have done. This I consider frightening. (Reaction of subjects, 1962)

In addition to issues of self-image, some participants reported disturbed 20 sleep and being upset for an extended period of time. "That night I couldn't sleep the test bothered me – even the next day I was upset" (Reaction of subjects, 1962). This view was echoed by another participant who commented on how difficult it was to shake off the traumatic impact of the experience: "I wouldn't want to do another experiment like that again for any amount of money. I'm still sorry I went to do it. It took me a couple of weeks before I was able to forget about it. I don't think it's right to put someone through such a nervous tension" (Reaction of subjects, 1962). The traumatic character of the experience was remarked upon by another participant who said that "the experiment left such an effect on me that I spent the night in a cold sweat and nightmares because of the fear that I might have killed the man in the chair" (Reaction of subjects, 1962). Another participant who identified himself as an alderman in the city of New Haven spoke of an intense but delayed reaction to his participation: "I sort of had a delayed reaction on this because I more or

---

a *[discussed ... exp.]* This material was given in square brackets in the original article.

less forgot about it, but after speaking with Dr. Milgram ... I think I was really – I don't know how to describe it – just completely depressed for a while" (Errera, 1963b). For some participants, the intensity of the experiment carried over into the workplace and one participant reported that it led indirectly to him losing his job:

> By coincidence a fellow employee had taken part in the same experiment before me ... Later we compared notes and during one of our discussions, I got hot under the collar and said things I normally never would say, especially during working hours. Directly, I made a very bad impression on some people within hearing distance because I used some vulgar words and shortly thereafter because of other conditions I lost my job. (Reaction of subjects, 1962)

It would of course be unfair to lay all of the blame for this man's loss of employment at the feet of Milgram. However, it is clear from the archival record that Milgram did traumatize many of his participants and there is evidence that in several cases he did not help his participants adequately deal with what they had experienced. Indeed in several cases, Milgram does not appear to have debriefed his participants at all; he simply sent them home. It is clear in hindsight, that Baumrind (1964) was entirely justified in calling Milgram to task for brutalizing his participants while failing, at least in some instances, to deal with the consequences in even the most minimal of ways. [...]

ETHICAL RISKS AND INTELLECTUAL REWARDS

Although Milgram never publicly disclosed the full extent of the harm he had inflicted, he did take on the ethical challenge of justifying the extreme nature of the experiment in relation to the intellectual benefits that accrued. Milgram claimed that the deception of participants and the "temporary" discomfort that many experienced was offset by the "revelation of certain difficult to get at truths" (1974, p. 198), namely that people have a tendency to obey any command that comes from a legitimate authority. While marshalling this argument, Milgram mobilized the ideal of the autonomous individual, while ignoring the powerful effect that context has on determining the meaning of any action. He insisted that his research was informed by the idea of the laboratory as a socially "neutral" space where each individual is free to act according to his or her conscience: "a person who comes to the laboratory is an active choosing adult capable of accepting or rejecting the prescriptions for action addressed to him" (Milgram, 1964, p. 852).

Here again, Milgram's vision of the experiment as a socially "neutral" test of autonomous individuals is undermined by the testimony of his participants.

Many of his participants indicated that the context of the psychological experiment had a very significant impact on their actions. They saw the experiment as a social contract – a "bargain" – in which they had an obligation as participants to follow instructions as best they could (Errera, 1963c, p. 21). In return, they expected to be treated in a manner that would respect their dignity and physical and psychological well-being. Milgram breached this contract in a highly dramatic and stressful fashion and then in effect stood back and denied that any such contract existed. He manipulated his trusting participants into doing something that they would never ordinarily do and then once the experiment was over he shifted all the responsibility from himself onto the participants insisting that they were "free moral agents" (as cited in Abse, 1973, p. 126) and that a majority of them had behaved in a "shockingly immoral way" (Milgram, 1964, p. 849). [...]

A key element in the social contract of a psychological experiment is trust 25 in the responsibility and competence of the scientific authority. Participants assume that the expert knows what he or she is doing and even if they have doubts, their lack of knowledge about the experimental situation combined with the repeated assurances of the expert is often enough to sustain compliance. As one obedient participant noted, "my main reason for not breaking off was the confidence I had in the experimenter knowing what he was doing" (Reaction of subjects, 1962). The participants had no reason not to trust the experimenter, a fact that Milgram was careful to exploit. When participants expressed concern for the "learner" and doubt as to whether they should proceed they were reassured in no uncertain terms: "although the shocks may seem painful, there is no permanent tissue damage, so please go on" (Milgram, 1963a, p. 374). It is impossible to know how many participants would have broken off the experiment without this reassurance, but it is likely that many took the experimenter at his word and continued on the basis of a declared commitment as to the safety of the procedure. "Giving the shocks did not upset me until the learner mentioned his heart, but I had faith in Yale that the doctor would stop the experiment if he thought it best" (Reaction of subjects, 1962).

In his analysis of the experiment, Milgram acknowledged that participants trust the experimenter and consequently transfer the responsibility for their behavior from themselves to the scientific authority. "In the laboratory, through a set of simple manipulations, ordinary people no longer perceived themselves as a responsible part of the causal chain leading to action against a person" (Milgram, 1974, p. 175). However, as we have seen, there was nothing "simple" about Milgram's manipulations. Participants were placed in an unfamiliar and disorienting environment with little to rely on apart from their own senses and the expertise of the authority figure. Their doubts and concerns were ignored and they were expressly told that all was well and that they must continue. In his published work, Milgram never conceded the unfairness of the scenario he had constructed. A psychological test licenses unusual or unexpected challenges for participants but it brings with it extraordinarily strong expectations of ethical conduct on the part of the experimenter. Participants in an experiment know that all may not be what it seems but at a minimum they assume – correctly as it turned out – that however strange, however seemingly inexplicable an experimental situation may appear, no one is going to be physically harmed, much less killed. As noted earlier, Milgram explicitly reaffirmed this common-sense understanding of the benign character of psychological experimentation and turned it to his advantage by instructing participants reluctant to continue with the shocks that there was "no permanent tissue damage" (Milgram, 1963a, p. 374). Some participants explicitly stated that their awareness of being in a "psychological experiment" influenced their actions. Explaining why he continued, one participant remarked that "I have faith in psychological experiments and suspected that the learner was not being hurt as badly as he pretended to be" (Errera, 1963c, p. 10). Even those participants who thought that the shocks were real were reassured by the professional context of the experiment: "We couldn't possibly conceive that anybody would allow any torture to go through Yale University" (Errera, 1963c, p. 10). [...]

Milgram's experiment demonstrated little beyond the commonplace observation that people in unfamiliar environments will trust authority figures and as a consequence can be deceived, manipulated, and taken advantage of. Milgram claimed to be divining dark and hitherto unexamined reaches of human nature when what he was really doing was toying with his participants' self-worth and physical well-being for the sake of a psychological study and his own professional advantage. Many participants spoke of their feelings of anger at having their trust betrayed in such a manipulative and potentially dangerous manner. "I never expected to be hoodwinked at an outstanding university such as Yale. It tends to make one cynical." One participant felt so mistreated that he considered legal action. "After the experiment was completed and I had left the laboratory, I felt that I should report your actions to the police" (Reaction of subjects, 1962).

Milgram kept such concerns to himself and he publically framed the experiment as something that actually benefitted his participants. The extreme circumstances of the experiment "provided people with an opportunity to learn something of importance about themselves" (Milgram, 1964, p. 850). To corroborate this claim, Milgram quoted several participants who framed their experience in a manner that reflected his own interpretation of the study as a moral test of character: Will the individual harm another? For these participants, the study helped "jar people out of complacency" by revealing obedient tendencies in themselves: "I think people should think more deeply about themselves and their relation to their world and to other people" (cited in Milgram, 1964, p. 850).

These were compelling testimonials and the majority of participants did indeed indicate that they had learned something through their participation. However, the self-knowledge was not always of the uplifting sort that Milgram reported. For many of Milgram's participants, the central test of the experiment was not "obedience to authority," but their own credulity. The experiment tested whether they were street-smart and savvy or easy "marks" who took too much on faith. Several participants indicated that the experiment taught them that they needed to be much more cynical in dealing with the world. "I've learned I am not as world-wise and sophisticated as I would like to believe. Your actors proved to me that I am a sucker for any two good con men. I now carry my wallet pinned in an inside pocket." Others commented on how instructive the study was for deepening their cynicism toward psychology. "Stay away from little men who try psychology without logic" one participant remarked (Reaction of subjects, 1962). For these participants, there was nothing especially edifying about being traumatized and manipulated and the study served only to underscore the importance of being mistrustful of everything – even something as seemingly benign as a study of "learning."

Torment in the Laboratory: The Seductive Appeal of Psi-power

Milgram misrepresented three important components of the Obedience research: (a) the extent of his debriefing procedures, (b) the risk posed by the experiment, and (c) the harm done to some of his participants. In his defense, proponents of the Obedience research have argued that any alleged ethical lapses are still offset by the enormous intellectual gains that have come as a consequence of these studies. Milgram famously framed the study in terms of the Holocaust, and while he later conceded that there were a number of important differences between his American laboratory participants and German SS[a] guards (Milgram, 1964), he insisted throughout his career that his work contained vital insights into Nazi crimes. Indeed, for Milgram (1974), explaining the Holocaust was a relatively modest undertaking that

30

---

a SS Abbreviation for *Shutzstaffel*, a Nazi organization. During the Holocaust, concentration camps and death camps were managed by a branch of the SS.

sold the experiment short: "To focus only on the Nazis, however despicable their deeds ... is to miss the point entirely" he insisted. "For the studies are principally concerned with the ordinary and routine destruction carried out by everyday people following orders" (p. 178). Thus, publically, the obedience research had a clear meaning and ethical justification for Milgram. It was a dramatization of the weakness of our "unique personalities" in relation to "larger institutional structures" (p. 188). For Milgram and his contemporary enthusiasts, the price of such knowledge was high, but it was and is a price worth paying (see Elms, 1982; Miller, 2009).

This is an appealing and well-rehearsed defense, however it is undermined by other records from Milgram's papers. While Milgram (1963a) was running the study and routinely and knowingly subjecting people to "uncontrollable seizures" and "violent convulsions," he admitted in his notebook that he had strong reservations about the ethical status and meaning of his work (p. 375). As a graduate student, he spoke emphatically against the use of deception in psychological research: "My own view is that deception for any purposes [in psychology] is unethical and tends to weaken the fabric of confidence so desirable in human relations" (Milgram, 1959). Insisting as a graduate student that "truth is never to be pursued irresponsibly" (as cited in Evans, 1980, p. 193), Milgram soon reconciled himself to the use of deception in his research, although the matter evidently tugged at his conscience, for in a manner similar to contemporary verbal contortions over the word "torture," he carefully avoided the use of the term "deception" in favor of the euphemism "technical illusions." [...]

In a moment of extraordinary candor, Milgram admitted that the principal motivation for all this torment had little to do with high-minded sentiment. It was driven by curiosity and something darker:

> Considered as a personal motive of the author – the possible benefits that might rebound to humanity – withered to insignificance alongside the stri-dent demands of intellectual curiosity. When an investigator keeps his eyes open throughout a scientific study he learns things about himself as well as his subjects and the conclusions do not always flatter. (Milgram, 1962b)

Milgram never specified what "unflattering" conclusions were to be drawn about himself concerning his role in these experiments, but it is not difficult to imagine the appeal of his experimental design to a young and physically small man. Most social psychological experiments place the experimenter in a position of power, but few confer upon the investigator the almost super-human capacity to psychologically crush other men, often individuals physically larger and older than oneself.[5] This fact was partly concealed by Milgram's emphasis on the "shockingly immoral" behavior of his participants, however a close reading of the initial published report clearly reveals the immense power that the experimental design conferred upon the young assistant professor. The design did not simply "stress" participants; it temporarily obliterated their agency[a] and self-possession leaving them utterly at the mercy of the experimenter. In his report, Milgram noted that on one occasion "we observed a seizure so violently convulsive that it was necessary to call a halt to the experiment" (Milgram, 1963a, p. 371). Such extraordinary manifestations of the experimenter's power were not rare, isolated occurrences. Milgram noted, in a matter-of-fact style, that extreme physiological stress reactions were "characteristic rather than exceptional responses to the experiment" (p. 375). And if that wasn't impressive enough, Milgram included a statement by an observer which summarized the experiment's psychologically destructive power:

> I observed a mature and initially poised businessman enter the laboratory smiling and confident. Within 20 minutes he was reduced to a twitching, stuttering wreck, who was rapidly approaching a point of nervous collapse. He constantly pulled an earlobe, and twisted his hands. At one point he pushed his fist into his forehead and muttered "Oh God, let's stop it." And yet he continued to respond to every word of the experimenter, and obeyed to the end. (cited in Milgram, 1963a, p. 377)

35

It is difficult to imagine how one could knowingly and regularly visit such anguish on so many innocent, unsuspecting people under any circumstances, let alone in the absence of a clear ethical goal. Milgram claimed that the torment was sustained by his "intellectual curiosity," but that just doesn't suffice as an explanation in the face of such unrelenting brutality. What seems more likely, although it must remain at the level of speculation, are the all too common feelings of power and self-importance that often come to individuals put in a position (however briefly) of total control.

The extraordinary power and feelings of self-aggrandizement provided by the social psychology laboratory were very much in evidence in a letter that Milgram sent to his research assistant Alan Elms in 1961. Milgram outlined some of the logistical considerations for the obedience experiments and he explained that one of Elms' jobs would be to "think of ways to deliver more people to the laboratory":

> This is a very important practical aspect of the research. I will admit it bears some resemblance to Mr. Eichmann's position, but you at least should have

---

a   *agency*   Ability to make choices and act freely.

no misconceptions of what we do with our daily quota. We give them a chance to resist the commands of a malevolent authority and assert their alliance with morality. (cited in Elms, 1995, p. 24)

The reference to Adolph Eichmann, senior SS officer and one of the architects of the Holocaust, was undoubtedly meant in jest, but it does suggest a level of ethical discomfort with what was about to unfold. Like Eichmann, Elms was to deliver innocent, unsuspecting people into an environment where they would be tormented. As we shall see later, inflicting anguish on the innocent did trouble Milgram slightly, but as this passage makes clear his extraordinary sense of moral entitlement was enough to allay such concerns. As a social psychologist, he saw himself as having the means and the authority to determine the ethical integrity of others according to a "test" of his own devising.

40   Milgram had set out to make his mark on the world and to develop the "boldest and most significant experimental research" he could think of (as cited in Blass, 2009, p. 40). He quickly discovered that the sustained torment and the temporary obliteration of his participants as autonomous moral agents was a pathway to professional success. Professional ambitions notwithstanding, there is evidence to indicate that Milgram enjoyed the brutality of the experiment and the sufferings of the participants. He never made so damning an admission in print (apart from his cryptic reference to learning things about himself that "do not always flatter"), but as a person, Milgram was known, even among friends, for his "off-putting, domineering, and prima donna-ish ways" (Blass, 2004, p. 185). Ever conscious of authority, Milgram's colleagues and former students noted that he was "routinely cruel to graduate students" (Pettigrew, 2005, p. 1778) with his "razor tongue" that "cut to the quick ... with a caustic, even devastating surgical precision" (Takooshian, 2000, p. 13). Always dominant in the classroom, he struck terror in the hearts of his students, challenging them by name "while openly deriding shallow comments" (Takooshian, 2000, p. 13). Known as an "equal opportunity insulter," Milgram brought an equally combative, hierarchal attitude to his relationships with colleagues (Blass, 2004, p. 182). A student recalled that Milgram was "egalitarian in his dislikes. He'll be as rude to other faculty and administrators as to students" (Tavris, 1974, p. 75).

The Obedience experiments provided plenty of opportunities for taking pleasure in the debasement of others. The special charm of the laboratory context for all this gruesomeness was the way it rendered the thrill of torment invisible. By tricking people in a laboratory instead of a "real world" setting such as a bank or school, ethical attention was shifted onto the actions of those being deceived rather than the people doing the manipulating. The laboratory setting cloaked the behavior of the experimenters themselves in the socially uplifting language of science while enabling them to strike a pose of self-sacrifice: the world weary scientist who torments others for their own good and that of humanity.

The ethical dubiousness of all this is apparent in the recollections of Alan Elms (1995). Sitting with Milgram behind a two-way mirror, Elms recalled watching with "fascination and with our share of tension" (p. 24) nearly 1000 gruesome displays of human suffering which saw, as a matter of routine, participants "sweat, tremble, stutter, bite their lips, groan, and dig their fingernails into their flesh" (Milgram, 1963a, p. 375). For Elms, this suffering quickly transformed the situation from an "artificially structured experiment" into "slice after slice of real life" with Milgram and Elms sitting as arbiters of the moral character of those they were manipulating. "We were distressed when some volunteers wept," Elms recalled and "appalled when others laughed as they administered shock after shock" (1995, p. 25). Elms and Milgram could delight in the dramatic intensity of the laboratory spectacle with a clear conscience – even though they had created the very conditions that facilitated the "aggressive," "immoral" behavior that they would subsequently condemn. It was of no matter since "science" was underwriting their conduct, the morality of using inhumane techniques to supposedly warn against the dangers of unquestioning inhumanity was seldom challenged.

In his published work, Milgram hid issues of power, sadism, and self-aggrandizement behind the Holocaust and the pose of disinterested science. However, Milgram's unpublished doubts concerning the ethics and meaning of the research suggest that darker, personal motivations were an important factor in the study. Milgram wanted to get ahead in psychology and short of physically harming someone he was clearly prepared to adopt the most extreme measures in order to achieve that goal. In pursuit of his goal to undertake the "boldest and most significant experimental research" (as cited in Blass, 2009, p. 40), he inflicted countless hours of at times gruesome torment on hundreds of unsuspecting innocent people, a fact which did at times weigh on him. In an unpublished reflection he remarked that "at times I have concluded that, although the experiment can be justified, there are still elements in it that are ethically questionable, that it is not nice to lure people into the laboratory and ensnare them into a situation that is stressful and unpleasant to them" (Milgram, n.d.). Publically however, Milgram does not appear to have been even mildly concerned about the ethics of actions. Indeed, he later indicated that he was "totally astonished" by ethical criticisms of Baumrind and others (Milgram cited in Evans, 1980, p. 193), a comment which in the context of his participants' feedback only serves to underscore his insensitivity and hubris. Milgram clearly felt that as a scientist he had the right to indulge his own intellectual interests, with little regard for the feelings of others, least of all his innocent participants. Quick to label his manipulated, "obedient" participants as nascent Nazis, he bristled at criticisms that his own work was excessive and complained that there was an "absence of any assumption of good will and good faith" (as cited in Evans, 1980, p. 194). In hindsight and in view of the archival record, it is apparent that Milgram's many critics in the 1960s and 1970s were entirely right to assume the worst.

Conclusion

[...] The experiment's enduring appeal is based in part on its theatricality – a fact readily conceded by Milgram himself (Nicholson, 2011). However, theatricality by itself is of little consequence; the study's remarkable capacity to fascinate comes from its ability to represent itself as a "scientific window" into graphic events of great historical consequence. Milgram famously framed the experiment as a kind of "mini-Holocaust," and the possibility of having even a portion of the Holocaust "laid-bare" in the comfort and safety of an American psychological laboratory has proven irresistible to many and an enormous boon to the legitimacy of social psychology. As Miller (2004) has noted, there is an "undeniable degree of prestige or intellectual status inherent in contributing significantly to an understanding of the Holocaust" (p. 227). More recently, the field has derived a similar prestige by applying Milgram to issues associated with the "War on Terror," most notably the torture of prisoners at Abu Ghraib by American military police. Burger (2009) put the matter succinctly when he noted that Milgram gives social psychology a place at the table in discussions of "atrocities, massacres, and genocide" (p. 10). With so much on the line, continued disciplinary interest in Milgram and a kind of "willed indifference" to its extensive ethical and intellectual shortcomings is hardly surprising.

45   "People are often blind to a loved one's faults," Wilson, DePaulo, Mook, and Klaaren (1993) remarked:

> It is a disturbing comment on the nature of scientific assessment that a similar process can occur when trained scientists evaluate research: Our infatuation with the importance of the topic of a study can make us overlook its imperfections. (p. 325)

What accounts for the wider public interest in the Obedience experiment? Beyond the obvious tabloid allure of people "killing" each other in the laboratory, the experiment references the human capacity for extreme violence and it holds an appealing moral: one should not always "obey" a directive just because it comes from a duly constituted authority. This is a message readily available from a variety of other sources, but Milgram renders the moral in a scientific idiom while fostering the comforting illusion that we are on the path to "knowing" what causes atrocities and thereby "preventing more Holocausts, Abu Ghraibs and other examples of wanton cruelty" (Cohen, 2008, 11th para.). The fact that neither Milgram nor any of his contemporary imitators are able to offer any plausible theoretical insights into these horrific events is beside the point. It is the moral message that matters along with the dream of a human nature brought under laboratory control (see Brannigan, 1997; Lemov, 2005).

Having devoted my time to criticizing Milgram's work, I would conclude with an ironic endorsement of its value in understanding how "domains of brutality" come to be established and sustained. Whatever "lasting value" the study has in this respect lies not with the behavior of the experimental participants but with the conduct of Milgram himself and that of his research associates. By invoking the ideal of scientific progress, Milgram was able to get a small team of researchers to deceive and torment nearly 1000 people. By all accounts, his research team were only too happy to comply; the rate of obedience for Milgram's research assistants was not 66% but 100%. Herein lies both the shortcoming and the value of Milgram's work. Taken at face value, the obedience experiments provide little insight into the Holocaust since the experience of Milgram's participants was quite unlike that of Nazi killers. Participants in the obedience study found themselves in an unfamiliar, ethically ambiguous context where they were manipulated into doing something that most did not wish to do, in the face of considerable pressure from a physically present authority figure and a promise that their actions were causing "no permanent tissue damage" (Milgram, 1963a, p. 374). In contrast, members of the Nazi killing machine were fully aware of the killing and destruction they were carrying out and most required no coercion or "encouragement" from an immediate superior. As Goldhagen (1996) has noted, German police battalions willingly rampaged across eastern Europe torturing and murdering thousands, often at their own discretion and with a viciousness and thoroughness that frequently exceeded what they had been ordered to do. Most gladly did so with a clear conscience, convinced of the ethical basis of their actions given the magnitude of the "Jewish threat." As one Nazi police official noted, members of the police battalions "were motivated by a great hatred against the Jews" and consequently "were, with few exceptions, quite happy to take part in shootings of Jews. They had a ball!" (cited in Goldhagen, 1996, p. 396).

While the conduct of the participants in the obedience experiments does little to "explain" the murderous rage of German police battalions, the attitude and actions of Milgram and his research team are consistent with the forms of justification noted by Goldhagen (1996). Unlike the obedience study's often bullied, misled, and reluctant experimental participants, Milgram and his team were, almost from the outset of the study, well aware of the torment that they were to inflict on others. Like members of the German police battalions, they believed in what they were doing and they invoked a "high" ideal to justify their actions (Mandel, 1998). What the obedience research thus reveals is not a fanciful "abandonment of humanity" or an equally outlandish "merger" of each individual's "unique personality into larger institutional structures" (Milgram, 1974, p. 188). The research unwittingly highlights instead the way in which an ideal – in this case the value of human experimentation – can underwrite a ready compliance with gratuitous torment.

50   If the recent revival of Milgram and his obedience work is any guide, some psychologists seem ready to reassert this "scientific" ideal and test its power anew. It is my hope that this foray into Milgram's archive demonstrates not only the folly of this undertaking but of the need to be much more self-critical when examining the ethical justifications of human experimentation in psychology.

(2011)

FUNDING

This research was supported by a grant from the Social Science and Research Council of Canada (grant no. 410-2002-1448).

NOTES

3. Burger's replication was featured on ABC News' January 3, 2007 broadcast of *Primetime*.
4. Australian psychologist and journalist Gina Perry (2008) interviewed some of the participants of the obedience experiment for a radio program "Beyond the Shock Machine" that was broadcast by the Australian Broadcasting Corporation.
5. For a detailed analysis of the role of masculinity in the Obedience experiments, see Nicholson (2011).

REFERENCES

Abse, D. (1973). *The dogs of Pavlov*. London, UK: Vallentine Mitchell.
American Psychological Association. (1959). Ethical standards of psychologists. *American Psychologist, 14*, 279–282.
Baron, R., Byrne, D., Branscombe, N., & Fritzley, H. (2011). *Mastering social psychology*. Toronto, Canada: Pearson.
Baumrind, D. (1964). Some thoughts on ethics of research: After reading Milgram's "behavioral study of obedience". *American Psychologist, 19*, 421–423.
Baumrind, D. (1985). Research using intentional deception. *American Psychologist, 40*, 165–174.
Bellak, G. (Producer), & Dubin, C. S. (Director). (1975). *The tenth level* [Motion picture]. United States: Paramount Pictures.

Benjamin, L., & Simpson, J. (2009). The power of the situation: The impact of Milgram's obedience studies on personality and social psychology. *American Psychologist, 64*, 12–19.

Blass, T. (2000). The Milgram paradigm after 35 years: Some things we now know about obedience to authority. In T. Blass (Ed.), *Obedience to authority: Current perspectives on the Milgram paradigm* (pp. 35–59). Mahwah, NJ: Lawrence Erlbaum Associates.

Blass, T. (2004). *The man who shocked the world: The life and legacy of Stanley Milgram.* New York, NY: Basic Books.

Blass, T. (2009). From New Haven to Santa Clara: A historical perspective on the Milgram obedience experiments. *American Psychologist, 64*, 37–45.

Brannigan, A. (1997). The postmodern experiment: Science and ontology in experimental social psychology. *British Journal of Sociology, 48*, 594–610.

Brown, D. (Creator), & Caron, B. (Director). (2006). *The heist.* [Television special] D. Brown, A. O'Connor, A. Owen, & R. Sandhu (Executive producers). London, UK: Channel 4 Broadcasting.

Burger, J. (2002). Four decades and counting [Review of the book *Obedience to authority: Current perspectives on the Milgram paradigm* by T. Blass (Ed.)]. *Contemporary Psychology, 47*, 665–667.

Burger, J. (2009). Replicating Milgram: Would people still obey today? *American Psychologist, 64*(1), 1–11.

Cohen, A. (2008, December 29). Four decades after Milgram, we're still willing to inflict pain. *New York Times*, A24. Retrieved from http://www.nytimes.com/2008/12/29/opinion/29mon3.html?scp=1&sq=&st=nyt

Darley, J. (1992). Social organization for the production of evil. *Psychological Inquiry, 3*, 199–218.

Elms, A. (1982). Keeping deception honest: Justifying conditions for social scientific research stratagems. In T. Beauchamp, R. Faden, R. J. Wallace, & L. Walters (Eds.), *Ethical issues in social science research* (pp. 232–245). Baltimore, MD: Johns Hopkins University Press.

Elms, A. (1995). Obedience in retrospect. *Journal of Social Issues, 51*, 21–31.

Elms, A. (2009). Obedience lite. *American Psychologist, 64*, 32–36.

Errera, P. (1963a, March 14). Defiant subjects. [Meeting conducted by Dr. Paul Errera]. Stanley Milgram Papers (Sanitized Data, Box 155A). Yale University Archives, New Haven, CT.

Errera, P. (1963b, April 11). Defiant subjects. [Meeting conducted by Dr. Paul Errera]. Stanley Milgram Papers (Sanitized Data, Box 155A). Yale University Archives, New Haven, CT.

Errera, P. (1963c, March 21). Obedient subjects. [Meeting conducted by Dr. Paul Errera]. Stanley Milgram Papers (Sanitized Data, Box 155A). Yale University Archives, New Haven, CT.

Errera, P. (1963d, April 4). Obedient subjects. [Meeting conducted by Dr. Paul Errera]. Stanley Milgram Papers (Sanitized Data, Box 155A). Yale University Archives, New Haven, CT.

Evans, R. (1980). *The making of social psychology.* New York, NY: Gardner Press.

Fenigstein, A. (1998). Were obedience pressures a factor in the Holocaust? *Analyse & Kritik, 20*, 54–73.

Fiske, S., & Harris, L. (2004). Why ordinary people torture enemy prisoners. *Science, 306*, 1482–1483.

French TV contestants made to inflict "torture". (2010, March 18). *BBC News.* Retrieved from: http://news.bbc.co.uk/1/hi/8571929.stm

Gilovich, T., Keltner, D., & Nisbet, R. (2006). *Social psychology.* New York, NY: Norton.

Goldhagen, D. (1996). *Hitler's willing executioners.* New York, NY: Knopf.

Harris, B. (1988). Key words: A history of debriefing in social psychology. In J. Morawski (Ed.), *The rise of experimentation in American psychology* (pp. 188–212). New Haven, CT: Yale University Press.

Helm, C., & Morelli, M. (1985). Obedience to authority in a laboratory setting: Generalizability and context dependency. *Political Studies, 33*, 610–627.

Henley, J. (2007, December 13). A glosssary of US military torture euphemisms. *The Guardian*. Retrieved from http://www.guardian.co.uk/world/2007/dec/13/usa.humanrights/print

Horgan, J. (2010, October 4). Are war crimes caused by bad apples or bad barrels? *Scientific American*. Retrieved from http://www.scientificamerican.com/blog/post.cfm?id=are-war-crimes-caused-by bad-apples-2010-10-04&sc=WR_20101006

Johnson, G. (1943). *American heroes and hero worship*. New York, NY: Harper & Brothers.

Lemov, R. M. (2005). *World as laboratory: Experiments with mice, mazes, and men* (1st ed.). New York, NY: Hill and Wang.

Mandel, D. (1998). The obedience alibi: Milgram's account of the Holocaust reconsidered. *Analyse & Kritik, 20*, 74–94.

Milgram, S. (1959). Note to self. Stanley Milgram Papers (Series 1, Box 23, Folder 383). Yale University Archives, New Haven, CT.

Milgram, S. (1962a). *Evaluation of obedience research: Science or art?* Unpublished manuscript, Yale University, New Haven, CT.

Milgram, S. (1962b, August). Note to self. Stanley Milgram Papers (Series II, Box 46, Folder 173). Yale University Archives, New Haven, CT.

Milgram, S. (1963a). Behavioral study of obedience. *Journal of Abnormal & Social Psychology, 67*, 371–378.

Milgram, S. (1963b, April 17). Letter to Dorwin Cartwright. Stanley Milgram Papers (Series 1, Box 1a, Folder 7). Yale University Archives, New Haven, CT.

Milgram, S. (1963c, September 27). Letter to Joan Daves. Stanley Milgram Papers (Series 1, Box 1a, Folder 9). Yale University Archives, New Haven, CT.

Milgram, S. (1964). Issues in the study of obedience: A reply to Baumrind. *American Psychologist, 19*, 848–852.

Milgram, S. (1965). Some conditions of obedience and disobedience to authority. *Human Relations, 18*, 57–76.

Milgram, S. (1974). *Obedience to authority: An experimental view*. London, UK: Tavistock.

Milgram, S. (n.d.). An experimenter's dilemma. Stanley Milgram Papers (Series II, Box 46, Folder 173). Yale University Archives, New Haven, CT.

Miller, A. G. (1986). *The obedience experiments: A case study of controversy in social science*. New York, NY: Praeger.

Miller, A. (2004). What can the Milgram obedience experiments tell us about the Holocaust? In A. Miller (Ed.), *Social psychology of good and evil* (pp. 193–239). New York, NY: Guilford.

Miller, A. (2009). Reflections on "Replicating Milgram" (Burger, 2009) [Peer commentary on the paper "Replicating Milgram: Would people still obey today?" by J. M. Burger]. *American Psychologist, 64*, 20–27.

Miller, A., Collins, B., & Brief, D. (1995). Perspectives on Obedience to Authority: The legacy of the Milgram experiments. *Journal of Social Issues, 51*, 1–19.

Mixon, D. (1972). Instead of deception. *Journal for the Theory of Social Behavior, 2*, 145–174.

Myers, D., & Smith, S. (2009). *Exploring social psychology*. Toronto, Canada: McGraw Hill Ryerson.

Navarick, D. (2009). Reviving the Milgram obedience paradigm in the era of informed consent. *The Psychological Record, 59*, 155–170.

Nicholson, I. (2011). "Shocking" masculinity: Stanley Milgram, "Obedience to Authority," and the crisis of manhood in Cold War America. *ISIS, 102*, 238–268.

Pancer, S. M. (1997). Social psychology: The crisis continues. In D. Fox & I. Prilleltensky (Eds.), *Critical psychology* (pp. 150–165). London, UK: Sage.

Patten, S. (1977a). The case that Milgram makes. *Philosophical Review, 86*, 350–364.

Patten, S. (1977b). Milgram's shocking experiments. *Philosophy, 52*, 425–440.

Perry, G. (2008, October 11). Beyond the shock machine [Radio broadcast]. Australian Broadcasting Corporation.

Pettigrew, T. (2005). [Review of the book *The Man Who Shocked the World*, by T. Blass]. *Social-Forces, 83*, 1778–1779.

Reaction of subjects. (1962). Stanley Milgram Papers (Series II, Box #44). Yale University Archives, New Haven, CT.

Russell, N. (2010). Milgram's obedience to authority experiments: Origins and early evolution. *British Journal of Social Psychology, 49*, 1–23.

Samelson, F. (1974). History, origin myth and ideology: "Discovery" of social psychology. *Journal of the Theory of Social Behavior, 4*(2), 218–231.

Schuller, H. (1982). *Ethical problems in psychological research.* New York, NY: Academic Press.

Sheppard, J., & Young, M. (2007). The routes of moral development and the impact of exposure to the Milgram obedience study. *Journal of Business Ethics, 75*, 315–333.

Silverman, I. (1977). Why social psychology fails. *Canadian Psychological Review, 18*, 353–358.

Smith, M. B. (1972). Is experimental social psychology advancing? *Journal of Experimental Social Psychology, 8*, 86–96.

Stam, H., Lubek, I., & Radtke, H. L. (1998). Repopulating social psychology texts: Disembodied "subjects" and embodied subjectivity. In B. Bayer & J. Shotter (Eds.), *Reconstructing the psychological subject: Bodies, practices and technologies* (pp. 153–186). London, UK: Sage.

Subjects' conversation. (1963, February 28). Stanley Milgram Papers (Series II, Box #44). Yale University Archives, New Haven, CT.

Takooshian, H. (2000). How Stanley Milgram taught about obedience and social influence. In T. Blass (Ed.), *Obedience to authority: Current perspectives on the Milgram paradigm* (pp. 9–24). Mahwah, NJ: Lawrence Erlbaum.

Tavris, C. (1974). The frozen world of the familiar stranger: An interview with Stanley Milgram. *Psychology Today, 8*, 71–73, 76–78, 80.

Wilson, T., DePaulo, B., Mook, D., & Klaaren, K. (1993). Scientists' evaluations of research: The biasing effects of the importance of the topic. *Psychological Science, 4*, 322–325.

Zimbardo, P. G. (2007). *The Lucifer effect: Understanding how good people turn evil* (1st ed.). New York, NY: Random House.

Questions:

1. What are the article's two main arguments, and where does the article explicitly state what they are? Does the slight delay in providing this thesis help or hinder your reading process?

2. Using Milgram's experiment as an example, critique or support the argument that ethical safeguards interfere with science.

3. Outline the differences between the concepts of "informed consent" and "retrospective consent."

4. Discuss Milgram's assertion that the "participant must be the ultimate source of judgment." Do you think this claim adequately justifies his experimental process?

5. Consider the article's use of primary, archival research. Does this evidence change your perception of Milgram's experiment? Why or why not?

6. This paper has 73 references. How does the sheer number of references influence your reading? Read through the references and describe what types of library databases and search terms you think Nicholson might have used.

7. What problems with the experiment does the article document the participants as identifying? After reading the extensive quotations from the participants given in this article, do you feel any more sympathetic toward them? Why or why not?

8. The article suggests that Milgram's work confused two types of obedience. What are the two types, and, according to Nicholson's article, where does the confusion occur?

9. The tone of the article shifts as the article progresses. Why do you think this might be? Do the personal attacks on Milgram help or hinder the article's effectiveness? Is there any justification for the *ad hominem* approach here?

10. Does the speculation about Milgram's motivation belong in this type of social scientific paper? Why do you think it is included here?

11. This article is careful about identifying exactly where all source material ideas and quotations are used. Identify one framed paraphrase and one direct quotation which is correctly integrated into one of the article's sentences. Now, rewrite the direct quotation as a fully framed paraphrase.

12. Argue for or against the article's assertion that Milgram's work demonstrates only that a subject can easily be manipulated "in unfamiliar circumstances."

13. Argue for or against the article's concluding argument that Milgram's work actually shows that the ideology of the absolute value of science justifies abuse of subjects for some researchers. The article asserts that "the rate of obedience for Milgram's research assistants was not 66% but 100%"; do you find this reinterpretation of the experiment convincing as part of a discussion of how "domains of brutality" develop?

14. Introduction to Psychology courses almost always require students to participate in psych experiments to obtain course credit. What sort of code of conduct would you expect researchers to observe if you were one of the students required to participate in an experiment?

fourteen

JOSEPH DIMOW

Resisting Authority: A Personal Account of the Milgram Obedience Experiments

In this article from the magazine Jewish Currents, *journalist and activist Joseph Dimow reflects on his experience as a participant in Stanley Milgram's "Obedience to Authority" experiment.*

☙

In 1961, I participated in a famous experimental study about obedience and authority – although I and other participants were led to believe it was a study of memory and learning. The experiment was designed by a Yale University professor of social psychology, Stanley Milgram, and resulted in a book, *Obedience to Authority*, which is still widely used in sociology courses.

Like many others in the New Haven area, I answered an ad seeking subjects for the experiment and offering five dollars, paid in advance, for travel and time. At the Yale facility, I met a man who looked very professorial in a white coat and horn-rimmed glasses. He led me into a room filled with an impressive display of electrical equipment. A second man was introduced to me as another subject for the experiment, and together we were told that the experiment was to test the widely held belief that people learn by punishment. In this case, one of us would be a "learner" and the other a "teacher." The teacher would read a list of paired words to the learner and then repeat the first word of the pair. If the learner did not respond with the correct second word, the teacher would deliver a "mild" electric shock to the learner as punishment.

This struck me as bizarre, and although the instructions were in accord with what we had been told, I wondered if something else was going on.

The "professor" said we would draw straws to see which of us would be the learner. He offered the straws to the other man, then announced that he had drawn the short straw and would be the learner. I hadn't seen either straw, and my doubts became suspicions that I was being deceived.

5   The learner, said the professor, would be in an adjoining room, out of my sight, and strapped to a chair so that his arms could not move – this so that the learner could not jump around and damage the equipment or do harm to himself. I was to be seated in front of a console marked with lettering colored yellow for "Slight Shock" (15 volts) up to purple for "Danger: Severe Shock" (450 volts). The shocks would increase by 15-volt increments with each incorrect answer.

I was very suspicious and asked a number of questions: Isn't it dangerous? How do you know the learner doesn't have a bad heart and can't take the shocks? What if he wants to stop, can he get out of the chair? The professor assured me that the shocks were not painful or harmful since the amperage was lowered as the voltage increased. He let me feel what a 45-volt shock would be like: a slight tickle. I asked the learner if he was willing to do this and why he didn't have any questions. He said, "Let's try it." With some trepidation on my part, we began the experiment. After a few shocks, the learner let out an "Ouch!" and I asked if he was okay. He said he was, but after the next shock, his complaint became louder. I said I would stop. The "professor" told me to continue, and the learner said he was ready to go on, too. I went on for two or three more shocks. With each, the learner's cry of pain became louder – and then he asked to stop, and I refused to go any further. The professor became very authoritative. He said that I was costing them valuable time, it was essential for me to continue, I was ruining the experiment. He asserted that he was in charge, not me. He reminded me that I had been paid and insisted that I continue. I refused, offered to give him back the five dollars, and told him that I believed the experiment to be really about how far I would go, that the learner was an accomplice, and that I was determined not to continue.

At that point, the professor gave up and his demeanor changed. Instead of being authoritative and assertive, he was detached and polite as he asked if I would answer some questions about what had taken place. I agreed, and he asked a series of questions about who was responsible for what had happened.

He showed me pictures and asked for my reactions. One was a painting of a boy looking like a cornered rat, cowering before a young teacher with a cane who looked reluctant to whip the boy but was overseen, in turn, by an older man who appeared to be a judge or headmaster and was ordering the young teacher to proceed. This seemed a crude way of getting me to identify with the teacher; I even asked if the cowering boy was drawn to look ugly so that he would attract no sympathy.

Then I was shown a pie chart on which I was asked to draw lines distributing responsibility for the experiment and its aftermath. I allotted one half the pie to the experimenter and one quarter each for the two subjects – then I drew more lines to show that as the experiment went on and the learner became an unwilling participant, unable to stop the experiment, his portion decreased while the experimenter's and the teacher's portions increased.

After several such questions, I asked if my suspicions were correct and   10 if the whole experiment was designed to see if ordinary Americans would obey immoral orders, as many Germans had done during the Nazi period. The professor declined to answer, but asked what had made me think that the experiment was not what had been described to me at the beginning. I told him that my suspicions had been aroused by the way the straws had been handled, by the idea that they would risk shocking a stranger, and by the fact that he, the professor, had been in the area with me the whole time and had never gone to observe the learner. He did not respond to my comments, but said I would receive a report when the experiment was completed.

Then the most disturbing part of the entire experience occurred: The professor brought in the learner and I was flabbergasted. His face was covered in tears and he looked haggard. He offered his hand and thanked me for stopping the experiment, saying that the shocks hadn't really hurt but anticipating them had been dreadful. I was confused as to whether he was in earnest or acting. I left unsure, and waited outside for the learner so I could discuss it with him. After about a half hour he had not appeared, and I was convinced that he was an actor and that my suspicions about the experiment had been correct. The report that I received confirmed that the experiment was designed to see how far subjects would go in obeying orders to administer pain to others. It had arisen out of the desire to understand the widespread obedience to horrendous and brutal orders in Nazi Germany. The report also confirmed that the professor and learner were indeed actors, although not professionals – and I have always thought that they deserved Academy Awards anyway.

After these many years, I don't remember what I did with the report (probably gave it to a sociology student), but I have been able to find information about the experiment from Milgram's and other books. Of forty participants in Milgram's first experiment, fifteen refused to continue at some point, while twenty-five went all the way to 450 volts – a "shock" that they administered three times before the experiment was ended by the professor. (There was no actual shock, of course. The actor playing the part of the learner reacted with a cry of pain to a red light, which lit whenever there was a supposed shock.)

The experiment was repeated in other venues away from the university in Hartford and Cambridge. Results were worst (that is, the highest percentage of testers went all the way to 450 volts) with a group of nurses in Bridgeport. The experiments were also repeated in Princeton, Munich, Rome, South

Africa, and Australia, with levels of obedience registering even higher than in New Haven. (In Munich, 85 percent of the subjects were obedient to the end.) Before the experiments began, Stanley Milgram sought predictions from a variety of people, including psychiatrists and college faculty and graduate students. "With remarkable similarity," he wrote, "they predicted that virtually all the subjects would refuse to obey the experimenter. The psychiatrists, specifically, predicted that most subjects would not go beyond 150 volts, when the victim makes his first explicit demand to be freed. They expected that only four percent would reach 300 volts, and that only a pathological fringe of about one in a thousand would administer the highest shock on the board" ("The Perils of Obedience").

Many social psychologists were critical of the experiment. Diana Baumrind, in the *American Psychologist*, 1964, complained that there was no informed consent and that even if valuable information were gleaned, it would "not justify the risk that real [emotional] harm [would be] done to the subject." Milgram maintained that a followup questionnaire showed that 84 percent of the subjects were glad to have been involved, 15 percent were neutral and only 1.3 percent were sorry or had negative feelings. Milgram also had a psychiatrist interview subjects thought most likely to have suffered consequences, and the doctor found no evidence of traumatic reactions. I don't remember if I answered this questionnaire, but I would have been in the neutral group.

15   In retrospect, I believe that my upbringing in a socialist-oriented family steeped in a class struggle view of society taught me that authorities would often have a different view of right and wrong than mine. That attitude stayed with me during my three and one half years of service in the army, in Europe, during World War II. Like all soldiers, I was taught to obey orders, but whenever we heard lectures on army regulations, what stayed with me was that we were also told that soldiers had a right to refuse illegal orders (though what constituted illegal was left vague).

In addition, in my position during the late 1940s as a staff member of the Communist Party, in which I held positions as chairman in New Haven and Hartford, I had become accustomed to exercising authority and having people from a variety of backgrounds and professions carry out assignments I gave them. As a result, I had an unorthodox understanding of authority and was not likely to be impressed by a white lab coat.

In the early 1950s, I was harassed and tailed by the FBI, and in 1954, along with other leaders of the Communist Party in Connecticut, I was arrested and tried under the Smith Act on charges of "conspiracy to teach and advocate the overthrow of the government by force and violence." We were convicted, as expected, and I was about to go to jail when the conviction was overturned on appeal. I believe these experiences also enabled me to stand up to an authoritative "professor."

This is not to say that membership in the Communist Party made me or anyone else totally independent. Many of us, in fact, had become accustomed to carrying out assignments from people with higher positions in the Party, even when we had doubts. Would I have refused to follow orders had the experimental authority figure been a "Party leader" instead of a "professor"? I like to think so, as I was never a stereotypical "true believer" in Party doctrine. This was one of the reasons, among others, that I left the Party in the late 1950s. In any event, I believe that my political experience was an important factor in determining my skeptical behavior in the Milgram experiment.

I think the experiment had only limited relevance to our understanding of the actions of the German people under Nazi rule. In the experiment, the professor had no power to enforce his orders. In Nazi Germany, the enforcement powers went from simple reprimand all the way to imprisonment and death. In addition, the role of the learner in the experiment was markedly different from the victimized Jews, Gypsies, gay men and others under Nazism, who had not volunteered to be in an "experiment" and had no ability to stop their suffering.

The results of the Milgram experiment should not surprise us. Most 20 people unquestioningly obey orders from authorities, and refusal is unusual. As children, after all, we are taught to obey our parents, teachers, employers and law enforcement officers. Perhaps that is why examples of refusal to obey immoral orders excite my admiration. This is especially so of the Israeli soldiers and pilots who are currently defending the morality of their country and of the Jewish people by refusing to serve in the Occupied Territories.[a] Some are paying a high price in lost jobs, destroyed careers and jail time for their actions. Their situation cannot be directly compared to that of us who were tested in the Milgram experiment, for they are not being explicitly ordered to injure innocent people (although they know that the "collateral damage" of their military actions includes innocents), nor are the people being killed and injured voluntarily participating in an experiment. Still, the results of the Milgram experiment demonstrate how rare and heroic is the "Courage to Refuse" (as one of their organizations is named). These are people who deserve to be honored.

(2004)

---

a   *Israeli soldiers ... Occupied Territories*   Some members of the Israeli military, many of them affiliated with the organization "Courage to Refuse," willingly participate in missions they consider part of their nation's defense but refuse to fight in territories Israel has occupied since the Six-Day War (1967). According to the UN, these territories are occupied in violation of international law.

Questions:

1. In what ways does the article mimic the tone and approach of Milgram's "Behavioral Study"? What aspects are different?

2. What particular writing elements of the article function as markers of the author's sincerity and truthfulness?

3. Why does the article provide so much background on Milgram's experiments? What audience does the article address?

4. In what ways does the author's experience of the study differ from the subject experiences described in "Behavioral Study"?

5. In what ways does the article present Milgram's experiment as poorly designed?

6. What does the wording "Milgram maintained" suggest? Point out other loaded words or phrasing.

7. What is the "most disturbing part" of the experiment for Dimow? What ethical guidelines seem to have been violated, according to this description?

8. What does the description of the picture and the pie chart contribute to your understanding of how Milgram's experiment was conducted? Does the description of Dimow's experience following the experiment make you reassess the claims made in "Issues in the Study of Obedience"? If so, how?

9. What does the experimenter's assertion that "shocks were not painful or harmful since the amperage was lowered as the voltage increased" mean?

10. What reasons are given for the article's skepticism about the relevance of Milgram's experiment to a discussion of what motivated Nazis? Look again at Milgram's explanations about the immediate historical background to his experiments. Which version do you find more convincing? Why?

11. The article provides an extended discussion of the author's personal background. What rhetorical or explanatory function does the material serve?

12. Look again at "Behavioral Study." What sort of background information on subjects is provided in the published study? What does that material tell a reader about the relationship between personal experience and resistance to unjust authority?

III

Medical Research: Experimenting with Drugs and Unsuspecting Populations

A.J. Wakefield, S.H. Murch, A. Anthony, J. Linnell, D.M. Casson, M. Malik, M. Berelowitz, A.P. Dhillon, M.A. Thomson, P. Harvey, A. Valentine, S.E. Davies, and J.A. Walker-Smith

RETRACTED: Ileal-Lymphoid-Nodular Hyperplasia, Non-Specific Colitis, and Pervasive Developmental Disorder in Children

This 1998 article by Andrew Wakefield et al., published in the UK medical journal The Lancet, *led to widespread fears that the measles, mumps, and rubella (MMR) vaccine could cause autism spectrum disorders. The article has since been shown to be fraudulent, and there is now a strong scientific consensus that vaccines do not cause autism;* The Lancet *retracted the article in 2010 after an investigation found Wakefield to have acted unprofessionally and unethically. See the selection following this article for a discussion of Wakefield's misconduct.*

ॐ

SUMMARY

Background
We investigated a consecutive series of children with chronic enterocolitis and regressive developmental disorder.[a]

Methods
12 children (mean age 6 years [range 3–10], 11 boys) were referred to a paediatric gastroenterology[b] unit with a history of normal development

---

a *chronic enterocolitis* Persistent or recurring inflammation of the small intestine and colon; *regressive developmental disorder* Disorder in which a child develops normally, and then displays a change or loss in previously acquired skills, such as language, and the beginning of behaviors characteristic of a pervasive development disorder, such as autism.

b *gastroenterology* Study of diseases affecting the stomach and intestines.

followed by loss of acquired skills, including language, together with diarrhoea and abdominal pain. Children underwent gastroenterological, neurological, and developmental assessment and review of developmental records. Ileocolonoscopy and biopsy sampling, magnetic-resonance imaging (MRI), electroencephalography (EEG), and lumbar puncture[a] were done under sedation. Barium follow-through radiography[b] was done where possible. Biochemical, haematological and immunological[c] profiles were examined.

Findings

Onset of behavioural symptoms was associated, by the parents, with measles, mumps, and rubella vaccination in eight of the 12 children, with measles infection in one child, and otitis media[d] in another. All 12 children had intestinal abnormalities, ranging from lymphoid nodular hyperplasia to aphthoid ulceration.[e] Histology showed patchy chronic inflammation in the colon in 11 children and reactive ileal lymphoid hyperplasia in seven, but no granulomas.[f] Behavioural disorders included autism (nine), disintegrative psychosis (one), and possible postviral or vaccinal encephalitis (two).[g] There were no focal neurological abnormalities[h] and MRI and EEG tests were normal. Abnormal laboratory results were significantly raised urinary

---

a *Ileocolonoscopy* Imaging of the ileum (the final section of the small intestine) done with the use of a long, flexible fiber optic tube inserted through the rectum; *biopsy sampling* Removal of a small piece of body tissue for examination under a microscope; *magnetic-resonance imaging* Method of recording an image of body tissues using powerful magnets; *electroencephalography* Visual tracing of the electrical activity of the brain; *lumbar puncture* Removal of fluid from the spine through a hollow needle inserted into the lower back.

b *Barium follow-through radiography* Medical imaging technique in which a patient swallows a solution containing barium sulphate. Barium sulphate shows as an opaque white area on x-rays, which allows the examiner to view the stomach and intestines as the solution passes through. X-rays are taken at regular intervals until the solution enters the large intestine.

c *haematological* Of or related to blood; *immunological* Of or related to the immune system.

d *otitis media* Infection in the middle ear.

e *lymphoid nodular hyperplasia* Rapid growth of lymph cells in areas where they are not normally found; *aphthoid ulceration* Blisters in the gastrointestinal tract or mouth.

f *Histology* Microscopic examination of the structure, composition, and function of tissues; *reactive ... hyperplasia* Rapid growth of lymph cells in the ileal area; *granulomas* Small masses of immune cells, usually produced as a result of inflammation or when the body tries to eliminate infection or a foreign substance.

g *disintegrative psychosis* Rapid change and subsequent decline of normal development in young children, usually progressing to severe developmental impairment and loss of social, verbal, and motor skills. Its cause is unknown; *encephalitis* Inflammation of the brain.

h *focal neurological abnormalities* Neurological problems resulting in impaired movement in a particular part of the body, speech problems, or sensory difficulties.

methylmalonic acid compared with age-matched controls (p=0.003), low haemoglobin in four children, and a low serum IgA in four children.[a]

Interpretation

We identified associated gastrointestinal disease and developmental regression in a group of previously normal children, which was generally associated in time with possible environmental triggers.

INTRODUCTION

We saw several children who, after a period of apparent normality, lost acquired skills, including communication. They all had gastrointestinal symptoms, including abdominal pain, diarrhoea, and bloating and, in some cases, food intolerance. We describe the clinical findings, and gastrointestinal features of these children.

PATIENTS AND METHODS

12 children, consecutively referred to the department of paediatric gastroenterology with a history of a pervasive developmental disorder with loss of acquired skills and intestinal symptoms (diarrhoea, abdominal pain, bloating and food intolerance), were investigated. All children were admitted to the ward for 1 week, accompanied by their parents.

Clinical investigations

We took histories, including details of immunisations and exposure to infectious diseases, and assessed the children. In 11 cases the history was obtained by the senior clinician (JW-S[b]). Neurological and psychiatric assessments were done by consultant staff (PH, MB) with HMS-4 criteria.[1] Developmental histories included a review of prospective developmental records from parents, health visitors, and general practitioners. Four children did not undergo psychiatric assessment in hospital; all had been assessed professionally elsewhere, so these assessments were used as the basis for their behavioural diagnosis.

After bowel preparation, ileocolonoscopy was performed by SHM or MAT under sedation with midazolam and pethidine.[c] Paired frozen and

---

a   *urinary methylmalonic acid*   Methylmalonic acid participates in various metabolic processes in the body; its unusually high presence in urine can indicate a vitamin B12 deficiency or a metabolic disorder;   *haemoglobin*   Protein in red blood cells that binds with oxygen and transports it throughout the body;   *serum IgA*   levels of Immunoglobulin A, an antibody present in the body's mucosal linings.

b   *JW-S*   J.A. Walker-Smith, one of the paper's co-authors. Throughout this article, the co-author responsible for a particular portion of the research is identified by her or his initials in parentheses.

c   *midazolam*   Drug used to treat seizures and for sedation;   *pethidine*   Pain reliever.

formalin-fixed mucosal biopsy samples were taken from the terminal ileum; ascending, transverse, descending, and sigmoid colons,[a] and from the rectum. The procedure was recorded by video or still images, and were compared with images of the previous seven consecutive paediatric colonoscopies (four normal colonoscopies and three on children with ulcerative colitis[b]), in which the physician reported normal appearances in the terminal ileum. Barium follow-through radiography was possible in some cases.

5   Also under sedation, cerebral magnetic-resonance imaging (MRI), electroencephalography (EEG) including visual, brain stem auditory, and sensory evoked potentials[c] (where compliance made these possible), and lumbar puncture were done.

Laboratory investigations

Thyroid function, serum long-chain fatty acids, and cerebrospinal-fluid lactate[d] were measured to exclude known causes of childhood neurodegenerative disease. Urinary methylmalonic acid was measured in random urine samples from eight of the 12 children and 14 age-matched and sex-matched normal controls, by a modification of a technique described previously.[2] Chromatograms[e] were scanned digitally on computer, to analyse the methylmalonic-acid zones from cases and controls. Urinary methylmalonic-acid concentrations in patients and controls were compared by a two-sample t test.[f] Urinary creatinine was estimated by routine spectrophotometric assay.[g]

---

a   *formalin-fixed ... samples*   Small tissue samples taken from a mucus membrane and preserved with formalin;   *terminal ileum*   Last part of the ileum, which forms the beginning of the connection between the small and large intestines;   *ascending ... colons*   Refers to, in order, the segments of the large intestine from connection with the small intestine to the rectum.

b   *ulcerative colitis*   Recurring acute inflammation of the colon with extensive ulceration, primarily of the lining of the colon.

c   *evoked potentials*   Neurological activity detected when a subject is exposed to a stimulus, such as a flashing light or a sound.

d   *Thyroid*   Endocrine gland located at the base of the neck. The thyroid secretes hormones that regulate growth and metabolism;   *serum long-chain fatty acids*   Fatty acids present in blood. Their measurement can help to diagnose conditions including adrenoleukodystrophy, a genetic neurological disorder;   *cerebrospinal-fluid lactate*   Levels of lactic acid in spinal fluid, measured to diagnose conditions affecting the brain and spinal cord, such as encephalitis or meningitis.

e   *Chromatograms*   Visual representations of a mixture's composition, created by separating the mixture into various components.

f   *two-sample t test*   Method of analyzing data in order to determine whether a difference between two groups is significant or random.

g   *Urinary creatinine*   Level of the metabolism byproduct creatinine in urine; a measure of kidney function;   *spectrophotometric assay*   Method of determining the contents or activity of a biological substance by measuring the amount of light the substance absorbs.

Children were screened for antiendomyseal antibodies and boys were screened for fragile-X[a] if this had not been done before. Stool samples were cultured for *Campylobacter* spp, *Salmonella* spp, and *Shigella* spp and assessed by microscopy for ova[b] and parasites. Sera were screened for antibodies to Yersinia *enterocolitica*.[c]

TABLE 1
Clinical Details and Laboratory, Endoscopic, and Histological Findings

Child	Age (years)	Sex	Abnormal laboratory tests	Endoscopic findings	Histological findings
1	4	M	Hb 10.8, PCV 0.36, WBC 16.6 (neutrophilia), lymphocytes 1.8, ALP 166	Ileum not intubated; aphthoid ulcer in rectum	Acute caecal cryptitis and chronic non-specific colitis
2	9.5	M	Hb 10.7	LNH of T ileum and colon; patchy loss of vascular pattern; caecal aphthoid ulcer	Acute and chronic non-specific colitis; reactive ileal lymphoid hyperplasia
3	7	M	MCV 74, platelets 474, eosinophils 2.68, IgE 114, IgG_1 8.4	LNH of T ileum	Acute and chronic non-specific colitis: reactive ileal and colonic lymphoid hyperplasia
4	10	M	IgE 69, IgG_1 8.25, IgG_4 1.006, ALP 474, AST 50	LNH of T ileum; loss of vascular pattern in rectum	Chronic non-specific colitis: reactive ileal and colonic lymphoid hyperplasia
5	8	M		LNH of T ileum; proctitis with loss of vascular pattern	Chronic non-specific colitis: reactive ileal lymphoid hyperplasia
6	5	M	Platelets 480, ALP 207	LNH of T ileum; loss of colonic vascular pattern	Acute and chronic non-specific colitis: reactive ileal lymphoid hyperplasia

a *antiendomyseal antibodies* Antibodies present in Celiac disease, a hypersensitivity to gluten; *fragile-X* Genetic condition that makes a particular X chromosome unusually vulnerable to damage and can cause autism and other disorders.

b *Campylobacter spp ... Shigella spp* Bacteria that can cause infections in the human digestive system; *ova* Eggs.

c *Yersinia enterocolitica* Bacterium that can cause enterocolitis or ileitis in humans.

7	3	M	Hb 9.4, WBC 17.2 (neutro-philia), ESR 16, IgA 0.7	LNH of T ileum	Normal
8	3.5	F	IgA 0.5, IgG 7	Prominent ileal lymph nodes	Acute and chronic non-specific colitis: reactive ileal lymphoid hyperplasia
9	6	M		LNH of T ileum; patchy erythema at hepatic flexure	Chronic non-specific colitis: reactive ileal and colonic lymphoid hyperplasia
10	4	M	IgG$_1$ 9.0	LNH of T ileum and colon	Chronic non-specific colitis: reactive ileal lymphoid hyperplasia
11	6	M	Hb 11.2, IgA 0.26, IgM 3.4	LNH of T ileum	Chronic non-specific colitis
12	7	M	IgA 0.7	LNH on barium follow-through; colonoscopy normal; ileum not intubated	Chronic non-specific colitis: reactive colonic lymphoid hyperplasia

LNH=lymphoid nodular hyperplasia; T ileum=terminal ileum. Normal ranges and units: Hb=haemoglobin 11·5–14·5 g/dL; PCV=packed cell volume 0·37–0·45; MCV=mean cell volume 76–100 pg/dL; platelets 140–400 10^9/L; WBC=white cell count 5·0–15·5 10^9/L; lymphocytes 2·2–8·6 10^9/L; eosinophils 0–0·4 10^9/L; ESR=erythrocyte sedimentation rate 0–15 mm/h; IgG 8–18 g/L; IgG$_1$ 3·53–7·25 g/L; IgG$_4$ 0·1–0·99 g/L; IgA 0·9–4·5 g/L; IgM 0·6–2·8 g/L; IgE 0–62 g/L; ALP=alkaline phosphatase 35–130 U/L; AST=aspartate transaminase 5–40 U/L.

Histology

Formalin-fixed biopsy samples of ileum and colon were assessed and reported by a pathologist (SED). Five ileocolonic biopsy series from age-matched and site-matched controls whose reports showed histologically normal mucosa were obtained for comparison. All tissues were assessed by three other clinical and experimental pathologists (APD, AA, AJW).

Ethical approval and consent

Investigations were approved by the Ethical Practices Committee of the Royal Free Hospital NHS Trust, and parents gave informed consent.

RESULTS

Clinical details of the children are shown in tables 1 and 2. None had 10 neurological abnormalities on clinical examination; MRI scans, EEGs, and cerebrospinal-fluid profiles were normal; and fragile X was negative. Prospective developmental records showed satisfactory achievement of early milestones in all children. The only girl (child number eight) was noted to be a slow developer compared with her older sister. She was subsequently found to have coarctation of the aorta.[a] After surgical repair of the aorta at the age of 14 months, she progressed rapidly, and learnt to talk. Speech was lost later. Child four was kept under review for the first year of life because of wide bridging of the nose. He was discharged from follow-up as developmentally normal at age 1 year.

In eight children, the onset of behavioural problems had been linked, either by the parents or by the child's physician, with measles, mumps, and rubella vaccination. Five had had an early adverse reaction to immunisation (rash, fever, delirium; and, in three cases, convulsions). In these eight children the average interval from exposure to first behavioural symptoms was 6.3 days (range 1–14). Parents were less clear about the timing of onset of abdominal symptoms because children were not toilet trained at the time or because behavioural features made children unable to communicate symptoms.

One child (child four) had received monovalent[b] measles vaccine at 15 months, after which his development slowed (confirmed by professional assessors). No association was made with the vaccine at this time. He received a dose of measles, mumps, and rubella vaccine at age 4.5 years, the day after which his mother described a striking deterioration in his behaviour that she did link with the immunisation. Child nine received measles, mumps, and rubella vaccine at 16 months. At 18 months he developed recurrent antibiotic-resistant otitis media and the first behavioural symptoms, including disinterest in his sibling and lack of play.

Table 2 summarises the neuropsychiatric diagnoses; the apparent precipitating events; onset of behavioural features; and age of onset of both behaviour and bowel symptoms.

---

a *coarctation of the aorta* Condition in which a portion of the aorta, the major artery leading away from the heart, is abnormally narrow.

b *monovalent* I.e., intended to immunize against only one disease.

TABLE 2
Neuropsychiatric Diagnosis

Child	Behavioural diagnosis	Exposure identified by parents or doctor	Interval from exposure to first behavioural symptom	Features associated with exposure	Age at onset of first symptom	
					Behaviour	Bowel
1	Autism	MMR	1 week	Fever/delirium	12 months	Not known
2	Autism	MMR	2 weeks	Self injury	13 months	20 months
3	Autism	MMR	48 h	Rash and fever	14 months	Not known
4	Autism? Disintegrative disorder?	MMR	Measles vaccine at 15 months followed by slowing in development. Dramatic deterioration in behaviour immediately after MMR at 4.5 years	Repetitive behaviour, self injury, loss of self-help	4.5 years	18 months
5	Autism	None – MMR at 16 months	Self-injurious behaviour started at 18 months		4 years	
6	Autism	MMR	1 week	Rash & convulsion; gaze avoidance & self-injury	15 months	18 months
7	Autism	MMR	24 h	Convulsion, gaze avoidance	21 months	2 years
8	Post-vaccinal encephalitis?	MMR	2 weeks	Fever, convulsion, rash & diarrhoea	19 months	19 months

9	Autistic spectrum disorder	Recurrent otitis media	1 week (MMR 2 months previously)	Disinterest; lack of play	18 months	2 years
10	Post-viral encephalitis?	Measles (previously vaccinated with MMR)	24 h	Fever, rash & vomiting	15 months	Not known
11	Autism	MMR	1 week	Recurrent "viral pneumonia" for 8 weeks following MMR	15 months	Not known
12	Autism	None – MMR at 15 months	Loss of speech development and deterioration in language skills noted at 16 months			Not known

MMR=measles, mumps, and rubella vaccine.

Laboratory tests

All children were antiendomyseal-antibody negative and common enteric pathogens were not identified by culture, microscopy, or serology.[a] Urinary methylmalonic-acid excretion was significantly raised in all eight children who were tested, compared with age-matched controls (p=0.003; figure 1). Abnormal laboratory tests are shown in table 1.

Endoscopic findings[b]

The caecum[c] was seen in all cases, and the ileum in all but two cases. Endo- 15
scopic findings are shown in table 1. Macroscopic colonic appearances were reported as normal in four children. The remaining eight had colonic and

a *enteric* Of the small intestine; *pathogens* Disease-causing agents or microorganisms; *serology* Diagnostic testing of the serum component of blood, particularly concerned with immune response to pathogens.

b *Endoscopic findings* Results of endoscopy, an examination conducted by inserting a fiber-optic cable or other tube-shaped device into a space in the body.

c *caecum* Dilated pouch forming the first part of the large intestine and located at the meeting of the small and large intestines.

rectal mucosal abnormalities including granularity, loss of vascular pattern, patchy erythema,[a] lymphoid nodular hyperplasia, and in two cases, aphthoid ulceration. Four cases showed the "red halo" sign around swollen caecal lymphoid follicles, an early endoscopic feature of Crohn's disease.[b][3] The most striking and consistent feature was lymphoid nodular hyperplasia of the terminal ileum which was seen in nine children (figure 2), and identified by barium follow-through in one other child in whom the ileum was not reached at endoscopy. The normal endoscopic appearance of the terminal ileum (figure 2) was seen in the seven children whose images were available for comparison.

Histological findings

Histological findings are summarised in table 1.

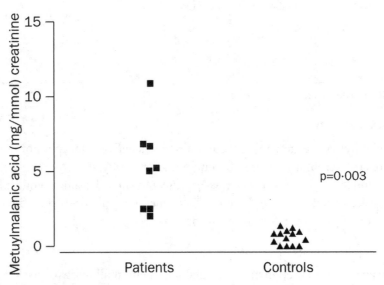

Figure 1.
Urinary methylmalonic-acid excretion in patients and controls.
p=Significance of mean excretion in patients compared with controls.

---

a *granularity* Presence of small grains or particles; *vascular pattern* Presence of blood vessels; *erthyema* Redness of skin or mucous membrane due to congestion of the blood capillary vessels, usually from injury, infection, or inflammation.

b *caecal lymphoid follicles* Sacs of lymph tissue in the caecum; *Crohn's disease* Chronic or recurring inflammation of the gastrointestinal tract, particularly affecting the intestines, most commonly the ileum, but often reaching into the colon.

Terminal ileum

A reactive lymphoid follicular hyperplasia[a] was present in the ileal biopsies of seven children. In each case, more than three expanded and confluent lymphoid follicles with reactive germinal centres[b] were identified within the tissue section (figure 3). There was no neutrophil infiltrate[c] and granulomas were not present.

Colon

The lamina propria was infiltrated by mononuclear cells (mainly lymphocytes and macrophages)[d] in the colonic-biopsy samples. The extent ranged in severity from scattered focal collections of cells beneath the surface epithelium[e] (five cases) to diffuse infiltration of the mucosa (six cases). There was no increase in intraepithelial[f] lymphocytes, except in one case, in which numerous lymphocytes had infiltrated the surface epithelium in the proximal colonic biopsies. Lymphoid follicles in the vicinity of mononuclear-cell infiltrates

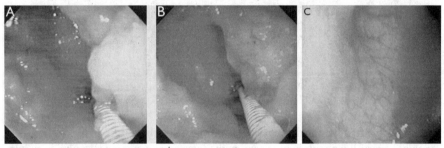

FIGURE 2.
Endoscopic view of terminal ileum in child three and in a child with endoscopically and histologically normal ileum and colon. Greatly enlarged lymphoid nodule in right-hand field of view. A and B=child three; C=normal ileum. Remainder of mucosal surface of terminal ileum is a carpet of enlarged lymphoid nodules.

---

a *lymphoid follicular hyperplasia* Inflammatory condition of unknown causes, characterized by the rapid growth of microscopically visible lymph tissue.

b *germinal centres* Locations of actively growing and changing lymphocytes (a type of white blood cell) within the center of a lymph nodule.

c *neutrophil infiltrate* Accumulation of neutrophils (a type of white blood cell), which can cause inflammation.

d *lamina propria* Highly vascular connective layer of the mucous membrane; *mononuclear cells* Cells with only one nucleus; *macrophages* Large phagocytic (disease-fighting) cells that can mobilize and travel to areas of infection, but otherwise are immobile within tissues.

e *focal* Concentrated; *epithelium* Thin layer of tissue that lines the gastrointestinal tract, blood vessels, and other small cavities in the body, as well as the outer surface of the body.

f *intraepithelial* Within the epithelium.

showed enlarged germinal centres with reactive changes that included an excess of tingible body macrophages.[a]

There was no clear correlation between the endoscopic appearances and the histological findings; chronic inflammatory changes were apparent histologically in endoscopically normal areas of the colon. In five cases there was focal acute inflammation with infiltration of the lamina propria by neutrophils; in three of these, neutrophils infiltrated the caecal (figure 3) and rectal-crypt epithelium.[b] There were no crypt abscesses. Occasional bifid[c] crypts were noted but overall crypt architecture was normal. There was no goblet-cell depletion but occasional collections of eosinophils[d] were seen in the mucosa. There were no granulomata. Parasites and organisms were not seen. None of the changes described above were seen in any of the normal biopsy specimens.

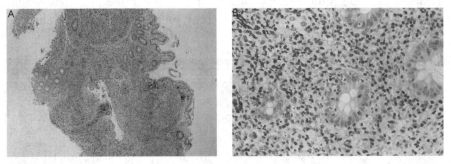

FIGURE 3.
Biopsy sample from terminal ileum (top) and from colon (bottom).
A=child three; lymphoid hyperplasia with extensive, confluent lymphoid nodules.
B=child three; dense infiltration of the lamina propria crypt epithelium by neutrophils and mononuclear cells. Stained with haematoxylin and eosin.

DISCUSSION

20 We describe a pattern of colitis and ileal-lymphoid-nodular hyperplasia in children with developmental disorders. Intestinal and behavioural pathologies may have occurred together by chance, reflecting a selection bias in a self-referred group; however, the uniformity of the intestinal pathological

---

a   *tingible body macrophages*   Macrophages primarily located in germinal centers and containing consumed and dying cells that can be stained for microscopic viewing.

b   *caecal and ... epithelium*   Epithelial layer covering the crypts, or glands, present in the lining of the caecum and rectum.

c   *bifid*   Divided into two parts.

d   *goblet-cell*   Cell that secretes an important component of mucus and is found many places in the body, including the intestinal crypt;   *eosinophils*   Type of white blood cell.

changes and the fact that previous studies have found intestinal dysfunction in children with autistic-spectrum disorders, suggests that the connection is real and reflects a unique disease process.

Asperger first recorded the link between coeliac disease and behavioural psychoses.[4] Walker-Smith and colleagues[5] detected low concentrations of alpha-1 antitrypsin in children with typical autism, and D'Eufemia and colleagues[6] identified abnormal intestinal permeability, a feature of small intestinal enteropathy,[a] in 43% of a group of autistic children with no gastro-intestinal symptoms, but not in matched controls. These studies, together with our own, including evidence of anaemia[b] and IgA deficiency in some children, would support the hypothesis that the consequences of an inflamed or dysfunctional intestine may play a part in behavioural changes in some children.

The "opioid excess" theory of autism, put forward first by Panksepp and colleagues[7] and later by Reichelt and colleagues[8] and Shattock and colleagues[9] proposes that autistic disorders result from the incomplete break-down and excessive absorption of gut-derived peptides from foods, including barley, rye, oats, and casein[c] from milk and dairy produce. These peptides may exert central-opioid effects, directly or through the formation of ligands with peptidase enzymes required for breakdown of endogenous central-nervous-system opioids,[9] leading to disruption of normal neuroregulation and brain development by endogenous encephalins and endorphins.[d]

One aspect of impaired intestinal function that could permit increased permeability to exogenous peptides is deficiency of the phenyl-sulphur-transferase systems,[e] as described by Waring.[10] The normally sulphated glycoprotein matrix[f] of the gut wall acts to regulate cell and molecular

a *alpha-1 antitrypsin* Protease inhibitor that helps protect tissues from the effects of inflammation; *enteropathy* Any disease of the intestine.
b *anaemia* Hemoglobin or red blood cell deficiency.
c *opioid* Chemical that stimulates the nervous system's opioid receptors, thus having effects on the body similar to those of opium; *peptides* Amino acid chains; *casein* Primary protein found in milk.
d *ligands* Molecules that have bonded to other molecules; *peptidase enzymes* Enzymes that catalyze the chemical breakdown of peptide links; *endogenous* Manufactured or caused by the body; *encephalins* Compounds found in the brain that are related to endorphins, and affect similar changes; *endorphins* Peptide hormones released in the nervous system; an important effect of endorphins is pain relief via stimulation of opiate receptors.
e *exogenous* From or having its cause outside the body; *phenyl-sulphur-transferase systems* Systems involving enzymes that metabolize sulphur and help the body break down the bonds of certain chemicals in order to remove toxins.
f *sulphated ... matrix* Sulphur-containing support structure for cells, providing protection from physical stresses and playing a role in communication between cells.

trafficking.[11] Disruption of this matrix and increased intestinal permeability, both features of inflammatory bowel disease,[17] may cause both intestinal and neuropsychiatric dysfunction. Impaired enterohepatic sulphation and consequent detoxification of compounds such as the phenolic amines[a] (dopamine, tyramine, and serotonin)[12] may also contribute. Both the presence of intestinal inflammation and absence of detectable neurological abnormality in our children are consistent with an exogenous influence upon cerebral function. Lucarelli's observation that after removal of a provocative enteric antigen children achieved symptomatic behavioural improvement, suggests a reversible element in this condition.[13]

Despite consistent gastrointestinal findings, behavioural changes in these children were more heterogeneous. In some cases the onset and course of behavioural regression was precipitous, with children losing all communication skills over a few weeks to months. This regression is consistent with a disintegrative psychosis (Heller's disease), which typically occurs when normally developing children show striking behaviour changes and developmental regression, commonly in association with some loss of coordination and bowel or bladder function.[14] Disintegrative psychosis is typically described as occurring in children after at least 2–3 years of apparently normal development.

25 Disintegrative psychosis is recognised as a sequel to measles encephalitis, although in most cases no cause is ever identified.[14] Viral encephalitis can give rise to autistic disorders, particularly when it occurs early in life.[15] Rubella virus is associated with autism and the combined measles, mumps, and rubella vaccine (rather than monovalent measles vaccine) has also been implicated. Fudenberg[16] noted that for 15 of 20 autistic children, the first symptoms developed within a week of vaccination. Gupta[17] commented on the striking association between measles, mumps, and rubella vaccination and the onset of behavioural symptoms in all the children that he had investigated for regressive autism. Measles virus[18,19] and measles vaccination[20] have both been implicated as risk factors for Crohn's disease and persistent measles vaccine-strain virus infection has been found in children with autoimmune hepatitis.[b 21]

We did not prove an association between measles, mumps, and rubella vaccine and the syndrome described. Virological studies are underway that may help to resolve this issue.

---

a   *enterohepatic sulphation*   Presence of sulphur in the circulating liver secretions traveling from the liver, to the intestine, to the blood, and back to the liver;   *phenolic amines*   Compounds that contain the organic compound phenol.

b   *autoimmune hepatitis*   Inflammation of the liver as a result of the production of antibodies against naturally occurring substances in the body.

If there is a causal link between measles, mumps, and rubella vaccine and this syndrome, a rising incidence might be anticipated after the introduction of this vaccine in the UK in 1988. Published evidence is inadequate to show whether there is a change in incidence[22] or a link with measles, mumps, and rubella vaccine.[23] A genetic predisposition to autistic-spectrum disorders is suggested by over-representation in boys and a greater concordance rate in monozygotic than in dizygotic twins.[a] [15] In the context of susceptibility to infection, a genetic association with autism, linked to a null allele of the complement (C) 4B gene located in the class III region of the major-histocompatibility complex,[b] has been recorded by Warren and colleagues.[24] C4B-gene products are crucial for the activation of the complement pathway[c] and protection against infection: individuals inheriting one or two C4B null alleles may not handle certain viruses appropriately, possibly including attenuated strains.

Urinary methylmalonic-acid concentrations were raised in most of the children, a finding indicative of a functional vitamin B12 deficiency. Although vitamin B12 concentrations were normal, serum B12 is not a good measure of functional B12 status.[25] Urinary methylmalonic-acid excretion is increased in disorders such as Crohn's disease, in which cobalamin excreted in bile[d] is not reabsorbed. A similar problem may have occurred in the children in our study. Vitamin B12 is essential for myelinogenesis[e] in the developing central nervous system, a process that is not complete until around the age of 10 years. B12 deficiency may, therefore, be a contributory factor in the developmental regression.[26]

We have identified a chronic enterocolitis in children that may be related to neuropsychiatric dysfunction. In most cases, onset of symptoms was after measles, mumps, and rubella immunisation. Further investigations are needed to examine this syndrome and its possible relation to this vaccine.

(1998)

---

a *monozygotic* Term applied to identical twins (who come from a single ovum); *dizygotic twins* Non-identical twins (who come from two different ova).

b *null allele* Gene or group of genes that does not perform its normal function; *complement (C)* Human protein that has a relationship to localized inflammation. It is encoded by the C4B gene; *class ... complex* Part of the major histocompatibility complex, a set of genes associated with immune function.

c *complement pathway* Set of specific antibody responses controlled by a group of blood proteins.

d *cobalamin* Vitamin B12; *bile* Fluid secreted by the liver that assists digestion.

e *myelinogenesis* Growth of the myelin sheath, the protective covering of the nerve fibers in the central nervous system.

Addendum: Up to Jan 28, a further 40 patients have been assessed; 39 with the syndrome.

Contributors: A J Wakefield was the senior scientific investigator. S H Murch and M A Thomson did the colonoscopies. A Anthony, A P Dhillon, and S E Davies carried out the histopathology. J Linnell did the B12 studies. D M Casson and M Malik did the clinical assessment. M Berelowitz did the psychiatric assessment. P Harvey did the neurological assessment. A Valentine did the radiological assessment. JW-S was the senior clinical investigator.

Acknowledgments: This study was supported by the Special Trustees of Royal Free Hampstead NHS Trust and the Children's Medical Charity. We thank Francis Moll and the nursing staff of Malcolm Ward for their patience and expertise; the parents for providing the impetus for these studies; and Paula Domizo, Royal London NHS Trust, for providing control tissue samples.

REFERENCES

1. Diagnostic and Statistical Manual of Mental Disorders (DSM-IV). 4th edn. Washington DC, USA: American Psychiatric Association, 1994.
2. Bhatt HR, Green A, Linnell JC. A sensitive micromethod for the routine estimations of methyl-malonic acid in body fluids and tissues using thin-layer chromatography. *Clin Chem Acta* 1982; **118**: 311–21.
3. Fujimura Y, Kamoni R, Iida M. Pathogenesis of aphthoid ulcers in Crohn's disease: correlative findings by magnifying colonoscopy, electromicroscopy, and immunohistochemistry. *Gut* 1996; **38**: 724–32.
4. Asperger H. Die Psychopathologie des coeliakakranken kindes. *Ann Paediatr* 1961; **197**: 146–51.
5. Walker-Smith JA, Andrews J. Alpha-1 antitrypsin, autism and coeliac disease. *Lancet* 1972; **ii**: 883–84.
6. D'Eufemia P, Celli M, Finocchiaro R, et al. Abnormal intestinal permeability in children with autism. *Acta Paediatrica* 1996; **85**: 1076–79.
7. Panksepp J. A neurochemical theory of autism. *Trends Neurosci* 1979; **2**: 174–77.
8. Reichelt KL, Hole K, Hamberger A, et al. Biologically active peptide-containing fractions in schizophrenia and childhood autism. *Adv Biochem Psychopharmacol* 1993; **28**: 627–43.
9. Shattock P, Kennedy A, Rowell F, Berney TP. Role of neuropeptides in autism and their relationships with classical neurotransmitters. *Brain Dysfunction* 1991; **3**: 328–45.
10. Waring RH, Ngong JM. Sulphate metabolism in allergy induced autism: relevance to disease aetiology, conference proceedings, biological perspectives in autism, University of Durham, NAS 35–44.
11. Murch SH, MacDonald TT, Walker-Smith JA, Levin M, Lionetti P, Klein NJ. Disruption of sulphated glycosaminoglycans in intestinal inflammation. *Lancet* 1993; **341**: 711–41.
12. Warren RP, Singh VK. Elevated serotonin levels in autism: association with the major histo-compatibility complex. *Neuropsychobiology* 1996; **34**: 72–75.
13. Lucarelli S, Frediani T, Zingoni AM, et al. Food allergy and infantile autism. *Panminerva Med* 1995; **37**: 137–41.
14. Rutter M, Taylor E, Hersor L. In: Child and adolescent psychiatry. 3rd edn. London: Blackwells Scientific Publications: 581–82.
15. Wing L. The Autistic Spectrum. London: Constable, 1996: 68–71.
16. Fudenberg HH. Dialysable lymphocyte extract (DLyE) in infantile onset autism: a pilot study. *Biotherapy* 1996; **9**: 13–17.

17. Gupta S. Immunology and immunologic treatment of autism. *Proc Natl Autism Assn Chicago* 1996; 455–60.
18. Miyamoto H, Tanaka T, Kitamoto N, Fukada Y, Takashi S. Detection of immunoreactive antigen with monoclonal antibody to measles virus in tissue from patients with Crohn's disease. *J Gastroenterol* 1995; **30**: 28–33.
19. Ekbom A, Wakefield AJ, Zack M, Adami H-O. Crohn's disease following early measles exposure. *Lancet* 1994; **344**: 508–10.
20. Thompson N, Montgomery S, Pounder RE, Wakefield AJ. Is measles vaccination a risk factor for inflammatory bowel diseases? *Lancet* 1995; **345**: 1071–74.
21. Kawashima H, Mori T, Takekuma K, Hoshika A, Hata A, Nakayama T. Polymerase chain reaction detection of the haemagglutinin gene from an attenuated measles vaccines strain in the peripheral mononuclear cells of children with autoimmune hepatitis. *Arch Virol* 1996; **141**: 877–84.
22. Wing L. Autism spectrum disorders: no evidence for or against an increase in prevalence. *BMJ* 1996; **312**: 327–28.
23. Miller D, Wadsworth J, Diamond J, Ross E. Measles vaccination and neurological events. *Lancet* 1997; **349**: 730–31.
24. Warren RP, Singh VK, Cole P, et al. Increased frequency of the null allele at the complement C4B locus in autism. *Clin Exp Immunol* 1991; **83**: 438–40.
25. England JM, Linnell JC. Problems with the serum vitamin B12 assay. *Lancet* 1980; **ii**: 1072–74.
26. Dillon MJ, England JM, Gompertz D, et al. Mental retardation, megaloblastic anaemic, homocysteine metabolism due to an error in B12 metabolism. *Clin Sci Mol Med* 1974; **47**: 43–61.

Questions:

1. Describe an ileocolonoscopy and a lumbar puncture. What are the possible adverse consequences of these procedures? When do you think ileocolonoscopies and lumbar punctures should be done on children? Ileocolonoscopy and lumbar puncture did not serve a diagnostic function for the children included in this study; do you think it was ethical to do them for the purposes of the study?

2. Does the experiment, as described, support the conclusions it asserts in the Discussion? Why or why not?

3. How does the article characterize the recruitment of these patients? How does the description of the enrolment process change between the Methods section and the Discussion section? What is the significance of this change?

4. List everything you know about the control subjects used in this experiment. What, if any, other information should have been provided?

5. The gold standard for research of this type involves double-blinding, comparator arms (control), prospective study, sequential enrolment, large enrolment, and randomization to the study group. Which, if any, of these standards does this experiment meet?

6. Which "history" dominates in this article: parental recollection or documented occurrence? Which *should* dominate in an article of this type?

7. Look carefully at the citations for this paper. How many times do these researchers cite themselves as authorities?

8. How current were the sources for this 1998 article?

9. A standard peer-reviewed scientific article is usually 5 (minimum) to 10 single-spaced pages. What is the average page length of the documents cited here? What sort of publication do you think these short pieces might be?

10. The article claims that D'Eufemia et al.'s article (reference 6) found 43% of a group of autistic children had "abnormal intestinal permeability." Look up the summary of D'Eufemia's article online. How many children constituted this 43%?

11. Review the Discussion section carefully, taking into account just where the sources are unreliable or very dated or Wakefield et al.'s own word. Characterize the reliability of the discussion.

12. "Disintegrative psychosis" is not autism and affects approximately 1 in 100,000 individuals. Does the Discussion's material on "disintegrative psychosis" have a strong relation to the discussion of this investigation? To what extent does this material actually support the article's contentions?

sixteen

Brian Deer

How the Case against the MMR Vaccine Was Fixed

*The late 1990s and early 2000s saw the spread of a belief – now
fully discredited – that some vaccines could cause the sudden onset
of regressive autism and other conditions in children. This article by
journalist Brian Deer investigates the 1998 study by researcher Andrew
Wakefield et al. (included in this anthology) that prompted the scare.*

e

W hen I broke the news to the father of child 11, at first he did not believe
me. "Wakefield told us my son was the 13th child they saw," he said,
gazing for the first time at the now infamous research paper which linked
a purported new syndrome with the measles, mumps, and rubella (MMR)
vaccine.[1] "There's only 12 in this."

That paper was published in the *Lancet* on 28 February 1998. It was
retracted on 2 February 2010.[2] Authored by Andrew Wakefield, John Walker-
Smith and 11 others from the Royal Free Hospital and School of Medicine,
London, it reported on 12 developmentally challenged children, and triggered
a decade long public health scare.

"Onset of behavioural symptoms was associated by the parents with
measles, mumps, and rubella vaccination in eight of the 12 children," began the paper's "findings." Adopting these claims as fact, its results section
added: "In these eight children the average interval from exposure to first
behavioural symptoms was 6.3 days (range 1-14)."

Mr. 11, an American engineer, looked again at the paper: a five page case
series of 11 boys and one girl, aged between 3 and 9 years. Nine children, it
said, had diagnoses of "regressive" autism,[a] while all but one were reported
with "non-specific colitis."[b] The "new syndrome" brought these together,
linking brain and bowel diseases. Child 11 was the penultimate case.

---

a   *"regressive" autism*   Form of autism in which a child does not begin to show symptoms until 15
    to 30 months of age.

b   *colitis*   Inflammation of the colon.

5   Running his finger across the paper's tables, over coffee in London, Mr. 11 seemed reassured by his anonymised son's age and other details. But then he pointed at table 2 – headed "neuropsychiatric diagnosis" – and for a second time objected.

"That's not true."

Child 11 was among the eight whose parents apparently blamed MMR. The interval between his vaccination and the first "behavioural symptom" was reported as 1 week. This symptom was said to have appeared at age 15 months. But his father, whom I had tracked down, said this was wrong.

"From the information you provided me on our son, who I was shocked to hear had been included in their published study," he wrote to me, after we met again in California, "the data clearly appeared to be distorted."

He backed his concerns with medical records, including a Royal Free discharge summary. Although the family lived 5000 miles from the hospital, in February 1997 the boy (then aged 5) had been flown to London and admitted for Wakefield's project, the undisclosed goal of which was to help sue the vaccine's manufacturers.

WAKEFIELD'S "SYNDROME"

10   Unknown to Mr. 11, Wakefield was working on a lawsuit,[3] for which he sought a bowel-brain "syndrome" as its centrepiece. Claiming an undisclosed £150 (€180; $230) an hour through a Norfolk solicitor named Richard Barr, he had been confidentially put on the payroll for two years before the paper was published, eventually grossing him £435 643, plus expenses.[4]

Curiously, however, Wakefield had already identified such a syndrome before the project that would reputedly discover it. "Children with enteritis/ disintegrative disorder [an expression he used for bowel inflammation and regressive autism[5]] form part of a new syndrome," he and Barr explained in a confidential grant application to the UK government's Legal Aid Board,[6] before any of the children were investigated. "Nonetheless the evidence is undeniably in favour of a specific vaccine induced pathology."

The two men also aimed to show a sudden onset "temporal association" – strong evidence in product liability. "Dr Wakefield feels that if we can show a clear time link between the vaccination and onset of symptoms," Barr told the legal board, "we should be able to dispose of the suggestion that it's simply a chance encounter."[7]

But child 11's case must have proved a disappointment. Records show his behavioural symptoms started *too soon*. "His developmental milestones were normal until 13 months of age," notes the discharge summary. "In the period 13-18 months he developed slow speech patterns and repetitive hand movements. Over this period his parents remarked on his slow gradual deterioration."

That put the first symptom two months earlier than reported in the *Lancet,* and a month before the boy had MMR. And this was not the only anomaly to catch the father's eye. What the paper reported as a "behavioural symptom" was noted in records as a chest infection.

"Please let me know if Andrew W has his doctor's license revoked," wrote 15 Mr. 11, who is convinced that many vaccines and environmental pollutants may be responsible for childhood brain disorders. "His misrepresentation of my son in his research paper is inexcusable. His motives for this I may never know."

The father need not have worried. My investigation of the MMR issue exposed the frauds behind Wakefield's research. Triggering the longest ever UK General Medical Council fitness to practise hearing, and forcing the *Lancet* to retract the paper, last May it led to Wakefield and Walker-Smith being struck off the medical register.[8-10]

Wakefield, now 54, who called no witnesses, was branded "dishonest," "unethical," and "callous."[8-10] Walker-Smith, now 74, the senior clinician in the project, was found to have presided over "high risk"[11] research without clinical indication or ethical approval. The developmentally challenged children of often vulnerable parents were discovered to have been treated like the doctors' guinea pigs.[10]

Lawsuit Test Case

But Mr. 11 was not the first parent with a child in the study whom I interviewed during my investigation. That was Mrs. 2: the first of the parents to approach Wakefield. She was sent to him by an anti-vaccine campaign called JABS. Her son had regressive autism,[12] longstanding problems with diarrhoea,[13] and was the prime example of the purported bowel and brain syndrome – still unsubstantiated 14 years later.[14] This boy would appear in countless media reports, and was one of the four "best" cases in Barr's lawsuit.

I travelled to the family home, 80 miles northeast of London, to hear about child 2 from his mother. That was in September 2003, when the lawsuit fell apart after counsel representing 1500 families said that, on the evidence, Barr's autism claims would fail.[15] By that time, Mrs. 2 had seen her son's medical records and expert reports, written for her case at trial.

Her concerns about MMR had been noted by her general practitioner 20 when her son was 6 years old. But she told me the boy's troubles began after his vaccination, which he received at 15 months. "He'd scream all night, and he started head banging, which he'd never done before," she explained.

"When did that begin, do you think?" I asked.

"That began after a couple of months, a few months afterward, but it was still, it was concerning me enough, I remember going back."

"Sorry. I don't want to be, like, massively pernickety, but was it a few months, or a couple of months?"

"It was more like a few months because he'd had this, kind of, you know, slide down. He wasn't right. He wasn't right. Before he started."

25 "Not quicker than two months, but not longer than how many months? What are we talking about here?"

"From memory, about six months, I think."

The next day, she complained to my editors. She said my methods "seemed more akin to the gutter press." But I was perplexed by her story, since there was no case in the *Lancet* that matched her careful account.

According to the paper, child 2 had his "first behavioural symptom" two weeks, not six months, after MMR. This was derived from a Royal Free medical history (citing "head banging" and "screaming" as the start) taken by Mark Berelowitz, a child psychiatrist and a coauthor of the paper. He saw Mrs. 2 during the boy's admission, at age 8, after she had discussed her son's story with Wakefield.[10]

As I later discovered, each family in the project was involved in such discussions before they saw the hospital's clinicians. Wakefield phoned them at home, and must have at least suggestively questioned them, potentially impacting on later history taking. But I knew little of such things then, and shared my confusion with Walker-Smith, who I met shortly after Mrs. 2.

30 "There is no case in the paper that is consistent with the case history [Mrs. 2] has given me," I told him. "There just isn't one."

"Well that could be true," the former professor of paediatric gastroenterology[a] replied, disarmingly. He knew the case well, having admitted the boy for the project and written reports for Barr, who paid him £23 000.[16]

"Well, so either what she is telling me is not accurate, or the paper's not accurate."

"Well I can't really comment," he said. "You really touch on an area which I don't think should be debated like this. And I think these parents are wrong to discuss such details, where you could be put in a position of having a lot of medical details and then try to match it with this, because it is a confidential matter."

It was not merely medically confidential, it was also legally protected: a double screen against public scrutiny. But responding to my first MMR reports

---

a *gastroenterology* Branch of medicine dedicated to the digestive system.

in the *Sunday Times,* in February 2004,[17] the GMC[a] decided to investigate the cases and requisitioned the children's records.

The regulator's main focus was whether the research was ethical. Mine was whether it was true. So as a five member disciplinary panel trawled through the records, with five Queen's counsel and three defendant doctors, I compared them with what was published in the journal.[18]

Multiple Discrepancies

The paper gave the impression that the authors had been scrupulous in documenting the patients' cases. "Children underwent gastroenterological, neurological, and developmental assessment and review of developmental records," it explained, specifying that *Diagnostic and Statistical Manual of Mental Disorders IV (DSM-IV)*[b] criteria were used for neuropsychiatric diagnoses. "Developmental histories included a review of prospective developmental records from parents, health visitors, and general practitioners."

When the details were dissected before the panel, however, multiple discrepancies emerged. A syndrome necessarily requires at least some consistency, but, as the records were laid out, Wakefield's crumbled.

First to crack was "regressive autism," the bedrock of his allegations.[3] "Bear in mind that we are dealing with regressive autism in these children, not of classical autism where the child is not right from the beginning," he later explained, for example, to a United States congressional committee.[19]

But only one – child 2 – clearly had regressive autism.[20] Three of nine so described clearly did not. None of these three even had autism diagnoses, either at admission or on discharge from the Royal Free.

The paper did not reveal that two of this trio were brothers, living 60 miles south of the hospital. Both had histories of fits and bowel problems recorded before they received MMR. The elder, child 6, aged 4 years at admission, had Asperger's syndrome,[c] [21] which is distinct from autism under DSM-IV, is not regressive,[22] and was confirmed on discharge.[10] His brother, child 7, was admitted at nearly 3 years of age without a diagnosis,[10] and a post-discharge letter from senior paediatric registrar and *Lancet* coauthor David Casson summarised: "He is not thought to have features of autism."

The third of this trio, child 12, was enrolled on the advice of the brothers' mother – reported in media as a JABS activist, who had herself "only

---

a *GMC* General Medical Council, an organization responsible for managing the national list of registered doctors and enforcing standards of medical practice in the United Kingdom.

b *Diagnostic and ... (DSM-IV)* Reference guide published by the American Psychological Association providing standard criteria for the classification of mental disorders.

c *Asperger's syndrome* Developmental disorder that shares some characteristics with autism.

relatively recently" blamed the vaccine. Child 12 was aged 6 at admission and had previously been assessed for possible Asperger's syndrome at Guy's Hospital, London, by a renowned developmental paediatrician. She diagnosed "an impairment in respect of language" – an opinion left undisturbed by Berelowitz.[10]

Mrs. 12 was a GMC witness at its mammoth hearing, which between July 2007 and May 2010 ran for 217 days. She explained that the brothers' mother had made her suspicious of MMR and gave her Barr's and Wakefield's names. Mrs. 12 approached them and filed a statement for legal aid before her son was referred.

"It was like a jigsaw puzzle – it suddenly seemed to fit into place," she told the panel, describing how she concluded, four years after the boy was vaccinated, that MMR was to blame for his problems. "I had this perfectly normal child who, as I could see, for no apparent reason started to not be normal."

The 12 children were admitted between July 1996 and February 1997, and others had connections not revealed in the paper, almost as striking as the trio's. The parents of child 9 and child 10 were contacts of Mrs. 2, who ran a group that campaigned against MMR. And child 4 and child 8 were admitted – without outpatient appointments[10] – for ileocolonoscopy[a] and other invasive procedures, from one Tyneside general practice, 280 miles from the Royal Free, after advice from anti-MMR campaigners.

PRE-EXISTING PROBLEMS

45 Both child 4 and child 8 were among the eight whose parents were reported to have blamed the vaccine. But although the paper specified that all 12 children were "previously normal," both had developmental delays, and also facial dysmorphisms,[b] noted before MMR vaccination.

In the case of child 4, who received the vaccine at 4 years, Wakefield played down problems, suggesting that early issues had resolved. "Child four was kept under review for the first year of life because of wide bridging of the nose," he reported in the paper. "He was discharged from follow-up as developmentally normal at age 1 year."

But medical records, presented by the GMC, give a different picture for this child. Reports from his pre-MMR years were peppered with "concerns over his head and appearance," "recurrent" diarrhoea, "developmental delay,"

---

a *ileocolonoscopy* Examination of the gastrointestinal tract.

b *facial dysmorphisms* Abnormally shaped faces. Facial dysmorphism can be associated with several developmental disorders.

"general delay," and restricted vocabulary. And although before his referral to Wakefield his mother had inquired about vaccine damage compensation, his files include a report of a "very small deletion within the fragile X gene," and a note of the mother's view that her concerns about his development began when he was 18 months old.

"In general, his mother thinks he developed normally initially and subsequently his problems worsened, and he lost some of his milestones, but he subsequently improved on a restrictive exclusion diet," wrote his general practitioner, William Tapsfield, referring the boy, then aged 9, after a phone conversation with Wakefield. "The professionals who have known [child 4] since birth don't entirely agree with this, however, and there is a suggestion that some of his problems may have started before vaccination."

Similarly with child 8, who was also described in the *Lancet* as having overcome problems recorded before MMR. "The only girl ... was noted to be a slow developer compared with her older sister," the paper said. "She was subsequently found to have coarctation of the aorta.[a] After surgical repair of the aorta at the age of 14 months, she progressed rapidly, and learnt to talk. Speech was lost later."

But Wakefield was not a paediatrician. He was a former trainee gastrointestinal surgeon with a non-clinical medical school contract.[10] And his interpretation differed from that of local consultants (including a developmental paediatrician and a geneticist) who had actually looked after the girl. Her doctors put the coarctation side by side with the developmental delay and dysmorphism, and noted of her vocabulary that, before MMR at 18 months, she "vocalised" only "two or three words." 50

"[Child 8's] mother has been to see me and said you need a referral letter from me in order to accept [child 8] into your investigation programme," the general practitioner, Diana Jelley, wrote to Wakefield at referral, when the girl was aged 3 and a half years. "I would simply re-iterate ... that both the hospital and members of the primary care team involved with [child 8] had significant concerns about her development some months before she had her MMR."

The girl's general practice notes also provide insight into the background to the 12 children's referrals. After person(s) unknown told Mrs. 8 that her daughter may have inflammatory bowel disease, Jelley wrote: "Mum taking her to Dr Wakefield, Royal Free hospital for CT scans/gut bi-

---

a *coarctation of the aorta* Condition in which a portion of the aorta, the major artery leading away from the heart, is abnormally narrow.

opsies ?Crohn's[a] – will need ref letter – Dr Wakefield to phone me. Funded through legal aid."

THE CHILD WAS "PALE"

The remaining five children served Wakefield's claims no better. There was still no convincing MMR syndrome.

Child 1, aged 3 years when he was referred to London, lived 100 miles from the Royal Free and had an older brother who was diagnosed as autistic. Child 1's recorded story began when he was aged 9 months, with a "new patient" note by general practitioner Andrea Barrow. One of the mother's concerns was that her son could not hear properly – which might sound like a hallmark presentation of classical autism, the emergence of which is often insidious. Indeed, a Royal Free history, by neurologist and coauthor Peter Harvey, noted "normal milestones" until "18 months or so."

55   This boy was vaccinated at 12 months of age, however. Thus neither 9 nor 18 months helped Wakefield's case. But in the *Lancet*, the "first behavioural symptom" was reported to have occurred "1 week" after the injection, holding the evidence for the lawsuit on track.

Step 1 to achieve this: two and a half years after the child was vaccinated, Walker-Smith took an outpatient history. Although the mother apparently had no worries following her son's vaccination, the professor elicited that the boy was "pale" 7-10 days after the shot. He also elicited that the child "possibly" had a fever, and "may" have been delirious, as well as pale.

"It's difficult to associate a clear historical link with the MMR and the answer to autism," Walker-Smith wrote to the general practitioner, with a similar letter to Wakefield, "although [Mrs. 1] does believe that [child 1] had an illness 7-10 days after MMR when he was pale, ?fever, ?delirious, but wasn't actually seen by a doctor."

Step 2: for the *Lancet* Wakefield dropped the question marks, turning Walker-Smith's queries into assertions. And, although Royal Free admission and discharge records refer to "classical" autism, step 3, the former surgeon reported "delirium" as the first "behavioural symptom" of *regressive* autism, with, step 4, a "time to onset" of 7 days.

So here – behind the paper – is how Wakefield evidenced his "syndrome" for the lawsuit, and built his platform to launch the scare.

---

a *?Crohn's* Referring to Crohn's Disease, chronic inflammation of the intestines or other part of the gastrointestinal tract.

"It is significant that this syndrome only appeared with the introduction 60 of the polyvalent MMR vaccine in 1988 rather than with the monovalent[a] measles vaccine introduced in 1968," he claimed in one of a string of patents he filed for businesses to be spun from the research.[23] "This indicates that MMR is responsible for this condition rather than just the measles virus."

Three of the four remaining children were seen in outpatients on the same day – in November 1996. None of their families were reported in the paper as blaming the vaccine. Child 5, from Berkshire, aged 7 at admission, had received MMR at 16 months. The paper reported concerns at 18 months, but the medical records noted fits and parental worries at 11 months. Child 9, aged 6, from Jersey, also had MMR at 16 months. His mother dated problems from 18-20 months. Child 10, aged 4, from south Wales, contracted a viral infection, which was suspected by parents and doctors to have caused his disorder, four months after his vaccination.

"Behavioural changes included repetitive behaviour, disinterest in play or head banging," said a question and answer statement issued by the medical school, concerning the *Lancet* 12, on the day of the paper's publication.

Another discrepancy to emerge during the GMC hearing concerned the number of families who blamed MMR. The paper said that eight families (1, 2, 3, 4, 6, 7, 8, and 11) linked developmental issues with the vaccine. But the total in the records was actually 11. The parents of child 5, 9, and 12 were also noted at the hospital as blaming the vaccine, but their stated beliefs were omitted from the journal.

CASE SELECTION

The frequency of these beliefs should not have surprised Wakefield, retained as he was to support a lawsuit. In the month that Barr engaged him – two years before the paper was published – the lawyer touted the doctor in a confidential newsletter to his MMR clients and contacts. "He has deeply depressing views about the effect of vaccines on the nation's children," Barr said.[24] "He is also anxious to arrange for tests to be carried out on any children ... who are showing symptoms of possible Crohn's disease. The following are signs to look for. If your child has suffered from all or any of these symptoms could you please contact us, and it may be appropriate to put you in touch with Dr Wakefield."

The listed symptoms included pain, weight loss, fever, and mouth ulcers. 65 Clients and contacts were quickly referred. Thus, an association between autism, digestive issues, and worries about MMR – the evidence that launched

---

a *polyvalent* I.e., targeting more than one disease; *monovalent* I.e., targeting only one disease.

the vaccine scare – was bound to be found by the Royal Free's clinicians because this was how the children were selected.

Moreover, through the omission from the paper of some parents' beliefs that the vaccine was to blame, the time link for the lawsuit sharpened. With concerns logged from 11 of 12 families, the maximum time given to the onset of alleged symptoms was a (forensically unhelpful) four months. But in a version of the paper circulated at the Royal Free six months before publication, reported concerns fell to nine of 12 families but with a still unhelpful maximum of 56 days.[25] Finally, Wakefield settled on 8 of 12 families, with a maximum interval to alleged symptoms of 14 days.

Between the latter two versions, revisions also slashed the mean time to alleged symptoms – from 14 to 6.3 days. "In these children the mean interval from exposure to the MMR vaccine to the development of the first behavioural symptom was six days, indicating a strong temporal association," he emphasised, in a patent for, among other things, his own measles vaccine,[26] eight months before the *Lancet* paper.

This leaves child 3. He was 6½ and lived on Merseyside: 200 miles from the hospital. He received MMR at 14 months, with the first concerns recorded in his GP notes 15 months after that. His mother – who 4 years later contacted Wakefield on the advice of JABS[27] – told me that her son had become aggressive towards a brother, and records say that his vocabulary had not developed.

"We both felt that the MMR needle had made [child 3] go the way he is today," the parents wrote to a local paediatric neurologist, Lewis Rosenbloom, 18 months before their son's referral to London. They told him they wanted "justice" from the vaccine's manufacturer and that they had been turned down for legal aid. "Although it is said that the MMR has never been proven to make children to be autistic, we believe that the injection has made [child 3] to be mentally delayed, which in turn may have triggered off the autism."

I visited this family twice. Their affected son was now a teenager and a challenge both to himself and to others. His mother said his diagnosis was originally "severe learning difficulties with autistic tendencies," but that she had fought to get it changed to autism.

As for a connection with MMR, there was only suspicion. I don't think his family was sure, one way or the other. When I asked why they took him to the Royal Free, his father replied: "We were just vulnerable, we were looking for answers."

What was unquestionably true was that child 3 had serious bowel trouble: intractable, lifelong, constipation. This was the most consistent feature among the 12 children's symptoms and signs[28] but, being the opposite of an expected finding in inflammatory bowel disease,[29] was nowhere mentioned in the paper. This young man's symptoms were so severe that he was dosed at his special school, his mother said, with up to five packets of laxative a day.

"You always knew when his stomach was hard," she told me, in terms echoed over the years by many parents involved with Wakefield. "He would start headbutting, kicking, breaking anything in the house. Then he would go to the toilet and release it."

For the Royal Free team, however, when reporting on these patients, such motility symptoms[30] were sidelined in the hunt for Wakefield's syndrome. In almost all the children, they noted commonly swollen glands in the terminal ileum, and what was reported as "non-specific colitis."[31,32] In fact, as I revealed in the *BMJ* last April,[33] the hospital's pathology service found the children's colons to be largely normal, but a medical school "review" changed the results.

In this evolution of the gut pathology to what was published in the *Lancet*, child 3's case was a prime example. After ileocolonoscopy (which GMC prosecution and defence experts agreed was not clinically indicated), the hospital's pathologists found all colonic samples to be "within normal histological limits." But three months after the boy was discharged, Walker-Smith recalled the records and changed the diagnosis to "indeterminate ileocolitis."[34]

"I think, sadly, this was the first child who was referred, and the long-term help we were able to give in terms of dealing with constipation was not there," he told the GMC panel. "However, we had excluded Crohn's disease and we had done our best to try and help this child, but in the end we did not."

So that is the *Lancet* 12: the foundation of the vaccine scare. No case was free of misreporting or alteration. Taken together, the NHS records cannot be reconciled with what was published, to such devastating effect, in the journal (table).

Wakefield, however, denies wrongdoing, in any respect whatsoever.[35] He says he never claimed the children had regressive autism, nor that he said they were previously normal. He never misreported or changed any findings in the study, and never patented a vaccine for measles. None of the children were Barr's clients before referral to the hospital, and he never received huge payments from the lawyer. There were no conflicts of interest. He is the victim of a conspiracy.[36,37] He never linked autism with MMR.

"Mr. Deer's implications of fraud against me are claims that a trained physician and researcher of good standing had suddenly decided he was going to fake data for his own enrichment," he said in a now abandoned complaint against me to the UK Press Complaints Commission. "The other authors generated and 'prepared' all the data that was reported in the *Lancet*. I merely put their completed data in tables and narrative form for the purpose of submission for publication."

Comparison of Three Features of the 12 Children in the Lancet Paper with Features Apparent in the NHS Records, Including Those from the Royal Free Hospital

Child No	Regressive autism		Non-specific colitis		First symptoms days after MMR		All three features	
	Lancet	Records*	Lancet	Records[†]	Lancet	Records[‡]	Lancet	Records
1	Yes	?	Yes	Yes	Yes	No	Yes	No
2	Yes	Yes	Yes	Yes	Yes	No	Yes	No
3	Yes	?	Yes	No	Yes	?	Yes	No
4	Yes	?	Yes	No	Yes	No	Yes	No
5	Yes	?	Yes	No	No	No	No	No
6	Yes	No	Yes	Yes	Yes	?	Yes	No
7	Yes	No	No	No	Yes	No	No	No
8	No	No	Yes	No	Yes	No	No	No
9	No	No	Yes	No	No	No	No	No
10	No	No	Yes	No	No	No	No	No
11	Yes	?	Yes	No	Yes	No	Yes	No
12	Yes	No	Yes	No	No	No	No	No
Total	9/12	?6/12	11/12	3/12	8/12	?2/12	6/12	0/12

See supplementary data on bmj.com for a version of this table with detailed footnotes.

*Regressive developmental disorder – autism.

[†]Royal Free hospital pathology service.

[‡]First behavioural symptoms ≤ 14 days after MMR.

80   But, despite signing up to claim credit for a paper in the *Lancet*, his co-authors Walker-Smith and Murch did not even know which case was which. Walker-Smith said he had "trusted" Wakefield. "When I signed that paper, I signed with good intent," he told the GMC panel. Denying any wrongdoing, he argued that the published report was not even about MMR, but merely described a new "clinico-pathological entity." He said that the admissions to the Royal Free were "entirely related to gastroenterological illness" and how the children were sourced was "irrelevant" and "immaterial." His lawyers said that he was appealing against the panel's decision and on these grounds they had advised him not to respond to my questions.

The journal, meanwhile, took 12 years to retract the paper, by which time its mischief had been exported. As parents' confidence slowly returned in Britain, the scare took off around the world, unleashing fear, guilt, and infectious diseases – and fuelling suspicion of vaccines in general. In addition to measles outbreaks, other infections are resurgent, with Mr. 11's home state of California last summer seeing 10 babies dead from whooping cough, in the worst outbreak since 1958.[38] Wakefield, nevertheless, now apparently self employed and professionally ruined, remains championed by a sad rump of disciples. "Dr Wakefield is a hero," is how one mother caught their mood in a recent *Dateline NBC* television investigation, featuring the story of the doctor and me. "I don't know where we would be without him."[39]

(2011)

Funding: Brian Deer's investigation, which led to the General Medical Council inquiry, was funded by the Sunday Times of London and the Channel 4 television network. Reports by Deer in the BMJ were commissioned and paid for by the journal. No other funding was received, apart from legal costs paid to Deer by the Medical Protection Society on behalf of Andrew Wakefield.

Competing interests: The author has completed the unified competing interest form at www.icmje. org/coi_disclosure.pdf (available on request from him) and declares no support from any organisation for the submitted work; no financial relationships with any organisation that might have an interest in the submitted work in the previous three years; BD's investigation led to the GMC proceedings referred to in this report, including the charges. He made many submissions of information but was not a party or witness in the case, nor involved in its conduct.

Provenance and peer review: Commissioned; externally peer reviewed.

REFERENCES

1. Wakefield AJ, Murch SH, Anthony A, Linnell, Casson DM, Malik M, et al. Ileal lymphoid nodular hyperplasia, non-specific colitis, and pervasive developmental disorder in children. *Lancet* 1998;351:637-41 [retracted].

2. Editors of the Lancet. Retraction: ileal lymphoid nodular hyperplasia, non-specific colitis, and pervasive developmental disorder in children. *Lancet* 2010;375:445.

3. MMR and MR vaccine litigation: Sayers and others v Smithkline Beecham plc and others - [2007] All ER (D) 30 (Jun).

4. Deer B. MMR doctor given legal aid thousands. *Sunday Times* 2006 December 31. www. timesonline.co.uk/tol/news/uk/article1265373.ece/.

5. Wakefield A. "Introduction to the rationale, aims and potential therapeutic implications of the investigation of children with disintegrative disorder (regressive autism); Heller's disease and intestinal symptomatology." (Document issued by Wakefield and mailed to doctors and parents who approached the Royal Free, dated 3 February 1997.)

6. Barr R, Wakefield A. Proposed protocol and costing proposals for testing a selected number of MR and MMR vaccinated children (and attached specification). Submitted to the Legal Aid Board 6 June 1996. [GMC fitness to practise panel hearing in the case of Wakefield, Walker-Smith and Murch. Day 11.]

7. Richard Barr. Letter to the Legal Aid Board. 22 November 1996. Day 11.

8. General Medical Council. Dr Andrew Wakefield: determinations on serious professional misconduct and sanctions. 24 May 2010. Wakefield: www.gmc-uk.org/Wakefield_SPM_and_ SANCTION.pdf_32595267.pdf.

9. General Medical Council. Professor John Walker-Smith: determinations on serious professional misconduct and sanctions, 24 May 2010. www.gmc-uk.org/Professor_Walker_Smith_SPM.pdf_32595970.pdf.

10. General Medical Council. Fitness to practise panel. Findings of fact. 28 January 2010. www.gmc-uk.org/static/documents/content/Wakefield__Smith_Murch.pdf.

11. British Paediatric Association. Guidelines for the ethical conduct of medical research involving children. *Bull Med Ethics* 1992;80:13-20.

12. Thomas N. Evidence to the panel. Day 107.

13. Cartmel R. Evidence to the panel. Day 14.

14. Buie T, Campbell DB, Fuchs GJ, Furata GT, Levy J, VandeWater J, et al. Evaluation, diagnosis, and treatment of gastrointestinal disorders in individuals with ASDs: a consensus report. *Pediatrics* 2010;125(suppl 1): s1-18.

15. Lord Justice May. Judgment in the court of appeal, London. R on the application of "H" v the Legal Services Commission. 28 February 2006.

16. Deer B. Revealed: Undisclosed payments to Andrew Wakefield at the heart of vaccine alarm. http://briandeer.com/wakefield/legal-aid.htm.

17. Deer B. Revealed: MMR research scandal. *Sunday Times* 2004 Feb 22. www.timesonline.co.uk/tol/life_and_style/health/article1027636.ece.

18. Deer B. MMR doctor fixed data on autism. *Sunday Times* 2009 Feb 8. www.timesonline.co.uk/tol/life_and_style/health/article5683671.ece.

19. Wakefield A. Evidence to the House of Representatives committee on government reform. 25-26 April 2001.

20. Rutter M. Evidence to the panel. Day 37. Day 39.

21. Walker-Smith J. Letter to Andrew Wakefield, 4 October 1996. Day 41.

22. Filipek PA. Autistic spectrum disorders. In: Swaiman KF, Ashwal S, eds. Pediatric Neurology, Principles and Practice. 3rd ed. Mosby, 1999.

23. Patent Office. Pharmaceutical composition for regressive behavioural disease. UK patent GB 2 325 856 A. Priority date 6 June 1997. Publication date 9 December 1998.

24. Dawbarns. Newsletter. February 1996.

25. Deer B. It's all change as MMR paper reveals key differences from published Lancet study. http://briandeer.com/mmr/lancet-versions.htm.

26. Patent Office. Filing receipt. 6 June 1997, published at Deer B. Revealed: the first Wakefield MMR patent describes "safer measles vaccine". http://briandeer.com/wakefield/vaccine-patent.htm.

27. Walker-Smith J. Letter to Ajjegowda Shantha. 4 April 1996.

28. Murch S, Thomson M, Walker-Smith J. Author's reply [letter]. *Lancet* 1998;351:908.

29. Squires RH, Colletti RB. Indications for pediatric gastrointestinal endoscopy: a medical position statement of the North American Society for Pediatric Gastroenterology and Nutrition. *J Pediatr Gastroenterol Nutr* 1996;23:107-10.

30. Afzal N, Murch S, Thirrupathy K, Berger L, Fagbemi A, Heuschkel R. Constipation with acquired megarectum in children with autism. *Pediatrics* 2003;112:939-42.

31. Deer B. Wakefield MMR-autism sign was recognized for years: as benign finding in children. http://briandeer.com/wakefield/ileal-hyperplasia.htm.

32. Turunen S, Karttonen TJ, Kokkonen J. Lymphoid nodular hyperplasia and cow's milk hypersensitivity in children with chronic constipation. *J Pediatrics* 2004;145:606-11.

33. Deer B. Wakefield's "autistic enterocolitis" under the microscope. *BMJ* 2010;340:c1127.

34. Walker-Smith J. Letter to Ajjegowda Shantha, with revised discharge summary. 31 December 1996.

35. Wakefield A. Complaint to the Press Complaints Commission. March 2009 (suspended on 10 February 2010 on grounds of non-pursuit by the complainant). http://briandeer.com/solved/wakefield-complaint.pdf.

36. Deer B. Vaccine victim: Andrew Wakefield invents a bizarre conspiracy [video]. http://briandeer.com/solved/tall-story.htm.

37. Profile: Andrew Wakefield, the man at the centre of the MMR scare. *Times* 2010 May 24. www.timesonline.co.uk/tol/news/uk/article7135099.ece.

38. California Department of Public Health. Pertussis report. 15 December 2010. http://www.cdph.ca.gov/programs/immunize/Documents/PertussisReport2010-12-15.pdf.

39. NBC News. A dose of controversy. *Dateline NBC, with Matt Lauer*. 30 August 2009, repeated and updated 30 May 2010.

The version of this article on bmj.com contains full footnotes.

Questions:

1. This article was commissioned by, and published in, the UK's most significant medical journal, *BMJ* (British Medical Journal). How does the tone of this article compare to the tone articles in medical research journals usually use? Given that Deer almost single-handedly exposed, through primary, investigative research, the flaws in Wakefield et al.'s paper, and that Wakefield unsuccessfully pursued Deer for libel, is the tone justified? Would another tone be more convincing?

2. Most science research articles use some variation of the following arrangement: Abstract, Introduction/Background, Methodology, Results, Discussion. What arrangement does Deer's article use? Why does the article start with Mr. 11 and his child?

3. What effect do the parent testimonials that feature so prominently in Deer's article have on your view of Wakefield et al.'s study?

4. Why do you think Walker-Smith makes the assertion that it is wrong to violate scientific or medical anonymity? Does Deer's article violate the anonymity principle?

5. Does the inductive and narrative nature of the article help or harm the paper? Do you think the paper is most fully aimed at a lay audience or at an audience of specialists? Why?

6. Look at the article's citations. Where does most of the information come from? How often does the article cite one of Deer's other investigative works as an authority? Does this affect your perception of the article's reliability?

7. Extended endnotes for this article are included in its online version, available at bmj.com. Note 18 of those extended notes includes the following information:

Ten children (1, 2, 3, 5, 6, 7, 8, 9, 10, 12) were found to have been subjected to invasive investigations for research purposes without ethical approval. In seven cases (1, 2, 3, 5, 8, 9, 12) this was found to be contrary to the child's clinical interests. Eight children (1, 2, 3, 4, 5, 8, 9, 12) were caused to undergo colonoscopies which were not clinically indicated. Seven children (1, 2, 3, 5, 8, 9, 12) were caused to undergo barium meals and follow throughs which were not clinically indicated. Three children (3, 9, 12) were caused to undergo lumbar punctures which were not clinically indicated.

Does Wakefield et al.'s conduct, as described in this endnote, seem ethical to you? Why do you think this information is not included in the body of Deer's article? What would be the effect of moving it up into the article and discussing it at length?

8. Look up the MMR scare on Wikipedia or on the Guardian newspaper's website (www.guardian.co.uk). What have the consequences of the scare been?

9. Does the article's comparison chart (between Wakefield et al.'s paper and the NHS records) help you to understand the article's point about the various diagnosis, MMR, and "autism" discrepancies? Make up a chart that presents the article's information in an alternative fashion. Explain why you chose that particular approach.

10. Mr. 11 states: "[Wakefield's] motives for this I may never know." In what ways does the fact that an "undisclosed goal" of Wakefield et al.'s study "was to help sue the [vaccine's] manufacturers" suggest the ethical and/or scientific validity of Wakefield's study is compromised?

11. List the failures of the research and research publication processes documented in this paper.

12. Look up the *Lancet*'s retraction of Wakefield's paper. Do you think it is adequate?

13. Why is peer review a blinded process on both sides? Do you think there might be occasions when the peer reviewers should be identified?

seventeen

Nancy F. Olivieri, Gary M. Brittenham,
Christine E. McLaren, Douglas M. Templeton,
Ross G. Cameron, Robert A. McClelland,
Alastair D. Burt, and Kenneth A. Fleming

Long-Term Safety and Effectiveness of Iron-Chelation Therapy with Deferiprone for Thalassemia Major

This scientific article from The New England Journal of Medicine *examines the effect of the drug deferiprone on patients with hemolytic anemia, a condition characterized by the unusually rapid destruction of red blood cells. Olivieri's work on deferiprone became a subject of controversy when the pharmaceutical company Apotex, which had provided funding for the study, attempted to prevent Olivieri from publishing the results; the selections following this article address the controversy.*

℮

Abstract

Background

Deferiprone is an orally active iron-chelation agent[a] that is being evaluated as a treatment for iron overload in thalassemia major.[b] Studies in an animal model showed that prolonged treatment is associated with a decline in the effectiveness of deferiprone and exacerbation of hepatic fibrosis.[c]

Methods

Hepatic iron stores were determined yearly by chemical analysis of liver-biopsy specimens, magnetic susceptometry,[d] or both. Three hepatopatholo-

---

a   *iron-chelation agent*   Drug that binds with excess iron in the body.
b   *thalassemia major*   Type of genetically inherited form of hemolytic anemia characterized by abnormalities in the red blood cells.
c   *hepatic* Relating to the liver; *fibrosis* Accumulation of scar tissue.
d   *magnetic susceptometry* Measurement of magnetic properties using induced voltage.

gists[a] who were unaware of the patients' clinical status, the time at which the specimens were obtained, and the iron content of the specimens examined 72 biopsy specimens from 19 patients treated with deferiprone for more than one year. For comparison, 48 liver-biopsy specimens obtained from 20 patients treated with parenteral deferoxamine[b] for more than one year were similarly reviewed.

Results

Of the 19 patients treated with deferiprone, 18 had received the drug continuously for a mean (±SE[c]) of 4.6±0.3 years. At the final analysis, 7 of the 18 had hepatic iron concentrations of at least 80 μmol[d] per gram of liver, wet weight (the value above which there is an increased risk of cardiac disease and early death in patients with thalassemia major). Of 19 patients in whom multiple biopsies were performed over a period of more than one year, 14 could be evaluated for progression of hepatic fibrosis; of the 20 deferoxamine-treated patients, 12 could be evaluated for progression. Five deferiprone-treated patients had progression of fibrosis, as compared with none of those given deferoxamine (P=0.04).[e] By the life-table[f] method, we estimated that the median time to progression of fibrosis was 3.2 years in deferiprone-treated patients. After adjustment for the initial hepatic iron concentration, the estimated odds of progression of fibrosis increased by a factor of 5.8 (95 percent confidence interval, 1.1 to 29.6) with each additional year of deferiprone treatment.

Conclusions

Deferiprone does not adequately control body iron burden in patients with thalassemia and may worsen hepatic fibrosis.

---

a *hepatopathologists* Clinical scientists and doctors who specialize in diseases of the liver.

b *parenteral* Introduced into the body through a route other than the mouth, usually through infusion or injection; *deferoxamine* At the time of the study, this drug was the only iron-chelation agent to be approved for clinical use.

c *SE* Standard error.

d *μmol* Micromoles, a unit of measure used in chemistry and related disciplines.

e *P* Probability, here an indication of the likelihood of getting the same experimental results as the ones observed if there were no relationship between the variables being studied.

f *life-table* Table that shows the probability of mortality and other health-related occurrences in a given year based on age and other factors.

ARTICLE

Transfusions[a] and iron-chelation therapy have dramatically improved the lives of patients with thalassemia major.[1] Transfusions can prevent death and promote normal development, but the iron in the transfused red cells accumulates and eventually damages the liver, heart, and other organs. Deferoxamine, the only iron-chelating agent approved for clinical use, prolongs survival and ameliorates iron-induced organ damage.[2,3] Unfortunately, to be effective, treatment with deferoxamine requires prolonged parenteral infusion, making compliance difficult. Considerable effort has been devoted to finding alternative treatments.[4] One candidate, deferiprone (1,2-dimethyl-3-hydroxypyridin-4-one, or L1), was initially evaluated with an indirect indicator of therapeutic effectiveness – the serum ferritin[b] concentration.[5-11] Subsequently, a direct quantitative assessment of body iron burden demonstrated a favorable effect of deferiprone on iron balance.[12] Recognized adverse effects of deferiprone include embryotoxicity, teratogenicity, arthritis, severe neutropenia, and agranulocytosis.[c] Recently, concern about long-term treatment with deferiprone was aroused by studies of an animal model of iron overload[13] in which long-term treatment with a closely related compound, 1,2-diethyl-3-hydroxypyridin-4-one, was associated with a loss of effectiveness and an exacerbation of hepatic fibrosis.

To determine whether the effects of deferiprone are sustained during long-term therapy, we measured hepatic iron during continued treatment of patients in whom body iron had been measured during short-term therapy.[12] To assess whether long-term therapy was associated with progression of hepatic fibrosis, a panel of hepatopathologists evaluated the liver-biopsy specimens obtained during this trial.

METHODS

Patients

Of 21 previously studied patients who received deferiprone for a mean (±SE) of 3.1±0.3 years,[12] 19 continued to receive deferiprone at a dose of 75 mg per kilogram of body weight per day, while undergoing repeated biopsies for hepatic iron measurements. Long-term effectiveness could be evaluated in 18

---

a *Transfusions* Additions of blood or blood components into the circulatory system.

b *ferritin* Protein indicative of the total amount of iron stored in the body.

c *embryotoxicity* Toxic effects on an embryo; *teratogenicity* Causation of problems in embryonic development; *neutropenia* Abnormal low levels of neutrophils, a type of white blood cell; *agranulocytosis* Abnormal low levels of granulocytes, another type of white blood cell.

patients, who had received the drug continuously for 4.6±0.3 years; 1 patient had stopped taking deferiprone shortly after the previous analysis, but the results of biopsies up to the discontinuation of therapy are included in the histologic[a] analysis. Beginning in year 3 of the seven-year study, hepatic iron was measured by magnetic susceptometry in vivo.[b] [14] Because this technique does not provide histologic information, the follow-up period for effectiveness (range, 2 to 7 years) was longer than that for histologic analysis (range, 2 to 6 years).

Patients received regular transfusions. The objective of transfusion was to maintain the hemoglobin[c] concentration above 9.5 g per deciliter. From November 1989 to November 1993, deferiprone was synthesized at the University of Toronto and encapsulated by NovaPharm Pharmaceuticals (Toronto). After November 1993, Apotex (Weston, Ont., Canada) supplied deferiprone tablets. The equivalence of the two formulations was not evaluated.

5   Body iron was evaluated in tissue obtained at biopsy, as described previously, and by magnetic susceptometry (Biomagnetic Technologies, San Diego, Calif.).[14,15] The magnetic measurements have been validated previously.[14] We converted the concentration of iron in dried samples to a wet weight, assuming a liver water content of 70 percent; chemical and magnetic values were used interchangeably.[12]

In the original trial design, each biopsy specimen, obtained primarily to monitor therapeutic effectiveness, was histologically reviewed, but serial biopsy specimens were not prospectively compared. Concern about hepatotoxicity[13,16] prompted a retrospective review of these results and of those from a comparison group of 20 deferoxamine-treated patients. The comparison group included all patients eight years of age or older for whom the results of two or more biopsies performed one year apart during continuous deferoxamine treatment were available.

EFFICACY MONITORING

A hepatic iron concentration of less than 80 μmol per gram of liver, wet weight, was considered to indicate effective iron-chelation therapy, and a concentration of 80 μmol or more per gram of liver, wet weight, was considered to indicate ineffective therapy.[12] These criteria, derived from a long-term trial in deferoxamine-treated patients,[17] were used to evaluate the short-term effectiveness of deferiprone in our previous study[12] and were applied in an

---

a   *histologic*   Relating to the microscopic structure of tissues.

b   *in vivo*   Latin: in the living; i.e., in live subjects.

c   *hemoglobin*   Iron-containing protein in red blood cells that carries oxygen.

identical manner in the present long-term study. Similarly, a serum ferritin concentration of less than 2500 µg per liter was considered to indicate effective iron-chelation therapy, and higher values ineffective therapy.[12,18]

HISTOLOGIC EVALUATION

An independent initial review of the biopsy specimens was carried out by the two study investigators who are hepatopathologists and a consultant. Before this review, the 72 biopsy slides were randomly arranged and each slide was assigned a unique number. Each pathologist graded the findings according to the system summarized in Table 1.[19] Each was unaware of the patients' clinical status, the date each sample was obtained, and the hepatic iron content of each biopsy specimen. After the completion of the initial review, the two study investigators conducted a consensus review, in which all biopsy specimens were examined jointly, after standards regarding sample adequacy and definitions of progression and regression of fibrosis had been agreed on. The results of this evaluation were subsequently reviewed with the consultant, and a final decision was made with regard to the adequacy and stage of each biopsy specimen.

TABLE 1.
System of Evaluating Hepatic-Biopsy Specimens for Architectural Changes, Fibrosis, and Cirrhosis*

Finding	Score[†]
No fibrosis	0
Fibrous expansion of some portal areas, with or without short fibrous septa	1
Fibrous expansion of most portal areas, with or without short fibrous septa	2
Fibrous expansion of most portal areas, with occasional portal-to-portal bridging	3
Fibrous expansion of portal areas, with marked portal-to-portal bridging as well as portal-to-central bridging	4
Marked portal-to-portal bridging as well as portal-to-central bridging, with occasional nodules (incomplete cirrhosis)	5
Probable or definite cirrhosis	6

* The criteria were obtained, with modifications, from Ishak et al.[19] The following additional features were noted but not scored: intra-acinar fibrosis, perivenular ("chicken-wire") fibrosis, and phlebosclerosis of terminal hepatic venules.

[†]The maximal score is 6.

Clinically significant progression of fibrosis was considered to have occurred if there was a change in the histologic score from 0 (no fibrosis) to 1 or greater, from 1 or 2 to 3 or greater, or from 3 or 4 to 5 or 6. Changes in the score from 1 to 2 or from 3 to 4 were not considered clinically significant. Similarly, a change in the score from 5 (incomplete cirrhosis) to 6 (probable or definite cirrhosis) was not considered clinically significant; hence, patients whose initial biopsy specimen showed cirrhosis could not be evaluated for progression of fibrosis. A biopsy specimen was considered adequate if two or more portal tracts[a] were present. This was considered the absolute minimum necessary for the assessment of fibrosis and cirrhosis, since the ability to identify these processes varies considerably depending on the size of the biopsy specimen.

MONITORING OF TOXICITY AND COMPLIANCE

10 Other types of safety monitoring in this study have been described previously.[12] Sexually active patients were asked to use reliable methods of contraception. We assessed compliance by monitoring the frequency with which pill bottles were opened, using bottles with microprocessors in the caps.[12,20]

STATISTICAL ANALYSIS

Data are presented as means ±SE. Medians and ranges are given for continuous variables, and proportions are given for dichotomous variables.[b] Pretreatment variables were compared between treatment groups by the Mann-Whitney test for continuous variables and by the Fisher-Irwin exact test[c] for dichotomous variables.[21] The Wilcoxon signed-rank test[d] was used to compare pretreatment and post-treatment values for continuous variables and to assess whether there was a change in compliance during the last two years of deferiprone therapy. The Kaplan-Meier product-limit method[e] was used to estimate the probability

---

a *portal tracts*   Cellular arrangements in the liver that conduct blood and other fluids.

b *continuous variables* Variables that can have any value within a given numerical range; *dichotomous variables* Variables with two potential values, used to separate information into two groups.

c *Mann-Whitney test* Mathematical process that can be used to determine if the difference in data between two groups is statistically significant; *Fisher-Irwin exact test* Mathematical process that can be used to determine if there is a statistically significant relationship between two dichotomous variables.

d *Wilcoxon signed-rank test* Mathematical process that can be used to compare two groups of data.

e *Kaplan-Meier product-limit method* Mathematical process used to create a "survival curve," which shows the probability of a system breakdown (such as death or, in this case, progression of fibrosis) over a period of time.

that each patient would not have progression of fibrosis for a specified period. The log-rank test[a] was used to compare differences in the length of time to the progression of fibrosis in the treatment groups.[22] Because the only patients with progression of fibrosis were in the deferiprone group, it was not possible to estimate the risk or odds of progression to fibrosis on the basis of the type of chelating therapy, the dichotomous predictor variable. Thus, multivariate logistic-regression models[b] were formed to examine the relation between the dependent variable, progression of hepatic fibrosis, and predictor variables, including the duration of deferiprone therapy, age at initial biopsy, sex, the presence of antibody to hepatitis C virus, initial hepatic iron concentration, and the amount of blood transfused.[23] Stepwise analysis[c] and an analysis of all possible subgroups were performed to choose the most parsimonious[d] model with statistically significant predictors. All tests were two-tailed;[e] a P value of 0.05 was considered to indicate statistical significance. The BMDP (BMDP Statistical Software, Los Angeles) and S-PLUS (Statistical Sciences, version 3.3 for Windows, Seattle) statistical computer packages were used for computations.

The study was approved by the human subjects committee of the Hospital for Sick Children, Toronto, and the Health Protection Branch of Health Canada. Written informed consent was obtained from each patient or the patients' parents.

RESULTS

Effectiveness of Deferiprone

Among the 18 patients in whom the effectiveness of deferiprone could be evaluated, the mean ($\pm$SE) hepatic iron concentration decreased from 88.7$\pm$12.1 to 65.5$\pm$7.9 µmol per gram of liver, wet weight (normal value, about 1.6), after a mean of 4.6$\pm$0.3 years of therapy (range, 2 to 7); this decrease of 23.2$\pm$10.9 µmol of iron per gram of liver, wet weight, was not significant (P=0.07). Initial and final hepatic iron concentrations are shown in Figure 1. In seven patients, hepatic iron concentrations at the end of treatment met or exceeded the threshold value of 80 µmol per gram of liver, wet weight, which is associated with an increased risk of cardiac disease and early death.[17]

---

a *log-rank test* Mathematical process used to compare two survival curves.

b *multivariate logistic-regression models* Mathematical models showing the probability relationship between a set of variables and one dichotomous variable.

c *Stepwise analysis* Mathematical process used to look for a relationship between one dependent variable and more than one independent variable by incorporating variables one by one in an optimal order.

d *most parsimonious* I.e., simplest possible.

e *two-tailed* Looking for both significantly high and significantly low results.

The serum ferritin concentration decreased from 4455±841 μg per liter at the beginning of treatment to 2831±491 μg per liter at the end of treatment. Expressed logarithmically, this decrease was significant (P=0.03). In nine of the patients, the serum ferritin concentration exceeded 2500 μg per liter, the threshold used to distinguish effective from ineffective chelation therapy.[12,18]

Data on compliance were available for all 18 patients for the last two full years of treatment. The median rate of compliance each year was 98 percent, with ranges of 90 to 100 percent for the penultimate year and 87 to 100 percent for the final year (P=0.70).

Histologic Analysis

15 Of the 72 biopsy specimens available from the patients treated with deferiprone, 17 (24 percent) were judged inadequate for evaluation. Histologic changes could not be evaluated in five patients: two did not have two adequate biopsy specimens that had been obtained at least one year apart, and three had

FIGURE 1.

Initial and Final Hepatic Iron Concentrations in 18 Patients with Thalassemia Major Treated with Deferiprone

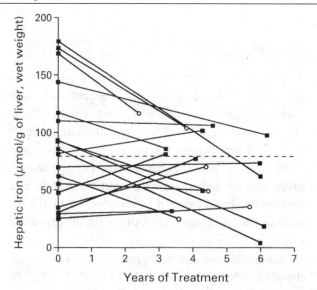

The dashed line indicates the value of 80 μmol of iron per gram of liver tissue, wet weight, above which there is an increased risk of cardiac disease and early death due to iron loading during long-term treatment with deferoxamine.[17] Squares indicate concentrations determined by liver biopsy, and circles concentrations determined by magnetic susceptometry.

cirrhosis at the initial evaluation. Thus, a total of 55 biopsy specimens from 14 patients were examined (Table 2).

TABLE 2.
Characteristics of the 14 Deferiprone-Treated Patients and the 12 Deferoxamine-Treated Patients in Whom Progression of Hepatic Fibrosis Could Be Evaluated

Characteristic	Deferiprone Group (N=14)	Deferoxamine Group (N=12)	P Value
Progression of hepatic fibrosis – no. (%)	5(36)	0	0.04
Age at initial biopsy – yr			
Median	18.2	13.9	0.1
Range	10.5-23.7	8.7-31.5	
Male sex – no. (%)	3 (21)	6 (50)	0.2
Antibody to hepatitis C – no (%)	6 (43)	5 (42)	1.0
Initial hepatic iron concentration – μmol/g of liver, wet weight			
Median	80.9	35.2	0.01
Range	24.2-180.1	10.8-226.4	
Duration of treatment – yr*			
Median	2.3	3.2	0.4
Range	1.3-4.0	1.3-4.3	
Amount of blood transfused – ml packed cells/kg/yr			
Median	77	82	0.5
Range	56-114	57-109	

*The period under consideration is the interval between the initial biopsy and the final biopsy. Data were available for 13 patients in the deferiprone group and 10 in the deferoxamine group.

Five patients treated with deferiprone had evidence of progression of fibrosis. The estimated median time to progression was 3.2 years. Figure 2 shows the initial and final histologic stages, hepatic iron concentrations, and duration of therapy for these five patients. In four patients, hepatic iron concentrations stabilized during therapy; in the fifth, the iron concentration decreased substantially. Figure 3 shows representative photomicrographs[a] of liver specimens from the initial and final biopsies in these patients. In one other patient, there was an improvement in the histologic stage over a period of 3.5 years.

---

a *photomicrographs*   Photographs or digital images taken through a microscope or magnifying lens.

FIGURE 2.
Changes in histologic findings and hepatic iron concentrations in the five Deferiprone-treated patients with progression of hepatic fibrosis. The worst possible fibrosis score is 6; the staging system is described in Table 1.

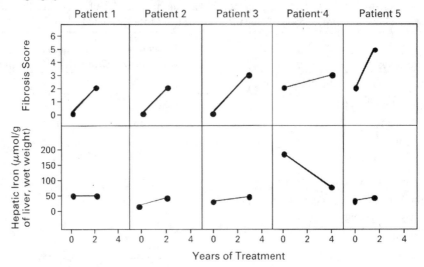

FIGURE 3.
Changes in histologic findings in the five Deferiprone-treated patients with progression of hepatic fibrosis. The top panels show the initial biopsy specimens, and the bottom panels the last biopsy specimens that could be evaluated (Masson's trichrome, x100).[a]

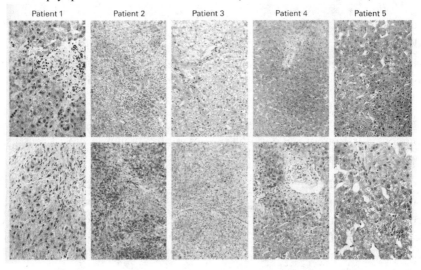

a *Masson's trichrome* Method for coloring microscopic images of tissue; *x100* Viewed at 100 times normal size.

TABLE 3.

Characteristics of the Patients with Progression of Hepatic Fibrosis during Treatment with Deferiprone and Those without Progression

Characteristic	Progression of Hepatic Fibrosis (N=5)	No Progression of Hepatic Fibrosis (N=9)	P Value
Age at initial biopsy – yr			
Median	21	16	0.03
Range	18-24	10-22	
Male sex – no. (%)	2 (40)	1 (11)	0.50
Antibody to hepatitis C – no (%)	4 (80)	2 (22)	0.09
Initial hepatic iron concentration – μmol/g of liver, wet weight			
Median	29.6	93.6	0.07
Range	24.2-180.1	55.4-174.2	
Duration of treatment – yr*			
Median	2.1	2.4	0.80
Range	1.6-4.0	1.3-3.4	
Amount of blood transfused – ml packed cells/kg/yr			
Median	77	77	0.60
Range	56-89	64-114	

*The period under consideration is the interval between the initial biopsy and the final biopsy. Data were available for five patients with progression of hepatic fibrosis and eight with no progression.

Table 3 shows that deferiprone-treated patients with progression of fibrosis were older than those without progression (P=0.03). There were no other significant differences (sex, prevalence of hepatitis C infection, initial hepatic iron concentration, or the duration of therapy or amount of blood transfused between the initial biopsy and the final biopsy) between the two groups.

Of the 48 biopsy specimens available from the 20 patients treated with deferoxamine, 8 (17 percent) were judged inadequate for evaluation. Histologic changes could not be evaluated in eight patients: six did not have two adequate biopsy specimens that had been obtained at least one year apart, and two had cirrhosis at the initial evaluation. Thus, a total of 31 biopsy specimens from 12 patients were examined (Table 2). None of the specimens showed evidence of progression of fibrosis. In one patient, there was an improvement in the histologic stage over a two-year period.

After adjustment for initial hepatic iron concentrations, multivariate logistic-regression analysis[a] showed that the estimated odds of progression to fibrosis increased by a factor of 5.8 (95 percent confidence interval, 1.1 to 29.6) with each additional year of deferiprone treatment. The deferiprone-treated patients had a significantly higher mean initial hepatic iron concentration than the deferoxamine-treated patients (P=0.01) (Table 2), but no significant differences between the two groups were identified with respect to age at initial biopsy, sex, prevalence of hepatitis C infection, or the duration of therapy or amount of blood transfused between the initial biopsy and the final biopsy.

Other Adverse Effects

20 Deferiprone therapy was not associated with clinically significant hematologic[b] changes, as evidenced by regular blood counts. No characteristic abnormalities in liver function were observed, although many patients had the small elevations in serum aminotransferase[c] concentrations that are commonly found during iron overload. In no patient did heart disease requiring medication develop during the study.

DISCUSSION

These results indicate that deferiprone is not an effective means of iron-chelation therapy in patients with thalassemia major and may be associated with worsening of hepatic fibrosis, even in patients whose hepatic iron concentrations have stabilized or decreased. After a mean of 4.6 years of deferiprone therapy, body iron burden was at concentrations associated with a greatly increased risk of cardiac disease and early death[17] in 7 of 18 patients (39 percent). Other investigators have recently reported that hepatic iron exceeded this threshold in 58 percent of patients who were treated with deferiprone for one to four years.[24] In our patients, differences in objectively determined rates of compliance or the rate of iron loading could not account for the lack of effectiveness of deferiprone.

The results of our review of liver-biopsy specimens suggest that extended deferiprone therapy may be associated with a worsening of hepatic fibrosis. Fibrosis progressed in 5 of the 14 patients (36 percent) in whom it could be evaluated, despite the stabilization of or a marked reduction in hepatic iron concentrations in all 5 patients. The estimated median time to progression of

---

a *multivariate logistic-regression analysis* Statistical method showing the probability relationships between multiple variables.

b *hematologic* Related to the blood.

c *aminotransferase* Enzyme involved in the production of some amino acids.

fibrosis was 3.2 years, and after adjustment for initial hepatic iron concentrations, the estimated odds of progression of fibrosis increased by a factor of 5.8 (95 percent confidence interval, 1.1 to 29.6) with each additional year of deferiprone treatment.

The worsening of hepatic fibrosis in deferiprone-treated patients is in contrast to the arrest of fibrosis regularly observed with deferoxamine therapy. In a seminal study, long-term therapy with deferoxamine halted the progression of hepatic fibrosis in patients with thalassemia major.[25-27] In these patients, progression was arrested despite a regimen of deferoxamine (0.5 g per day intramuscularly, six days per week)[25] now considered suboptimal because it merely stabilizes, rather than reduces, hepatic iron concentrations. Moreover, fibrosis was halted despite a final mean hepatic iron concentration (139 μmol per gram of liver, wet weight)[25] that was more than twice that in our deferiprone-treated patients with progression of fibrosis (49.5 μmol per gram of liver, wet weight). Subsequent studies[28] and the results in our comparison group of deferoxamine-treated patients confirm that modern regimens of parenteral deferoxamine arrest fibrosis.

Our findings virtually recapitulate those in a gerbil model of iron overload in which administration of a closely related drug (1,2-diethyl-3-hydroxypyridin-4-one) was associated with initial loss of efficacy, worsening of hepatic fibrosis, and cardiac fibrosis.[13] There has been concern that hydroxypyridinones may exacerbate iron-related tissue damage.[29-31] Deferiprone is a bidentate[a] chelator, and three molecules are needed to occupy the six coordination sites of a single atom of iron. In contrast, one molecule of the hexadentate[b] deferoxamine binds a single atom of iron; the chelate[c] (ferrioxamine) is virtually inert biologically. At low concentrations of deferiprone relative to the concentrations of available iron, partially bound forms of iron (bound to only one or two molecules of deferiprone) appear in which the unoccupied coordination sites remain reactive and able to catalyze the formation of hydroxyl radical or other reactive oxygen species.[d] [32] There is increasing evidence to suggest that these reactive oxygen species are involved in the pathogenesis of hepatic fibrosis. Recently, the potential cellular toxicity of deferiprone has been shown in erythrocytes[33] and cultured myocytes.[e] [34]

---

a *bidentate* Having two atoms that can bind to a metal.
b *hexadentate* Having six atoms that can bind to a metal.
c *chelate* Substance produced by the binding of a chelation agent with a metal.
d *reactive oxygen species* Types of molecules containing oxygen that are normally present in the body but can cause cell damage at high concentrations.
e *erythrocytes* Red blood cells; *myocytes* Muscle cells.

25   The limitations of our histopathological analysis should be emphasized: our analysis was retrospective, the number of patients studied was small, and the patients treated with deferoxamine do not constitute a true control population. Furthermore, histologic assessment was based on relatively few biopsy specimens, although this was true in both the deferiprone group and the deferoxamine group. Because this study was observational rather than randomized and the effect of hepatic fibrosis was not confirmed by challenge after discontinuation of the drug, the relation between deferiprone and fibrosis cannot be considered definite or proved. Nonetheless, we could identify no other causes of the accelerated fibrosis.

The patients with progression of fibrosis did not differ significantly from those without progression with respect to sex, prevalence of hepatitis C[a] infection, initial hepatic iron concentrations, duration of therapy, or the rate of iron accumulation. The consensus of the pathologists was that there was no difference between groups in the type of inflammatory changes. Nonetheless, we cannot rule out the possibility of an interaction between deferiprone and hepatitis C infection. The patients with progression of fibrosis were older than those without progression, and the likelihood of progression of fibrosis was greater in patients with lower hepatic iron concentrations. Because we are unable to identify definite risk factors for accelerated fibrosis, we have discontinued deferiprone therapy in all patients, including those who are unable or unwilling to use deferoxamine in standard regimens.

Despite their limitations, the results of our analysis, together with theoretical considerations and findings in animal studies, indicate that deferiprone may worsen hepatic fibrosis. Before it can be considered for clinical use, even in patients who are unwilling or unable to use deferoxamine in standard regimens, prospective clinical trials are mandatory to evaluate the possibility of irreversible hepatic damage.

(1998)

Supported in part by research grants from the Medical Research Council of Canada, the Ontario Heart and Stroke Foundation, the Ontario Thalassemia Foundation, the Cooley's Anemia Foundation, Apotex, and the National Institutes of Health (DK49108, HL58203, and HL61219). Dr. Olivieri is supported in part by a Scientist Award from the Medical Research Council of Canada.

We are indebted to Professor Peter Scheuer, London, for his independent analysis of the liver-biopsy slides; to David Nathan, David Weatherall, Michael Baker, Antonio Cao, Helen Chan, John Dick, Peter Durie, Brenda Gallie, Marc Giacomelli, John Harris, David Kern, F.A. Olivieri, Robert Phillips, Eliot Phillipson, Sergio Piomelli, Alvin Zipursky, and Stanley Zlotkin for ongoing encouragement, advice, and support; and to Naomi Klein, Maria Muraca, Allyson Muroff, and Helen Schinkel for assistance in data management.

a   *hepatitis C*   Virus that can cause hepatic fibrosis.

REFERENCES

1. Cooley's Anemia Progress Review Committee. Cooley's anemia: progress in biology and medicine. Bethesda, Md.: Division of Blood Diseases and Resources, 1995.
2. Fosburg MT, Nathan DG. Treatment of Cooley's anemia. Blood 1990;76:435-44.
3. Olivieri NF, Brittenham GM. Iron-chelating therapy and the treatment of thalassemia. Blood 1997;89:739-61. [Erratum, Blood 1997;89:2621.]
4. Bergeron RJ, Brittenham GM, eds. Development of iron chelators for clinical use. Boca Raton, Fla.: CRC Press, 1994.
5. Kontoghiorghes GJ, Aldouri MA, Sheppard LN, Hoffbrand AV. 1,2- dimethyl-3-hydroxypyrid-4-one, an orally active chelator for treatment of iron overload. Lancet 1987;1:1294-5.
6. Kontoghiorghes GJ, Bartlett AN, Hoffbrand AV, et al. Long-term trial with the oral iron chelator 1,2-dimethyl-3-hydroxypyrid-4-one (L1). I. Iron chelation and metabolic studies. Br J Haematol 1990;76:295- 300.
7. al-Refaie FN, Wonke B, Hoffbrand AV, Wickens DG, Nortey P, Kontoghiorghes GJ. Efficacy and possible adverse effects of the oral iron chelator 1,2-dimethyl-3-hydroxypyrid-4-one (L1) in thalassemia major. Blood 1992;80:593-9.
8. al-Refaie FN, Hershko C, Hoffbrand AV, et al. Results of long-term deferiprone (L1) therapy: a report by the International Study Group on Oral Iron Chelators. Br J Haematol 1995;91:224-9.
9. Olivieri NF, Koren G, Hermann C, et al. Comparison of oral iron chelator L1 and desferrioxamine in iron-loaded patients. Lancet 1990;336:1275-9.
10. Tondury P, Kontoghiorghes GJ, Ridolfi-Luthy AR, et al. L1 (1,2-dimethyl-3-hydroxypyrid-4-one) for oral iron chelation in patients with beta-thalassaemia major. Br J Haematol 1990;76:550-3.
11. Agarwal MB, Gupte SS, Viswanathan C, et al. Long-term assessment of efficacy and safety of L1, an oral iron chelator, in transfusion dependent thalassaemia: Indian trial. Br J Haematol 1992;82:460-6.
12. Olivieri NF, Brittenham GM, Matsui D, et al. Iron-chelation therapy with oral deferiprone in patients with thalassemia major. N Engl J Med 1995;332:918-22.
13. Carthew P, Smith AG, Hider RC, Dorman B, Edwards RE, Francis JE. Potentiation of iron accumulation in cardiac myocytes during the treatment of iron overload in gerbils with the hydroxypyridinone iron chelator CP94. Biometals 1994;7:267-71.
14. Brittenham GM, Farrell DE, Harris JW, et al. Magnetic-susceptibility measurement of human iron stores. N Engl J Med 1982;307:1671-5.
15. Overmoyer BA, McLaren CE, Brittenham GM. Uniformity of liver density and nonheme (storage) iron distribution. Arch Pathol Lab Med 1987;111:549-54.
16. Wong A, Alder V, Robertson D, et al. Liver iron depletion and toxicity of the iron chelator deferiprone (L1, CP20) in the guinea pig. Biometals 1997;10:247-56.
17. Brittenham GM, Griffith PM, Nienhuis AW, et al. Efficacy of deferoxamine in preventing complications of iron overload in patients with thalassemia major. N Engl J Med 1994;331:567-73.
18. Olivieri NF, Nathan DG, MacMillan JH, et al. Survival in medically treated patients with homozygous b-thalassemia. N Engl J Med 1994;331:574-8.
19. Ishak K, Baptista A, Bianchi L, et al. Histological grading and staging of chronic hepatitis. J Hepatol 1995;22:696-9.
20. Cramer JA, Mattson RH, Prevey ML, Scheyer RD, Ouellette VL. How often is medication taken as prescribed? A novel assessment technique. JAMA 1989;261:3273-7. [Erratum, JAMA 1989;262:1472.]
21. Fleiss JL. Statistical methods for rates and proportions. 2nd ed. New York: John Wiley, 1981.

22. Kaplan EL, Meier P. Nonparametric estimation from incomplete observations. J Am Stat Assoc 1958;53:457-81.

23. Peto R, Pike MC, Armitage P, et al. Design and analysis of randomized clinical trials requiring prolonged observation of each patient. II. Analysis and examples. Br J Cancer 1977;35:1-39.

24. Hoffbrand AV, Al-Refaie F, Davis B, et al. Long-term trial of deferiprone in 51 transfusion-dependent iron overloaded patients. Blood 1998;91:295-300.

25. Barry M, Flynn DM, Letsky EA, Risdon RA. Long-term chelation therapy in thalassaemia major: effect on liver iron concentration, liver histology, and clinical progress. BMJ 1974;2:16-20.

26. Risdon RA, Flynn DM, Barry M. The relation between liver iron concentration and liver damage in transfusional iron overload in thalassaemia and the effect of chelation therapy. Gut 1973;14:421.

27. Risdon RA, Barry M, Flynn DM. Transfusional iron overload: the relationship between tissue iron concentration and hepatic fibrosis in thalassaemia. J Pathol 1975;116:83-95.

28. Aldouri MA, Wonke B, Hoffbrand AV, et al. Iron state and hepatic disease in patients with thalassaemia major, treated with long term subcutaneous desferrioxamine. J Clin Pathol 1987;40:1353-9.

29. Halliwell B. Drug antioxidant effects: a basis for drug selection? Drugs 1991;42:569-605.

30. Halliwell B. Iron, oxidative damage, and chelating agents. In: Bergeron RJ, Brittenham GM, eds. Development of iron chelators for clinical use. Boca Raton, Fla.: CRC Press, 1994:33-56.

31. Nathan DG. An orally active iron chelator. N Engl J Med 1995;332:953-4. [Erratum, N Engl J Med 1995;332:1315.]

32. Motekitis RJ, Martell AE. Stabilization of the iron (III) chelates of 1,2-dimethyl-3-hydroxypyrid-4-ones and related ligands. Inorg Chim Acta 1991;183:71-80.

33. Cragg L, Hebbel RP, Solovey A, Miller WJ, Enright H. The iron chelator L1 potentiates iron-mediated oxidative DNA damage. Blood 1996; 88:Suppl 1:646a. abstract.

34. Hershko C, Link G, Pinson A, Konijn AM. Deferiprone (L1) fails to mobilize iron and promotes iron cardiotoxicity at suboptimal L1/iron concentrations. Blood 1997;90:Suppl 1:11a. abstract.

Questions:

1. Does the article provide a rationale for the investigation? If so, where?

2. Most human clinical drug trials in the past have used short study periods. What dangers might be associated with this approach?

3. In recent years, post-market reporting of drug side-effects has been put into place. Do you know which federal body you should report such side-effects to? If so, what is it? If not, how would you go about finding out?

4. This study was published in the face of significant resistance from the Canadian patent-holder for deferiprone. Does anything in the tone of the article suggest that this is the case? If so, what? If not, why do you think the authors chose this particular tone?

5. What does it mean that "chemical and magnetic values were used interchangeably"? What does this type of phrasing suggest about the intended audience of the paper?

6. Which of the illustrations do you find assist in understanding the study's findings? Why would pictures be included?

7. Report in clear language what the following phrase means: "Because the only patients with progression of fibrosis were in the deferiprone group, it was not possible to estimate the risk or odds of progression to fibrosis on the basis of the type of chelating therapy, the dichotomous predictor variable."

8. The article's last paragraph is fairly strongly worded for a scientific study. Rewrite it, maintaining the strength of the wording but using language and tone appropriate to either a news article or a blog post.

9. Studies such as this one that show negative results (meaning they do not, for example, support the use of a new treatment) are less likely to be published. What are some of the problems associated with this?

10. Does a researcher have a responsibility to reveal negative study findings (findings that suggest a treatment is not working) to a study participant while the study is in progress? Why or why not? What safeguards should be provided to study participants?

11. Many studies in the past have refused to enroll female participants. What putative arguments were given for the exclusion? How does this study address one of those arguments?

12. Are the sub-headings given in the abstract necessary for clarity? Rate the clarity of the abstract.

eighteen

DAVID G. NATHAN AND
DAVID J. WEATHERALL

Academic Freedom in
Clinical Research

This editorial article, originally published in The New England Journal
of Medicine, *examines the conflict between Apotex, a Canadian drug
manufacturer, and Nancy Olivieri, a clinical researcher. From its
consideration of the controversy, the article draws lessons regarding
the relationship between scientists and industry.*

❦

"Is the university–industrial complex out of control?" The editorial
that appeared under this eye-catching title in *Nature* in January 2001
came to the conclusion that links between academia and industry are of
increasing concern to academics and to society at large and that the sectors
involved must review and revise their policies in order to sustain the public
accountability and academic freedom of universities.[1] It ended with a list
of New Year's resolutions for actions that would be required to maintain
public trust in higher education and publicly funded research. But with the
notable exception of the decision of many leading medical and scientific
journals to tighten up regulations regarding conflicts of interest among their
authors,[2] these resolutions seem to have gone the way of most New Year's
resolutions.

A recently published independent review sponsored by the Canadian Association of University Teachers (CAUT) of a long-standing dispute involving Nancy Olivieri, a clinical researcher at the Hospital for Sick Children
(Toronto) and the University of Toronto, provides an impetus to revisit this
issue.[3] Since the remarkable advances in the biomedical sciences stemming
from the human genome project are likely to reach their potential for clinical
application only through an increasingly effective partnership between academia and industry that preserves academic freedom, ensuring such freedom
is a matter of great urgency.

THE DISPUTE

The Olivieri case involves the search for a safe and effective orally active iron chelator.[a] This quest has dominated clinical research on thalassemia[b] and other conditions characterized by iron overload for decades.[4,5] Of many compounds that have been considered, only one, a hydroxypyridin-4-one with the generic name deferiprone, has entered clinical trials. The events surrounding clinical trials of this drug at the Hospital for Sick Children and the University of Toronto represent a modern nadir in the relations among academic investigators, their institutions, and the pharmaceutical industry.

Deferiprone, a bivalent iron chelator, was initially synthesized by Robert Hider and his colleagues.[6] It was briefly licensed to Ciba–Geigy (now Novartis) but was abandoned by the company in 1993 because of its low therapeutic index[c] in animals without iron overload,[7] its poor stoichiometry[d] (three molecules of drug are required for binding of each iron atom[7]), and its rapid removal from the circulation.[8,9] The efficacy of the drug is therefore critically dependent on its poorly maintained concentration in body fluids.

Deferiprone was first investigated in uncontrolled clinical trials by a group at the Royal Free Hospital in London.[10] After reading the report of these trials, Olivieri and her colleagues produced enough of the drug to initiate clinical studies. Encouraged by her preliminary results, she and Gideon Koren, a former colleague of hers at the Hospital for Sick Children, entered a collaboration with Apotex, a Canadian manufacturer of generic drugs. Apotex produced the drug, and Olivieri, who is the director of the largest hemoglobinopathy[e] clinic in North America, began a relatively short-term, uncontrolled clinical trial of deferiprone in patients with thalassemia who had iron overload. The trial was sponsored by a grant from Apotex to the Hospital for Sick Children. To obtain the grant, Olivieri and Koren (then the associate director of the Hospital for Sick Children Research Institute) signed a confidentiality agreement that was compatible at that time with the policy at both the University of Toronto and the hospital, and the hospital accepted the grant.

---

a   *iron chelator*   Substance used to remove excess iron from the body.
b   *thalassemia*   Genetic condition in which red blood cells are abnormal and so are destroyed unusually quickly.
c   *therapeutic index*   Ratio between the dosage necessary for effective treatment and, in animals, a fatal dosage.
d   *stoichiometry*   Ratio between substances involved in a chemical reaction.
e   *hemoglobinopathy*   Genetic problem relating to hemoglobin, the protein in red blood cells that carries oxygen.

The initial results were published in 1995 and were encouraging.[11] The drug appeared to reduce or maintain liver iron levels in patients with thalassemia who had undergone multiple transfusions. Although an editorial accompanying the report warned that much more time would be required to determine the efficacy of deferiprone,[12] there was high expectation among researchers and physicians who treat patients with thalassemia that the drug would prove useful in the management of the iron overload associated with long-term transfusion. In the process of trying to gather the data to answer this important question, Olivieri, Apotex, the Hospital for Sick Children, Koren, and the University of Toronto became embroiled in a nasty controversy.

Olivieri started a second prospective trial (for which she did not sign a confidentiality agreement) in which deferiprone treatment was compared with therapy with the standard drug, deferoxamine. Patients who take deferoxamine regularly have a steady decline in hepatic[a] iron levels. In a substantial proportion of patients in Olivieri's second study, deferiprone either failed to reduce hepatic iron levels below their starting points or actually increased them to a value that was substantially above their starting points.[13] In addition, in some of the patients who received deferiprone but in none of those who received deferoxamine (who had much lower liver iron levels, on average), increased hepatic fibrosis appeared to have developed, as judged by four independent pathologists who were blinded to the treatment-group assignments. An editorial[14] accompanying the published report stated that deferiprone might be ineffective or even toxic in some patients and that further trials were required. The question of whether deferiprone induces hepatic fibrosis remains unresolved.

When she became aware of the findings, Olivieri thought it was her responsibility to report these adverse events to her institutional review board, present them to a scientific meeting, and submit them for publication. In response, Apotex stopped all clinical trials involving Olivieri and threatened to take legal action for her violation of the confidentiality agreement that she had signed before the first trial if she released to the public the information gained in the second trial. Within a few years, two lawsuits totaling $20 million were formally lodged against her. Olivieri defied Apotex by submitting the material for publication and presenting it at a scientific meeting.[15]

CURRENT STATUS OF DEFERIPRONE

It takes years to demonstrate whether an iron chelator is clinically effective. In the case of deferoxamine, for example, more than two decades of

---

a *hepatic* Relating to the liver.

observation were required to show that the incidence of cardiac disease, the most common cause of death in patients with thalassemia, could be reduced by the drug.[16-19] As of this writing, the safety and efficacy of deferiprone have not been established; it is beyond the scope of this article to review the scientific data concerning the use of the drug in any detail. Published papers suggest that deferiprone does a poor job of removing iron from hepatic stores in a substantial proportion of treated patients.[20-23] The issue of its safety with respect to hepatic fibrosis has not been resolved. Suffice it to say, when the dispute began, Olivieri had good reason to believe that deferiprone was neither safe nor effective.

THE ETHICAL STRUGGLE

The series of events that followed the disagreement between Olivieri and Apotex and the company's attempt to prevent the investigator from publishing her results are described in the CAUT report, an exhaustively researched and annotated 540-page document.[3] The report describes years of harassment and the generation of misinformation about Olivieri, as well as providing even more worrisome accounts of a large donation that the University of Toronto was negotiating with Apotex.[3] The report makes clear that Olivieri's academic freedom to present her concerns to her peers was abridged. More important, although the Hospital for Sick Children and the University of Toronto knew that this freedom was under attack, Olivieri received harassment instead of support from the hospital and ineffectual support from the university in her legal stand against Apotex.

The CAUT report is not the only one that has been produced on this subject. About two years ago a panel sponsored by the Hospital for Sick Children found fault with Olivieri's handling of events after the termination of the trial.[24] The CAUT report provides strong evidence that the panel's conclusion was based at least in part on incorrect information that was fed to it by Koren, who, the panel knew, was in the midst of his own bitter dispute with Olivieri.[3] Subsequent to the release of the panel's report, which is now known to be flawed, the hospital's medical advisory committee inquired further into Olivieri's conduct – a process that culminated in her being referred for research misconduct to the College of Physicians and Surgeons of Ontario,[3] the medical licensing board of the province. Information about the humiliating referral was made widely available to the public by both the Hospital for Sick Children and the University of Toronto, even though release of such confidential information was in violation of the then-current policy of the university. Recently, the College of Physicians and Surgeons issued a statement concluding that none of the allegations have any basis and completely vindicating Olivieri on all scores.[25]

CONCLUSIONS

Although the Olivieri debacle is complicated by personal animosity, poor administrative judgment, and bad behavior among academic colleagues, the case report raises a number of fundamental questions about the research interface among teaching hospitals, academic clinical departments of universities, and industry. As the authors of the CAUT report state in their summary, these are issues that affect the entire biologic-research community.

What is central to this particular case is the principle of academic freedom. It does not matter whether, in the end, Olivieri was right or wrong in her assessment of the efficacy and safety of deferiprone or whether she is a pleasant or difficult colleague. What matters is that, at the time of the dispute, Olivieri was concerned enough about the safety of the drug to convey her doubts to a scientific meeting and to a peer-reviewed journal so that members of the medical community could judge the issue for themselves.

But the CAUT report also underscores the inadequacies of the University of Toronto and the Hospital for Sick Children in resolving this vexing problem at the interface among the university, industry, and the investigator. In our opinion, the hospital had an inadequate control mechanism for clinical research, and its leadership managed the case poorly. At the same time, administrators at the university were unable or unwilling to persuade the hospital or Apotex to provide more effective and stable management for what became a public crisis. In part, the university's response might have been influenced by the prospect of a large gift from Apotex, but in our opinion, most of the failure was due to excessive legalism and an unwillingness to force a high-stakes and very public confrontation with the Hospital for Sick Children or Apotex.

15 The Toronto episode probably represents an extreme case, engendered in part by deep hostilities within the ranks of the faculty at the Hospital for Sick Children and the determination of Apotex to win at any cost. We would like to believe that most universities would support faculty members who were exerting their right to academic freedom in the face of angry and disappointed industrial sponsors of trials that did not go their way. Apotex is probably an unusual company. Few pharmaceutical companies, even those with severe hubris, would so zealously pursue a drug like this one – recently described by the dean of the University of Toronto as an agent with "uneven efficacy and uncertain toxicity"[26] – to the point of litigation and adverse publicity.

But as unusual as this case may be, it is not unique. There are other examples of serious failure of institutional support in similar circumstances.[27-29] There can be no useful interactions between industry and academic physicians if lawsuits are to result from publication of data that may not suit the business plans of the sponsor. Institutions must protect the right and duty of their faculty members to publish their research. No sponsor can be permitted to stand in the way of that right and duty, because they represent the central ethos of university life.

What lessons can be learned from the recent events in Toronto and the growing evidence that things may not be much better, albeit less bizarre, elsewhere? It appears that at the time when Olivieri entered into her contractual agreement with Apotex, neither the Hospital for Sick Children nor the University of Toronto had adequate procedures for arrangements of this kind with industry – a situation that has since been corrected.[3,26] Most contracts now allow investigators to publish information arising from joint studies of this type after a fixed period, although the specifics vary widely among contracts. But is this provision sufficient? The enormous legal and financial power of the pharmaceutical industry puts clinical investigators in a very difficult position if there is a major controversy about the outcome of a particular study. Clinical scientists and, incidentally, industry itself need a fail-safe mechanism for cases of this kind in order to protect academic freedom, the safety of patients, and the rights of industrial sponsors. Such a mechanism could be created if, as part of the original contract, an independent external review panel were established, with members who were acceptable to both parties, to mediate in cases of scientific disagreement about the outcome of a particular study. Indeed, this approach might be more effective if a national panel were established by a respected body such as the National Institutes of Health, the Institute of Medicine, or equivalent national research councils in other countries in order to deal with problems of this kind. It would be hoped that such a standing committee would rarely need to be activated, but its very existence might serve as a deterrent against events like those that occurred in Toronto.

Universities will have to decide on the extent to which they wish to become commercialized and will have to monitor the effect that such commercialization has on the pattern of their research, on public confidence in research, and on academic freedom.[30] They need to reexamine every aspect of their contracts with industry, ways of preventing dangerous relationships between faculty members and industry, and ways of evolving standards for research practice that will protect scientists when difficulties arise. Universities must develop much clearer understandings with their teaching hospitals about the standing of academic clinicians; in the Toronto affair, neither party seemed to know what the other was doing and no one was in overall control.[3]

We now have the potential to enter one of the most productive periods in biomedical research, the success of which will depend to no small degree on an increasingly close partnership between universities and industry. It is vital, therefore, that the problems of this interface be recognized and corrected. No doubt, leaders of medical schools and teaching hospitals, or their equivalents, who read the CAUT report about events in Toronto – and they should read it – will want to tell themselves that such things could never happen in their universities. Unfortunately, they are happening, although they generally take milder and less public forms. All of us in academic medicine must look carefully at our own houses and set standards that protect the rights of faculty members to express their opinions in scholarly settings and journals. The Olivieri case represents an important warning that academic freedom can disappear if we do not protect it. How tempting and comforting it would be to believe that the case is unique. And how much wiser it would be to conclude instead that there, but for the grace of God, go we.

(2002)

Supported in part by a grant (HL54785-07) from the National Heart, Lung, and Blood Institute.

REFERENCES

1. Is the university-industrial complex out of control? Nature 2001;409:119-119
2. Davidoff F, DeAngelis CD, Drazen JM, et al. Sponsorship, authorship, and accountability. N Engl J Med 2001;345:825-827
3. Thompson J, Baird P, Downie J. Report of the Committee of Inquiry on the case involving Dr. Nancy Olivieri, the Hospital for Sick Children, the University of Toronto, and Apotex Inc. Toronto: Canadian Association of University Teachers, 2001.
4. Pippard MJ, Weatherall DJ. Oral iron chelation therapy for thalassaemia: an uncertain scene. Br J Haematol 2000;111:2-5
5. Porter JB. Practical management of iron overload. Br J Haematol 2001;115:239-252
6. Hider RC, Singh S, Porter JB, Huehns ER. The development of hydroxypyridin-4-ones as orally active iron chelators. Ann N Y Acad Sci 1990;612:327-338
7. Berdoukas V, Bentley P, Frost H, Schnebli HP. Toxicity of oral iron chelator L1. Lancet 1993;341:1088-1088
8. Singh S, Epemolu RO, Dobbin PS, et al. Urinary metabolic profiles in human and rat of 1,2-dimethyl- and 1,2-diethyl-substituted 3-hydroxypyridin-4-ones. Drug Metab Dispos 1992;20:256-261
9. Choudhury R, Singh S. Effect of iron overload on the metabolism and urinary recovery of 3-hydroxypyridin-4-one chelating agents in the rat. Drug Metab Dispos 1995;23:314-320
10. Kontoghiorghes GJ, Aldouri MA, Hoffbrand AV, et al. Effective chelation of iron in beta thalassaemia with the oral chelator 1,2-dimethyl-3-hydroxypyrid-4-one. Br Med J (Clin Res Ed) 1987;295:1509-1512

11. Olivieri NF, Brittenham GM, Matsui D, et al. Iron-chelation therapy with oral deferiprone in patients with thalassemia major. N Engl J Med 1995;332:918-922

12. Nathan DG. An orally active iron chelator. N Engl J Med 1995;332:953-954[Erratum, N Engl J Med 1995;332:1315.]

13. Olivieri NF, Brittenham GM, McLaren CE, et al. Long-term safety and effectiveness of iron-chelation therapy with deferiprone for thalassemia major. N Engl J Med 1998;339:417-423

14. Kowdley KV, Kaplan MM. Iron-chelation therapy with oral deferiprone – toxicity or lack of efficacy? N Engl J Med 1998;339:468-469

15. Olivieri NF. Long-term followup of body iron in patients with thalassemia major during therapy with the orally active iron chelator deferiprone (L1). Blood 1996;88:Suppl 1:310a-310a abstract.

16. Letsky EA, Miller F, Worwood M, Flynn DM. Serum ferritin in children with thalassaemia regularly transfused. J Clin Pathol 1974;27:652-655

17. Wolfe L, Olivieri N, Sallan D, et al. Prevention of cardiac disease by subcutaneous deferoxamine in patients with thalassemia major. N Engl J Med 1985;312:1600-1603

18. Olivieri NF, Nathan DG, MacMillan JH, et al. Survival in medically treated patients with homozygous β-thalassemia. N Engl J Med 1994;331:574-578

19. Brittenham GM, Griffith PM, Nienhuis AW, et al. Efficacy of deferoxamine in preventing complications of iron overload in patients with thalassemia major. N Engl J Med 1994;331:567-573

20. Tondury P, Zimmermann A, Nielsen P, Hirt A. Liver iron and fibrosis during long-term treatment with deferiprone in Swiss thalassaemic patients. Br J Haematol 1998;101:413-415

21. Del Vecchio GC, Crollo E, Schettini F, Schettini F, Fischer R, De Mattia D. Factors influencing effectiveness of deferiprone in a thalassaemia major clinical setting. Acta Haematol 2000;104:99-102

22. Mazza P, Amurri B, Lazzari G, et al. Oral iron chelating therapy, a single center interim report on deferiprone (L1) in thalassemia. Haematologica 1998;83:496-501

23. Hoffbrand AV, Al-Refaie F, Davis B, et al. Long-term trial of deferiprone in 51 transfusion-dependent iron overloaded patients. Blood 1998;91:295-300

24. Naimark A, Knoppers BM, Lowry FH. Clinical trial of L1 (deferiprone) at the Hospital for Sick Children in Toronto: a review of facts and circumstances. Toronto: Hospital for Sick Children, 1998.

25. Cranton J. Letter to Dr. Laurence Becker. Toronto: College of Physicians and Surgeons of Ontario, December 19, 2001.

26. Naylor CD. The deferiprone controversy: time to move on. CMAJ 2002;166:452-453

27. Kern D, Crausman RS, Durand KT, Nayer A, Kuhn C III. Flock worker's lung: chronic interstitial lung disease in the nylon flocking industry. Ann Intern Med 1998;129:261-272[Erratum, Ann Intern Med 1999;130:246.]

28. Davidoff F. New disease, old story. Ann Intern Med 1998;129:327-328[Erratum, Ann Intern Med 1999;130:246.]

29. Rennie D. Thyroid storm. JAMA 1997;277:1238-1243[Erratum, JAMA 1997;277:1762.]

30. Somerville MA. A postmodern moral tale: the ethics of research relationships. Nat Rev Drug Discov 2002;1:316-320

Questions:

1. What is academic freedom? Who should have academic freedom? Does this differ from who effectively *does* have academic freedom?

2. What role(s) should universities play in interfacing between their researchers and pharmaceutical companies?

3. Look up other high-profile cases in which university researchers have had difficult interactions with pharmaceutical companies (e.g. references 27-29). Choose one and write a very short summary presentation of the case.

4. In what ways should a drug company be accountable to the public?

5. In what ways should a university researcher be accountable to the public?

6. Do you think the proposed mediation process would work? Why or why not?

7. Characterize Olivieri and Apotex as they are presented in this article. Characterize Nathan and Weatherall; what sort of authorial character (or ethos) is presented in this article? What writing and communication strategies are used in these representations?

8. Of the two CAUT and University of Toronto reports, which do you feel is likely to be most unbiased (most trustworthy)? What characteristic of either group influenced your choice the most?

9. The article begins by citing an editorial in *Nature*. What is *Nature*? Does a reference to it carry special weight? Why or why not?

10. Why is the human genome project mentioned in this article? Is the mention a *non sequitur*? Why or why not?

11. Provide sub-headings for this editorial article. Would the addition of these subheadings constitute an improvement? Why or why not?

12. Describe the tone of paragraph 15 (beginning "The Toronto episode ...."). Look up the classical rhetoric term *occupatio*. Do you think it applies here? How else could you describe the tone or effect of this statement?

nineteen

George Constantinou, Stavros Melides, Bernadette Modell, Michael Spino, Fernando Tricta, David G. Nathan, and David J. Weatherall

The Olivieri Case

Included below are two letters to the editor of The New England Journal of Medicine *regarding David Nathan and David Weatherall's article "Academic Freedom in Clinical Research" (included in this anthology), as well as Nathan and Weatherall's response to these letters of concern.*

℮

To the Editor:

In their article on the dispute over the iron-chelating agent[a] deferiprone (Oct. 24 issue),[1] Nathan and Weatherall focus on the ethical implications for the relationships among scientists, their institutions, and industry. Their article ignores recent evidence on the safety and efficacy of the drug[2-5] and does nothing to resolve the uncertainty that has made deferiprone unavailable to the majority of patients with thalassemia[b] worldwide.

This very public controversy has generated fear, uncertainty, and doubt among such patients and their physicians, fomented mistrust among clinicians and researchers, and undermined patients' confidence in their doctors. Such doubts have been heightened by the high respect in which North American science is held. The Food and Drug Administration has not licensed deferiprone,[c] since its decisions are made in a North American context and cannot remain uninfluenced by such a high-profile dispute. Consequently, regulatory authorities in many countries have also declined to license deferiprone. As a result, although it has been used safely by thousands of patients for as long as 10 years and is marketed cheaply by Cipla in India, doctors in

---

a   *iron-chelating agent*   Drug that binds with excess iron in the body.

b   *thalassemia*   Genetic condition in which red blood cells are abnormal and so are destroyed unusually quickly.

c   *The Food ... licensed deferiprone*   The FDA approved deferiprone for thalassemia patients in 2011.

neighboring countries have been unable to use it, despite the fact that it offers the only hope for their patients with thalassemia. For every year this situation continues, at least 2000 to 3500 patients with thalassemia who receive regular transfusions[a] die worldwide from untreated iron overload – more than the total number of such patients living in North and South America combined. Patients cannot wait 10 years for a better iron-chelating agent. They need deferiprone now because it is affordable, tolerable, and adequately safe and effective.

George Constantinou
Stavros Melides
United Kingdom Thalassaemia Society, London N14 6PH, United Kingdom
george@constantine58.freeserve.co.uk
Bernadette Modell, Ph.D., F.R.C.P.
University College London, London N19 5LW, United Kingdom

REFERENCES

1. Nathan DG, Weatherall DJ. Academic freedom in clinical research. N Engl J Med 2002;347:1368-1371
2. Wonke B, Wright C, Hoffbrand AV. Combined therapy with deferiprone and desferrioxamine. Br J Haematol 1998;103:361-364
3. Giardina PJ, Grady RW. Chelation therapy in β-thalassaemia: an optimistic update. Semin Hematol 2001;38:360-366
4. Wanless IR, Sweeney G, Dhillon AP, et al. Lack of progressive hepatic fibrosis during long-term therapy with deferiprone in subjects with transfusion-dependent beta-thalassaemia. Blood 2002;100:1566-1569
5. Anderson LJ, Wonke B, Prescott E, Holden S, Walker JM, Pennell DJ. Comparison of effects of oral deferiprone and subcutaneous desferrioxamine on myocardial iron concentrations and ventricular function in beta-thalassaemia. Lancet 2002;360:516-520

To the Editor:

Nathan and Weatherall misrepresent facts related to Apotex and deferiprone. Moreover, one of the authors failed to disclose his relationship with a company that is developing a competing drug[1] and, in so doing, violated recent *Journal* guidelines for authorship and accountability.[2] We take issue with many of the statements made by Nathan and Weatherall.

The facts as seen by Apotex are as follows. In 1992, Olivieri and a coin-vestigator requested that Apotex support the development of deferiprone (L1), advocating the need for this oral drug in treating iron overload in patients with thalassemia major. Although it was needed by patients, deferiprone would not

---

a   *transfusions*   Additions of blood or blood components into the circulatory system.

be viewed by a pharmaceutical company as a profitable investment because there are fewer than 700 patients with thalassemia major in the United States, there is a risk of agranulocytosis,[a] and there would be patent protection for only a short time. Notwithstanding these limitations, Apotex made a commitment to develop deferiprone because of its potential for enhancing the survival of a segment of the population of patients with thalassemia major.

In 1993, Apotex initiated a program to determine the safety and efficacy of deferiprone, leading to its evaluation in more than 800 patients in Europe, North America, and other parts of the world. In 1995, Olivieri and a colleague of hers in Toronto expressed concern about a loss of response in six of their patients.[3] The scientists at Apotex reviewed the data and concluded that they did not support this interpretation. To obtain an independent assessment, Apotex distributed the raw data to all the other investigators studying deferiprone. The investigators were unanimous in their disagreement with Olivieri's interpretation. Apotex then convened an international committee of experts to review the raw data and Olivieri's interpretation. The committee concluded as follows: "Specifically, the Committee does not find a trend toward a loss of effectiveness of therapy in patients treated with L1 [deferiprone] on a long-term basis. There are no sudden, unexpected changes in regard to failure of therapy."[4]

By that time, Apotex had decided not to renew Olivieri's contract and discontinued the study in Toronto, while continuing the studies at all other sites. Nathan and Weatherall characterize the termination of Olivieri's contract as an attempt to stop her from divulging her views, but the systematic approach to disclosure and review demonstrates that such a perspective is invalid.

Three months after her contract had been terminated, Olivieri submitted an abstract for presentation at a scientific conference,[3] not disclosing that a review of the data had been conducted and that two committees had rejected her conclusions. Only after her contract had been terminated did she allege that the use of deferiprone might be associated with an increase in the risk of hepatic fibrosis.[b] [5] We think these claims have been disproved.[6]

Nathan and Weatherall demonstrate a lack of current knowledge regarding the clinical safety of deferiprone, its efficacy, and its role in the treatment of thalassemia major. Readers who wish to learn about deferiprone may refer to recent reports from independent[6-10] and Apotex-sponsored[11] studies. Deferiprone, now approved in 24 countries, is the only lifesaving alternative for patients with thalassemia major who will not or cannot take deferoxamine

---

a   *agranulocytosis* Abnormal low levels of granulocytes, a type of white blood cell.
b   *hepatic* Relating to the liver; *fibrosis* Accumulation of fibrous scar tissue.

and for whom no other treatment is approved; Nathan and Weatherall would deny them this treatment.

In addition, Nathan and Weatherall falsely contend that within "a few years" after Apotex stopped the trials, "two lawsuits totaling over $20 million were formally lodged against" Olivieri. At no time did Apotex initiate any lawsuit against Olivieri. The truth may be found in the records of the Court of Ontario. It was Olivieri who filed numerous lawsuits against those who disagreed with her actions, including other scientists, the media, and Apotex. As part of its defense, Apotex filed a counterclaim.

10   In summary, the record shows a picture remarkably different from that portrayed by Nathan and Weatherall. What is academic freedom, and what evidence can an author hide from a journal or the scientific community? Is it acceptable to suppress the fact that conclusions have been challenged, particularly when extensive peer review has resulted in the rejection of the author's conclusions? What are the responsibilities that accompany academic freedom, and how are the patients' interests best served? There is a responsibility that comes with the right to academic freedom, and that is the duty of full disclosure and scientific objectivity.

Michael Spino, Pharm.D.
Fernando Tricta, M.D.
Apotex, Toronto, ON M9L 1T9, Canada
mspino@apotex.com

REFERENCES

1. Nisbet-Brown E, Olivieri NF, Giardina PJ, et al. ICL670a, a tridentate orally-active iron chelator, provides net negative iron balance and increased serum iron binding capacity in iron-overloaded patients with thalassemia. Blood 2001;98:747a-747a abstract.

2. Davidoff F, DeAngelis CD, Drazen JM, et al. Sponsorship, authorship, and accountability. N Engl J Med 2001;345:825-827

3. Olivieri NF. Long-term followup of body iron in patients with thalassemia major during therapy with the orally active iron chelator deferiprone (L1). Blood 1996;88:Suppl 1:310a-310a abstract.

4. Schwartz E, Blumer J, Corey M, Wonke B. Report: expert advisory panel on L1 efficacy. Toronto: Hospital for Sick Children, July 12–23, 1996.

5. Olivieri NF, Brittenham GM, McLaren CE, et al. Long-term safety and effectiveness of iron-chelation therapy with deferiprone for thalassemia major. N Engl J Med 1998;339:417-423

6. Wanless IR, Sweeney G, Dhillon AP, et al. Lack of progressive hepatic fibrosis during long-term therapy with deferiprone in subjects with transfusion-dependent beta-thalassemia. Blood 2002;100:1566-1569

7. Anderson LJ, Wonke B, Prescott E, Holden S, Walker JM, Pennell DJ. Comparison of effects of oral deferiprone and subcutaneous desferrioxamine on myocardial iron concentrations and ventricular function in beta-thalassaemia. Lancet 2002;360:516-520

8. Ceci A, Baiardi P, Felisi M, et al. The safety and effectiveness of deferiprone in a large-scale, 3-year study in Italian patients. Br J Haematol 2002;118:330-336

9. Maggio A, D'Amico G, Morabito A, et al. Deferiprone versus deferoxamine in patients with thalassemia major: a randomized clinical trial. Blood Cells Mol Dis 2002;28:196

10. Hershko C, Link G, Konijn AM, Huerta M, Rosenmann E, Reinus C. The iron-loaded gerbil model revisited: effects of deferoxamine and deferiprone treatment. J Lab Clin Med 2002;139:50-58

11. Cohen AR, Galanello R, Piga A, Dipalma A, Vullo C, Tricta F. Safety profile of the oral iron chelator deferiprone: a multicentre study. Br J Haematol 2000;108:305-312

Author/Editor Response

As physicians who have cared for patients with thalassemia for more than 40 years, we completely understand the anxieties expressed by Mr. Constantinou and colleagues. They desperately want an effective oral chelator; so do we. It took many years of careful clinical studies to prove that deferoxamine can save the lives of patients with thalassemia.[1,2] Before patients and governments in developing countries give up this drug and accept deferiprone as an effective agent for treating thousands of patients, a long-term trial must be performed to prove that it can control body iron levels. Sadly, as we understand it, the only study of this kind was stopped by the manufacturers of deferiprone six years ago. Had it not been stopped, we would have this information by now; there are no shortcuts to assessing control of the slow accumulation of iron in patients with thalassemia.

Although the current Apotex lawsuit against Olivieri may be defensive, Spino and Tricta fail to mention that their company threatened her repeatedly with legal action if she publicized her concerns about the efficacy and safety of deferiprone and even threatened the American Society of Hematology in the same way if it allowed her to present her results at its annual meeting. Furthermore, if the company's advisors were so confident that Olivieri was incorrect and that body iron levels were being controlled, why did they immediately stop the trial? Had they not done so, or had they at least ensured that adequate data on iron levels were made available from the toxicity studies that were continued, we would now have this vital information.

Spino and Tricta also suggest that we are denying the drug to those "who will not or cannot take deferoxamine." Patients in whom deferoxamine results in adequate control of the level of iron accumulation should survive for a long time without cardiac or other complications.[1,2] We ask only that similar controlled data be collected for deferiprone. In fact, a recent, peer-reviewed article reviewing the field of chelation concluded, on the basis of currently available evidence, that deferiprone does not control iron accumulation in a substantial number of cases.[3] Although the more recent studies of toxicity and cardiac function cited by Spiro and Tricta are interesting, ultimately it is iron accumulation that kills patients with thalassemia. Very few "cannot" take deferoxamine on medical grounds; those who "will not" should be reminded of the uncertainty of the long-term efficacy of deferiprone.

Finally, Spino and Tricta call into question the objectivity of our review by suggesting that one of us has a conflict of interest, presumably because of recent work on an oral chelator manufactured by Novartis. We would remind them, however, that we first asked for adequately controlled trials of deferiprone in 1995[4] – long before this new agent was on the scene – and have asked many times since. We have never had a financial interest in Novartis (Novartis Oncology has made a grant to the Dana-Farber Cancer Institute for the development of kinase inhibitors[a] for cancer; it has nothing to do with chelation research, and we are not supported by it).

Our review of the Olivieri case was written because we are concerned about problems at the interface between academia and industry and about the importance of open scientific debate and academic freedom.[5] The fact that Spino and Tricta choose to interpret the article as the expression of a wish on our part to deny patients lifesaving drugs suggests that our fears are well founded.

David G. Nathan, M.D.
Harvard Medical School, Boston, MA 02115
David J. Weatherall, M.D.
University of Oxford, Oxford OX3 9DS, United Kingdom

REFERENCES

1. Olivieri FN, Nathan DG, MacMillan JH, et al. Survival in medically treated patients with homozygous β-thalassemia. N Engl J Med 1994;331:574-578
2. Brittenham GM, Griffith PM, Nienhuis AW, et al. Efficacy of deferoxamine in preventing complications of iron overload in patients with thalassemia major. N Engl J Med 1994;331:567-573
3. Porter JB. Practical management of iron overload. Br J Haematol 2001;115:239-252
4. Nathan DG. An orally active iron chelator. N Engl J Med 1995;332:953-954[Erratum, N Engl J Med 1995;332:1315.]
5. Thompson J, Baird P, Downie J. Report of the Committee of Inquiry on the case involving Dr. Nancy Olivieri, the Hospital for Sick Children, the University of Toronto, and Apotex Inc. Toronto: Canadian Association of University Teachers, 2001.

(2003)

---

a   *kinase inhibitors*   Drugs that treat cancer by inhibiting the action of improperly functioning kinases (a type of enzyme).

Questions:

1. Look at both letters and Nathan and Weatherall's reply. Which do you find most convincing? Why?

2. Outline the major claims of one of the letters. In what ways, if at all, can these claims be seen as addressing the concerns expressed in the original editorial?

3. Even if all the claims made in the letter by Apotex were true, would this mean that Olivieri's article (included in this anthology) should not have been published? Why or why not?

4. What is wrong with the claim that a drug should be put on the market because "patients cannot wait"?

5. Who approves a drug for use for humans in Canada? If you were part of that department, what sorts of things would you take into account in approving a drug? Rank these things from most to least important.

6. What do you make of the Apotex claim that "Deferiprone, now approved in 24 countries, is the only lifesaving alternative for patients with thalassemia major who will not or cannot take deferoxamine and for whom no other treatment is approved; Nathan and Weatherall would deny them this treatment"? What sort of rhetorical strategies are being used here?

7. Assess the tone of the Apotex letter. Rewrite a paragraph to make it more balanced in approach.

8. The Apotex letter says Nathan and Weatherall "suppress" information. What does suppressing mean? Does the Apotex letter "suppress" (or not include) any information that you know about?

9. Characterize Nathan and Weatherall's responses to the two letters; how do they respond to each letter differently?

twenty

F.E. Vera-Badillo, R. Shapiro, A. Ocana, E. Amir, and I.F. Tannock

from Bias in Reporting of End Points of Efficacy and Toxicity in Randomized, Clinical Trials for Women with Breast Cancer

This 2012 article from Annals of Oncology *examines biased reporting in published clinical trials of breast cancer treatments, finding that both the safety and the efficacy of drugs were frequently misrepresented in the articles studied.*

❧

ABSTRACT

Background. Phase III[a] randomized, clinical trials (RCTs) assess clinically important differences in end points[b] that reflect benefit to patients. Here, we evaluate the quality of reporting of the primary end point (PE) and of toxicity in RCTs for breast cancer.

Methods. PUBMED[c] was searched from 1995 to 2011 to identify RCTs for breast cancer. Bias in the reporting of the PE and of toxicity was assessed using pre-designed algorithms. Associations of bias with the Journal Impact Factor (JIF), changes in the PE compared with information in ClinicalTrials. gov and funding source were evaluated.

---

a *Phase III* Term applied to large-scale clinical trials; usually, a drug will be authorized for sale only after Phase III trials have demonstrated its safety and effectiveness.
b *end points* Objects of measurement in a clinical trial, such as survival or toxicity.
c *PUBMED* Search engine of the life sciences and biomedical research database MEDLINE. Both PUBMED and MEDLINE are maintained by the United States National Library of Medicine.

Results. Of 164 included trials, 33% showed bias in reporting of the PE and 67% in the reporting of toxicity. The PE was more likely to be reported in the concluding statement of the abstract when significant differences favoring the experimental arm[a] were shown; 59% of 92 trials with a negative PE used secondary end points to suggest benefit of experimental therapy. Only 32% of articles indicated the frequency of grade 3 and 4 toxicities in the abstract. A positive PE was associated with under-reporting of toxicity.

Conclusion. Bias in reporting of outcome is common for studies with negative PEs. Reporting of toxicity is poor, especially for studies with positive PEs.

INTRODUCTION

Phase III randomized, clinical trials (RCTs) are designed to detect or exclude clinically important differences between experimental and control groups in end points that reflect benefit to patients.[1] Such trials provide the gold standard to evaluate the efficacy and toxicity of new drugs before approval by regulatory authorities.[2,3]

Appropriate design and objective reporting of RCTs in journals are essential to inform clinicians about the activity and safety of new medical interventions. It is good practice to design RCTs with no more than three outcomes for which hypothesis testing is planned.[4] Otherwise multiple significance testing may lead to apparently significant results that occur by chance. These outcomes should normally include at least one end point reflecting potential benefit and at least one reflecting potential harm (e.g. grade 3 and 4 adverse events[b]). Reviews have shown that a substantial proportion of clinical trials have suboptimal reporting of harm – a large number of trials have shown deficiency in the report of toxicity, especially severe toxicity, graded as 3 and 4.[5] Guidelines such as Consolidated Standards of Reporting Trials can improve the quality of reporting of clinical trials.[6]

Bias in reporting of clinical trials and selective publication can create false perceptions of drug efficacy and safety. There is evidence for selective reporting of favorable results and suppression of unfavorable data from publication, leading to inappropriate conclusions.[2,7] This may be influenced by publication bias – the association between positive results and acceptance of

---

a   *experimental arm*   Group of subjects who have been treated with the experimental method; the other group of subjects is the "control arm."

b   *grade 3 ... adverse events*   Severe, or life-threatening physical changes that are experienced by a patient during a medical treatment; such occurrences are considered "adverse events" regardless of whether they are caused by the treatment.

reports for publication.[4,8] Selection bias can affect not only the interpretation of the trial itself but also the interpretation of subsequent systematic reviews or overviews, producing inaccurate summaries of research[2,9] and misrepresentation of toxicity.[10] Reporting of harms may be viewed as discrediting the reporting of benefits.

Spin, a type of bias, is defined as use of reporting strategies to highlight that the experimental treatment is beneficial, despite a statistically non-significant difference in the primary outcome, or to distract the reader from statistically non-significant results.[11] It is important to recognize the presence of bias and spin in reports of clinical trials, and to evaluate their importance when placing a RCT in context and ascribing a level of credibility.[12]

5 Here, we review the papers reporting RCTs for breast cancer to quantify the extent of biased reporting, and to guide readers in judging the credibility of their conclusions. Because busy clinicians often read only the abstracts of publications,[13] we have emphasized accurate reporting of the primary endpoint (PE) and toxicity in the abstract. We hypothesized that despite the availability of guidelines to minimize bias in reporting, this remains prevalent.

METHODS

Literature Search and Study Selection. We carried out an electronic search of MEDLINE (Host: PubMed) for publications from January 1995 to August 2011 using the following MeSH[a] terms: randomized clinical trial, RCT, Phase III and breast neoplasms[b] or breast cancer. The inclusion criteria were human studies published in English and including patients aged ≥18 years. We excluded trials with sample size less than 200 patients as they were unlikely to be definitive studies and more likely to have higher levels of bias. Furthermore, the focus of this study was to assess reporting of clinical trials that potentially change clinical practice. Other exclusion criteria included trials where the PE was not a time to event end point, commentaries, review articles, observational studies, meta-analyses, ongoing studies and articles for which only the abstract was available.

Data Extraction and Analysis. The following data were extracted independently from each RCT by two authors (FV and RS): setting of treatment (adjuvant versus metastatic[c]), sponsorship (industry versus non-industry or

---

a *MeSH* Medical Subject Headings, used to index articles for the database MEDLINE.
b *neoplasms* Abnormal masses of tissue; tumors.
c *adjuvant* Intended to prevent recurrence of a cancer; an example would be post-surgery chemotherapy; *metastatic* Term applied to treatments used for secondary tumor development distant from original cancer site.

not stated), year of publication, impact factor of the journal where the trial was published (as a continuous variable[a]), the primary and secondary end points [overall survival (OS), progression-free survival (PFS), disease-free survival (DFS), response rate, toxicity or quality of life], whether the PE was defined in the abstract and/or in the paper and whether the secondary end points were reported in the abstract and in the paper. If data on PFS or DFS were not reported, similar end points such as time to progression, time-to-treatment failure or event-free survival[b] were extracted instead. For papers with more than two arms, the assessment of efficacy was not considered if at least one arm was positive (see Figure 1); if all were negative, it was included for analysis. For toxicity, all trials were included regardless of the number of study arms. The journal impact factor (JIF) was extracted from the Gerstein Science Information Center of the University of Toronto through the Web of Science (Host:BIOSIS) and was retrieved from the Journal Citation reports up to 2012.

Figure 1.
Decision tree for assessment of reporting of the primary end point in the concluding statement of the abstract

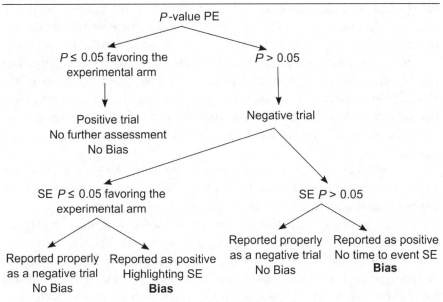

PE = primary end point; SE = secondary end point.

---

a *continuous variable* Variable that can have any value within a given numerical range.
b *event-free survival* Length of time until a subject experiences the particular symptoms or biological events that a treatment was intended to remove.

For recent trials, we also explored whether the PE listed in the trial registry ClinicalTrials.gov was the same as that reported in abstracts or papers reporting the same trial. We initially searched articles for any reference to trial registration. For those not reporting such data, we manually searched the ClinicalTrials.gov database for trial-related information.

End Points. The primary analysis included the assessment of the prevalence of bias and spin in reporting the PE of the study, and in reporting toxicity; the secondary analysis evaluated predictors of bias and spin.

Bias was defined as inappropriate reporting of the PE and toxicity, with emphasis on reporting of these outcomes in the abstract. A decision tree was used to assess whether the PE was reported with bias, and whether a secondary end point was used to imply benefit of the experimental arm (Figure 1). Studies where multiple PEs were reported and where at least one end point was positive were not considered for assessment of bias.

Bias in reporting of toxicity was assessed using a hierarchy scale from 1 (excellent) to 7 (very poor) to indicate whether reporting of grade 3 and 4 toxicities occurred in the concluding statement of the abstract, elsewhere in the abstract, in the results section of the paper, only in a table or not at all, with lower scores if they were also included in the discussion section of the paper (Figure 2). We defined reporting of grade 3 and 4 toxicities as poor if they were not mentioned in the abstract (scale of 5–7 in our hierarchy), and good (scale 1–2) if they were mentioned in the concluding statement of the abstract. When there were no statistically significant differences in toxicity, a general statement in the abstract was deemed to be sufficient; when statistically significant differences were seen, it was expected that they would be reported in the abstract.

Spin was defined as the use of words in the concluding statement of the abstract to suggest that a trial with a negative PE was positive based on some apparent benefits shown in one or more secondary end points.

Predictors of bias included the impact of source of funding on reporting of the PE, the relationship between the quality of reporting of the PE and of toxicity and the frequency of a change in the PE from the original protocol (when this could be obtained from ClinicalTrials.gov) to a published paper. Also whether the JIF, setting of the trial (adjuvant versus metastatic), definitive versus surrogate end point (e.g. OS compared with PFS or DFS) and modification of the PE influenced the prevalence of bias or spin.

Statistical Analysis. Data were presented descriptively as means or medians. Predictors of bias were assessed by the chi-squared test and by univariable logistic regression (categorical variables) or univariable linear regression[a] (continuous variables). Correlations between variables were tested using Spearman's correlation and the magnitude of association was assessed as described by Burnand et al.[14] All statistical analyses were conducted using SPSS statistical software version 17 (IBM Corp, Arkmon, New York). All significance tests were two-sided[b] using an alpha level of 0.05. No correction was applied for multiple statistical testing.

FIGURE 2.
Hierarchy scale for reporting of adverse events

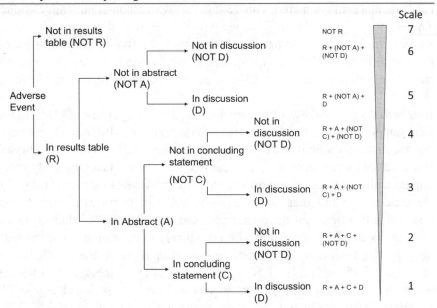

Not R = not reported in results table; Not D = not reported in discussion; Not A = not reported in abstract; Not C = not reported in concluding statement; R = reported in results table; D = reported in discussion; A = reported in abstract; C = reported in concluding statement.

---

a *chi-squared test* Statistical test used to determine the probability that a set of data reflects a significant relationship between variables; *univariable logistic regression* Statistical method showing the probability relationship between one variable and a given result; *categorical variables* Variables that can have any of a set of specific values, sometimes including non-numerical values; *linear regression* Statistical method that can be used to highlight a relationship between a categorical variable and a continuous variable, presenting that relationship in a form that can be shown as a line on a graph.

b *two-sided* Looking for both significantly high and significantly low results.

Results

15 A total of 568 articles were identified initially and 164 RCTs (148 for systemic therapy,[a] 11 for radiation therapy and 5 for surgical therapy) were eligible for analysis[....] Eighty-one trials (49.4%) were conducted in the adjuvant setting and 83 (50.6%) evaluated experimental therapy for women with metastatic breast cancer. OS was the PE in only 27 trials (16.5%) and DFS or PFS was the PE in 137 studies (83.5%). Only 30 trials (18%) were identified as included in ClinicalTrials.gov. Among these studies, the PE was changed in the final report in seven (23.3%) studies. Seventy-two (43.9%) studies were positive with a significant P-value for the difference in primary endpoint favoring the experimental arm compared with 92 (56.1%) with a non-significant P-value. The majority, 150 trials (91.4%), were published in medium or high impact journals; the calculated median impact factor was 19. Bias in reporting efficacy (Spearman's rho = 0.27 and chi-squared $P = 0.30$)[b] and toxicity (Spearman's rho = 0.46 and chi-squared $P = 0.06$) were not influenced by date of publication.

Bias in the Reporting of the Primary End Point. Fifty-four trials (32.9% of the total sample) were reported as positive, based on a non-PE, despite not finding a statistically significant difference in the PE. These reports were biased and used spin in an attempt to conceal that bias. When assessing only those reports with a non-significant difference in the PE between the arms ($N = 92$), the incidence of this bias increased to 59.0%. Compared with studies with a statistically significant difference between arms in the PE, studies with a non-significant difference showed a statistically significant association with not reporting the PE in the concluding statement of the abstract (27% versus 7%, OR = 5.15, 95% CI = 1.86–14.26, $P = 0.001$). Compared with studies where the PE did not change, there was a trend for trials with a change of PE to report a statistically significant difference for (the new) PE (OR = 2.29, 95% CI = 0.37–14.32, $P = 0.47$). There was no association between the JIF and bias (Spearman's rho = −0.10, $P = 0.20$).

There were no apparent differences in the probability of bias in trials conducted in the adjuvant or metastatic settings (chi-squared $P = 0.146$). There was also no association between bias and the type of PE, (OS versus DFS or PFS, chi-squared $P = 0.23$).

---

a *systemic therapy* Treatment that impacts the whole body via the bloodstream; chemotherapy is an example of systemic therapy for cancer.

b *Spearman's rho* Type of correlation coefficient, which indicates how close the relationship between the variables is to that of a linear equation; *P* Probability, here an indication of the likelihood of getting the same experimental results as the ones observed if there were no relationship between the variables being studied.

Bias in Reporting of Toxicity. A total of 110 (67.1%) papers met our definition of biased reporting of toxicity. Distribution of bias according to the hierarchy scale is reported in Table 2. There was a statistically significant association between biased reporting of toxicity and observation of a statistically significant difference in the arms for the PE (OR = 2.00, 95% CI = 1.02–3.94, P = 0.044). There was no association between biased reporting of toxicity and biased reporting of efficacy (chi-squared P = 0.43), or with change of the PE (OR = 0.58, 95% CI = 0.1–3.2, P = 0.17). The JIF was not associated with biased reporting of toxicity (Spearman's rho = −0.153, Chi squared P = 0.73). Bias in the reporting of toxicity was significantly associated with the use of OS as the PE (OR = 3.30, 95% CI = 1.1–10.1, P = 0.028). Reporting of toxicity was not influenced by the setting of the trial (adjuvant versus metastatic, OR = 1.68, 95% CI = 0.9 to 3.3, P = 0.12).

TABLE 2.
Distribution of bias in reporting toxicity

Toxicity hierarchy scale	Number of trials N=164	%	Positive PE (%)	Negative PE (%)
1	17	10	7 (4)	10 (6)
2	7	4	2 (1)	5 (3)
3	20	12	4 (2)	16 (10)
4	10	6	5 (3)	5 (3)
5	21	13	11 (7)	10 (6)
6	55	34	31 (19)	24 (15)
7	34	21	12 (7)	22 (13)

Higher numbers and shaded area refer to poor reporting of toxicity. PE, primary end point.

Influence of Funding on Results. Funding from industry partners was reported in 103 (62.8%) studies, 32 (19.5%) studies were funded by academic or governmental grants and in 29 (17.7) studies the source of funding was not stated. Three studies reported this in the abstract,[15–17] while all others reported this information in the body of the manuscript. Success in finding a significant difference between the arms for the PE and bias in the reporting of this end point were not influenced by source of funding (chi-squared P = 0.78 and P = 0.71, respectively). Similarly, industry funding was not associated with biased reporting of toxicity (chi-squared P = 0.71). There was no effect of industry funding on odds of change in the PE (OR = 3.20, CI = 0.3–31.4, P = 0.41). There was a significant, but weak association between funding from the industry and higher JIF of the published study (Spearman's rho = 0.39 and chi-squared P = 0.05).

DISCUSSION

20 Several papers have evaluated the frequency and characteristics of bias in the reporting of efficacy, [3,4,18–21] but these reports have tended to focus on heterogeneous[a] medical conditions and not on cancer clinical trials. Furthermore, there are limited data in the literature about bias in the reporting of toxicity.[11] Here, we have explored the frequency of bias in reporting of efficacy and toxicity in randomized trials evaluating treatments of breast cancer. We focus our research on breast cancer, given that it is the most common malignancy in women, has substantial mortality[22] and is a cancer site with a large number of trials.

The objective of a phase III RCT is to detect or exclude differences in end points that will reflect benefit to patients. The chosen end points should be measures of patient benefit (i.e. improved efficacy or better safety and tolerability). OS is the gold standard for the assessment of benefit: it is unambiguous and is not subject to investigator interpretation.[23] PFS or DFS may be suitable end points if these measures are valid surrogates for OS,[24] or possibly in trials where there is a high rate of cross-over to the experimental arm which confounds interpretation of OS. For women with breast cancer, neither DFS nor PFS have been shown to be adequate surrogates for OS[25,26] but 83.5% of our cohort of trials used these end points.

Bias in the reporting of the PE was prevalent, especially when statistical significance of the difference in the PE between the arms was not found. You et al.[27] evaluated reports of RCTs published between 2005 and 2009, and found that there was misinterpretation of the PE in 21.6% of the trials; this included non-significance in a superiority trial interpreted as showing treatment equivalence, study conclusion based on end points other than PE, study considered positive despite a non-significant P-value, and study conclusions based only on one end point when there were co-PEs. We found a higher incidence of inappropriate reporting of the PE in RCTs for breast cancer that increased dramatically when only the trials with a non-significant P-value were assessed. Consequently, spin was used frequently to influence, positively, the interpretation of negative trials, by emphasizing the apparent benefit of a secondary end point. We found bias in reporting efficacy and toxicity in 32.9% and 67.1% of trials, respectively, with spin and bias used to suggest efficacy in 59% of the trials that had no significant difference in their PE. These results are similar to those in other areas of medicine.[3] In contrast to those data where bias in the reporting of toxicity was less frequent when the PE was positive, we found that bias in the reporting of toxicity was higher

---

a *heterogeneous* Diverse, varied.

when the trial had a significant *P*-value for the difference in the PE between experimental and control arms. A possible explanation for this finding may be that investigators and/or sponsors then focus on efficacy as the basis of registration and downplay toxicity to make the results more attractive.

To avoid selection for publication of positive trials, and/or publication of a subset of the original recorded outcomes on the basis of the results, registration of trials is now mandatory. Due to our period of evaluation (1995–2011), only 18% of our trials were registered in ClinicalTrials.gov. In some of these trials the PE was changed between the time of registration and reporting of their results. Among these trials, there was a trend towards change of the PE being associated with positive results, suggesting that it may be a strategy to make a negative trial appear positive. This may be on the basis of a low likelihood of observing enough events for this end point to be statistically significant or even a lack of effect of the experimental therapy to modify the original PE (usually OS). Trial registration does not necessarily remove bias in reporting outcome, although it does make it easier to detect.[18]

The pharmaceutical industry is increasingly influential in clinical trial sponsorship with data showing an increase in industry sponsorship of phase III RCTs from 24 to 72% over a 30 year period.[3,19,20] In our cohort of trials, 67% were industry sponsored, but we found no association between industry sponsorship and biased reporting of either efficacy or toxicity, and no association of for-profit sponsorship with change of the PE between that listed in trial registries and the final publication.

There are some limitations to our study. First, we searched only breast cancer trials and we cannot extrapolate our findings to published reports of trials for other types of cancer. Second, including studies with less than 200 patients would be likely to increase the level of bias, but the clinical impact of such studies is low. Third, we utilized subjective measures for some of our outcome measures such as the presence of spin. Fourth, our scales used to assess bias in reporting of efficacy and toxicity were based on our interpretation of the characteristics that a paper has to accomplish to be considered unbiased, but they have not been validated. Fifth, many of our included trials were not available at ClinicalTrials.gov. This database was established in 2002[28] and many trials initiated before this date were not included. Furthermore, many European trials were not initially included in the US-based ClinicalTrials.gov database and European Clinical Trials Registries do not have easily searchable databases.[29] Our analysis of change in the PE should, therefore, be interpreted with caution.

In conclusion, bias in the reporting of efficacy and toxicity remains prevalent. Clinicians, reviewers, journal editors and regulators should apply a critical eye to trial reports and be wary of the possibility of biased reporting. Guidelines are necessary to improve the reporting of both efficacy and toxicity.

(2012)

Disclosure: The authors have declared no conflicts of interest.

REFERENCES

1. Ocana A, Tannock IF. When are 'positive' clinical trials in oncology truly positive? J Natl Cancer Inst 2010; 103: 16–20.
2. Dwan K, Altman DG, Arnaiz JA et al. Systematic review of the empirical evidence of study publication bias and outcome reporting bias. PLoS One 2008; 3: e3081.
3. Chan AW, Hrobjartsson A, Haahr MT et al. Empirical evidence for selective reporting of outcomes in randomized trials: comparison of protocols to published articles. JAMA 2004; 291: 2457–2465.
4. Kirkham JJ, Altman DG, Williamson PR. Bias due to changes in specified outcomes during the systematic review process. PLoS One 2010; 5: e9810.
5. Ioannidis JP. Adverse events in randomized trials: neglected, restricted, distorted, and silenced. Arch Intern Med 2009; 169: 1737–1739.
6. Ioannidis JP, Evans SJ, Gotzsche PC et al. Better reporting of harms in randomized trials: an extension of the CONSORT statement. Ann Intern Med 2004; 141: 781–788.
7. Williamson PR, Gamble C, Altman DG et al. Outcome selection bias in metaanalysis. Stat Methods Med Res 2005; 14: 515–524.
8. Krzyzanowska MK, Pintilie M, Tannock IF. Factors associated with failure to publish large randomized trials presented at an oncology meeting. JAMA 2003; 290: 495–501.
9. Ioannidis JP. Why most published research findings are false. PLoS Med 2005; 2: e124.
10. Cuervo LG, Clarke M. Balancing benefits and harms in health care. BMJ 2003; 327: 65–66.
11. Pitrou I, Boutron I, Ahmad N et al. Reporting of safety results in published reports of randomized controlled trials. Arch Intern Med 2009; 169: 1756–1761.
12. Ioannidis JP. Limitations are not properly acknowledged in the scientific literature. J Clin Epidemiol 2007; 60: 324–329.
13. Barry HC, Ebell MH, Shaughnessy AF et al. Family physicians' use of medical abstracts to guide decision making: style or substance? J Am Board Fam Pract 2001; 14: 437–442.
14. Burnand B, Kernan WN, Feinstein AR. Indexes and boundaries for 'quantitative significance' in statistical decisions. J Clin Epidemiol 1990; 43: 1273–1284.
15. Gianni L, Eiermann W, Semiglazov V et al. Neoadjuvant chemotherapy with trastuzumab followed by adjuvant trastuzumab versus neoadjuvant chemotherapy alone, in patients with HER2-positive locally advanced breast cancer (the NOAH trial): a randomised controlled superiority trial with a parallel HER2-negative cohort. Lancet 2010; 375: 377–384.
16. Cortes J, O'Shaughnessy J, Loesch D et al. Eribulin monotherapy versus treatment of physician's choice in patients with metastatic breast cancer (EMBRACE): a phase 3 open-label randomised study. Lancet 2011; 377: 914–923.
17. Martin M, Segui MA, Anton A et al. Adjuvant docetaxel for high-risk, node-negative breast cancer. N Engl J Med 2010; 363: 2200–2210.
18. Kirkham JJ, Dwan KM, Altman DG et al. The impact of outcome reporting bias in randomised controlled trials on a cohort of systematic reviews. BMJ 2010; 340: c365.
19. Boutron I, Dutton S, Ravaud P et al. Reporting and interpretation of randomized controlled trials with statistically nonsignificant results for primary outcomes. JAMA 2010; 303: 2058–2064.

20. Smyth RM, Kirkham JJ, Jacoby A et al. Frequency and reasons for outcome reporting bias in clinical trials: interviews with trialists. BMJ 2011; 342: c7153.
21. Rising K, Bacchetti P, Bero L. Reporting bias in drug trials submitted to the Food and Drug Administration: review of publication and presentation. PLoS Med 2008; 5: e217; discussion e217.
22. Siegel R, Ward E, Brawley O et al. Cancer statistics, 2011: the impact of eliminating socioeconomic and racial disparities on premature cancer deaths. CA Cancer J Clin 2011; 61: 212–236.
23. Pazdur R. Endpoints for assessing drug activity in clinical trials. Oncologist 2008; 13(Suppl 2): 19–21.
24. Booth CM, Eisenhauer EA. Progression-free survival: meaningful or simply measurable? J Clin Oncol 2012; 30(10): 1030–1033.
25. Ocana A, Tannock IF. When are 'positive' clinical trials in oncology truly positive? J Natl Cancer Inst 2011; 103: 16–20.
26. Amir E, Seruga B, Kwong R, Tannock IF, Ocaña A. Poor correlation between progression-free and overall survival in modern clinical trials: are composite endpoints the answer? Eur J Cancer 2012; 48(3): 385–388.
27. You B, Gan HK, Pond G et al. Consistency in the analysis and reporting of primary end points in oncology randomized controlled trials from registration to publication: a systematic review. J Clin Oncol 2012; 30: 210–216.
28. http://clinicaltrials.gov/ct2/info/about. (16 March 2012, date last accessed).
29. https://www.clinicaltrialsregister.eu/. (16 March 2012, date last accessed).

Questions:

1. What is a Phase III human trial? What phases does a drug for human consumption need to go through in order to be legally saleable in Canada?

2. Look up one of the following drugs and see if you can find information about the data that has had to be added since the drug was approved for widespread human use: prozac (or SSRIs in general); antipsychotics (try searching for these drugs in combination with the terms "elderly" and "black box warning"); lipitor.

3. Outline what a "gold-standard" RCT looks like according to this article.

4. Rewrite the introduction as though you were either a newspaper reporter or Ben Goldacre (see an excerpt from his book *Bad Pharma* in this anthology).

5. Look up Consolidated Standards of Reporting Trials. According to the source you found, why are consolidated standards necessary?

6. List the types of bias this article addresses. What other types of investigative and reporting bias can you think of?

7. What is an impact factor? Why is it important? Why do you think the article did not explain why it was using impact factor as a variable?

8. Look up ClinicalTrials.gov. Who has to register trials here?

9. Why does it matter if an experiment's PE is changed once the results are in? Why does it matter if toxicity is downplayed? Write two or three lines that could be added to the article to present the importance of the article's findings more explicitly.

10. Consider the article's stated limitations. Why do you think the article suggests its results should be treated with caution?

11. Grade the reliability of the sources used in this article and defend your ranking.

12. The article states that the "clinical impact" of trials with less than 200 participants is low. Does it provide any justification for this assertion? In what ways could these trials be made to have significant clinical impacts?

13. Why do you think pharmaceutical companies now sponsor 72% of these clinical trials, as opposed to 24% three decades ago? Should this be the case? Why or why not?

twenty-one

BEN GOLDACRE

Ghostwriters in the Sky

This September 2010 article, written for the Guardian *by epidemiologist and popular science writer Ben Goldacre, discusses the problem of medical ghostwriting. The article focuses on a court case advanced against Wyeth, a large pharmaceutical company that employed a medical ghostwriting company to create academic articles on one of Wyeth's medications.*

❧

If I tell you that Katie Price[a] did not, necessarily, write her own book, this is not a revelation. From academics I have slightly higher expectations, but now the legal system has spat out another skip full of documents: this time, we get a new insight into the strange phenomenon of medical ghost-writing.

Attributed authorial assistance is one thing. This is different, and more cynical. A commercial medical writing company is employed by a drug company to produce a programme of academic papers that can be rolled out in academic journals to build a brand message. After copywriters produce the articles, in collaboration with the drug company, to their specifications, the ghostwriting company finds some academics who are willing to put their names to them, perhaps after a few modest changes.

The latest documents come from a court case brought against Wyeth by around 14,000 patients who developed breast cancer while taking their hormone replacement therapy,[b] Prempro. The open access journal PLoS Medicine, acting with the *New York Times*, argued successfully in court that 1500 documents from the case which detailed the ghostwriting should be placed in the public domain, because they represent important information on a potential threat to public health. Now, PLoS has published the first academic analysis of these documents, which is free to access online.

HRT, we should remember, has had a rocky history. Initially the panacea to all ills, by 1998 the HERS[c] trial showed it didn't prevent cardiovascular events after all, and by 2002 the Women's Health Initiative trial showed it

---

a   *Katie Price*   British celebrity who achieved fame as a model.
b   *hormone replacement therapy*   Drug given to reduce menopause symptoms.
c   *HERS*   Heart and Estrogen/Progestin Replacement Study.

also increased the risk of breast cancer and stroke. We now know it increases the risk of dementia and incontinence. Survey data shows that even today, many gynaecologists continue to have beliefs about the efficacy of HRT that are in excess of the evidence. Reading how the literature was engineered, it's not hard to see why.

5   The company DesignWrite boasts that over 12 years they have "planned, created, and/or managed hundreds of advisory boards, a thousand abstracts and posters, 500 clinical papers, over 10,000 speakers' bureau programs, over 200 satellite symposia, 60 international programs, dozens of websites, and a broad array of ancillary printed and electronic materials." They proposed a "planned publication program" to Wyeth, consisting of review articles, case reports, letters, editorials, commentaries and more, using the medical literature as a marketing tool.

DesignWrite wrote the first drafts, and sent them to Wyeth, who then advised on the creation of a second draft. Only then did the paper get sent to the academic who would appear as the "author." Review articles cost Wyeth $20,000. Abstracts are $4,000. The academics weren't paid cash, but they did get an easy publication in an academic journal for their CV.[a] And once the publication process was in train, the chap from Wyeth's marketing depart-ment helpfully provided comments and suggestions for the authors to use in response to peer reviewers' comments.

The PLoS documents show DesignWrite sold Wyeth more than 50 peer reviewed journal articles for HRT, and a similar number of conference posters, slide kits, symposia, and journal supplements. The analysis in PLoS (by an academic who appeared as a paid witness against the company in court) found that these publications variously promoted unproven and unlicensed benefits of Wyeth's HRT drug, undermined its competitors, and downplayed its harms.

You might imagine there are rules against this. There are not: there are traditions, good faith, and leaky regulations. It's illegal, for example, for a pharmaceutical company to promote its drug for "off label" use, which means selling it to treat a medical problem for which that drug has no license. In the case of Wyeth's HRT drug, that meant they couldn't market it for preventing Alzheimer's, Parkinson's, and wrinkles, to name but a few. The PLoS analy-sis finds that many articles produced by DesignWrite for Wyeth promoted the drug's use for exactly these conditions: but academic journal publications are not regarded as promotional activity, so this was all legal.

Worst of all is the complicity of the academics, and in very large numbers. There is no possible way they could persuade themselves that what they were doing was correct. "Research shows high clinician reliance on journal articles

---

a   *CV*   Curriculum vitae, a résumé of academic and professional accomplishment.

for credible product information," said DesignWrite, in their initial pitch. They're right, and that's for a reason: when you read an academic paper, you trust it was written by the person whose name is on it.

There are very simple solutions. If a commercial writer employed by a 10 pharmaceutical company has the idea for a paper, and writes it, then their name and their company's name should be on the paper. If the first, last, and corresponding authors on a paper didn't write or lead on it, they should say so, loudly and clearly. Universities, which are currently inconsistent on ghost writing, could take a lead, but they don't, and so these serious problems in academia will persist, because they are slightly complicated, and hidden from public scrutiny. That's why you should read about them in PLoS, talk about them, crane your neck over, scratch your chin, and mutter in astonishment. Nobody in a regulatory role is interested. Our only hope is the power of shame.

(2010)

Questions:

1. Compare this piece with the following excerpt from Goldacre's book *Bad Pharma*. Is the writing "voice" the same in each? Is the essential audience the same for each piece? If there are differences, how do the differences in writing style reflect differences in audience?

2. Katie Price is a British model and reality TV actor known for having extensive plastic surgery. Why do you think this article uses her (ghost-written) biography as its opening example?

3. How pervasive do you think the medical ghostwriting described here is? Find some evidence to support your argument. What is the source of that evidence and can it be trusted? Why or why not?

4. See if you can find information on how prevalent HRT prescription is at a national level. Taking into account the information presented in "Ghostwriters," do you think this level of prescription is justified?

5. Should academic/pharmaceutical writing services such as DesignWrite be regulated? If so, by whom?

6. Does your university or college have a policy against academics claim-ing credit for ghostwritten articles? What sorts of ethical problems can you think of that are associated with academics achieving, for instance, tenure, on the basis of research they have not done? How could your university or college prevent such problems?

7. Add one or two suggestions of your own to the solutions proposed here.

8. Why do you think regulators are not interested in regulating ghostwrit-ing? Give one reason why they should be.

twenty-two

Ben Goldacre

from *Bad Pharma*

Scientist and author Ben Goldacre's book Bad Pharma *argues that
the pharmaceutical industry manipulates and distorts the information
gathered through drug tests and trials. The following excerpt describes
and criticizes the practice of ghostwriting in medical journals and
recommends ways to curb this practice.*

❧

It goes without saying that when we're dealing with medical treatments,
which can be hugely harmful as well as helpful, it's vitally important that
all our information is reliable, and transparent. But there is another ethical
dimension, which often seems to be neglected.

These days, in most universities, we send a long and threatening document to every undergraduate student, explaining how every paragraph
of every essay and dissertation they submit will be put through a piece of
software called TurnItIn, expensively developed to detect plagiarism. This
software is ubiquitous, and every year its body of knowledge grows larger, as
it adds every student project, every Wikipedia page, every academic article,
and everything else it can find online, in order to catch people cheating. Every
year, in every university, students are caught receiving outside help; every
year, students are disciplined, with points docked and courses marked as
"failed." Sometimes they are thrown off their degree completely, leaving a
black mark of intellectual dishonesty on their CV[a] forever.

And yet, to the best of my knowledge, no academic anywhere in the world
has ever been punished for putting their name on a ghostwritten academic
paper. This is despite everything we know about the enormous prevalence of
this unethical activity and despite endless specific scandals around the world
involving named professors and lecturers, with immaculate legal documentation, and despite the fact that it amounts, in many cases, to something that is
certainly comparable to the crime of simple plagiarism by a student.

---

a *CV* Curriculum vitae, a résumé of academic and professional accomplishment.

Not one has ever been disciplined. Instead, they have senior teaching positions.

So, what do the regulations say about ghostwriting? For the most part, very little. A survey in 2010 of the top fifty medical schools in the United States found that all but thirteen had no policy at all prohibiting their academics putting their name to ghostwritten articles.[90] The International Committee of Medical Journal Editors, meanwhile, has issued guidelines on authorship, describing who should appear as a named author on a paper, in the hope that ghostwriters will have to be fully declared as a result. These are widely celebrated, and everyone now speaks of ghostwriting as if it has been fixed by the ICMJE. But in reality, as we have seen so many times before, this is a fake fix: the guidelines are hopelessly vague, and are exploited in ways that are so obvious and predictable that it takes only a paragraph to describe.

The ICMJE criteria require that someone is listed as an author if they fulfill the following three criteria: they contributed to the conception and design of the study (or data acquisition, or analysis and interpretation); they contributed to drafting or revising the manuscript; and they had final approval on the contents of the paper. This sounds great, but because you have to fulfill *all three* criteria to be listed as an author, it is very easy for a drug company's commercial medical writer to do almost all the work, but avoid being listed as an author. For example, a paper could legitimately have the name of an independent academic on it, even if they only contributed 10 per cent of the design, 10 per cent of the analysis, a brief revision of the draft, and agreed to the final contents. Meanwhile, a team of commercial writers employed by a drug company on the same paper would not appear in the author list, anywhere at all, even though they conceived the study in its entirety, did 90 per cent of the design, 90 per cent of the analysis, 90 per cent of the data acquisition, and wrote the entire draft.[91]

In fact, often the industry authors' names do not appear at all, and there is just an acknowledgment of editorial assistance to a company. And often, of course, even this doesn't happen. A junior academic making the same contribution as many commercial medical writers – structuring the write-up, reviewing the literature, making the first draft, deciding how best to represent the data, writing the words – would get their name on the paper, sometimes as first author. What we are seeing here is an obvious double standard. Someone reading an academic paper expects the authors to be the people who conducted the research and wrote the paper: that is the cultural norm, and that is why medical writers and drug companies will move heaven and earth to keep their employees' names off the authors list. It's not an accident, and there is no room for special pleading. They don't want commercial writers in the author list, because they know it looks bad.

Is there a solution? Yes: it's a system called "film credits," where everyone's contribution is simply described at the end of the paper: "X designed the study, Y wrote the first draft, Z did the statistical analysis," and so on. Apart from anything else, these kinds of credits can help ameliorate the dismal political disputes within teams about the order in which everyone's name should appear. Film credits are uncommon. They should be universal.

If I sound impatient about any of this, it's because I am. I like to speak with people who disagree with me, to try to change their behaviour, and to understand their position better: so I talk to rooms full of science journalists about problems in science journalism, rooms full of homeopaths about how homeopathy doesn't work, and rooms full of people from big pharma about the bad things they do. I have spoken to the members of the International Society of Medical Publications Professionals three times now. Each time, as I've set out my concerns, they've become angry (I'm used to this, which is why I'm meticulously polite, unless it's funnier not to be). Publicly, they insist that everything has changed, and ghostwriting is a thing of the past. They repeat that their professional code has changed in the past two years. But my concern is this. Having seen so many codes openly ignored and broken, it's hard to take any set of voluntary ideals seriously. What matters is what happens, and undermining their claim that everything will now change is the fact that nobody from this community has ever engaged in whistleblowing (though privately many tell me they're aware of dark practices continuing even today). And for all the shouting, this new code isn't even very useful: a medical writer could still produce the outline, the first draft, the intermediate drafts and the final draft, for example, with no problem at all; and the language used to describe the whole process is oddly disturbing, assuming – unthinkingly – that the data is the possession of the company, and that it will 'share' it with the academic.

10   But more than that, even if we did believe that everything has suddenly changed, as they claim, as everyone in this area always claims – and it will be half a decade, at least, as ever, before we can tell if they're right – not one of the longstanding members of the commercial medical writing community has ever given a clear account of why they did the things they described above with a clear conscience. They paid guest authors to put their names on papers they had little or nothing to do with; and they ghostwrote papers covertly, knowing exactly what they were doing, and why, and what effect it would have on the doctors reading their work. These are the banal, widespread, bread-and-butter activities of their industry. So, a weak new voluntary code

with no teeth from people who have not engaged in full disclosure – nor, frankly, offered an apology – is not, to my mind, any evidence that things have changed.

WHAT CAN YOU DO?

1. Lobby for your university to develop a strong and unambiguous code forbidding academic staff from being involved in ghostwriting. If you are a student, draw parallels with the plagiarism checks that are deployed on your own work.
2. Lobby for the following changes in all academic journals you are involved in:
    * A full description of "film credit" contributions at the end of every paper, including details of who initiated the idea for the publication.
    * A full declaration of the amount paid to any commercial medical writing firm for each paper, in the paper, and of who paid it.
    * Every person making a significant contribution should appear as a proper author, not "editorial assistance."
3. Raise awareness of the issue of ghostwriting, and ensure that everyone you know realizes that the people who appear as authors on an academic paper may have had little to do with it.
4. If you teach medical students, ensure that they are aware of this widespread dishonesty among senior figures in the academic medical literature.
5. If you are aware of colleagues who have accepted guest authorship, discuss the ethics of this with them.
6. If you are a doctor or an academic, lobby for your Royal College or academic society to have a strong code forbidding involvement in ghostwriting.

(2012)

REFERENCES

90. Lacasse JR, Leo J. Ghostwriting at Elite Academic Medical Centers in the United States. PLoS Med. 2010 Feb 2;7(2):e1000230.
91. Matheson A. How Industry Uses the ICMJE Guidelines to Manipulate Authorship – And How They Should Be Revised. PLoS Med. 2011;8(8):e1001072.

Questions:

1. Assess the effectiveness of the comparison made at the beginning of this excerpt. How does it work?

2. One of the supporting phrases this article uses is "to the best of my knowledge." Why do you think it is used, and do you find it a convincing rhetorical strategy? Why or why not?

3. Outline the argumentative structure used in this excerpt. Does it follow a science writing structure in any way?

4. Do you agree with this excerpt's characterization of the ICMJE's voluntary code of conduct and its lack of efficacy? What pressures can you think of that might prevent the ICMJE from taking a stronger stance on this activity?

5. Look at one of the biggest journals in your field. What sort of a research and writing credit approach does it require?

6. This article makes strong use of the personal voice in a public context. Rewrite a paragraph in a more formal language aimed at journal publication.

IV

Genetics: Inheritance, Measurement, and Pseudo-Measurement

twenty-three

STEPHEN JAY GOULD

Critique of *The Bell Curve*

*The following is a chapter from the second edition of evolutionary
biologist and popular science writer Stephen Jay Gould's book* The
Mismeasure of Man. *The chapter examines the central arguments in* The
Bell Curve: Intelligence and Class Structure in American Life, *a 1994
book discussing IQ differences between racial groups and arguing that
low IQ is related to problems such as poverty, unemployment, and crime.*

☙

THE BELL CURVE

*T*he Bell Curve by Richard J. Herrnstein and Charles Murray provides a
superb and unusual opportunity for insight into the meaning of experiment as a method in science. Reduction of confusing variables is the primary
desideratum[a] in all experiments. We bring all the buzzing and blooming confusion of the external world into our laboratories and, holding all else constant
in our artificial simplicity, try to vary just one potential factor at a time. Often,
however, we cannot use such an experimental method, particularly for most
social phenomena when importation into the laboratory destroys the subject
of our investigation – and then we can only yearn for simplifying guides in
nature. If the external world therefore obliges and holds some crucial factors
constant for us, then we can only offer thanks for such a natural boost to
understanding.

When a book garners as much attention as *The Bell Curve* has received,
we wish to know the causes. One might suspect content itself – a startling
new idea, or an old suspicion now verified by persuasive data – but the reason
might well be social acceptability, or just plain hype. *The Bell Curve* contains
no new arguments and presents no compelling data to support its anachro-
nistic social Darwinism. I must therefore conclude that its initial success in
winning such attention must reflect the depressing temper of our time – a
historical moment of unprecedented ungenerosity, when a mood for slashing

---

a *desideratum* Necessary or desired thing.

social programs can be so abetted by an argument that beneficiaries cannot be aided due to inborn cognitive limits expressed as low IQ scores.

The Bell Curve rests upon two distinctly different but sequential arguments, which together encompass the classical corpus of biological determinism as a social philosophy. The first claim (Chapters 1–12) rehashes the tenets of social Darwinism as originally constituted. ("Social Darwinism" has often been used as a general term for any evolutionary argument about the biological basis of human differences, but the initial meaning referred to a specific theory of class stratification within industrial societies, particularly to the idea that a permanently poor underclass consisting of genetically inferior people had precipitated down into their inevitable fate.)

This social Darwinian half of *The Bell Curve* arises from a paradox of egalitarianism. So long as people remain on top of the social heap by accident of a noble name or parental wealth, and so long as members of despised castes cannot rise whatever their talents, social stratification will not reflect intellectual merit, and brilliance will be distributed across all classes. But if true equality of opportunity can be attained, then smart people rise and the lower classes rigidify by retaining only the intellectually incompetent.

5   This nineteenth-century argument has attracted a variety of twentieth-century champions, including Stanford psychologist Lewis M. Terman, who imported Binet's[a] original test from France, developed the Stanford-Binet IQ test, and gave a hereditarian[b] interpretation to the results (one that Binet had vigorously rejected in developing this style of test); Prime Minister Lee Kuan Yew of Singapore, who tried to institute a eugenics[c] program of rewarding well-educated women for higher birthrates; and Richard Herrnstein, coauthor of *The Bell Curve* and author of a 1971 *Atlantic Monthly* article that presented the same argument without documentation. The general claim is neither uninteresting nor illogical, but does require the validity of four shaky premises, all asserted (but hardly discussed or defended) by Herrnstein and Murray. Intelligence, in their formulation, must be depictable as a single number, capable of ranking people in linear order, genetically based, and effectively immutable. If any of these premises are false, the entire argument collapses. For example, if all are true except immutability, then programs for early

---

a   *Binet*   Alfred Binet (1857–1911), a psychologist best known for developing the method of intelligence testing that would later evolve into the "IQ test." However, he argued that intelligence could not be reduced to a fixed number.

b   *hereditarian*   Related to the belief that inherited genetic traits are a significant factor in the development of personality and intelligence.

c   *eugenics*   Selective breeding intended to produce physically or mentally improved human beings. Eugenics has played a key role in legitimizing racist ideologies such as Nazism.

intervention in education might work to boost IQ permanently, just as a pair of eyeglasses may correct a genetic defect in vision. The central argument of *The Bell Curve* fails because most of the premises are false.

The second claim (Chapters 13–22), the lightning rod for most commentary, extends the argument for innate cognitive stratification by social class to a claim for inherited racial differences in IQ – small for Asian superiority over Caucasian, but large for Caucasians over people of African descent. This argument is as old as the study of race. The last generation's discussion centered upon the sophisticated work of Arthur Jensen[a] (far more elaborate and varied than anything presented in *The Bell Curve*, and therefore still a better source for grasping the argument and its fallacies) and the cranky advocacy of William Shockley.[b]

The central fallacy in using the substantial heritability of within-group IQ (among whites, for example) as an explanation for average differences between groups (whites vs. blacks, for example) is now well known and acknowledged by all, including Herrnstein and Murray, but deserves a restatement by example. Take a trait far more heritable than anyone has ever claimed for IQ, but politically uncontroversial – body height. Suppose that I measure adult male height in a poor Indian village beset with pervasive nutritional deprivation. Suppose the average height of adult males is 5 feet 6 inches, well below the current American mean of about 5 feet 9 inches. Heritability within the village will be high – meaning that tall fathers (they may average 5 feet 8 inches) tend to have tall sons, while short fathers (5 feet 4 inches on average) tend to have short sons. But high heritability within the village does not mean that better nutrition might not raise average height to 5 feet 10 inches (above the American mean) in a few generations. Similarly the well-documented 15-point average difference in IQ between blacks and whites in America, with substantial heritability of IQ in family lines within each group, permits no conclusion that truly equal opportunity might not raise the black average to equal or surpass the white mean.

Since Herrnstein and Murray know and acknowledge this critique, they must construct an admittedly circumstantial case for attributing most of the black-white mean difference to irrevocable genetics – while properly stressing that the average difference doesn't help at all in judging any particular person because so many individual blacks score above the white mean in

---

a   *Arthur Jensen*   American psychologist (1923–2012) best known for arguing that genetics was partially responsible for the differences in average IQ between racial groups. Gould critiques Jensen's arguments elsewhere in *The Mismeasure of Man*.

b   *William Shockley*   American inventor (1910–89) who became infamous as an advocate of eugenics.

IQ. Quite apart from the rhetorical dubriety[a] of this old ploy in a shopworn genre – "some-of-my-best-friends-are-group-x" – Herrnstein and Murray violate fairness by converting a complex case that can only yield agnosticism into a biased brief for permanent and heritable difference. They impose this spin by turning every straw on their side into an oak, while mentioning but downplaying the strong circumstantial case for substantial malleability and little average genetic difference (impressive IQ gains for poor black children adopted into affluent and intellectual homes; average IQ increases in some nations since World War II equal to the entire 15-point difference now separating blacks and whites in America; failure to find any cognitive differences between two cohorts of children born out of wedlock to German women, and raised in Germany as Germans, but fathered by black and white American soldiers).

Disturbing as I find the anachronism of *The Bell Curve,* I am even more distressed by its pervasive disingenuousness. The authors omit facts, misuse statistical methods, and seem unwilling to admit the consequences of their own words.

DISINGENUOUSNESS OF CONTENT

10 The ocean of publicity that has engulfed *The Bell Curve* has a basis in what Murray and Herrnstein (*New Republic,* October 31, 1994) call "the flashpoint of intelligence as a public topic: the question of genetic differences between the races." And yet, since the day of publication, Murray has been temporizing and denying that race is an important subject in the book at all; instead, he blames the press for unfairly fanning these particular flames. He writes with Herrnstein (who died just a month before publication) in the *New Republic*: "Here is what we hope will be our contribution to the discussion. We put it in italics; if we could we would put it in neon lights: *The answer doesn't much matter.*"

Fair enough in the narrow sense that any individual may be a rarely brilliant member of an averagely dumb group (and therefore not subject to judgment by the group mean), but Murray cannot deny that *The Bell Curve* treats race as one of two major topics, with each given about equal space; nor can he pretend that strongly stated claims about group differences have no political impact in a society obsessed with the meanings and consequences of ethnicity. The very first sentence of *The Bell Curve*'s preface acknowledges equality of treatment for the two subjects of individual and group differences: "This book is about differences in intellectual capacity among people and

---

a *dubriety* Doubtfulness.

groups and what these differences mean for America's future." And Murray and Herrnstein's *New Republic* article begins by identifying racial difference as the key subject of interest: "The private dialogue about race in America is far different from the public one."

DISINGENUOUSNESS OF ARGUMENT

The Bell Curve is a rhetorical masterpiece of scientism,[a] and the particular kind of anxiety and obfuscation that numbers impose upon nonprofessional commentators. The book runs to 845 pages, including more than 100 pages of appendices filled with figures. So the text looks complicated, and reviewers shy away with a knee-jerk claim that, while they suspect fallacies of argument, they really cannot judge. So Mickey Kaus writes in the *New Republic* (October 31): "As a lay reader of *The Bell Curve*, I'm unable to judge fairly," as does Leon Wieseltier in the same issue: "Murray, too, is hiding the hardness of his politics behind the hardness of his science. And his science for all I know is soft.... Or so I imagine. I am not a scientist. I know nothing about psychometrics."[b] Or Peter Passell in the *New York Times* (October 27, 1994): "But this reviewer is not a biologist, and will leave the argument to experts."

In fact, *The Bell Curve* is extraordinarily one-dimensional. The book makes no attempt to survey the range of available data, and pays astonishingly little attention to the rich and informative history of this contentious subject. (One can only recall Santayana's dictum, now a cliché of intellectual life: "Those who cannot remember the past are condemned to repeat it.") Virtually all the analysis rests upon a single technique applied to a single set of data – all probably done in one computer run. (I do agree that the authors have used the most appropriate technique – multiple regression – and the best source of information – the National Longitudinal Survey of Youth – though I shall expose a core fallacy in their procedure below. Still, claims as broad as those advanced in *The Bell Curve* simply cannot be adequately defended – that is, either properly supported or denied – by such a restricted approach.)

The blatant errors and inadequacies of *The Bell Curve* could be picked up by lay reviewers if only they would not let themselves be frightened by numbers – for Herrnstein and Murray do write clearly and their mistakes are both patent and accessible. I would rank the fallacies in two categories: omissions and confusions, and content.

---

a   *scientism*   I.e., exploitation of undue belief in the authority of science.
b   *psychometrics*   Scientific measurement and analysis of mental abilities and other psychological information.

15   1. *Omissions and confusions:* While disclaiming on his own ability to judge, Mickey Kaus (in the *New Republic*) does correctly identify "the first two claims" that are absolutely essential "to make the pessimistic 'ethnic difference' argument work": "(1) that there is a single, general measure of mental ability; (2) that the IQ tests that purport to measure this ability ... aren't culturally biased."

Nothing in *The Bell Curve* angered me more than the authors' failure to supply any justification for their central claim, the *sine qua non*,[a] of their entire argument: the reality of IQ as a number that measures a real property in the head, the celebrated "general factor" of intelligence (known as *g*) first identified by Charles Spearman in 1904. Murray and Herrnstein simply proclaim that the issue has been decided, as in this passage from their *New Republic* article: "Among the experts, it is by now beyond much technical dispute that there is such a thing as a general factor of cognitive ability on which human beings differ and that this general factor is measured reasonably well by a variety of standardized tests, best of all by IQ tests designed for that purpose."

Such a statement represents extraordinary obfuscation, achieved by defining "expert" as "that group of psychometricians working in the tradition of *g* and its avatar IQ." The authors even admit (pp. 14–19) that three major schools of psychometric interpretation now contend, and that only one supports their view of *g* and IQ – the classicists as championed in *The Bell Curve* ("intelligence as a structure"), the revisionists ("intelligence as information processing"), and the radicals ("the theory of multiple intelligences").

This vital issue cannot be decided, or even understood without discussing the key and only rationale that *g* has maintained since Spearman invented the concept in 1904 – factor analysis. The fact that Herrnstein and Murray barely mention the factor analytic argument (the subject receives fleeting attention in two paragraphs) provides a central indictment and illustration of the vacuousness in *The Bell Curve*. How can authors base an eight-hundred-page book on a claim for the reality of IQ as measuring a genuine, and largely genetic, general cognitive ability – and then hardly mention, either pro or con, the theoretical basis for their certainty? Various clichés like "*Hamlet* without the Prince of Denmark" come immediately to mind.

Admittedly, factor analysis is a difficult and mathematical subject, but it can be explained to lay readers with a geometrical formulation developed by L.L. Thurstone in the 1930s and used by me in Chapter 7 of *The Mismeasure of Man*. A few paragraphs cannot suffice for adequate explanation, so,

a   *sine qua non*   Necessary element or condition.

although I offer some sketchy hints below, readers should not question their own IQs if the topic still seems arcane.

In brief, a person's performances on various mental tests tend to be posi- 20 tively correlated – that is, if you do well on one kind of test, you tend to do well on the others. This result is scarcely surprising, and is subject to either purely genetic (the innate thing in the head that boosts all scores) or purely environmental interpretation (good books and good childhood nutrition to enhance all performances). Therefore, the positive correlations say nothing in themselves about causes.

Charles Spearman used factor analysis to identify a single axis – which he called *g* – that best identifies the common factor behind positive correlations among the tests. But Thurstone later showed that *g* could be made to disappear by simply rotating the factor axes to different positions. In one rotation, Thurstone placed the axes near the most widely separated of attributes among the tests – thus giving rise to the theory of multiple intelligences (verbal, mathematical, spatial, etc., with no overarching *g*). This theory (the "radical" view in Herrnstein and Murray's classification) has been supported by many prominent psychometricians, including J.P. Guilford in the 1950s, and most prominently today by Howard Gardner. In this perspective, *g* cannot have inherent reality, for *g* emerges in one form of mathematical representation for correlations among tests, and disappears (or at least greatly attenuates) in other forms that are entirely equivalent in amounts of information explained. In any case, one can't grasp the issue at all without a clear exposition of factor analysis – and *The Bell Curve* cops out completely on this central concept.

On Kaus's second theme of "cultural bias," *The Bell Curve*'s presentation matches Arthur Jensen's, and that of other hereditarians, in confusing a technical (and proper) meaning of bias (I call it "S-bias" for "statistical") with the entirely different vernacular concept (I call it "V-bias") that agitates popular debate. All these authors swear up and down (and I agree with them completely) that the tests are not biased – in the statistician's definition. Lack of S-bias means that the same score, when achieved by members of different groups, predicts the same consequence – that is, a black person and a white person with an identical IQ score of 100 will have the same probabilities for doing anything that IQ is supposed to predict. (I should hope that mental tests aren't S-biased, for the testing profession isn't worth very much if practitioners can't eliminate such an obvious source of unfairness by careful choice and framing of questions.)

But V-bias, the source of public concern, embodies an entirely different issue that, unfortunately, uses the same word. The public wants to know whether blacks average 85 and whites 100 because society treats blacks unfairly – that is, whether lower black scores record biases in this social sense. And this crucial question (to which we do not know the answer) cannot be addressed by a demonstration that S-bias doesn't exist (the only issue treated, however correctly, by *The Bell Curve*).

2. *Content:* As stated above, virtually all the data in *The Bell Curve* derive from one analysis – a plotting, by a technique called multiple regression, of the social behaviors that agitate us, such as crime, unemployment, and births out of wedlock (treated as dependent variables), against both IQ and parental socioeconomic status (treated as independent variables). The authors first hold IQ constant and consider the relationship of social behaviors to parental socioeconomic status. They then hold socioeconomic status constant and consider the relationship of the same social behaviors to IQ. In general, they find a higher correlation with IQ than with socioeconomic status; for example, people with low IQ are more likely to drop out of high school than people whose parents have low socioeconomic status.

25 But such analyses must engage two issues – form *and strength* of the relationship) – and Herrnstein and Murray only discuss the issue that seems to support their viewpoint, while virtually ignoring (and in one key passage almost willfully and purposely hiding) the other factor that counts so profoundly against them. Their numerous graphs only present the *form* of the relationships – that is, they draw the regression curves of their variables against IQ and parental socioeconomic status. But, in violation of all statistical norms that I've ever learned, they plot *only* the regression curve and do not show the scatter of variation around the curve, so their graphs show nothing about the *strength* of the relationship – that is, the amount of variation in social factors explained by IQ and socioeconomic status.

Now why would Herrnstein and Murray focus on the form and ignore the strength? Almost all of their relationships are very weak – that is, very little of the variation in social factors can be explained by either IQ or socioeconomic status (even though the form of this small amount tends to lie in their favored direction). In short, IQ is not a major factor in determining variation in nearly all the social factors they study – and their vaunted conclusions thereby collapse, or become so strongly attenuated that their pessimism and conservative social agenda gain no significant support.

Herrnstein and Murray actually admit as much in one crucial passage on page 117, but then they hide the pattern. They write: "It almost always explains less than 20 percent of the variance, to use the statistician's term, usually less than 10 percent and often less than 5 percent. What this means in English is that you cannot predict what a given person will do from his IQ score.... On the other hand, despite the low association at the individual level, large differences in social behavior separate groups of people when the groups differ intellectually on the average." Despite this disclaimer, their remarkable next sentence makes a strong causal claim: "We will argue that intelligence itself, not just its correlation with socioeconomic status, is responsible for these group differences." But a few percent of statistical determination is not equivalent to causal explanation (and correlation does not imply cause in any case, even when correlations are strong – as in the powerful, perfect, positive correlation between my advancing age and the rise of the national debt). Moreover, their case is even worse for their key genetic claims – for they cite heritabilities of about 60 percent for IQ, so you must nearly halve the few percent explained if you want to isolate the strength of genetic determination by their own criteria!

My charge of disingenuousness receives its strongest affirmation in a sentence tucked away on the first page of Appendix 4, page 593, where the authors state: "In the text, we do not refer to the usual measure of goodness of fit for multiple regressions, R^2, but they are presented here for the cross-sectional analysis." Now why would they exclude from the text, and relegate to an appendix that very few people will read or even consult, a number that, by their own admission, is "the usual measure of goodness of fit." I can only conclude that they did not choose to admit in the main text the extreme weakness of their vaunted relationships.

Herrnstein and Murray's correlation coefficients are generally low enough by themselves to inspire lack of confidence. (Correlation coefficients measure the strength of linear relationships between variables; positive values run from 0.0 for no relationship to 1.0 for perfect linear relationship.) Although low figures are not atypical in the social sciences for large surveys involving many variables, most of Herrnstein and Murray's correlations are very weak – often in the 0.2 to 0.4 range. Now, 0.4 may sound respectably strong, but – and now we come to the key point – R^2 is the square of the correlation coefficient, and the square of a number between 0 and 1 is less than the number itself, so a 0.4 correlation yields an r-squared of only 0.16. In Appendix 4, then, we discover that the vast majority of measures for R^2, excluded from the main body of the text, have values less than 0.1. These very low values of R^2 expose the true weakness, in any meaningful vernacular sense, of nearly all the relationships that form the heart of *The Bell Curve*.

DISINGENUOUSNESS OF PROGRAM

Like so many conservative ideologues who rail against a largely bogus ogre of suffocating political correctness, Herrnstein and Murray claim that they only seek a hearing for unpopular views so that truth will out. And here, for once, I agree entirely. As a cardcarrying First Amendment (near) absolutist, I applaud the publication of unpopular views that some people consider dangerous. I am delighted that *The Bell Curve* was written – so that its errors could be exposed, for Herrnstein and Murray are right in pointing out the difference between public and private agendas on race, and we must struggle to make an impact upon the private agendas as well.

But *The Bell Curve* can scarcely be called an academic treatise in social theory and population genetics. The book is a manifesto of conservative ideology, and its sorry and biased treatment of data records the primary purpose – advocacy above all. The text evokes the dreary and scary drumbeat of claims associated with conservative think tanks – reduction or elimination of welfare, ending of affirmative action in schools and workplaces, cessation of Head Start[a] and other forms of preschool education, cutting of programs for slowest learners and application of funds to the gifted (Lord knows I would love to see more attention paid to talented students, but not at this cruel and cynical price).

The penultimate chapter presents an apocalyptic vision of a society with a growing underclass permanently mired in the inevitable sloth of their low IQs. They will take over our city centers, keep having illegitimate babies (for many are too stupid to practice birth control), commit more crimes, and ultimately require a kind of custodial state, more to keep them in check (and out of our high IQ neighborhoods) than with any hope for an amelioration that low IQ makes impossible in any case. Herrnstein and Murray actually write (p. 526): "In short, by custodial state, we have in mind a high-tech and more lavish version of the Indian reservation for some substantial minority of the nation's population, while the rest of America tries to go about its business."

The final chapter then tries to suggest an alternative, but I have never read anything so feeble, so unlikely, so almost grotesquely inadequate. They yearn romantically for the "good old days" of towns and neighborhoods where all people could be given tasks of value and self-esteem could be found for all steps in the IQ hierarchy (so Forrest Gump[b] might collect the clothing for the church raffle, while Mr. Murray and the other bright folks do the planning and keep the accounts. Have they forgotten about the town Jew and the dwellers on the other side of the tracks in many of these idyllic villages?). I do believe in this concept of neighborhood, and I will fight for its return. I grew up in such a place within that mosaic known as Queens, New York City, but

---

a *Head Start* American preschool program for children from low-income families.
b *Forrest Gump* Fictional character (in *Forrest Gump*, a 1986 novel and 1994 film) whose IQ score is significantly below average.

can anyone seriously find solutions (rather than important palliatives) to our social ills therein?

However, if Herrnstein and Murray are wrong about IQ as an immutable thing in the head, with humans graded in a single scale of general capacity, leaving large numbers of custodial incompetents at the bottom, then the model that generates their gloomy vision collapses, and the wonderful variousness of human abilities, properly nurtured, reemerges. We must fight the doctrine of *The Bell Curve* both because it is wrong and because it will, if activated, cut off all possibility of proper nurturance for everyone's intelligence. Of course we cannot all be rocket scientists or brain surgeons (to use the two current slang synecdoches[a] for smartest of the smart), but those who can't might be rock musicians or professional athletes (and gain far more social prestige and salary thereby) – while others will indeed serve by standing and waiting.

I closed Chapter 7 in *The Mismeasure of Man* on the unreality of g and the fallacy of regarding intelligence as a single innate thing-in-the-head (rather than a rough vernacular term for a wondrous panoply of largely independent abilities) with a marvelous quote from John Stuart Mill, well worth repeating to debunk this generation's recycling of biological determinism for the genetics of intelligence:

> The tendency has always been strong to believe that whatever received a name must be an entity or being, having an independent existence of its own. And if no real entity answering to the name could be found, men did not for that reason suppose that none existed, but imagined that it was something particularly abstruse and mysterious.

How strange that we would let a single false number divide us, when evolution has united all people in the recency of our common ancestry – thus undergirding with a shared humanity that infinite variety which custom can never stale. *E pluribus unum.*[b]

(1996)

---

a *synecdoches* Poetic devices in which the name of a part of something is substituted for the whole.

b *E pluribus unum* Latin: Out of many, one; a motto appearing on the Seal of the United States.

Questions:

1. Look at the chapter's use of the term "we." Who is the article's "we"? How do you know?

2. How does the chapter define "social Darwinism"?

3. This piece takes the form of an extended book review (or what would, in a journal, be called a review article). Contrast it with Gladwell's "None of the Above" (included in this anthology), which is also a book review. How is the tone different? What sorts of authority claims does each make?

4. What is "biological determinism"? Give another example of it.

5. How does the article show Herrnstein and Murray as outmoded, even antiquated, thinkers?

6. What are the four shaky premises on which *The Bell Curve*'s argument rests, and how are they linked so that if any one is false, the entire argument collapses?

7. What is cognitive stratification? How is it related to "inherited racial differences in IQ"?

8. This article functions, in part, through analogy and example. Find one example of each. In each case, how effective do you find the strategy identified?

9. Why would a "white mean" be the chosen standard for a social discussion of IQ? If you were going to test an abstract skill, what level of ability would it be best to grade it against?

10. Stephen Jay Gould was a New Yorker from a secular Jewish family. He worked at Harvard and is usually referred to as a "paleontologist, evolutionary biologist, and historian of science." Which (if any) elements of that description make a difference to your assessment of the article's reliability? Why?

11. In paragraph 12, Gould calls *The Bell Curve* "a rhetorical masterpiece of scientism." What evidence does he give in support of this characterization?

12. Why does it matter that Herrnstein and Murray's argument does not address the strength of the regression correlations?

13. The work makes a distinction between non-biased statistics and "vernacular concept" bias. What is the distinction and why is it important?

14. What reasons can you think of to explain why the article ends with the Latin statement "*E pluribus unum*"?

twenty-four

MALCOLM GLADWELL

None of the Above: What IQ Doesn't Tell You about Race

Is IQ genetically determined? This article by a noted journalist explores the meaning of IQ and its relationship to race, class, and culture.

℮

One Saturday in November of 1984, James Flynn, a social scientist at the University of Otago, in New Zealand, received a large package in the mail. It was from a colleague in Utrecht, and it contained the results of IQ tests given to two generations of Dutch eighteen-year-olds. When Flynn looked through the data, he found something puzzling. The Dutch eighteen-year-olds from the nineteen-eighties scored better than those who took the same tests in the nineteen-fifties – and not just slightly better, *much* better.

Curious, Flynn sent out some letters. He collected intelligence-test results from Europe, from North America, from Asia, and from the developing world, until he had data for almost thirty countries. In every case, the story was pretty much the same. IQs around the world appeared to be rising by 0.3 points per year, or three points per decade, for as far back as the tests had been administered. For some reason, human beings seemed to be getting smarter.

Flynn has been writing about the implications of his findings – now known as the Flynn effect – for almost twenty-five years. His books consist of a series of plainly stated statistical observations, in support of deceptively modest conclusions, and the evidence in support of his original observation is now so overwhelming that the Flynn effect has moved from theory to fact. What remains uncertain is how to make sense of the Flynn effect. If an American born in the nineteen-thirties has an IQ of 100, the Flynn effect says that his children will have IQs of 108, and his grandchildren IQs of close to 120 – more than a standard deviation higher. If we work in the opposite direction, the typical teen-ager of today, with an IQ of 100, would have had grandparents with average IQs of 82 – seemingly below the threshold necessary to graduate from high school. And, if we go back even farther, the Flynn effect puts the average IQs of the schoolchildren of 1900 at around 70, which is to suggest, bizarrely, that a century ago the United States was populated largely by people who today would be considered mentally retarded.

For almost as long as there have been IQ tests, there have been IQ fundamentalists. H.H. Goddard, in the early years of the past century, established the idea that intelligence could be measured along a single, linear scale. One of his particular contributions was to coin the word "moron." "The people who are doing the drudgery are, as a rule, in their proper places," he wrote. Goddard was followed by Lewis Terman, in the nineteen-twenties, who rounded up the California children with the highest IQs, and confidently predicted that they would sit at the top of every profession. In 1969, the psychometrician Arthur Jensen argued that programs like Head Start, which tried to boost the academic performance of minority children, were doomed to failure, because IQ was so heavily genetic; and in 1994 Richard Herrnstein and Charles Murray, in "The Bell Curve," notoriously proposed that Americans with the lowest IQs be sequestered in a "high-tech" version of an Indian reservation, "while the rest of America tries to go about its business."[1] To the IQ fundamentalist, two things are beyond dispute: first, that IQ tests measure some hard and identifiable trait that predicts the quality of our thinking; and, second, that this trait is stable – that is, it is determined by our genes and largely impervious to environmental influences.

5   This is what James Watson, the co-discoverer of DNA, meant when he told an English newspaper recently that he was "inherently gloomy" about the prospects for Africa. From the perspective of an IQ fundamentalist, the fact that Africans score lower than Europeans on IQ tests suggests an ineradicable cognitive disability. In the controversy that followed, Watson was defended by the journalist William Saletan, in a three-part series for the online magazine *Slate*. Drawing heavily on the work of J. Philippe Rushton – a psychologist who specializes in comparing the circumference of what he calls the Negroid brain with the length of the Negroid penis – Saletan took the fundamentalist position to its logical conclusion. To erase the difference between blacks and whites, Saletan wrote, would probably require vigorous interbreeding between the races, or some kind of corrective genetic engineering aimed at upgrading African stock. "Economic and cultural theories have failed to explain most of the pattern," Saletan declared, claiming to have been "soaking [his] head in each side's computations and arguments." One argument that Saletan never soaked his head in, however, was Flynn's, because what Flynn discovered in his mailbox upsets the certainties upon which IQ fundamentalism rests. If whatever the thing is that IQ tests measure can jump so much in a generation, it can't be all that immutable and it doesn't look all that innate.

The very fact that average IQs shift over time ought to create a "crisis of confidence," Flynn writes in "What Is Intelligence?," his latest attempt to puzzle through the implications of his discovery. "How could such huge gains be intelligence gains? Either the children of today were far brighter than their parents or, at least in some circumstances, IQ tests were not good measures of intelligence."

The best way to understand why IQs rise, Flynn argues, is to look at one of the most widely used IQ tests, the so-called WISC (for Wechsler Intelligence Scale for Children). The WISC is composed of ten subtests, each of which measures a different aspect of IQ. Flynn points out that scores in some of the categories – those measuring general knowledge, say, or vocabulary or the ability to do basic arithmetic – have risen only modestly over time. The big gains on the WISC are largely in the category known as "similarities," where you get questions such as "In what way are 'dogs' and 'rabbits' alike?" Today, we tend to give what, for the purposes of IQ tests, is the right answer: dogs and rabbits are both mammals. A nineteenth-century American would have said that "you use dogs to hunt rabbits."

"If the everyday world is your cognitive home, it is not natural to detach abstractions and logic and the hypothetical from their concrete referents," Flynn writes. Our great-grandparents may have been perfectly intelligent. But they would have done poorly on IQ tests because they did not participate in the twentieth century's great cognitive revolution, in which we learned to sort experience according to a new set of abstract categories. In Flynn's phrase, we have now had to put on "scientific spectacles," which enable us to make sense of the WISC questions about similarities. To say that Dutch IQ scores rose substantially between 1952 and 1982 was another way of saying that the Netherlands in 1982 was, in at least certain respects, much more cognitively demanding than the Netherlands in 1952. An IQ, in other words, measures not so much how smart we are as how *modern* we are.

This is a critical distinction. When the children of Southern Italian immigrants were given IQ tests in the early part of the past century, for example, they recorded median scores in the high seventies and low eighties, a full standard deviation below their American and Western European counterparts. Southern Italians did as poorly on IQ tests as Hispanics and blacks did. As you can imagine, there was much concerned talk at the time about the genetic inferiority of Italian stock, of the inadvisability of letting so many second-class immigrants into the United States, and of the squalor that seemed endemic to Italian urban neighborhoods. Sound familiar? These days, when talk turns to the supposed genetic differences in the intelligence of certain races, Southern Italians have disappeared from the discussion. "Did their genes begin to mutate somewhere in the 1930s?" the psychologists Seymour Sarason and John Doris ask, in their account of the Italian experience. "Or is it possible that somewhere in the 1920s, if not earlier, the sociocultural history of Italo-Americans took a turn from the blacks and the Spanish Americans which permitted their assimilation into the general undifferentiated mass of Americans?"

10   The psychologist Michael Cole and some colleagues once gave members of the Kpelle tribe, in Liberia, a version of the WISC similarities test: they took a basket of food, tools, containers, and clothing and asked the tribesmen to sort them into appropriate categories. To the frustration of the researchers, the Kpelle chose functional pairings. They put a potato and a knife together because a knife is used to cut a potato. "A wise man could only do such-and-such," they explained. Finally, the researchers asked, "How would a fool do it?" The tribesmen immediately re-sorted the items into the "right" categories. It can be argued that taxonomical categories are a developmental improvement – that is, that the Kpelle would be more likely to advance, tech-nologically and scientifically, if they started to see the world that way. But to label them less intelligent than Westerners, on the basis of their performance on that test, is merely to state that they have different cognitive preferences and habits. And if IQ varies with habits of mind, which can be adopted or discarded in a generation, what, exactly, is all the fuss about?

When I was growing up, my family would sometimes play Twenty Ques-tions on long car trips. My father was one of those people who insist that the standard categories of animal, vegetable, and mineral be supplemented with a fourth category: "abstract." Abstract could mean something like "whatever it was that was going through my mind when we drove past the water tower fifty miles back." That abstract category sounds absurdly difficult, but it wasn't: it merely required that we ask a slightly different set of questions and grasp a slightly different set of conventions, and, after two or three rounds of practice, guessing the contents of someone's mind fifty miles ago becomes as easy as guessing Winston Churchill. (There is one exception. That was the trip on which my old roommate Tom Connell chose, as an abstraction, "the Unknown Soldier" – which allowed him legitimately and gleefully to answer "I have no idea" to almost every question. There were four of us playing. We gave up after an hour.) Flynn would say that my father was teaching his three sons how to put on scientific spectacles, and that extra practice probably bumped up all of our IQs a few notches. But let's be clear about what this means. There's a world of difference between an IQ advantage that's genetic and one that depends on extended car time with Graham Gladwell.

Flynn is a cautious and careful writer. Unlike many others in the IQ debates, he resists grand philosophizing. He comes back again and again to the fact that IQ scores are generated by paper-and-pencil tests – and making sense of those scores, he tells us, is a messy and complicated business that requires something closer to the skills of an accountant than to those of a philosopher.

For instance, Flynn shows what happens when we recognize that IQ is not a freestanding number but a value attached to a specific time and a specific test. When an IQ test is created, he reminds us, it is calibrated or "normed" so that the test-takers in the fiftieth percentile – those exactly at the median – are assigned a score of 100. But since IQs are always rising, the only way to keep that hundred-point benchmark is periodically to make the tests more difficult – to "renorm" them. The original WISC was normed in the late nineteen-forties. It was then renormed in the early nineteen-seventies, as the WISC-R; renormed a third time in the late eighties, as the WISC III; and renormed again a few years ago, as the WISC IV – with each version just a little harder than its predecessor. The notion that anyone "has" an IQ of a certain number, then, is meaningless unless you know which WISC he took, and when he took it, since there's a substantial difference between getting a 130 on the WISC IV and getting a 130 on the much easier WISC.

This is not a trivial issue. IQ tests are used to diagnose people as mentally retarded, with a score of 70 generally taken to be the cutoff. You can imagine how the Flynn effect plays havoc with that system. In the nineteen-seventies and eighties, most states used the WISC-R to make their mental-retardation diagnoses. But since kids – even kids with disabilities – score a little higher every year, the number of children whose scores fell below 70 declined steadily through the end of the eighties. Then, in 1991, the WISC III was introduced, and suddenly the percentage of kids labeled retarded went up. The psychologists Tomoe Kanaya, Matthew Scullin, and Stephen Ceci estimated that, if every state had switched to the WISC III right away, the number of Americans labeled mentally retarded should have doubled.

That is an extraordinary number. The diagnosis of mental disability is one of the most stigmatizing of all educational and occupational classifications – and yet, apparently, the chances of being burdened with that label are in no small degree a function of the point, in the life cycle of the WISC, at which a child happens to sit for his evaluation. "As far as I can determine, no clinical or school psychologists using the WISC over the relevant 25 years noticed that its criterion of mental retardation became more lenient over time," Flynn wrote, in a 2000 paper. "Yet no one drew the obvious moral about psychologists in the field: They simply were not making any systematic assessment of the IQ criterion for mental retardation."

Flynn brings a similar precision to the question of whether Asians have a genetic advantage in IQ, a possibility that has led to great excitement among IQ fundamentalists in recent years. Data showing that the Japanese had higher

IQs than people of European descent, for example, prompted the British psychometrician and eugenicist Richard Lynn to concoct an elaborate evolutionary explanation involving the Himalayas, really cold weather, premodern hunting practices, brain size, and specialized vowel sounds. The fact that the IQs of Chinese-Americans also seemed to be elevated has led IQ fundamentalists to posit the existence of an international IQ pyramid, with Asians at the top, European whites next, and Hispanics and blacks at the bottom.

Here was a question tailor-made for James Flynn's accounting skills. He looked first at Lynn's data, and realized that the comparison was skewed. Lynn was comparing American IQ estimates based on a representative sample of schoolchildren with Japanese estimates based on an upper-income, heavily urban sample. Recalculated, the Japanese average came in not at 106.6 but at 99.2. Then Flynn turned his attention to the Chinese-American estimates. They turned out to be based on a 1975 study in San Francisco's Chinatown using something called the Lorge-Thorndike Intelligence Test. But the Lorge-Thorndike test was normed in the nineteen-fifties. For children in the nineteen-seventies, it would have been a piece of cake. When the Chinese-American scores were reassessed using up-to-date intelligence metrics, Flynn found, they came in at 97 verbal and 100 nonverbal. Chinese-Americans had slightly lower IQs than white Americans.

The Asian-American success story had suddenly been turned on its head. The numbers now suggested, Flynn said, that they had succeeded not because of their *higher* IQs, but despite their *lower* IQs. Asians were overachievers. In a nifty piece of statistical analysis, Flynn then worked out just how great that overachievement was. Among whites, virtually everyone who joins the ranks of the managerial, professional, and technical occupations has an IQ of 97 or above. Among Chinese-Americans, that threshold is 90. A Chinese-American with an IQ of 90, it would appear, does as much with it as a white American with an IQ of 97.

There should be no great mystery about Asian achievement. It has to do with hard work and dedication to higher education, and belonging to a culture that stresses professional success. But Flynn makes one more observation. The children of that first successful wave of Asian-Americans really did have IQs that were higher than everyone else's – coming in somewhere around 103. Having worked their way into the upper reaches of the occupational scale, and taken note of how much the professions value abstract thinking, Asian-American parents have evidently made sure that their own children wore scientific spectacles. "Chinese Americans are an ethnic group for whom high achievement preceded high IQ rather than the reverse," Flynn concludes, reminding us that in our discussions of the relationship between

IQ and success we often confuse causes and effects. "It is not easy to view the history of their achievements without emotion," he writes. That is exactly right. To ascribe Asian success to some abstract number is to trivialize it.

Two weeks ago, Flynn came to Manhattan to debate Charles Murray at a forum sponsored by the Manhattan Institute. Their subject was the black-white IQ gap in America. During the twenty-five years after the Second World War, that gap closed considerably. The IQs of white Americans rose, as part of the general worldwide Flynn effect, but the IQs of black Americans rose faster. Then, for about a period of twenty-five years, that trend stalled – and the question was why. [20]

Murray showed a series of PowerPoint slides, each representing different statistical formulations of the IQ gap. He appeared to be pessimistic that the racial difference would narrow in the future. "By the nineteen-seventies, you had gotten most of the juice out of the environment that you were going to get," he said. That gap, he seemed to think, reflected some inherent difference between the races. "Starting in the nineteen-seventies, to put it very crudely, you had a higher proportion of black kids being born to really dumb mothers," he said. When the debate's moderator, Jane Waldfogel, informed him that the most recent data showed that the race gap had begun to close again, Murray seemed unimpressed, as if the possibility that blacks could ever make further progress was inconceivable.

Flynn took a different approach. The black-white gap, he pointed out, differs dramatically by age. He noted that the tests we have for measuring the cognitive functioning of infants, though admittedly crude, show the races to be almost the same. By age four, the average black IQ is 95.4 – only four and a half points behind the average white IQ. Then the real gap emerges: from age four through twenty-four, blacks lose six-tenths of a point a year, until their scores settle at 83.4.

That steady decline, Flynn said, did not resemble the usual pattern of genetic influence. Instead, it was exactly what you would expect, given the disparate cognitive environments that whites and blacks encounter as they grow older. Black children are more likely to be raised in single-parent homes than are white children – and single-parent homes are less cognitively complex than two-parent homes. The average IQ of first-grade students in schools that blacks attend is 95, which means that "kids who want to be above average don't have to aim as high." There were possibly adverse differences between black teen-age culture and white teen-age culture, and an enormous number of young black men are in jail – which is hardly the kind of environment in which someone would learn to put on scientific spectacles.

Flynn then talked about what we've learned from studies of adoption and mixed-race children – and that evidence didn't fit a genetic model, either. If IQ is innate, it shouldn't make a difference whether it's a mixed-race child's mother or father who is black. But it does: children with a white mother and a black father have an eight-point IQ advantage over those with a black mother and a white father. And it shouldn't make much of a difference where a mixed-race child is born. But, again, it does: the children fathered by black American G.I.s in postwar Germany and brought up by their German mothers have the same IQs as the children of white American G.I.s and German mothers. The difference, in that case, was not the fact of the children's blackness, as a fundamentalist would say. It was the fact of their *Germanness* – of their being brought up in a different culture, under different circumstances. "The mind is much more like a muscle than we've ever realized," Flynn said. "It needs to get cognitive exercise. It's not some piece of clay on which you put an indelible mark." The lesson to be drawn from black and white differences was the same as the lesson from the Netherlands years ago: IQ measures not just the quality of a person's mind but the quality of the world that person lives in.

(2007)

Correction

1.   In fact, Herrnstein and Murray deplored the prospect of such 'custodialism' and recommended that steps be taken to avert it. We regret the error.

Questions:

1.   Does the article give the reader enough information to evaluate whether Flynn's interpretation of the data is superior to the other interpretations that are mentioned? If so, how do you know? If not, what information is missing?

2.   The article states that Flynn says making "sense of" IQ scores "requires something closer to the skills of an accountant than to those of a philosopher." What does this statement mean? What differentiates Flynn's interpretation of the numbers from those of the "IQ fundamentalists"?

3.   What are some real-world implications of the way society defines and interprets IQ? Draw examples from Gladwell's article, the ideas of H.H. Goddard (discussed therein), and/or your own experience.

4.   Summarize this article's thesis. Where does the thesis statement occur?

5.   What argumentative strategies does the article use to support this thesis? Which ones are the most effective?

6. Outline Gladwell's article in terms of argument function (thesis, background, explication, summary, etc.). Can you think of any other way to arrange the material more effectively?

7. Does the story about Gladwell's father help you to understand cognitive abstraction? What does the story contribute to the article?

8. According to the article, what does IQ actually measure? How important do you think this measurement is?

9. Considering the article's arguments, do you think that IQ should be used in the diagnosis of mental disability? Why or why not?

10. In part, IQ tests purport to measure one's ability to apply abstract taxonomical categories. According to the article, "it can be argued" that the ability to apply these categories is "a developmental improvement" that might lead to technological and scientific advancement. Defend or argue against this proposition.

11. Look at Stephen Jay Gould's "Critique of *The Bell Curve*" (included in this anthology). What elements of Gladwell's article are popularized or paraphrased material from Gould's work?

12. Gladwell is Canadian, though he now lives in the United States and writes regularly for *The New Yorker*. Does knowledge of his cultural background make a difference to how you respond to this article? Should it have any bearing?

13. Gladwell identifies as black. Does knowledge of his racial background make a difference to how you respond to this article? Should it have any bearing?

14. Look up eugenics. Look up the Pioneer Fund, on the board of which "British psychometrician and eugenicist Richard Lynn" sits. Does knowing about the nature of the fund change your view of his argument in any way?

twenty-five

Richard C. Strohman

Linear Genetics, Non-Linear Epigenetics: Complementary Approaches to Understanding Complex Diseases

This article by biologist Richard Strohman examines the limitations of genetic analysis as a means of predicting an organism's traits. The article explores non-linear epigenetics as a source of insight with implications for our understanding of disease, evolution, and the process of gene expression.

☙

ABSTRACT

Recent discoveries and rediscoveries in molecular and cell biology, in population and evolutionary biology, and in disease natural history raise new doubts about the ability of genetic analysis alone to predict multifactorial[a] (polygenic) human diseases and other complex phenotypes.[b] These doubts serve to redirect our attention to epigenetic regulation[c] as a second informational system in parallel with the genome. Epigenetic regulation is now viewed by many biologists as a process that includes mechanisms capable of constraining the genome and providing for new patterns of gene expression. Epigenetic networks, both intra- and inter-cellular, provide a basis for nonlinear and chaotic views of cellular and tissue level differentiation and organization, and thus provide a more dynamic approach to understanding

---

a   *multifactorial*   Caused by a combination of multiple genetic factors.
b   *phenotypes*   Characteristics in an organism, determined by the interaction of genetics and environment.
c   *epigenetic regulation*   Heritable effects on gene expression caused by anything other than changes to DNA sequence.

the creation of complex phenotypes, even from isogenic[a] conditions. The reality of regulatory networks within cells inserted, as it were, between genome and phenome,[b] also helps explain the difficulties now encountered when prediction and diagnosis of complex disease omits epigenetic considerations and depends entirely on gene causality.

INTRODUCTION

The Human Genome Project (HGP) is a superb technology capable of generating a high density map of perhaps all 100,000 human genes. As one of the leading architects of this project has said, "... this will also allow polygenic diseases and traits to be resolved into Mendelian[c] components and thereby mapped" (Hood 1992). However, while such a map is within the technical reach of the HGP it will not be possible to extract from a reading of the map information that, by itself, will be sufficient in the diagnosis or prediction of polygenic diseases and traits. It is not possible because the logic of polygenic diseases, as contrasted with monogenic diseases,[d] is not to be found in the genome. Rather, that logic is encoded in a cellular epigenetic network of genes, gene products, and environmental signaling. This network is characterized by enormous complexity and informational redundancy from which is generated unexpected outcomes (phenotypes) driven by small changes in boundary conditions and environments (Strohman 1993, 1994). Continued emphasis on linear genetic logic, and discounting of epigenetic approaches, presents serious problems for the future of biotechnology in the health/medical arena.

In what follows below I will try to convince you of the problems inherent in current directions within the Human Genome Project, as well as in biotechnology in general. Analysis of these problems has been set out in prior publications, which should be consulted for details and complete reference citations (Strohman 1993, 1994).

Informational redundancy is now well known at the level of the genome (Tautz 1992) and, in and of itself, constitutes a serious threat to the uniqueness equation underlying the HGP (Brennert 1990). This equation:

$$\text{Unique Genes} \to \text{Unique Effects}$$

---

a *isogenic*   Genetically similar or the same.
b *phenome*   All phenotypic traits of an organism.
c *Mendelian*   I.e., associating specific individual genes with specific characteristics.
d *monogenic diseases*   Inherited diseases caused by a single gene.

defines saturation mutagenesis[a] as the technological basis of efforts to associate specific genes with specific diseases (Wilkins 1993). However, in the presence of redundant genes[b] the uniqueness equation is severely compromised and is liable to serious error with equally serious consequences in clinical practice. In summary, the assumptions of the uniqueness equation are essential in the following areas: medical genetics, which seeks isomorphic[c] mapping of human diseases to Mendelian genes; molecular biology, which seeks to identify unique genetically based mechanisms driving cellular processes; and developmental biology, which presupposes (1) genetic programs, (2) additivity of gene effects,[d] and (3) the ability to map complex developmental stages to additive programmatic sequences in DNA.

These assumptions and presuppositions are being questioned at all levels of basic experimental biology […]. New, and old, research findings are leading to a rejection of genetic determinism[e] as the major paradigm of modern biology. Scientists are struggling with the apparent limits of genetic reductionism[f] and are attempting to restructure genetic mechanisms within a larger context of "decision making" within cells and organisms (Nijhout 1990; Goodwin 1985). This context is sometimes seen to be ruled by principles of non-linear dynamics and chaos theory,[g] but for this presentation at least, it is described under the more traditional heading of epigenetic regulation. The main point here is that while aspects of basic biological research are moving away from linear genetic determinism and toward non-linear complexity, applied biomedical technology remains unaffected by this fundamental shift in research emphasis. Thus, a rift is seen developing between basic and applied biology – a rift that could be dangerous to our public health. [...]

---

a   *saturation mutagenesis* · Method of identifying the functions of specific portions of DNA by causing mutations in those portions.

b   *redundant genes*   Multiple genes that perform the same role.

c   *isomorphic*   Parallel in form.

d   *genetic programs*   Term referring to the metaphorical understanding of DNA as encoding a linear sequence of instructions similar to the code in a straightforward computer program; *additivity of gene effects*   I.e., the idea that the effects of individual genes can simply be added together, as opposed to interacting with and affecting each other.

e   *genetic determinism*   Theory that genes ultimately determine all physical and behavioral human traits.

f   *genetic reductionism*   Theory that a single gene corresponds to a single trait.

g   *chaos theory*   Field of study in mathematics which observes the chaotic results of dynamic systems that are sensitive to differences in initial conditions.

Epigenetic Aspects of Cell Regulation

What most biologists have assumed for years, but have never really formal- 5
ized, is that every cell contains not one, but two, informational systems; the
first is genetic and the second *epigenetic*.

The familiar genetic system of:

$$DNA \rightarrow RNA^a \rightarrow Protein \rightarrow Phenotype$$

is applicable to a small range of human phenotypes. In biomedicine it
is restricted to monogenic diseases like Duchenne muscular dystrophy,
hemophilia,[b] and a host of other diseases. However, these diseases remain a
small percentage of our disease load and account for less than 2 percent of the
total (Weatherall 1982; McKeown 1988; and below).

The *epigenetic informational system* in cells is depicted in figure 1.
The sets of (a) interactive genes (epistasis[c]), (b) interactive genes and gene
products (epistasis, pleiotropy[d]), and (c) interactive gene products and environment (polygenic and pleiotropic effects) define an unstable *epigenetic*
system of great complexity inserted between unitary genetic elements and
the final phenotype. In fact, this is a chaotic system with the major characteristic that, while a detailed map may be generated of all components, it
will be impossible for mutational analysis alone to predict a unique outcome.
As an example, a mutant gene may be redundant as seen in the case of an
angiotensin[e] II pathway in heart tissue (see Strohman 1994). A single mutation in this pathway does not predict heart disease since the pathway contains
many alternative genetic elements, all of which perform identical functions.
Alternatively, even without redundancy at the gene level, many examples are
found in cellular metabolic[f] (epigenetic) networks, where the network will
simply be able to reset itself when given appropriate signals. The outcome
of this resetting is often positive but unpredictable, as found in adaptation of
master runners to aerobic stress. These older men show epigenetic regulation

---

a *RNA* Ribonucleic acid, which manufactures proteins based on the genetic information contained in DNA.

b *Duchenne ... dystrophy* Recessive genetic disorder of males that causes muscle degeneration and eventual death in boys; *hemophilia* Genetic disorder limiting the body's ability to form blood clots and prevent excessive bleeding.

c *epistasis* Suppression or alteration of one gene's effects through interaction with another gene at a different location in the chromosome.

d *pleiotropy* Influence of one gene on multiple phenotypic traits.

e *angiotensin* Protein that raises blood pressure.

f *cellular metabolic* Relating to the biochemical processes that occur within a cell.

of several key glycolytic and mitochondrial enzymes so that overall oxygen utilization is enhanced, as compared to younger men who achieve the same endpoint but through a different pathway (Strohman 1994). This response may even involve changes in gene expression since physical activity and electrical stimulation are known to repress and activate genes coding for isoenzymes[a] in skeletal muscle cells (see Strohman 1994). Thus, the context for patterns of gene expressions is found, not in the genome, but in interactive epigenetic networks.

FIGURE 1.

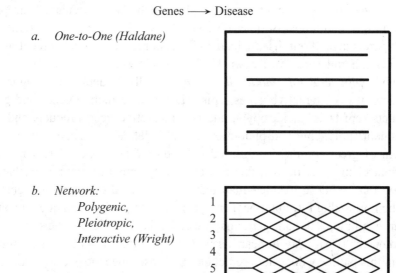

Genes ⟶ Disease

a. One-to-One (Haldane)

b. Network:
   Polygenic,
   Pleiotropic,
   Interactive (Wright)

A biomedical genetics view (a) sees a linear connection between genes and disease. This is a correct view for those monogenic diseases that make up 2% of our disease load. The epigenetic view (b) for the common, polygenic diseases (98% of our health problem) sees disease influenced by many genes interacting with one another and with environmental signals. For example, if gene 3 in the figure is mutant or missing it would be difficult, if not impossible, to trace its effect through the maze of interaction shown. Indeed, its effect may be neutralized since gene 3 may have a backup or substitute gene (redundancy at the genome level). Alternatively, whatever process is affected by gene 3 is also affected by many other genes acting in a network or circuit that changes on signaling from the environment (redundancy at the epigenetic level). For the polygenic diseases genetic diagnosis will, therefore, provide unreliable information.

---

a *isoenzymes* Different enzymes that catalyze the same chemical reaction.

An epigenetic system may be said to be chaotic in that, while it is impossible to predict which alternative pathway will be used, it will be possible to determine potential for adaptive change under precisely defined initial conditions (Skinner et al. 1992; Kaufmann 1993). The system is thus a *determinative chaotic system*, open to new approaches that combine linear genetic with non-linear complex system (epigenetic) analysis. Looked at in this way we may predict a new opportunity in biotechnology; viz., the definition of complex system parameters and specific environmental perturbations that elicit unique disease/health outcome (Sing & Reilly 1993).

We now look at specific instances in which mainstream genetic determinism has found itself in conflict with new (and old) findings from basic research in genetics and in other areas of molecular and cell biology. Most of these conflicts may be resolved by recasting them in epigenetic terms. [...]

POPULATION BIOLOGY CONFLICTS WITH GENETIC DETERMINISM

Under the uniqueness equation described in the equation above the HGP has 10 distilled a simplistic set of assumptions and goals:

1. Genes determine diseases.
2. Genes determine aging.
3. Genetic analysis provides diagnosis and therapy for disease and aging.

But fundamental rules governing population genetics stand in at least partial contradiction to the uniqueness equation and to the human genome project assumptions. This is a complex subject, but essentially the unique relationship between genes and phenotypes is flawed because most complex phenotypes (including diseases) have no unique genetic basis. Rather, the relationship between genome and phenome is characterized by great complexity involving interaction between many genes, gene products, and environmental signaling. This interaction may involve 10, 100, 1,000 or more genes for any common disease like cancer or the heart diseases. We don't know exactly how many genes interact in these cases, but as the number of interactive genes increases the effect of each gets smaller and more open to compensation (Wahlsten 1990). In addition, the interaction will be a function of an individual's natural history and environmental setting so that, even in simplified cases where genetic connections may be traced, genes will have different effects under different environments. The basic message from population genetics is that a precipitating environment is required to produce disease manifestation across the entire range of genetic variation (McKeown 1979). For problems like cardiovascular disease, most cancers, noninsulin-dependent diabetes, and most mental diseases, there is no evidence for single gene causality, and certainly none that would support the uniqueness equation fundamental to the human genome project.

Why is it that genetic diagnosis is predicted to fail in these cases? In brief, the argument is that the major statistical tool, analysis of variance, or ANOVA,[a] as developed by Fisher, is insensitive to heredity-environment interaction. This insensitivity is minimized in the agricultural breeding experiments for which ANOVA was designed because large sample size is normally the rule. In medical genetic studies (extended families) or in behavior genetics (twin studies), the sample sizes are small, so that error is large in detecting lack of interaction between heredity and environment. As Wahlsten (1994) points out, a newer statistical approach, multiple regression,[b] is replacing ANOVA; but for the kinds of studies we are discussing the two procedures are essentially equivalent. Experts in agricultural genetics generally accept significant interaction between genes and environment and are extremely cautious in applying heritability coefficients or in assigning any significant numerical value to genetic cause when dealing with complex traits. Their position is that if gene effects are interactive (not additive) with environmental effects, it is incorrect to use ANOVA for assessing genetic contribution to a particular phenotype across a range of environments. Medical geneticists, however, using the same ANOVA but with significantly smaller sample sizes, not surprisingly, do not find evidence for interaction and, therefore, assume that heredity and environment are additive. They then assign great significance to heritability coefficients and are confident that these numbers describe quantitatively the contribution of separate heredity and environment to any particular phenotype. We have a medical literature, then, that asserts with great confidence, but with serious theoretical reservations from sectors of population genetics, that this or that complex disease, while having an environmental component, also has a separate genetic component that can be discovered and utilized in pursuit of some hypothetical treatment strategy. It is beyond the scope of this review to enter this controversy fully; it is enough to state the minimum conclusion that medical genetics, with a linear view of gene-disease causality, finds itself in serious debate with a significant segment of its parent science of population genetics, which sees complex traits, including disease, as highly interactive and impossible to reduce to genetic elements alone (figure 1).

---

a *ANOVA* Method of determining whether there are significant differences between groups of data.

b *multiple regression* Method of estimating relationships between variables based on study data.

Disease Natural History Conflicts with Genetic Determinism

Here the issue is simple. There is a class of diseases for which the uniqueness equation is adequate. Diseases determined at fertilization, as Thomas McKeown (1979) has made clear, are based in genetic abnormalities of one kind or another. For simplicity we call these monogenic diseases. Examples are sickle cell anemia, cystic fibrosis,[a] and Duchenne muscular dystrophy. There are literally thousands of these diseases, but they occur within the human population at extremely low frequency and account for less than 2 percent of our total disease load. The basic message here is that 98 percent of the time our babies are well born, with genetic constitutions capable of supporting a life span of more than 100 years, with an average life expectancy of about 85 years, and an old age relatively free of morbidity (Freis & Crapol 1981; Tsai et al. 1978). In order for all of this to happen the human genome needs to find itself in an environment for which it has adequate representation – proper nutrition, housing, and sanitation, to name the obvious requirements.

There is a second level at which the biomedical paradigm is in conflict with actual disease distribution. It is assumed by the paradigm that complex traits like cancer and cardiovascular diseases have an important genetic component available not only to genetic analysis but also to genetic therapy. The reasoning is that these and other diseases attack people mostly in older (post-60) age groups. As such, the responsible genes would be beyond the reach of natural selection, which operates effectively at younger pre-reproductive ages. This being the case, it is argued that heart and cancer diseases are "old" entities, have always been with us (as have their genes), but show up significantly now because it is only recently that our population has aged sufficiently for them to become a problem. If this is true, then – so goes the argument – these are genetic diseases, pure and simple, and may be attacked as such (McKeown 1979, 1988).

But the natural history of our complex diseases shows that, in all prob- 15
ability, these are not genetic diseases, but are *diseases of civilization*. Of course, they have some genetic basis, but this basis is so broad as to be trivial with regard to providing precise genetic answers (figure 1). Like all polygenic traits, genes are necessary but not sufficient. Evidence that diseases of civilization are not simply genetic includes the following examples (McKeown 1988). First, twin studies show extremely low concordance for most cancers and heart disease. Second, these same diseases show remarkable variation in

---

a *sickle cell anemia* Mutation that distorts the shape of red blood cells and limits their ability to carry oxygen; *cystic fibrosis* Disorder causing the production of thick mucus in the body, which limits breathing and causes respiratory infection.

identical populations over time and over geographical and migratory patterns. These variations disclose, for example, that diseases tend to be place- (environment) specific and that, when people migrate, they tend to have those diseases common to their host population, not those that are common to the genes they brought with them, i.e., not common to their native population. These variations are reversible. Finally, these diseases are rare in populations that have not been influenced by Western habits. Natural history studies all indicate that our major premature killers are not genetic in any straightforward causal sense; they are diseases associated with changes in environment. That is the message from the past and present. That message, extended into the future, is that new diseases, their prevention and therapy, will also be associated with environmental change.

EVOLUTIONARY BIOLOGY CONFLICTS WITH GENETIC DETERMINISM

Most people, scientists included, are not aware of problems within evolutionary biology having to do with genetic mechanisms. These problems do not provide any weakening of the foundations supporting evolution; they do provide concern that we may have oversimplified the idea that evolution is to be explained by genetic mechanisms alone. Again, this is a complex area, but we can state the following premise. Genetic change (adaptation) is seen as one end point of evolution, and change in genes (mutations) is seen as one element providing a basis for phenotypic variance that may be acted upon by natural selection. But gene changes alone will not and cannot explain evolution. For a complete explanation we require an understanding of how individual organisms generate their phenotypes in the presence (or absence) of gene changes in a variety of environmental settings (Goodwin 1985). It is this understanding that we do not have at this time. Individual development is one missing link in our current theory of evolution; a link that is recognized, and one that the biological community is now struggling to supply and incorporate into a more complete picture of natural selection (Gottlieb 1992). We may provide a few examples of the conflict here. First, there is the fact of an absence of relationship between genetic and morphological[a] complexity of species. Some closely related species cannot be seen by expert examination to be different (have different morphology), yet they show great variation in complexity at both genetic and protein-sequence levels. Somehow organisms are able to take vastly different genomes and to construct nearly identical phenomes, and this fact cannot presently be explained by a simple linear

---

a *morphological* Having to do with the physical features, internal anatomy, and other observable characteristics of organisms.

genetic paradigm. Second, humans and chimps have a very different morphology, and yet humans do not differ genetically from chimps by more than 1-2 percent. Somehow we are able to construct very different organisms from very similar genomes, and this is currently not explained by genetic theory (see below for further discussion).

DEVELOPMENTAL CELL AND MOLECULAR BIOLOGY CONFLICT WITH GENETIC DETERMINISM

There are many conflicts here. First, genetic determinism for complex traits has assumed the notion of "gene programs" to help explain the causal linkage between genes and phenotypes. But this assumption has been found to be without experimental verification. There are no genetic programs (Nijhout 1990; Goodwin 1985). There are only genes that encode for proteins. Some of these genes, and their protein products, are extremely important. For example, some products bind to DNA and are involved in regulating development of morphological patterns (Wilkins 1993). When they are mutated or missing, the effects on a complex trait are profound. But these genes also exist in a cellular epigenetic context and depend on this context for the control of their expression in a species-specific manner.

A second conflict comes with the realization of the fact of *informational redundancy* in organisms, and especially within cells. The uniqueness equation is undermined in the presence of a determination that more than one gene can bring about a unique effect. It is further weakened when we realize that not only is there informational redundancy at the gene level, but at the epigenetic level as well. There are many examples in the current literature of experimental biology testifying to the ability of the organism to get along without what were thought to be crucial genes. The organism, when a gene is missing, finds other genes or finds new ways (epigenetic controls) for vast numbers of remaining genes to interact and to produce the same or highly similar phenotypes (Strohman 1994).

CONFLICT RESOLUTION

A major assumption of modern biomedical thinking is that genetic inheritance is the only inheritance. But biologists have always known this to be incomplete and we are now rediscovering the nature of our oversimplified paradigm (Sapp 1987). In modern developmental biology the idea of genetic programs as a script for phenotype is being abandoned. There is no isomorphic mapping of complex phenotypes to Mendelian factors (Stent 1987), and the mechanism by which the organism elicits phenotypic variability from isogenic or near-isogenic situations remains a profound mystery. The work on sibling species reveals that organisms may remain constant in morphology for millions of years, even while they are enormously divergent at the level of DNA (Nanny 1982). Humans and chimps are shown to be nearly identical in genetic terms, revealing that the organism is able to draw vastly different phenotypes from highly similar genotypes (Wilson et al. 1977). Thus, profound questions are raised concerning the assumption of gene programming. First, there appears to be less of a relationship between genetic and morphological complexity than we have thought. Second, if the program is not in the genes, and organisms clearly are programmed, then where is the program? These and newer variations of complex, non-linear themes tend to be suppressed by our nearly monolithic commitment to molecular genetic mechanisms. This review has suggested that the molecular reductionist program to explain life has serious limits and that new epigenetic approaches to genetic regulation will be crucial. What might these new approaches be?

20   John Maddox, the editor of *Nature*, has written that modern biology, in concentrating on mechanism, has neglected theoretical approaches that might provide structure to the enormous data base accumulated by strictly molecular inquiry (Maddox 1992), and he has suggested that such a conceptual structure might include a quantitative approach to dynamical cellular properties such as concentration fluxes of molecules that would control gene expression (Maddox 1992a). Numerical characterization of these properties might then provide a basis for theory construction concerning regulation at levels higher than the gene. Theoretical physicist-cum-biologist, Walter Elsasser, has in fact laid out a basic description of a holistic theoretical biology in which dynamical properties play the role of higher order regulation (Elsasser 1987). It is apparent that new research opportunities need to be created that will encourage work on these dynamical systems, and the theoretical structure hinted at by Maddox and Elsasser may lie, at least partially, in the theory of complex systems.

One might begin the merger of genetic reductionism and epigenetic complexity with those areas where multigenic systems are known to be coordinated by higher order cellular responses to environmental conditions. Nobel laureate, Barbara McClintock, who described mobile genetic elements[a] long before they were discovered by molecular biology, had always been preoccupied with mechanisms that rapidly reorganize the genome. In one of her last reviews she wrote of the significance of responses of the genome to challenge. McClintock ended the article by saying: "We know about the components of

---

a   *mobile genetic elements*   Portions of DNA that can change position.

genomes ... We know nothing, however, about how the cell senses danger and initiates responses to it that often are truly remarkable" (1984).

At the cell level Sing et al. offer an interesting epigenetic approach to complex analysis of heart disease, with multigenetic causality linked to interactive environments (1993). At levels above the cell – for complex physiological systems – chaos theory builds on epigenetic thinking and is already providing new ways to think about complex systems. This is particularly true for cardiac function, where sinus arrhythmia,[a] long thought to be low-level noise, or random fluctuation in heart rate, is now seen as high order chaos (Skinner et al. 1992). Coupling of heart rate to brain function, and thus to experience, has long been appreciated as an observable patterned occurrence but was mostly inexplicable through standard physiological experiment (Bond et al. 1973). Chaos theory is able to provide a method of revealing generic pattern in what was thought to be random variation. Recognition of these patterns allows new insights into brain-heart physiology and may even allow prediction of sudden cardiac death among patients at risk (Skinner et al. 1992).

It is here, at this interface between cell/organism and external world, that new research efforts might be focused. Initial cellular responses are epigenetic in nature and involve the selection of adaptive responses from a bewildering array of molecular possibilities. At cellular and higher levels we expect that evolution has worked to select not just single genes, but integrated behavior, or generic patterns of response, at all levels of biological organization (Kaufmann 1993). These patterns cannot be seen by linear analysis. It is at this level that the theory of complex systems might prove to be useful. Generic patterns, with some ultimate basis in genomic reorganization, changes in gene expression, and so on, would perhaps be open to theoretical structuring. Explanations and predictions of behavior leading to cancer or other cellular pathology, and to disease of the heart and other complex organs, would then not need to depend entirely on an apparently endless reductionistic analysis. They could rely more on understanding rules of higher level organization – rules that, themselves, have been selected and that control downstream mechanistic elements.

(1995)

NOTE

For a complete reference guide, see reviews by R.C. Strohman cited below.

---

a   *arrhythmia*   Heart condition that results in an irregular beat rhythm.

References

Bond, W.C., Bohs, C., Ebey, J., & Wolf, S. (1973). Rhythmic heart rate variability (sinus arrhythmia) related to stages of sleep. *Conditional Reflex* 8(2): 98-107.

Brenner, S., Dove, W., Hewrskowitz, I., and Thomas, R. (1990). Genes and development: Molecular and logical themes. *Genetics* 126: 479-486.

Elsasser, W. (1987). *Reflections on the Theory of Organisms*. Quebec: Orbis Publishing.

Fries, I.F. and Crapo, L.M. (1981). *Vitality and Aging: Implications of the Rectangular Curve*. N.Y.: W.H. Freeman.

Goodwin, B.C. (1985). What are the causes of morphogenesis? *BioEssays* 3: 32-36.

Gottlieb, G. (1992). *Individual Development and Evolution: The Genesis of Novel Behavior*. Oxford: Oxford University Press.

Hood, L. (1992). In D. J. Kevles and L. Hood (Eds.), *The Code of Codes*, Cambridge, MA: Harvard University Press.

Kaufmann, S. (1993). *The Origins of Order*. N.Y.: Oxford University Press.

Maddox, J. (1992a). Finding wood among the trees. *Nature* 333: 11.

Maddox, J. (1992b). Is molecular biology yet a science? *Nature* 355: 201.

McClintock, B. (1984). The significance of responses of the genome to challenge. *Science* 226: 792-801.

Mckeown, T. (1979). *The Role of Medicine: Dream, Mirage or Nemesis?* Princeton, N.J.: Princeton University.

Mckeown, T. (1988). *The Origins of Human Disease*. N.Y.: Basil Blackwell.

Nanny, D.L. (1982). Genes and phenes in tetrahymena. *BioScience* 32(10): 783-788.

Nijhout, H.F. (1990). Metaphors and the role of genes in development. *BioEssays* 12: 441-446.

Sapp, J. (1987). *Beyond the Gene: Cytoplasmic Inheritance and the Struggle for Authority in Genetics*. Oxford: Oxford University Press.

Sing, C.F. and Reilly, S.L. (1993). Genetics of common diseases that aggregate but do not segregate in families. In Sing, C.F. and Hanis, C.L. (Eds.) *Genetics of Cellular, Individual, Family and Population Variability*. N.Y.: Oxford University Press, 140-161.

Skinner, J.E., Molnar, M., Vybiral, T., and Mitra, M. (1992). Application of chaos theory to biology and medicine. *Integrative Physiological and Behavioral Science* 27: 39-53.

Stent, G. (1981). Strength and weakness of the genetic approach to the development of the nervous system. *Ann. Rev. Neurosci.* 4: 163-194.

Strohman, R.C. (1993). Ancient genomes, wise bodies, unhealthy people: Limits of genetic thinking in biology and medicine. *Perspectives in Biology & Medicine* 37(1): 112-145.

Strohman, R.C. (1994). Epigenesis: The missing beat in biotechnology? *Biotechnology* 12(2): 156-164.

Tautz, D. (1992). Redundancies, development and the flow of information. *BioEssays* 14: 263-266.

Tsai, S.P., Lee, E.S., and Hardy, R.J. (1978). The effect of reduction in leading causes of death: Potential gains in life expectancy, *AJPH* 68: 966-971.

Wahlsten, D. (1990). Insensitivity of the analysis of variance to heredity-environment interaction. *Behav. and Brain Sci.* 13: 109-161.

Weatherall, D.J. (1982). The new genetics and clinical practice. Nuffield Provincial Hospitals Trust, London.

Wilkins, A.S. (1993). *Genetic Analysis of Animal Development*, 2d ed. N.Y.: Wiley-Liss Press.

Wilson, A.C., Carlson, S.S., and White, T.J. (1977). Biochemical evolution. Ann. Rev. *Biochem.* 46: 573-639.

Questions:

1. This piece was published in *Integrative Physiological and Behavioural Science*. Who do you think reads this journal? How does the article's language reflect that audience's level of knowledge and specific language expectations?

2. Do you think scientists find this argument unusual now? Compare it to that found in the article by Baverstock and Rönkkö (included in this anthology). In what ways does Baverstock and Rönkkö's argument update this one? Does their argument seem to you to be a mathematical modelling of what Strohman's article calls "chaos theory" or a "determinative chaotic system"?

3. Is this article any more or less easy to understand than Baverstock and Rönkkö's? Is it any more or less scientific? Explain.

4. How many human genome elements did there turn out to be, at least as currently understood? Why do you think the number 100,000 was discussed here?

5. Identify one example of loaded language in this paper. Do you think its use is justified?

6. This piece makes heavy reference to other works by Strohman, which apparently provide extensive coverage of the material he discusses here. Do you think this approach works? Why or why not?

7. Consider the claims, first made in the 1990s, that an individual's statistical likelihood of developing breast cancer can be identified in relation to possessing the BCRA1 or BCRA2 gene deletion. How might the theories proposed here problematize these assumptions? Do you think they actually do so?

8. Consider the statistical discussion in this paper. Do you think the discussion is aimed at statisticians? If not, what does that tell you about the paper's intended audience?

twenty-six

KEITH BAVERSTOCK AND MAUNO RÖNKKÖ

from Epigenetic Regulation[a] of the Mammalian Cell

This 2008 article from PLOS ONE *describes the epigenetic processes that occur in cells, arguing that epigenetics plays a significant role in the inheritance of characteristics, the development of species through evolution, and the manifestation of cancer.*

☙

ABSTRACT

BACKGROUND

Understanding how mammalian cells are regulated epigenetically to express phenotype[b] is a priority. The cellular phenotypic transition, induced by ionising radiation,[c] from a normal cell to the genomic instability[d] phenotype, where the ability to replicate the genotype accurately is compromised, illustrates important features of epigenetic regulation. Based on this phenomenon and earlier work we propose a model to describe the mammalian cell as a self assembled open system operating in an environment that includes its genotype, neighbouring cells and beyond. Phenotype is represented by high dimensional attractors, evolutionarily conditioned for stability and robustness and contingent on rules of engagement between gene products encoded in the genetic network.

---

a *Epigenetic Regulation* Heritable effects on gene expression caused by anything other than DNA sequence.

b *phenotype* Observable characteristics of an organism determined by the interaction of genetics and environment.

c *ionising radiation* Particle radiation that uses energy to remove an electron from an atom or molecule.

d *genomic instability* High frequency of mutations within a genome.

METHODOLOGY/FINDINGS

We describe how this system functions and note the indeterminacy and fluidity of its internal workings which place it in the logical reasoning framework of predicative logic.[a] We find that the hypothesis is supported by evidence from cell and molecular biology.

CONCLUSIONS

Epigenetic regulation and memory are fundamentally physical, as opposed to chemical, processes and the transition to genomic instability is an important feature of mammalian cells with probable fundamental relevance to speciation and carcinogenesis.[b] A source of evolutionarily selectable variation, in terms of the rules of engagement between gene products, is seen as more likely to have greater prominence than genetic variation in an evolutionary context. As this epigenetic variation is based on attractor states[c] phenotypic changes are not gradual; a phenotypic transition can involve the changed contribution of several gene products in a single step.

INTRODUCTION

Today one of the most pressing issues in biology is to understand how the epigenetic aspects of the cell are regulated, that is, how the appropriate gene products are brought into action when and only when appropriate. Writing in 1958 Nanney[1] poses, under the heading "Epigenetic Control," the question of whether it is a "template replicating mechanism," i.e. DNA replication, or "some other" unspecified mechanism, which manifests phenotype at the cellular level. In essence Nanney was questioning whether all the then known empirical evidence about biological function, which he reviews, regarding the stability of phenotype could be accounted for as a result of "genetic regulation," or whether there was a need to invoke "epigenetic regulation" in addition. He concludes by nominating two separate mechanisms by which "homeostasis" could be achieved, namely a replicating template

---

a *predicative logic*   Type of logic in which all elements are reduced to symbols to produce logical equations.

b *speciation*   Process by which new species develop and arise; *carcinogenesis*   Creation of cancer.

c *attractor states*   States that a dynamic system is most likely to move toward, such that no matter what starting conditions occur, the system will eventually settle on one of its attractor states. For further definition of attractors and of other terms, please refer to the "Statement of hypothesis and definition of terms" provided later in the article.

mechanism or another, *"perhaps self-regulating metabolic[a] patterns"* as suggested by Delbrück at a Congress on Genetics in 1949. In the discussion following a paper that had attributed a specific phenomenon to the reproduction of genes that were favoured or inhibited by environmental conditions Delbrück noted that *"many systems in flux equilibrium are capable of several different equilibria under identical conditions. They can pass from one state to another under the influence of transient perturbations."*[2] Today we would refer to "flux equilibrium" as a dynamic steady state.[b]

In the event biology has invested heavily in the "template replicating mechanism" to the almost complete exclusion of any alternative. Prior to 1953 the concept of a gene was much more fluid than it is today being based primarily on empirical evidence of how it could be inherited and mutated. However, it can be argued that the case made by Schrödinger in 1943,[3] on quantum mechanical grounds, that the property "life" could not be based on statistical averaging, as is for example, temperature, and must therefore (because at that time there was no obvious alternative) be based on a mechanism (he used the analogy of a clock based on an aperiodic crystal)[c] was highly influential in the subsequent development of cell and molecular biology. The extraordinary elegance of DNA as a semi-conservative replicating mechanism[d] seems to have sealed the fate of the subject up to at least 2000.

Phenomena such as imprinting and the fact that a single genotype gives rise to more than 200 cellular phenotypes, could not, however, be explained without resort to some kind of "extra genetic" or epigenetic phenomenon. Indeed, assuming that all the information necessary to regulate the deployment of the code is encoded in the genotype leads to an infinite regression. Today there is a high degree of consensus that imprinting and other aspects of epigenetic regulation are controlled by chemical marking, methylation and acetelation[e] of DNA and the histones in chromatin[f] [4,5] and that these marks also constitute the epigenetic memory.[6] These it is generally assumed serve in a complex manner and in conjunction with sequences in the genome

---

a *metabolic* Relating to the biochemical processes that occur within a cell.

b *dynamic steady state* State that continually changes in small ways, but does not show large-scale changes because its small changes balance each other.

c *aperiodic crystal* Structure containing genetic information in covalent chemical bonds. The concept was introduced by Erwin Schrödinger in his non-fiction text *What Is Life?* (1943).

d *semi-conservative ... mechanism* Refers to the process by which DNA is replicated, which produces two DNA strands, each containing one original and one new strand.

e *methylation and acetelation* Attachment of a group of atoms (a methyl group or an acetyl group, respectively), to a molecule.

f *histones* Proteins that order and package DNA into sections called nucleosomes; *chromatin* Contents of the cell nucleus, a combination of DNA and histones.

associated with coding regions, to regulate the transcription[a] process. The study of the role of these "epigenetic marks" is now a major activity in cell and molecular biology.[7]

In parallel and in recognition of the fact that separating the genome into fragments for detailed study followed by re-synthesis has limits as a strategy for understanding biology, approaches under the heading of "systems biology" have burgeoned. However, as is made clear by O'Malley and Dupré[8] it is far from clear what exactly the term "systems biology" means. They define two main approaches, namely pragmatic (labelled type 1) and theoretic (type 2). The majority of systems biologists are of type 1 and *for them 'system' is a convenient but vague term that covers a range of detailed interactions with specifiable functions.*"[8] Type 2 systems biologists see a fundamental aspect to the term "system" along the lines of that advanced by Bertalanffy[9] as general systems theory. The essence of this approach is that the system is thermodynamically open[b] and that the high level properties of a system, such as phenotype, emerge from the global *interactions* of its component parts to give a result that is greater than the sum of those parts. This leads Huang[10] to distinguish between types 1 and 2 by the terms "localist" and "globalist."

Recently uncovered features of the cell would argue strongly for the global-ist perspective as the more likely to be relevant to understanding biology. [...]

Here we propose a hypothesis/model based on recognised features of the cell to describe the epigenetic regulation of the mammalian cell as a system somewhat similar to the concept Delbrück advanced in 1949,[2] namely based on dynamic steady states and thermodynamically open. We strive for realism in our assumptions recognising that the complexity of the model may make it computationally relatively intractable. However, we believe that the qualitative understanding of the way the cell operates would provide the most reliable basis for simplifying the model. We examine the evidence that supports the model and discuss its implications for understanding the processes that regulate cells.

We start with the phenomenon of genomic instability as induced by ionising radiation.[17] Previously we have drawn attention to the implications of the chemically friable[c] nature of DNA under physiological conditions[18] and subsequently described genomic instability as a stochastic[d] epigenetic

---

a *transcription*   Copying of a segment of DNA to produce RNA. Transcription is the first part of the process leading to the expression of a gene.

b *thermodynamically open*   Term applied to systems enclosed by a boundary that is permeable to the outside environment, such that matter can both enter and leave the system.

c *friable*   Easily reduced to smaller pieces.

d *stochastic*   Non-deterministic.

phenotypic transition between attractors, essentially specific patterns of gene products active in the cell, representing phenotype.[19] Subsequently, Huang et al[20] have identified, experimentally, such attractors as representatives of phenotype in the chemically induced differentiation of neutrophil precursors to the terminally differentiated state. Essentially, the chemical perturbation of the precursor attractor stimulates the transition to other attractors[21] and ultimately the terminally differentiated state.[a]

Attractors are components of a state space[b] with a dimension for each of the gene products coded for by the genotype, i.e., more than 100,000 in the human. A typical attractor might involve between 1000 and 10,000 active gene products. The attractors within the system are defined by rules of engagement between gene products and envisaged to be essentially point attractors, as opposed to limit cycle attractors, but they should be seen as elements of a limit cycle representing the cell cycle.[c]

In the case of radiation induced genomic instability the physical properties of energy deposition by ionising radiation and the low doses required to initiate genomic instability indicate that the transition to instability is not a genetic effect and must therefore be epigenetic in character.[22,23] It has been proposed[19] that the normally stable phenotype of a cell is represented by an evolutionarily conditioned or "home" attractor, that is, one that has been evolutionarily selected most importantly for two properties, namely robustness, or resistance to perturbation and fidelity in the replication of the genotype, or stability.

10   Exposure to ionising radiation, because it causes molecular damage to the genomic DNA and therefore genotype, which, if not repaired prior to cell division may compromise the genotype, thus places increased demands on the ongoing damage detection and repair processes in the cell, which are components of the home attractor. If that stress exceeds a critical value an irreversible, due to the high dimensionality of the attractor, transition to a variant and unconditioned attractor is stimulated. See Figure 1. If the cell can survive and divide at the variant attractor it will accrue the genotypic damage that characterises the instability phenotype by virtue of a lower level of fidelity in replication. The genomic instability phenotype is thus a mutator phenotype.[d]

---

a   *terminally differentiated state*   State in which a cell will no longer develop into a new type of cell.

b   *state space*   Set of all states that are possible in a given system.

c   *point attractors ... cycle attractors*   Attractors that represent an individual, specific state as opposed to a set of states that is cycled through repeatedly;   *cell cycle*   Process leading to cell division and replication.

d   *mutator phenotype*   Phenotype showing the mutation rate of cancer cells in comparison with the infrequent, random mutation rate of normal cells.

Figure 1.
Illustration of responses of the system to genotypic damage.

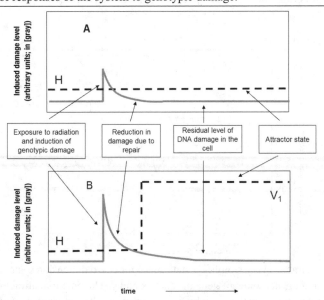

Panel A: A relatively small exposure to ionising radiation creates damage in the genomic DNA ([gray] trace) which is detected and repaired by the cell. The prevailing home attractor, H, is perturbed but not irreversibly so, i.e. the basin of attraction[a] is not exceeded and the system stays in the H attractor. Panel B: A larger exposure to ionising radiation but still within the capacity of the cell to repair causes the cell to exceed the basin of attraction of the home attractor, H, and stimulates the irreversible transition to the variant attractor V1.

Thus, genomic instability can be seen as the *loss*, at the cellular level, of the ability to replicate the genotype with the optimal level of integrity that was *gained* substantially through evolutionary conditioning subsequent to the origin of the species to which that cell belongs. In effect evolutionary conditioning minimises the residual damage in the dynamic steady state between DNA degradation and repair; the conditioning is thus a purely epigenetic evolutionary selection process.

It is the openness of the cell to its environment that is at the root of the instability phenomenon. Exposure to radiation, an extrinsic agent, causes the cell to respond to detect and repair the damage to the genotype. The consequent increased demand for the gene products responsible for detection and repair of the damage represents a perturbation of the attractor, which if

---

a   *basin of attraction*   Set of points in the space of system variables such that any initial conditions that fit in this set will evolve to reach a particular attractor.

sufficiently severe will exceed the basin of attraction in respect of one or more gene products and thus the adoption of the variant attractor. (See Figure 2.)

FIGURE 2.
Illustration of a state space.

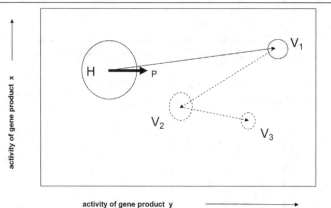

activity of gene product y

The figure illustrates a very simplified state space for a two dimension system, the coordinates indicating the activities of gene products x and y. The potential attractors are represented by circles, the diameters of which are proportional to their basins of attraction, the home attractor H being the largest because of environmental conditioning. A perturbation P of H beyond the basin of attraction due to an increase in gene product y causes the adoption of variant attractor V1. This is the initiation step of genomic instability. Subsequently, due to the relatively reduced robustness of variant, i.e., unconditioned, attractors, further transitions (dotted lines) to other variant attractors characterises the genomic instability phenotype.

METHODS

Statement of hypothesis and definition of terms

Our hypothesis is that the cellular phenotype of a mammalian cell is represented by a complex high dimensional dynamic attractor embedded in a state space with a dimension for each active gene product encoded in the genotype. The state space is therefore a proteomic[a] state space. The active gene products are assumed to interact selectively through rules of engagement, which are non-deterministic to allow for interactions with the environment, including other cells in the organism (one aspect of the openness). We further assume that the gene products are metabolised by the system (a second aspect of openness) and we assume there exists for each gene product a multi-compartmental dynamic steady state originating in the transcription of the

---

a *proteomic* Refers to the "proteome," all the proteins that are expressed by the genetic and epigenetic material in a given cell or group of cells.

coding sequences and terminating in the depletion of the active gene product, either as a result of it having been incorporated into the cellular architecture or selectively destroyed after use or being subject to spontaneous degradation. See Figure 3. This is referred to as the post-transcriptional[a] dynamic steady state. Prior to use gene products are stored or present in inactive forms, mRNA, tRNA, unfolded peptide[b] and inactive protein. Being an open system driven by attractors there is no continuum of stable states in the system; stable states are "quantised"[c] at the discrete high dimensional attractors.

FIGURE 3.
Illustration of the multi-compartmental dynamic steady state.

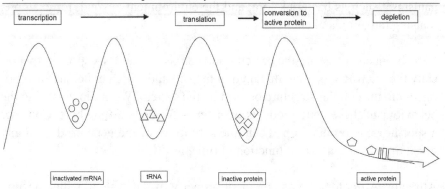

The figure illustrates the multi-compartmental dynamic steady state initiated by the transcription of coding sequences to mRNA, which is translated to peptide and finally yields active gene products which are depleted (block arrow) through use. There may be additional depletion (not indicated) by spontaneous decay from the product compartments.

The attractors available for occupation in the state space are an emergent property of the system determined by the rules of engagement, which also give rise to an architecture that influences transitions between attractors. The rules of engagement can be seen as the edges in a network, the nodes of which are the gene products and as does the genotype, they exhibit selectable variation. It is therefore assumed that they have been acted upon in evolutionary terms to increase the fitness of the architecture, including the attractor locations in the state space. Attractor transitions are equivalent to phenotypic

---

a   *post-transcriptional*   Referring to the point after a DNA sequence is transcribed by mRNA, but before the sequence is translated into a peptide by tRNA.

b   *mRNA*   Messenger ribonucleic acid, which copies genetic information from DNA and carries it out of the cell nucleus;   *tRNA*   Transfer RNA, which transports amino acids to a site of protein production;   *peptide*   Chemical compound consisting of two or more amino acids in a chain.

c   *"quantised"*   I.e., divided into distinct, separate values.

transitions and thus represent biological processes at the cellular level such as differentiation, carcinogenesis and evolution.

15 The system: It is important to be clear about the boundary between the system and its environment. In this case we define the system as the mammalian cell and all the material therein. However, we exclude the informational content (base sequence) of the genomic DNA but not the substance. Thus, the system is open in the sense that coding information derives along with other non-system "information" from the environment, including the neighbouring cells in the tissue and organism as well as, where appropriate, cohabiting organisms[24] such as bacterial flora, and the environment beyond, for example, ionising radiation and chemicals.

Gene products: these are the proteins and certain of the RNA species, specifically the microRNAs, manufactured in the cell and which either are incorporated into the cellular structure or used by the system. We are interested in the behaviour of these gene products with time. We denote time by t. Consider a specific gene product, gp. Then, the activity m of the gene product at any given time is captured as a function of time, $m_{gp}(t)$.

Attractor: Attractors are an emergent property of the system, which occupy a "point" or "volume" of the state space and are surrounded by a basin of attraction from which states drain into the attractor. It thus represents a domain of stability, albeit, limited by the boundary of the basin of attraction. Each gene product gp is governed by an attractor a_{gp}. The attractor determines a value range, a lower and an upper bound for the activity of the gene product. We denote the range of the attractor by $[low_{gp}, up_{gp}]$. In particular, if the activity of the gene product is within the attractor range, it remains there. In other words, if $low_{gp} \leq m_{gp}(t1) \leq up_{gp}$ holds for some $t1$, then also the condition $low_{gp} \leq m_{gp}(t2) \leq up_{gp}$ holds for any $t1 < t2$. Each attractor a_{gp} determines a basin b_{gp}. The basin of the attractor is a value range, indicating the minimum and maximum bounds for the activity of the gene product, such that if the activity of the gene product is within the range, it will eventually reach a value within the attractor range. We denote the range of the basin by $[min_{gp}, max_{gp}]$. Thus, formally, if $min_{gp} \leq m_{gp}(t1) \leq max_{gp}$ holds for some $t1$, then there exists also $t2$ such that $t1 < t2$ and $low_{gp} \leq m_{gp}(t2) \leq up_{gp}$ holds. In addition to the attractors and their basins of attraction, there exist volumes of state space through which the system transits during transitions and which exert some influence over the direction of migration.

Dynamic steady state: a condition in which two or more opposing processes are balanced to produce a stable state. Two categories are of particular interest, namely the DNA degradation under physiological conditions (due to, for example, hydrolysis[a]) opposed to the repair of that degradation by cellular repair processes, and the metabolic process generating gene products, commencing with their transcription opposed to their depletion through use, the post-translational steady state. (See Figure 3.)

Rule of engagement: The rules of engagement speak of the active gene products in time. Consider, for instance gene products gp_a and gp_b. Then a rule is of the generic form "*IF gp_a is active THEN gp_b is active*," stating that the activity of gp_a *implies* the activity of gp_b. Formally, the activity of a gene product is expressed with respect to some activity ranges, r_{gpa} and r_{gpb}, at points t1 and t2 in time. Then, a rule of engagement is a relation: $m_{gpa}(t1) \in r_{gpa} \Rightarrow m_{gpb}(t2) \in r_{gpb}$. For a gene product there are typically many rules of engagement and, thus, a gene product can be engaged with several other gene products. Consequently, a perturbation of any one gene product has the potential to perturb all those with which it is engaged.

Stability: within this context stability refers to the ability of the genome to replicate its genotype with maximum fidelity. DNA is an unstable compound under physiological conditions and thus is subject to ongoing repair. Degradation and repair are opposing processes which create a dynamic steady state of minimal residual damage in the system at any point in time.[18,19] This dynamic steady state is crucial for the long term stability of the system.

Robustness: the property of the system to resist perturbations of its stable states and of its transitions between stable states.

Homeostasis: is the property of an open system to regulate its internal state and maintain a stable condition.

Evolutionary conditioning: an evolutionary process whereby variations in the rules of engagement are selected particularly if they improve the integrity of the replication of the genotype, i.e. enhance stability, and/or expand the basin of attraction of the attractor, thus enhancing robustness.

---

a *hydrolysis* Destruction, through a reaction with water, of the molecular bonds in a chemical compound.

RESULTS

Description of the operation of the system

The system comprises two primary components, namely, the rules of engagement governing the regulation or deployment of the gene products and the material which is regulated to "build" the system. Placed in the environment, rather than the system, is the genotypic information that codes for the gene products. The reason for this is that the rules of engagement can be regarded as the formal causal component of the system (the syntax[a]). The residue, genomic coding sequence and environmental influences, are then regarded as the semantic[b] component. Separating them in this way, as does for example Rosen,[25,26] allows for a clearer logical definition[8] and treatment of the system.

25   Spatial distribution of gene coding sequences in the nucleus ensures that the gene products that are required by the current attractor are available to be drawn upon.[15,16] It is assumed that the gene coding sequences are transcribed stochastically as and when two conditions are met, namely that the chromatin structure is appropriate for the transcriptional apparatus to access the coding sequence and the sequence is activated for transcription. The transcribed products are stored (usually) in inactive forms in multi-compartmented dynamic steady states (one for each gene product) as a component of the routine metabolic activity of the cell. See Figure 3.

In general, if an attractor $\mathbf{a}_{gp}$ is perturbed, i.e., the value for $\mathbf{m}_{gp}(t) < \mathbf{min}_{gp}(t)$ or $\mathbf{m}_{gp}(t) > \mathbf{max}_{gp}(t)$ for one or more gene products, the system will exit the current attractor and adopt a variant attractor $\mathbf{v}_{gp}$.

Two circumstances in which the prevailing attractor can be perturbed are now considered. One can be regarded as *scheduled* within the system and its environment and thus part of its normal operation and the other *unscheduled* or stochastic and "forced" from the environment.

Differentiation[c] is the most common scheduled phenotypic transition between attractors at the cellular level. It can be initiated through signalling from its close environment or from within the system. It can be induced in the laboratory by specific drugs.[20] There are several ways of perturbing the existing state of the attractor. For example, a change in the level of activity of specific gene products can be induced by acting on the inactive protein

---

a   *syntax*   In linguistics, the grammatical rules structuring a language.

b   *semantic*   In linguistics, "semantics" refers to the meanings of words and sentences, as opposed to their structure.

c   *Differentiation*   Process in which different types of cells are produced to fulfill specific functions in the body.

of a specific gene product precursor to up-regulate[a] the gene product, or by transfer of active gene product from the cytoplasm[b] to the nucleus. In effect any perturbation of the activity of a gene product, $\mathbf{m_{gp}}$, up- or down-regulation, that places it outside the range of activities of the attractor and its basin, $\mathbf{m_{gp}(t) < min_{gp}}$, or $\mathbf{m_{gp}(t) > max_{gp}}$ will lead to a transition.

An example of an unscheduled phenotypic transition is the induction of genomic instability by ionising radiation. Here it is envisaged that stress on the post-transcriptional steady states of gene products dealing with the damage detection and repair of the genotype can cause the basin of attraction to be exceeded and the system to be released from its "normal" or "home" attractor. In this case the system will migrate to a variant attractor, which because it has not been occupied before has not been evolutionarily conditioned. It is thus likely to be less robust, i.e., a smaller basin of attraction and less stable, i.e., less proficient at error free replication of the genotype, than the normal attractor. A consequence of the loss of robustness will be that the variant attractor will be more prone to environmental perturbation and thus prone to migrate to other variant attractors. See Figure 2. For this reason the instability phenotype is best referred to as an *incomplete* phenotype. A second consequence will be due to the loss of stability resulting in a mutator phenotype.

Support for the hypothesis

We briefly review here the evidence that supports the idea that cellular regulation, both in mammalian cells and their evolutionary precursors, microorganisms, is essentially a physical process involving transitions between dynamic attractors, which are a product of self-organisation.[...]

In single celled organisms transcription of the coding sequence and regulation are more-or-less synonymous (although fission yeast is an exception). For multi-celled organisms, cooperating to form a tissue or organism, a more complex form of regulation is required. We therefore postulate that before multi-cellular growth could be established measures had to evolve to regulate the production of gene products much more closely and reduce the noise at the gene product level. This we propose is achieved through the development of post-translational processes, which serve to partly de-couple regulation from transcription. Transcription is stochastic but a post transcriptional steady state in which gene products in inactive forms are stored intervenes between transcription and regulation. (See Figure 3.)

---

a *up-regulate* Increase expression (of a gene).
b *cytoplasm* Substances surrounding the nucleus and enclosed by the cell membrane.

Huang et al[20] have provided the first experimental demonstration in mammalian cells showing that the drug induced *in vitro*[a] differentiation of a neutrophil precursor to the terminally differentiated state involved the transition between two high dimensional attractors representing phenotype. Human promyelocytic HL60 cells[b] *in vitro* can be reliably stimulated to differentiate to stable neutrophils with drugs, for example DMSO. Serial measurements of gene profiling[c] as a surrogate for the genomic state during the process induced by two drugs, showed that the differentiation pathways were dependent on the identity of the initiating drug. Thus, the concept of a single encoded differentiation pathway within the system was rejected. When the differentiation process is reversed by manipulating the drug concentration hysteresis[d] was observed.[21] This is interpreted by the authors as indicating the presence in the differentiation process of attractor states intermediate between the precursor and the terminally differentiated state. These experiments provide strong support for the concept of cellular processes being in essence transitions between attractors representing phenotypic states, the actual "route" of the transition being dependent on the conditions initiating the transition rather than an encoded pathway.

The proposal that the well established phenomenon of genomic instability induced by ionising radiation can be understood in terms of an epigenetic transition between dynamic attractors representing phenotype was advanced in 2000.[19] A prediction of the proposal is that once destabilised the genome will "wander" in the state space adopting variant attractors, Figure 2, and thus a destabilised clone will exhibit an increasing diversity of gene expression with time. [...]

[... T]here is clear support for the contention that dynamic attractors represent phenotype in mammalian cells and that this has been inherited from more primitive organisms. Attractors are robust to disruptive environmental influences to a degree but beyond a limit defined by the basin of attraction, which is a product of evolutionary conditioning, the system can be irreversibly perturbed adopting the instability phenotype. This we argue is a funda-

---

a   *in vitro*   Term referring to experimentation conducted on biological matter as opposed to whole living organisms.

b   *Human promyelocytic HL60 cells*   Cells cultured from a sample taken from a leukemia patient.

c   *gene profiling*   Measurement of the type and quantity of molecules produced when a particular gene is expressed.

d   *hysteresis*   Temporal gap between a cause and its effect (here, between the change in drug concentration and the change in cell behavior).

mentally important property of the epigenetic regulatory system that in germ cells plays a pivotal role in evolution and in somatic cells[a] in carcinogenesis.

DISCUSSION

Implications of the hypothesis

We now describe the principal implications entailed by the hypothesis/model: 35

- Epigenetic regulation and epigenetic memory are fundamentally physical processes deriving in part from the intrinsic rules of engagement between active gene products and in part from extrinsic influences. At mitosis, and fusion in germ cells,[b] the attractor is inherited to determine the phenotype of the offspring.
- Due to the influences from the environment the rules of engagement are indeterminate. Further, due to its openness the system operates far from equilibrium. This results in indeterminacy in the identities of the gene products. To deal with this inherent indeterminacy it is proposed that predicate logic systems, such as Refinement Calculus, are the most appropriate computational tools.
- Epigenetic variation exists in the form of attractors that are dormant (variant attractors) in the system but which can be occupied if the system is subject to an unscheduled expulsion from its normal attractor. When such a transition occurs there is a step change in phenotype, i.e., the change is not gradual. Epigenetic variation in germ cells could play a role in speciation.[c] In somatic cells the adoption of a variant attractor could be the initiation of carcinogenesis.

Each of these implications will be addressed in outline here and in more detail elsewhere.

Epigenetic regulation and memory

Epigenetic regulation can be seen as a physical process preceded by the 40 stochastic transcription of the appropriate coding sequences, dependent on the spatial ordering of the chromatin and the "status" of those sequences and contingent on the availability of the gene product precursors contained in multi-compartment post-translational steady states. Thus, although the

---

a *somatic cells* All cells, apart from reproductive cells.
b *mitosis* Process of cell division; *fusion in germ cells* Joining together of reproductive cells (i.e., fertilization).
c *speciation* Process by which new species develop and arise.

transcriptome[a] reflects the regulatory processes of the cell it is not as direct a reflection as the active proteome due to the buffering effect of the post-transcriptional steady states. For example, within a minute or two of damage being inflicted on the genotype by ionising radiation H2AX[b] labelling occurs at damage sites, checkpoints are instigated to delay replication and macro-scopically discernible foci of proteins assemble around the break[34-36] but it is not until tens of minutes later that the system responds with transcriptional responses.[37]

The epigenetic memory at mitosis involves the inheritance of the attractor by the offspring cells and thus is again a physical process. Following meiosis[c] and fusion in germ cells the situation, specifically for male cells, is more complicated.[38-41] In the final stages of spermatogenesis the last traces of cyto-plasm are expelled from the sperm thus precluding translation of transcripts to peptides, i.e. in effect interrupting the post translational processes. However, in principle the attractor could be sustained by the previously translated but inactive and stored precursors to gene products. It would seem reasonable to assume that attractors with low metabolic activity could thus survive the final stage of spermiogenesis[d] through to fusion.

There is ample evidence of epigenetic inheritance of genomic instability along the germ line and the subject has been extensively reviewed recently[42,43] so it will not be repeated here. It is important to recognise that the epigenetic inheritance of the GI phenotype is not Lamarckian[e] in character in so far as it is wholly without direction; the GI phenotype is a purely stochastic response to an environmental stimulus.

The current view, namely that epigenetic regulation is based on chromatin and DNA marking, certainly applies to, for example, imprinting. However, marking regulates at the transcription stage and it is evident that other more "immediate" processes are involved in the second by second regulation of the system. We therefore conclude that regulation is primarily a physical property of the attractor of which marking may be a consequence. Much the same argument applies to the epigenetic memory.

---

a   *transcriptome*   All RNA produced in one cell or group of cells.
b   *H2AX*   Histone that marks damaged genetic material for repair.
c   *meiosis*   Cell division for the specific purpose of sexual reproduction.
d   *spermiogenesis*   Creation and maturation of sperm cells.
e   *Lamarckian*   Reference to Jean-Baptiste Lamarck (1744–1829), whose now-rejected theory of evolution suggests that organisms acquire needed characteristics during their lifetimes, and that these characteristics are passed on through reproduction.

Indeterminacy

We predict that the operation of the system will be characterised by indeterminacy and thus there are implications for the computational approaches that appropriately address the system. As the system is open in respect of mass and energy flux it is far from equilibrium. Specific protein structures derived from a given peptide sequence result from the folding of the peptide and the characteristic structure is usually taken to be that with the lowest energy, i.e., the equilibrium structure. In the open environment of the cell such a restriction would not apply and many folded proteins could result from a single peptide, i.e., coding sequence. In addition many proteins have indeterminate structures[44,45] and in some cases can adopt a binding structure under the influence of the binding site.[46]

Thus, any computational approach has to be top-down and able to accommodate the inherent uncertainties. We suggest that Refinement Calculus will find an application here. Refinement Calculus[47] is a lattice-theoretic framework for reasoning. It was originally introduced as a tool for proving properties about specifications and computer algorithms, to be able to refine them into executable computer programs in a provably correct, stepwise manner. Because of this, Refinement Calculus is particularly suited for reasoning about open and complex systems, when there is only partial information available in the presence of non-determinism.

Because of its strong uniform formal foundation, built upon lattice theory and higher-order logic, Refinement Calculus bridges the gap between many popular reasoning styles, including agent based reasoning, contract based reasoning, and use of game theory. In other words, Refinement Calculus is at its best in reasoning about the precondition for reaching a certain state, when the interaction mechanisms are known only to some degree of certainty. Such a piece of information is crucial, if we wish to ensure that a set of specifications and claims about the system are consistent.

When considering a dynamically based system, the rules of engagement are seen as (partial) specifications in terms of Refinement Calculus. By measuring some of the attractors and the attractor ranges, we can then start proving the consistency of the rules, and infer other potential rules of engagement governing the system. It should be noted that due to the openness and complexity of the underlying system, the cell dynamics, there is very little hope of obtaining an algorithm-like, mechanical description of its functionality; rather, the system will most likely be described as a network of partial specifications, or rules of engagement, interacting in a non-deterministic manner. Then, Refinement Calculus provides a valuable tool, fixed-point

45

reasoning,[48] for understanding the potential outcome of those interactions. In particular, Refinement Calculus excels at finding out the governing state for some specific state to be reached by the network of rules of engagement.

However, there are indications that some measure of simplification can still result in meaningful models. [...]

Epigenetic variation

[...] An important feature of the adoption of epigenetic variants is that phenotypic change will not be gradual: the adoption of a variant attractor could involve a change in the contribution of several gene products in a single transition. This has implications for the evolutionary selection of epigenetic variation. Gradualism is universally accepted as fundamental to Darwinian theory.[51,52] According to Gould the term is a *"deductive intellectual consequence of asserting that natural selection acts as the creative mechanism of evolutionary change."* It has three meanings in the theory, namely as a means of distinguishing the theory from other so called theories such as Lamarckianism, as a means of refuting saltationism,[a] which it is argued would compete with natural selection as the creative force behind evolution and finally supporting the view that the demonstrable micro-evolutionary process (adaptation) that is central to Darwinism, would over geological timescales produce the full diversity of life that is observed today and in the fossil record. The theory of punctuated equilibrium[53] refutes this last meaning of gradualism, requiring that the process of evolution occurs in rapid spurts followed by long periods of "equilibrium" where no or very little change takes place, as the fossil record indicates.

50   It should be noted that the non-gradualism we are proposing, saltationism, does not challenge Darwin's "creative force" as the change it produces is subject to natural selection.

Depending on the specific circumstances, that is, relative loss of stability and robustness, and the extent of phenotypic change, such transitions to genomic instability in germ cells could co-evolve genetically and epigenetically, potentially resulting in evolutionary consequences ranging from minor adaptation to the origin of a new species. The initial stages in the case of speciation would be characterised by increased frequency of mutation, which over several generations would decline as integrity of replication increased and the new home attractor increased in robustness, both features that would be subject to selection for fitness.

---

a   *saltationism* Theory suggesting that evolution is not gradual, but proceeds by sudden developments.

Out of the two sources of variation it would seem that the epigenetic variation would make the more important contribution to a new species or the evolution of a new phenotypic feature, this by virtue of the non-gradual element inherent in this process. Consider the similarities in genotypes of mammalian species and the concurrent diversity in phenotypes. For example, mice have about the same number of protein coding genes as humans and over 90% of the mouse and human genomes can be partitioned into corresponding regions of conserved synteny,[a] that is, the order of genes has been conserved since the two species diverged from a common ancestor.[54] More than 99% of the proteins in the mouse genome are shared with other mammals and 98% with humans. Similarly the chimpanzee has a genome that differs from the human genome in only 4% of the bases overall and less than 1% in gene sequences coding for proteins.[55]

An overwhelmingly large fraction of the phenotypic differences between mammalian species relies on the arrangement, including scale, of a more or less common set of cellular phenotypes. Thus, in theory one could contemplate identical genotypes for mouse and man with only the rules of engagement defining the phenotypic differences.

The notion that non-gradualism underlies speciation has been discussed since Darwin's time. For example, as noted by Patrick Bateson,[56] Galton used the analogy of a "rough stone" with many facets that could, if sufficiently perturbed, make a jerky transition from resting on one facet to resting on another. This analogy captures the essence of the present model.

If the transition to genomic instability takes place in a somatic cell we suggest that the end result may be malignancy. Carcinogenesis, like genomic instability, is characterised by a mutator phenotype.[57] The relative loss of stability and robustness of the instability phenotype may result in changes in epigenetic regulation and the acquisition of mutations that a) give a selective growth advantage by, for example, the loss of a checkpoint and b) preclude the complete reversal of the process by modification of the state space due to the loss of or gain of dimensions (active gene products). Again a co-evolution of genetic and epigenetic variation may result in the instability phenotype resolving into a malignant phenotype. That there is an epigenetic component to carcinogenesis has been long recognised. Early experiments transplanting malignant cells into blastocysts[b] demonstrated that the malignant phenotype could be reversed.[58,59] Later, malignant nuclei from mice transplanted into enucleated eggs were grown into normal embryos.[60] The view of carcino-

55

---

a   *synteny*   Spatial overlap of genes within the same chromosome of an organism.

b   *blastocysts*   Early development structures containing embryonic cells.

genesis advanced here (see also),[19] while recognising the importance of mutations in achieving the "hallmarks of cancer,"[61] e.g., loss of senescence,[a] anchorage free growth, etc. sees such mutations as the consequence of an underlying and more fundamental epigenetic process that leads the system into a specific domain of the state space associated with malignancy, via a series of randomly adopted variant attractors, Figure 2. Thus, as is observed, the malignant phenotype is not well defined either in terms of the attractor that represents it or in terms of the mutations that it has acquired, although both may be "characteristic" of the disease.

CONCLUSION

In his Spinoza Lectures, the philosopher John Dupré[62] says "scientific modelling is not like building a scale model of a ship ... rather scientific models are successful to the extent that they identify the factors, or variables, that really matter." Regarding the cell as a material system driven by external "forces" in terms of the genotype, signals from neighbouring cells in the same organism and influences from the wider environment, including in some cases other organisms, is an attempt to extract those factors. Necessarily the detail that characterises the internal working of the cell, which is the subject of mainstream cell and molecular biology, is ignored. Walter Elsasser in 1981[63] sought principles, consistent with quantum mechanics, governing biology, where replicates at any level, organisms within a species to cells in a tissue, were characterised by intrinsic variability, and thus at odds with the concept of "mechanisms." He concluded biology relied on selection from a vast number of states and that [hereditary] reproduction rather than being duplication (possibly with errors) was better represented as "creativity with constraints," a process "released" by genes as operators or predicates. It is our contention that the evidence that can be garnered from the products of cell and molecular biology research since 1981 fully support Elsasser's prognosis.

(2008)

Acknowledgments: The authors gratefully acknowledge the comments of an anonymous referee and insightful and productive discussions over a number of years with Oleg Belyakov, Bob Cundall, Harold Hillman, Hooshang Nikjoo, Darius Leszczynski, Mike Thorne, Dillwyn Williams and especially Alwyn C. Scott who sadly died on 22 January 2007.

Author Contributions: Wrote the paper: KB. Other: Conceived and developed the ideas with the exception of the relvance of Refinement Calculus to the model/hypothesis proposed: KB. Contributed the insight that refinement Calculus was an appropriate modality to deal with the model/hypothesis proposed by the first author: MR.

---

a   *senescence*   Cell aging, cell death.

Funding: The authors have no support or funding to report.

Competing interests: The authors have declared that no competing interests exist.

REFERENCES

1. Nanney DL (1958) Epigenetic Control Systems. Proc Natl Acad Sci USA 44: 712–717.
2. Delbrück M (1949) translation of discussion following a paper by Sonneborn, T. M. and Beale, G.H. Unites Biologiques Douees de Continute Genetique:. pp. 33–35.
3. Schrödinger E (1944) What is life? Cambridge: Cambridge University Press.
4. Weinhold B (2006) Epigenetics: the science of change. Environ Health Perspect 114: a160–167.
5. Jaenisch R, Bird A (2003) Epigenetic regulation of gene expression: how the genome integrates intrinsic and environmental signals. Nat Genet 33: Suppl245–254.
6. Bird A (2002) DNA methylation patterns and epigenetic memory. Genes Dev 16: 6–21.
7. Bird A (2007) Perceptions of epigenetics. Nature 447: 396–398.
8. O'Malley MA, Dupré J (2005) Fundamental issues in systems biology. Bioessays 27: 1270–1276.
9. Bertalanffy L von (1950) An outline of general systems theory. British Journal for the Philosophy of Science 1: 134–165.
10. Huang S (2004) Back to the biology in systems biology: what can we learn from biomolecular networks? Brief Funct Genomic Proteomic 2: 279–297.
11. Felsenfeld G (1992) Chromatin as an essential part of the transcriptional mechanism. Nature 355: 219–224. doi:10.1038/355219a0.
12. Cremer T, Cremer C (2001) Chromosome territories, nuclear architecture and gene regulation in mammalian cells. Nat Rev Genet 2: 292–301.
13. Cremer M, von Hase J, Volm T, Brero A, Kreth G, et al. (2001) Non-random radial higher-order chromatin arrangements in nuclei of diploid human cells. Chromosome Res 9: 541–567.
14. Bolzer A, Kreth G, Solovei I, Koehler D, Saracoglu K, et al. (2005) Three-dimensional maps of all chromosomes in human male fibroblast nuclei and prometaphase rosettes. PLoS Biol 3: e157.
15. Fraser P, Bickmore W (2007) Nuclear organization of the genome and the potential for gene regulation. Nature 447: 413–417.
16. Kosak ST, Scalzo D, Alworth SV, Li F, Palmer S, et al. (2007) Coordinate gene regulation during hematopoiesis is related to genomic organization. PLoS Biol 5: e309.
17. Kadhim MA, Macdonald DA, Goodhead DT, Lorimore SA, Marsden SJ, et al. (1992) Transmission of chromosomal instability after plutonium alpha-particle irradiation. Nature 355: 738–740.
18. Baverstock KF (1991) DNA instability, paternal irradiation and leukaemia in children around Sellafield. Int J Radiat Biol 60: 581–595.
19. Baverstock K (2000) Radiation-induced genomic instability: a paradigm-breaking phenomenon and its relevance to environmentally induced cancer. Mutat Res 454: 89–109.
20. Huang S, Eichler G, Bar-Yam Y, Ingber DE (2005) Cell fates as high-dimensional attractor states of a complex gene regulatory network. Phys Rev Lett 94: 128701.
21. Chang HH, Oh PY, Ingber DE, Huang S (2006) Multistable and multistep dynamics in neutrophil differentiation. BMC Cell Biol 7: 11.
22. Morgan WF (2003) Non-targeted and delayed effects of exposure to ionizing radiation: I. Radiation-induced genomic instability and bystander effects in vitro. Radiat Res 159: 567–580.
23. Morgan WF (2003) Non-targeted and delayed effects of exposure to ionizing radiation: II. Radiation-induced genomic instability and bystander effects in vivo, clastogenic factors and transgenerational effects. Radiat Res 159: 581–596.

24. McFall-Ngai MJ (2002) Unseen forces: the influence of bacteria on animal development. Dev Biol 242: 1–14.

25. Rosen R (1991) Life Itself: a Comprehensive Inquiry into the Nature, Origin and Fabrication of Life. Allen THF, Roberts DW, editors. New York: Columbia University Press.

26. Rosen R (2000) Essays of Life Itself; Timothy EH, Roberts A, Roberts DW, editors. New York: Columbia University Press.

27. Kauffman SA (1993) The Origins of Order: Self Organisation and Selection in Evolution. Oxford: Oxford University Press.

28. Aldana M, Cluzel P (2003) A natural class of robust networks. Proc Natl Acad Sci USA 100: 8710–8714.

29. Aldana M, Balleza E, Kauffman S, Resendiz O (2007) Robustness and evolvability in genetic regulatory networks. J Theor Biol 245: 433–448.

30. Kashiwagi A, Urabe I, Kaneko K, Yomo T (2006) Adaptive response of a gene network to environmental changes by fitness-induced attractor selection. PLoS ONE 1: e49.

31. Davidich MI, Bornholdt S (2008) Boolean network model predicts cell cycle sequence of fission yeast. PLoS ONE 3: e1672.

32. Li F, Long T, Lu Y, Ouyang Q, Tang C (2004) The yeast cell-cycle network is robustly designed. Proc Natl Acad Sci USA 101: 4781–4786.

33. Falt S, Holmberg K, Lambert B, Wennberg A (2003) Long-term global gene expression patterns in irradiated human lymphocytes. Carcinogenesis 24: 1823–1845.

34. Shiloh Y (2003) ATM and related protein kinases: safeguarding genome integrity. Nat Rev Cancer 3: 155–168.

35. Zhou BB, Elledge SJ (2000) The DNA damage response: putting checkpoints in perspective. Nature 408: 433–439.

36. Rogakou EP, Boon C, Redon C, Bonner WM (1999) Megabase chromatin domains involved in DNA double-strand breaks in vivo. J Cell Biol 146: 905–916.

37. Watson A, Mata J, Bahler J, Carr A, Humphrey T (2004) Global gene expression responses of fission yeast to ionizing radiation. Mol Biol Cell 15: 851–860.

38. Krawetz SA (2005) Paternal contribution: new insights and future challenges. Nat Rev Genet 6: 633–642. doi:10.1038/nrg1654.

39. Rousseaux S, Caron C, Govin J, Lestrat C, Faure AK, et al. (2005) Establishment of male-specific epigenetic information. Gene 345: 139–153.

40. Miller D, Ostermeier GC, Krawetz SA (2005) The controversy, potential and roles of spermatozoal RNA. Trends Mol Med 11: 156–163.

41. Wykes SM, Krawetz SA (2003) The structural organization of sperm chromatin. J Biol Chem 278: 29471–29477.

42. Barber RC, Dubrova YE (2006) The offspring of irradiated parents, are they stable? Mutat Res 598: 50–60.

43. Bouffler SD, Bridges BA, Cooper DN, Dubrova Y, McMillan TJ, et al. (2006) Assessing radiation-associated mutational risk to the germline: repetitive DNA sequences as mutational targets and biomarkers. Radiat Res 165: 249–268.

44. Chen JW, Romero P, Uversky VN, Dunker AK (2006) Conservation of intrinsic disorder in protein domains and families: II. functions of conserved disorder. J Proteome Res 5: 888–898.

45. Romero P, Obradovic Z, Dunker AK (2004) Natively disordered proteins: functions and predictions. Appl Bioinformatics 3: 105–113.

46. Sugase K, Dyson HJ, Wright PE (2007) Mechanism of coupled folding and binding of an intrinsically disordered protein. Nature 447: 1021–1025.

47. Back R, Wright von J (1998) Refinement Calculus: A Systematic Introduction. New York: Springer-Verlag.

48. Tarski AA (1955) Lattice-Theoretical Fixed-point Theorem and Its Applications. Pacific J Math 5: 285–309.
49. Kitano H (2002) Computational systems biology. Nature 420: 206–210.
50. Krishnan A, Giuliani A, Tomita M (2007) Indeterminacy of reverse engineering of Gene Regulatory Networks: the curse of gene elasticity. PLoS ONE 2: e562.
51. Gould SJ (2002) The Structure of Evolutionary Theory. Cambridge Mass.: Harvard University Press.
52. Mayr E (2002) What Evolution Is. London: Weidenfeld and Nicholson.
53. Gould SJ, Eldredge N (1993) Punctuated equilibrium comes of age. Nature 366: 223–227.
54. Waterston RH, Lindblad-Toh K, Birney E, Rogers J, Abril JF, et al. (2002) Initial sequencing and comparative analysis of the mouse genome. Nature 420: 520–562.
55. Li WH, Saunders MA (2005) News and views: the chimpanzee and us. Nature 437: 50–51.
56. Bateson P (2002) William Bateson: a biologist ahead of his time. J Genet 81: 49–58.
57. Bielas JH, Loeb KR, Rubin BP, True LD, Loeb LA (2006) Human cancers express a mutator phenotype. Proc Natl Acad Sci USA 103: 18238–18242.
58. Mintz B, Illmensee K (1975) Normal genetically mosaic mice produced from malignant teratocarcinoma cells. Proc Natl Acad Sci USA 72: 3585–3589.
59. Illmensee K, Mintz B (1976) Totipotency and normal differentiation of single teratocarcinoma cells cloned by injection into blastocysts. Proc Natl Acad Sci USA 73: 549–553.
60. Li L, Connelly MC, Wetmore C, Curran T, Morgan JI (2003) Mouse embryos cloned from brain tumors. Cancer Res 63: 2733–2736.
61. Hanahan D, Weinberg RA (2000) The hallmarks of cancer. Cell 100: 57–70.
62. Dupré J (2008) The Constituents of Life. Amsterdam: Van Gorcum.
63. Elsasser WM (1981) Principles of a new biological theory: a summary. J Theor Biol 89: 131–150.

Questions:

1. Reflect on the first paragraph of the "statement of hypothesis." What are the key points discussed in this paragraph?

2. According to the article, how closely do genotypes relate to their resulting phenotypes? Do small changes in the environment in which the genotype is expressed result in small or large phenotypic variation? What examples support your conclusion?

3. The article provides definitions for key terms used in the article. How do these definitions add to your understanding of the article?

4. Why, according to the article, is the dynamic state of attractors "fundamentally important" to the epigenetic regulatory system? The article defines attractor states, but it uses the concept metaphorically, in architectural terms (as something that can be occupied, but in which the shape of the architecture changes). How does the metaphor assist your understanding of the process?

5. What role do computational approaches play in understanding the epigenetic regulatory system? Are they applied to anything specifically?

6. The article (briefly) uses linguistic theory in a metaphoric way. Where does it do so, and what does the use do in relation to the formulaic approach that occurs in the definitions section?

7. The article states, "it should be noted that the non-gradualism we are proposing, saltationism, does not challenge Darwin's 'creative force' as the change it produces is subject to natural selection." Why would the authors specifically address Darwin's theory? What influence might this statement have over the article's audience?

8. The article concludes with a quotation from philosopher John Dupré. What is the significance of this quotation in the context of this article?

9. The article defines homeostasis as "the property of an open system to regulate its internal state and maintain a stable condition." According to the article, what significance does the concept of homeostasis have in relation to the regulatory systems analysed?

10. In the section "Support for the Hypothesis," the article provides key supporting factors for the main argument. Is this what you would consider "evidence" in a strictly experimental sense, or is this article engaging in philosophy of science? Given that epigenetics is a very new field, is this approach appropriate?

11. In what ways does the article speak to a scientific reader in the field? In what ways does the article speak to a reader outside the field?

12. The article explains Refinement Calculus. What does the inclusion of this definition suggest about the intended audience? Do you think those who work with Refinement Calculus need this explanation? What, then, is the explanation doing here?

13. Reread the final paragraph of the "Epigenetic Variation" section. Explain how the article relates genomic instability to the generation of cancerous cells.

14. After reading the entire article, refer back to the article's abstract section (Background, Methodology/Findings, and Conclusions). Is this concise description of the article sufficient in order to understand the key points of the hypothesis? Why or why not?

15. Look at the Author Contributions section. Who wrote this article? Does the difference in contributions explain anything about the article's approaches?

V

Agricultural Science:
Humans and Other Animals

twenty-seven

JEFF DOWNING

from Non-Invasive Assessment of Stress in Commercial Housing Systems: A Report for the Australian Egg Corporation Limited

This March 2012 report by a respected veterinary researcher reaches some surprising conclusions about the degree to which egg-laying birds in various types of egg production systems in Australia – those labeled "conventional cage," "free range," and "barn" – undergo stress.

❧

FOREWORD

This project determined the corticosterone concentrations in albumen[a] of eggs collected from the three main production systems used in the Australian egg industry: conventional cages, free range and barn. Egg albumen corticosterone concentrations are used as a noninvasive measure of stress in laying hens because of their correlation with plasma[b] corticosterone concentrations.

This project was funded from industry revenue which is matched by funds provided by the Federal Government.

This report[c] is an addition to AECL's range of research publications and forms part of our R&D program, which aims to support improved efficiency,

---

a *corticosterone* Steroid hormone that serves several functions and is produced in greater quantities in response to stress; *albumen* Egg white or, more specifically, the proteins found within egg white.
b *plasma* The clear liquid portion of blood.
c The full report runs to 82 pages; the present excerpts are intended to convey its most important findings, but readers are encouraged to consult the full report, which is available through the Australian Egg Corporation site at www.aecl.org/dmsdocument/41. The report is copyright © 2012 Australian Egg Corporation Limited. "However," as is declared on the copyright page, "AECL encourages wide dissemination of its research provided the Corporation is clearly acknowledged."

sustainability, product quality, education and technology transfer in the Australian egg industry. [...]

Executive Summary

The basis of the current study was to evaluate corticosterone concentrations in albumen of eggs collected from hens maintained in the three main production systems used in the Australian egg layer industry (Conventional cages (CC), Free range (FR) and Barn (Bn)) at 24, 32, 42, 52, 62 and 72 weeks of age. Commercial producers were initially contacted by the Australian Egg Corporation Limited detailing the project to be undertaken and requesting their involvement. Following a positive response from individual producers, meetings were arranged and details of how the egg collections were to be made were discussed. In the study, five free range (FR 1-5), four conventional cage (CC 1-4) and three barn (Bn 1-3) flocks were sampled. From each of the participating farms, ninety eggs were collected at random when the hens were at the appropriate ages. For all except farm CC4, eggs were supplied by the producers who randomly selected the eggs from those laid on one day in the week of the specified flock age. Eggs were collected at the same ages for all systems but because flocks were placed at different times, the collections started in different months, ranging from July 2009 to March 2010.

Egg and albumen weights, egg albumen corticosterone concentration and total corticosterone in albumen were determined for all eggs collected. [...]

When the farms are grouped into production (housing) systems it was found that the type of system had no significant effect on egg albumen corticosterone concentration (P=0.78)[a] or on the total amount of egg albumen corticosterone (P=0.48). There was large variation in these measures for individual farms even in the same production system. [...]

[...] 1 General Introduction [...]

1.2 Physiological responses to stress [...]

A consistent non-specific response [to stress] is an increase in corticosterone secretion. [...] If the stress is continued for a chronic[b] period, the corticosterone-induced changes are detrimental to the hen. Pathological con-

---

a   *P*   Probability, here an indication of the likelihood of getting the same experimental results as the ones observed if there were no relationship between the variables being studied.

b   *chronic*   I.e., long-term.

sequences include ulcers, hypertension[a] and immunosuppression and effects can be permanent after removal of the stress and, if continued, can result in death. Short term stressors such as heat (Beuving, 1980), food and water deprivation (Beuving, 1980), transport (Broom and Knowles, 1989) and fear (Beuving et al., 1989) give rise to elevated corticosterone levels. [...]

1.13.2 EGG CORTICOSTERONE

Previous reports to the AECL (Downing and Bryden, 2002; 2005) identified the relationship between stressful events, plasma and egg albumen corticosterone concentrations. In these reports, examples are provided of how albumen corticosterone concentrations increase when hens are challenged with events that are known to be stressful.

Steroid hormones accumulate during egg formation (Rettenbacher et al., 2009). Royo et al. (2008) found that around about 80% of egg corticosterone was found in the yolk and 20% in albumen. During egg formation, yolk accumulation occurs over 7-12 days before ovulation (Johnson, 1986) whereas the albumen is accumulated over 4-6 hours the day before the egg is laid. Accumulation of corticosterone in yolk and albumen could provide measures of long-term and short-term stress, respectively. [...]

1.18 Project objectives

[...] The objective of the current study was to evaluate corticosterone concentrations in albumen from eggs collected from hens maintained in the three main production systems used in Australia – conventional cages (CC), free range (FR) and barn (Bn) – at 24, 32, 42, 52, 62 and 72 weeks of age. While a main aim was to compare the differences between systems, the pattern on individual farms would help to identify potential periods when hens were experiencing conditions which result in stress.

2 GENERAL MATERIALS AND METHODS

2.1 General methodology

2.1.1 DETERMINATION OF CORTICOSTERONE CONCENTRATIONS IN ALBUMEN

The corticosterone concentrations in egg albumen were determined by radio-immunoassay[b] (RIA). [...]

---

a   *hypertension*   High blood pressure.

b   *radioimmunoassay*   Laboratory procedure that can be used to test the quantity of a hormone or other antigen in a sample. A supply of the antigen is made radioactive and added to the sample, where it interacts with the sample; the interaction between the radioactive antigen and the sample is then analyzed to determine how much of the (non-radioactive) antigen was originally present.

2.2 Animal ethics

Hens maintained at the Poultry Research Unit, University of Sydney, Camden, NSW,[a] Australia, were part of a flock maintained for student teaching and research and were approved by the University of Sydney Animal Care and Ethics Committee and complied with the Australian Code of Practice for the use of Animals for Scientific Purposes. All other flocks were commercial flocks.

2.3 Egg collections

From each of the participating farms, ninety eggs were collected at random when the hens were 24, 32, 42, 52, 62 and 72 weeks of age. For all commercial farms, eggs were supplied by the producer. Producers were instructed to randomly sample the eggs from the egg collection belt on one day in the relative collection week. [...]

2.4 Description of participating farms

Individual producers were initially contacted by the Australian Egg Corporation Limited detailing the project to be undertaken and requesting their involvement. Following a positive response from producers, a meeting was arranged and details of how the collections would be made were discussed. At the end of the production cycle each producer was asked to fill in a questionnaire giving details about the facilities and flocks involved in the study. [The strain of hen was Isa Brown at all farms.]

2.4.1 FREE RANGE FARMS

Farm 1-FR1
Location: North Western Sydney Basin, NSW, Australia [...]
Shed capacity: 15,000 at placement at ten hens/m^2. The shed was divided into two roughly equal sections with a wire partition to prevent hens from "swarming" [...]
Mortality: 5.01% [...]

10 **Farm 2-FR2**
Location: Southern Tablelands of New South Wales, Australia [...]
Shed capacity: 13,000 at 10 birds/m^2. Hens reared on the floor with perches and transferred at 15 weeks and trained to use the slatted area and perches. Hens have access to the free range area from 20 weeks [...]
Mortality: 7.22% [...]

---

a *NSW* New South Wales, a state in Australia.

Farm 3-FR3
Location: Tamworth, NSW, Australia [...]
Shed capacity: 7,500 at ten birds/m^2 [...]
Mortality: 4% [...]

Farm 4-FR4
Location: North Western Sydney Basin, NSW, Australia [...]
Shed capacity: 10,000 at placement at ten hens/m^2 [...]
Mortality: 3.5% [...]

Farm 5-FR5
Location: North Western Sydney Basin, NSW, Australia [...]
Shed capacity: 15,000 at placement at ten hens/m^2 [...]
Mortality: 3.1% [...]

2.4.2 BARN FARMS

Farm 6-Bn1
Location: Western Sydney Basin, NSW, Australia [...]
Shed capacity: 2,500 at 10 birds/m^2 [...]
Mortality: Not available [...]

Farm 7-Bn2
Location: North Western Sydney Basin, NSW, Australia [...]
Shed capacity: 9,500 at placement at seven hens/m^2 [...]
Mortality: 4.0% [...]

Farm 8-Bn3
Location: Mid North Coast, NSW, Australia [...]
Shed capacity: 10,000 at 10 hens/m^2 [...]
Mortality: Not provided for commercial in confidence reasons [...]

2.4.3 CONVENTIONAL CAGE FARMS

Farm 9-CC1
Location: Western Sydney Basin, NSW, Australia [...]
Shed capacity: 44,000 at placement. Six tiers of five birds/cage [...]
Mortality: 1.75% [...]

Farm 10-CC2
Location: Western Sydney Basin, NSW, Australia [...]
Shed capacity: 22,500 at placement with five hens/cage with five tiers [...]
Mortality: 6.5% [...]

Farm 11-CC3
Location: North Coast of NSW
Shed capacity: 3,300 and single tier five birds/cage [...]
Mortality: 3% [...]

20 ## Farm 12-CC4
Location: University of Sydney, Camden, NSW, Australia
Shed capacity: 1200. Single bird cages – Eggs were collected from a sub sample of 100 birds maintained in the facility [...]
Mortality: 2.0% [...]

2.5 Statistical analysis

Egg weight, albumen weight, albumen corticosterone concentrations and total amount of corticosterone in albumen were response variables tested. [...]

3 RESULTS

Collections were made at the correct age for all farms except farm Bn1, where the 72 week collection was not made as the producer moulted the flock at 68 weeks. Collection of eggs was made after this moult, but the analysis was not considered to be part of the project and so have not been included.

3.1 Collection values for individual farms

3.1.1 FREE RANGE FARMS: ALBUMEN CORTICOSTERONE CONCENTRATION (NG/G)[a] [...][b]

TABLE 1.
Albumen Corticosterone Concentration (ng/g) – Free Range Farms

Age at Collection (weeks)	FR 1	FR 2	FR 3	FR 4	FR 5
24	0.52	1.30	1.38	1.23	1.31
32	0.97	1.50	1.10	0.97	0.83
42	0.79	1.06	1.21	0.73	0.85
52	0.96	0.87	0.92	0.85	0.74
62	1.00	0.89	0.83	0.87	0.74
72	0.91	0.92	0.73	0.76	1.09

a   *ng*   Nanogram; one nanogram is 10^{-9} grams.
b   [editors' note]   For reasons of space the information on albumen corticosterone concentrations and on range in egg corticosterone concentrations in the three types of farm is presented here in the form of charts, labeled Tables 1 through 6; in the full report the same information is presented over many pages, together with additional information.

3.1.2 BARN FARMS: ALBUMEN CORTICOSTERONE CONCENTRATION (NG/G) [...]

TABLE 2.
Albumen Corticosterone Concentration (ng/g) – Barn Farms

Age at Collection (weeks)	Bn 1	Bn 2	Bn 3
24	1.27	1.13	0.69
32	1.24	1.66	0.81
42	1.15	1.29	0.88
52	0.75	0.77	0.67
62	0.72	0.73	0.81
72	-	0.71	0.90

3.1.3 CONVENTIONAL CAGE FARMS: ALBUMEN CORTICOSTERONE CONCENTRATION (NG/G) [...]

TABLE 3.
Albumen Corticosterone Concentration (ng/g) – Conventional Cage Farms

Age at Collection (weeks)	CC 1	CC 2	CC 3	CC 4
24	0.81	1.27	1.38	0.72
32	1.23	1.13	0.95	0.78
42	0.90	1.05	0.70	0.73
52	0.85	0.94	0.75	0.61
62	0.92	0.70	0.68	0.74
72	0.83	0.83	0.64	0.83

3.4 The comparisons between different production systems [...]

3.4.3 CORTICOSTERONE CONCENTRATIONS

[...] The mean egg albumen concentration for the three production systems at the various collection ages is given in figure 3.10. The production system had no significant effect on egg albumen corticosterone concentration (P=0.78). The collection age had a significant effect (P=0.02), with the collection age x system interaction not significant (P=0.72). The albumen corticosterone concentrations were higher in the early stages of the production cycle. [...]

3.7 The range in egg corticosterone concentrations for all farms

[... Results are shown for] albumen corticosterone concentrations for all eggs collected from each of the 12 farms. An arbitrary value of 1.5 ng/g has been selected as the cut off point for the determination of the percentage of eggs with high corticosterone concentrations. This arbitrary value has been selected because for each housing system the farm with the lowest corticosterone concentrations are below this cut off value. [...]

Figure 3-10.

The mean egg albumen corticosterone concentrations for the three production systems.

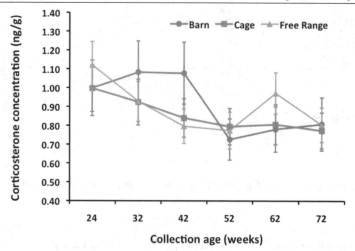

Conventional cages (n=4), Barn (n=3) and free range (n=5). The effect of production system was not significant (P=0.78).

3.7.1 Free Range Farms [...]

Table 4.

Range in Egg Corticosterone Concentrations:
Percentage of Eggs with Values above 1.5 ng/g – Free Range farms

Age at collection (weeks)	FR 1	FR 2	FR 3	FR 4	FR 5
24	0%	25.6%	41.1%	14.4%	20.0%
32	1.1%	41.1%	11.4%	3.3%	1.1%
42	0%	5.6%	17.8%	2.3%	6.8%
52	5.6%	1.1%	2.2%	0%	1.11%
62	6.7%	3.3%	1.1%	3.3%	0%
72	4.4%	2.2%	2.2%	0%	12.2%

3.7.2 Barn Farms [...]

Table 5.

Range in Egg Corticosterone Concentrations:
Percentage of Eggs with Values above 1.5 ng/g – Barn farms

Age at collection (weeks)	Bn 1	Bn 2	Bn 3
24	24.4%	13.3%	1.1%
32	18.9%	56.7%	0%
42	14.4%	21.1%	3.3%
52	0%	0%	0%
62	0%	1.1%	0%
72	-	0%	4.5%

3.7.3 CONVENTIONAL CAGE FARMS [...]

TABLE 6.
Range in Egg Corticosterone Concentrations:
Percentage of Eggs with Values above 1.5 ng/g – Conventional Cage farms

Age at collection (weeks)	CC 1	CC 2	CC 3	CC 4
24	0%	25.6%	33.3%	1.1%
32	20.0%	14.4%	2.2%	1.1%
42	3.3%	4.4%	0%	0%
52	0%	3.3%	0%	0%
62	3.3%	0%	1.1%	1.1%
72	1.1%	0%	0%	0%

3.7.4 THE MEAN PERCENTAGE OF EGGS WITH HIGH CORTICOSTERONE CONCENTRATIONS FOR THE THREE PRODUCTION SYSTEMS

The mean (± SEM)[a] percentage of hens with albumen corticosterone concentrations above 1.5 ng/g for the different production systems is shown in figure 3.28.

The effect of production system was not significant (P=0.39) while the effect of collection age was significant (P=0.001). [...]

FIGURE 3-28.
The mean (± SEM) percentage of hens with albumen corticosterone concentrations above 1.5 ng/g for the different production systems.

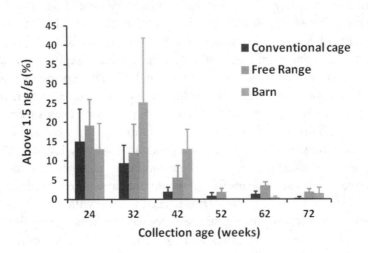

---

a *SEM* Standard error of the mean, an indicator of the mean's reliability.

4 Discussion

Stress is relevant to welfare as it indicates a condition requiring an adjustment by the animal to adverse stimuli. If the adaptation is quick and successful then the consequences to the animal are minimal. If the stress is persistent, even if the adverse stimulus doesn't remain the same but varies, then the consequences for the animal can be poorer welfare. [...]

Eggs were collected at the same ages for all farms but because flocks were placed at different times, the collections started in different months of the year. [...] There are two levels to evaluate the results from the current study, collectively as part of a prescribed housing system and on an individual farm basis. The changes occurring within a farm are relevant to physiological status of the hens on that farm, irrespective of the housing type employed. [...]

The type of production system had no significant effect on egg albumen corticosterone concentration (P=0.78) or on the total egg albumen corticosterone concentrations (P=0.48). With corticosterone concentrations used as a measure of stress in hens this would suggest that the hens from any particular system were not any more stressed than those in another system. However, the grouping of farms into production systems masks some of the individual farm differences. [...]

For some farms specific observations are worth noting. The corticosterone concentrations for farm FR1 increased at weeks 52, 62 and 72, and were as high as those seen at week 32. These elevated corticosterone concentrations corresponded to significant decreases in egg production in the week of the relevant egg collection[....] On investigation, it was noted that this flock of hens were infected with lice at this time. Interestingly, this health issue was associated with a decrease in production but also an increase in egg albumen corticosterone concentrations. For farm Bn3 an increase in corticosterone concentrations was recorded in weeks 62 and 72. This corresponded to an outbreak of cannibalism which forced the producer to reduce light intensity in an effort to eliminate the problem. These stress provoking events were related to elevated albumen corticosterone concentrations. Farm CC4 had high egg production throughout the production cycle and relatively low egg albumen corticosterone concentrations throughout the cycle. Hens on this farm were maintained in individual cages and so would not have experienced the potential social stresses hens on other farms needed to cope with.

Social interactions can be stressful for laying hens (Craig and Guhl, 1969; Hughes et al., 1997; Bilick and Keeling, 2000; Keeling et al., 2003). Group effects on hen welfare had received considerable attention with small group sizes allowing for formation of fairly stable social hierarchies (Keeling,

1995). In group housing, there is competition for space and important social interactions take place which can be stressful (Keeling, 1995). Social structures, familiarity and competition are key issues responsible for the level of aggressive behaviour in group housed hens. The extensive interactions that occur during establishment of social orders in small groups of caged hens could result in stress that is seen as elevated plasma corticosterone and consequently egg albumen corticosterone (Downing and Bryden, 2009). In the floor based systems, hens are continually faced with changing social interactions. Therefore, important considerations in assessing welfare of hens are: the size of the enclosure, number of hens in the facility and the availability of resources (Mench and Keeling, 2001). In alternative production systems (barn and free range) hens experience a range of densities in the enclosure (Channing et al., 2001). It has been reported that hens are limited to identifying about 100 individuals in a group (Nicol et al., 1999) and that they prefer to be with familiar hens rather than unfamiliar hens (Bradshaw, 1992; Freire et al., 1997). Nicol and co-workers (1999) observed that aggressive behaviour was more prominent in small and medium sized flocks compared to larger flocks. In large flocks, hens probably adopt strategies that reduce social contact (Oden et al., 2000; Mench and Keeling, 2001; Freire et al., 2003). In large flocks, victimisation of some hens could seriously compromise their welfare. Stressors are thought to have additive effects on plasma corticosterone concentration. The combination of changing environmental conditions and social interactions could have accounted for the higher albumen corticosterone concentration at the beginning of the production cycle. Such social interactions would not have taken place on farm CC4 (housed singly), where egg production was high and albumen corticosterone concentrations remained low. [...]

5 GENERAL CONCLUSIONS

When farms are grouped into production systems there were no differences in mean corticosterone concentrations. For each production system the variation between farms was large and limited the value of such a comparison. While the debate about the merits of different production systems will continue, what is happening on individual farms is very relevant to hen welfare.

For most commercial farms the albumen corticosterone concentrations were high at the start of the production cycle and this could reflect the extent of challenges that hens face early in the production cycle. The different patterns seen in albumen corticosterone concentrations thereafter could reflect the rate at which the hens are able to adapt to the challenges in their environment.

The different patterns of albumen corticosterone concentrations observed in the early part of the production provide evidence that:

35 • Farm management practices in the early phases of the production cycle could be relevant to the challenges hens need to deal with in their environment. Early rearing management could be designed to familiarise hens to the production housing and accommodate their adaptation to the housing transfer.

• The persistently low albumen corticosterone concentrations, low mortality, high egg production and large egg size recorded for farm CC4, where hens were housed individually, serves to illustrate the importance of group dynamics and social adaptation in laying hens.

• Commercial Farms FR1 and Bn3 were exceptions having persistently low albumen corticosterone concentrations through weeks 24-62. Clearly, large flocks in floor systems are capable of overcoming the issues of social stress and adaptation and can maintain low albumen corticosterone concentrations in the early phases of the production cycle. Again, it could reflect the role management or early rearing has in limiting the challenges faced by laying hens.

• Four of the five farms that had more persistent elevated albumen corticosterone concentrations […] in the early production period were large free range or barn flocks.

• The measurement of albumen corticosterone concentrations could be correlated with a lice infestation (Farm FR1) and an outbreak of cannibalism (Farm Bn3).

40 • In any flock there are likely to be some hens that perceive the challenges as more severe than others and have high corticosterone concentrations.

• The mean albumen corticosterone concentrations over the entire production cycle tended to be lower in flocks with lower mortality. However, further data is needed to establish a definitive relationship.

• The data suggest that the elevated albumen corticosterone concentrations in the early stages of the production cycle are likely correlated with reductions in performance. Again, further data collection is needed to verify this relationship.

At this stage, the measurement of albumen corticosterone concentrations has highlighted the importance of early adaptation to housing system, which is likely to be improved with further knowledge and attention to management.

The high degree of variation between farms in a commercial setting necessitates that more farm collections are needed to further validate the sensitivity of albumen corticosterone concentrations as a technique for use in assessing welfare. Validation of the relationship with production and mortality need to be further explored.

6 Recommendations

Continued assessment of egg albumen corticosterone concentrations with 45
particular interest to its correlation with production, body weight and mortality but also extend this to other measures of welfare such as plumage condition, health, feather pecking and cannibalism.

The comparison between production systems is not that useful because the variation between farms within a system can be large.

In a commercial context, the focus of welfare assessment should start at the flock level but eventually get to evaluating how individual hens cope within their environment.

More thorough practical benchmarking of body weights, plumage condition, body injury, other measures of stress [...] and mortality is required at the farm level.

The rearing management of pullets[a] in assisting them to adapt to new housing; group dynamics and social interactions requires more attention.

The use of egg albumen corticosterone concentrations be promoted 50
within the Poultry Science community as a technique that can further our understanding of bird welfare assessment.

(2012)

References

Beuving, G. (1980) Corticosteroids in laying hens: In *The Laying Hen and Its Environment*. R. Moss. (Ed). Martinus Nijhoff, The Hague. p. 65.

Bilick, B. and Keeling, L.J. (2000) Relationship between feather pecking and ground pecking in laying hens and the effect of group size. *Applied Animal Behaviour Science*, 68, 55.

Bradshaw, R.H. (1992) Conspecific discrimination and social preference in the laying hen. *Applied Animal Behaviour Science*, 33, 69.

Channing, C.E., Hughes, B.O. and Walker, A.W. (2001) Spatial distribution and behaviour of laying hens housed in an alternative system. *Applied Animal Behaviour Science*, 72, 335.

Downing, J.A. and Bryden, W.L. (2002). A non-invasive test of stress in laying hens. Report to the Rural Industries Research and Development Corporation, (RIRDC, Canberra), Project US-71A.

Downing, J.A. and Bryden, W.L. (2005). Non-invasive stress assessment of commercial egg industry practices. Report to the Rural Industries Research and Development Corporation, (RIRDC, Canberra), Project US-107A.

Downing, J.A. and Bryden, W.L. (2008) Determination of corticosterone concentrations in egg albumen: A non-invasive indicator of stress in laying hens. *Physiology and Behaviour*, 95, 381.

Downing, J.A. and Bryden, W.L. (2008) The effects of housing laying hens as groups in conventional cages on plasma and egg albumen corticosterone concentrations. Australian Poultry Science Symposium, 20, 157.

Freire, R., Appleby, M.C. and Hughes, B.O. (1997) Assessment of pre-laying motivation in the domestic hen using social interaction. *Animal Behavior*, 54, 313.

---

a *pullets* Young female chickens.

Freire, R., Wilkins, L.J., Short, F. and Nicol, C.J. (2003) Behaviour and welfare of individual laying hens in non-cage systems. *British Poultry Science*, 44, 22.

Hughes, B.O. and Black, A.J. (1976) The influence of handling on egg production, egg shell quality and avoidance behaviour of hens. *British Poultry Science*, 17, 135.

Hughes, B.O., Carmichael, N.L., Walker, A.W. and Grigor, P.N. (1997) Low incidence of aggression in large flocks of laying hens. *Applied Animal Behaviour Science*, 54, 215.

Keeling, L.J. (1995) Spacing behaviour and an ethological approach to assessing optimum space allocations for groups of laying hens. *Applied Animal Behaviour Science*, 44, 171.

Keeling, L.J., Estevez, I., Newberry, R.C. and Correia, M.G. (2003) Production-related traits of layers reared in different flock sizes: the concept of problematic intermediate group sizes. *Poultry Science*, 82, 1393-1396.

Mench, J. and Keeling, L.J. (2001) The social behaviour of domestic birds. In: *Social Behaviour in Farm Animals*. Keeling, L.J. and Gonyou, H.W. (eds.), CAB International, Wallingford, UK, 177.

Nicol, C.J., Gregory, N.G., Knowles, T.G., Parkman, I.D. and Wilkins, L.J. (1999) Differential effects of increased stocking density, mediated by increased flock size, on feather pecking and aggression in laying hens. *Applied Animal Behaviour Science*, 65, 137-152.

Oden, K., Vestergaard, K.S. and Algers, B. (2000) Space use and agonistic behaviour in sex composition in large flocks of laying hens. *Applied Animal Behaviour Science*, 67, 307.

Rettenbacher, S., Mostl, E. and Groothuis, T.G.G. (2009) Gestagens and glucocorticoids in chicken eggs. *General and Comparative Endocrinology*, 164, 125.

Royo, F., Mayo, S., Carlsson, H.E. and Hau, J. (2008) Egg corticosterone: A noninvasive measure of stress in egg-laying birds. *Journal of Avian Medicine and Surgery*, 22, 310.

Questions:

1. In the foreword, the authors note that this study was funded equally by industry and the federal government of Australia. Joint sponsoring of research by industry and government has become increasingly common in much of the world over the past few decades. What benefits might such cooperation have? What drawbacks? Comment with reference to this report.

2. Describe the differences in the way hens are kept in the "free range farms," "barn farms," and "conventional cage farms" involved in this study.

3. Research the egg-related regulations in your own jurisdiction. What conditions must be satisfied in that jurisdiction for eggs to be labeled "free range"?

4. The system known variously as "conventional cage" or "battery cage" originated in the 1950s, when producers found that stacking birds in rows of cages was more economical than allowing them to run about in a barnyard. By the late twentieth century, five or more birds were typically kept in a cage less than half a square meter in size. Comment on

the rhetorical effect of the terms used to describe this system: "battery cage" (the original term) and "conventional cage" (the term now more commonly used in the industry).

5. Figure 3.28 shows the mean percentage of birds with corticosterone concentrations above 1.5 ng/g – a level presented as indicating a high degree of stress. These specific cortisol measurements show that a lower percentage of conventionally housed birds are highly stressed, as compared to the free range farm birds. Do you find this result surprising? Why or why not?

6. Look at table 6. What effect does the inclusion of CC4 – a "single bird cage" farm – among the "conventional cage" farms have on the overall conventional results, as they are reported in 3.7.4 and shown in figure 3.28?

7. The second paragraph of general conclusions suggests that the high indications of stress "at the start of the production cycle […] could reflect the extent of challenges the hens face early in the production cycle." If the article had used the phrase "the conditions the hens are subjected to" instead of "the challenges the hens face," what would be the rhetorical effect of this change? What sort of rhetorical strategy is employed in the use of the phrase "the extent of challenges the hens face"? List the conditions under which one of the sets of hens was grown. Who is responsible for that set of conditions?

8. The general conclusions state that "the persistently low albumen corticosterone concentrations, low mortality, high egg production and large egg size recorded for farm CC4, where hens were housed individually, serves to illustrate the importance of group dynamics and social adaptation in laying hens." Read the long final paragraph in the "Discussion" section of Downing's report again; to what degree does that section of the report help to make Downing's conclusion about farm CC4 more clear? Would you draw the same conclusions Downing does from this evidence? Why or why not?

9. What (if anything) do you think the results from the other farms in the study, where the density of hens is typically 10 per square meter, have to say about "group dynamics and social interaction"?

twenty-eight

HUMANE SOCIETY INTERNATIONAL

Beyond Doubt: Intensive Farming Practices Results in Stressed Birds

This short article, which was posted on the Humane Society International website on March 26, 2012, strongly contests the findings of Jeff Downing's report (included in this anthology), suggesting that the conclusions which should be drawn from the evidence presented are quite different from the ones put forward by Downing.

❧

A study, "Non-invasive assessment of stress in commercial housing systems" by Dr. Jeff Downing, has proven beyond doubt that intensive farming practices in the egg industry, whether they are caged, barn, or labelled "free range," will ultimately result in stressed birds.

Conversely, this study is seriously flawed and the glaring omission of true free-range production systems from Dr. Downing's research points to an industry-contrived outcome.

However, Dr. Downing cannot be held accountable for this inconclusive study as Australian Egg Corporation, who has once again ignored the small producer farming eggs in true free-range fashion on open pastures, supplied the producers taking part in this research.

"We are suggesting that if any of our true free range producers had been included in this study there would have been a very different outcome," says Humane Choice Chief Operating Officer, Lee McCosker.

5   "What this study does point to is that we cannot have any faith in eggs labelled as free range until we have an enforceable definition. The five so-called 'free range' farms that took part in this study had very large flock sizes, extremely high stocking densities with high mortality rates and [the birds on these farms] were clearly just as stressed as caged birds, casting even more doubt on the free range claims made on supermarket eggs."

Last week Humane Choice represented our true free-range producers as a member of the Egg Labelling Forum facilitated by the NSW[a] Food Authority on behalf of the NSW Minister of Agriculture, Katrina Hodgkinson.

The scope of discussion is *Truth in Labelling* but animal welfare, the environment, and food safety or production systems are not within the scope of this discussion.

Until we have a legal definition for free-range, we will never have truth in labelling.

(2012)

Questions:

1. Identify the grammatical error in the title of this web article. What effect do errors of this sort have on the credibility of a publication? To what extent do you think that effect is justified?

2. The first paragraph of this article refers to Downing's study as having proven something "beyond doubt." Yet the third paragraph describes the study as "inconclusive." In what way(s) do you think the authors of the article regard the study as being "inconclusive"?

3. Based on your own assessment of the information provided in Downing's study, to what extent do you agree with the assertion made in the first paragraph of this article?

4. Paraphrase the arguments made by Humane Choice COO Lee McCosker in the fourth paragraph of this article. Based on your reading of both the Downing study and this selection, to what extent do you feel McCosker's criticisms of the Downing study are justified?

5. The tone of Downing's study gives an appearance of scientific objectivity, and the tone of this Humane Society web article is clearly intentionally partisan. Identify specific writing elements in each piece that contribute to their very different tones.

a   *NSW*   New South Wales, a state in Australia.

twenty-nine

CHRISTINE PARKER

The Truth about Free Range Eggs Is Tough to Crack

*In this August 6, 2013, article from the news and commentary
website* The Conversation, *Monash University law professor
Christine Parker endeavors to clear up some of the confusion
regarding standards in Australian egg production, and,
more generally, regarding what constitutes "free range."*

❦

Queensland recently changed its regulation of free range eggs, lifting the number of hens allowed per hectare from 1,500 to 10,000. This is more than a six-fold increase.

Choice and animal welfare and free range farming advocates are in an uproar about the changes. Queensland "free range" no longer means free range at all, they say.

The Queensland Department of Agriculture, Forestry and Fisheries says that the new figure is necessary, so that the Queensland egg industry won't be disadvantaged compared with other states.

In fact no other states have a legislated definition for free range, or minimum stocking density. The department says industry practice is to stock free range egg facilities well in excess of 1,500 birds per hectare.

5   So, how are we to know that our free range eggs are really free range?

STANDARDS GALORE

The Primary Industries departments of all Australian governments, state and federal, work together to set animal welfare guidelines for egg production in the Model Code of Practice for Poultry. The latest version was agreed in 2002 and is now under review. Currently it states that free range can mean up to 1,500 birds per hectare standard, but this could change.

The industry services body that represents producers, the Australian Egg Corporation, admitted last year that some free range egg production facilities stock up to 30 or 40,000 hens per hectare.

The Egg Corp has proposed an industry standard for free range of up to 20,000 birds per hectare. Their proposed standard was assessed by the Australian Competition and Consumer Commissioner as likely to mislead and deceive consumers.

In January 2013 supermarket giant Coles announced that it would only stock cage free eggs in its own brand range. For Coles, "free range" would mean a maximum of 10,000 hens per hectare outdoors.

The 10,000 figure was not based on any particular evidence or science. Rather it is based on a combination or balancing of what animal welfare requires, what industry say they can accomplish, and what Coles believes consumers feel they can afford based on extensive consumer research.

The new Queensland regulation is still better than other states. It set a limit on outdoor stocking density lower than the 20,000 per hectare proposed by the Egg Corp, and the currently unlimited industry practice.

It also states that production facilities can only go above 1500 up to the 10,000 if hens are moved around and the ground has 60 per cent vegetation cover.

So What Should Australia Do?

The big problem with the new Queensland regulation is that it seems to accept that supermarkets in consultation with industry can ultimately decide what animal welfare practices are acceptable. In the absence of government regulation, supermarkets decide what free range means and what choices are available to consumers.

In fact consumers who buy "free range" in supermarkets are actually buying something that would be more accurately described as "barn" or, in more Australian vernacular, "shed" laid. These hens live and eat in large crowded industrial sheds with some access to an outside ranging area that is often bare and uninteresting.

By contrast many consumers, animal welfare advocates and food activists probably think free range means eggs from hens that largely range outside in paddocks, with ground foliage and tree cover and access to an indoor area to nest and perch.

Rather than letting supermarkets and industry dictate what "free range" means in the absence of government regulation, all Australian states and territories should mandate compliance with at least the minimal animal welfare conditions in the Model Code of Practice.

They should also legislate definitions of cage, barn or shed, and free range that make it clear that what often currently counts as free range in the supermarkets is actually barn or shed with outside access, not a truly alternative free range production system.

Consumers need to recognise that the only true free range eggs currently available are premium products that cost more than supermarket brand free range eggs.

An Australian state that really wanted to help its egg industry might do more to help consumers get direct access to farmers outside of the supermarket system.

20   The South Australian government's recent proposal to introduce and support its own voluntary "South Australian Free Range" with more stringent standards is a step in this direction. What a pity Queensland chose to loosen its standards rather than market its difference.

(2013)

Questions:

1. The author, Christine Parker, is an academic at Monash University. Do you think this piece was written primarily for an academic audience? Why or why not?

2. To what extent does this article help to make the perhaps complicated Australian egg farming situation clear without oversimplifying? Are there areas in which it could do a better job of making things clear?

3. How long are the sentences and paragraphs in this article? What sort of an intended audience do these lengths suggest?

4. The last sentence of the article adopts a tone that is different from that of the rest of the article. Which tone do you think is more compelling – the tone of the last sentence, or the tone of the rest of the article? Why?

5. In many jurisdictions (including most Canadian and American ones), little or no regulation exists concerning cruelty to animals in an agricultural setting. To what extent should governments step in to limit cruelty in such situations? To what extent should humans be prepared to pay more in order to ensure that animals raised in an agricultural setting are not treated cruelly? Write a brief essay discussing these issues.

6. Why do you think this article "front-loads" all its evidence? Would it be more effective if it started with the "So What Should Australia Do?" section? Why or why not?

thirty

Toby G. Knowles, Steve C. Kestin, Susan M.
Haslam, Steven N. Brown, Laura E. Green,
Andrew Butterworth, Stuart J. Pope, Dirk
Pfeiffer, and Christine J. Nicol

from Leg Disorders in Broiler Chickens: Prevalence, Risk Factors, and Prevention

This article from PLOS ONE *investigates the leg health issues of chickens raised for meat on factory farms in the UK and finds that a high percentage of chickens have difficulty walking. The research identifies farming practices and other factors that may contribute to this problem.*

☙

ABSTRACT

Broiler (meat) chickens have been subjected to intense genetic selection. In the past 50 years, broiler growth rates have increased by over 300% (from 25 g per day to 100 g per day). There is growing societal concern that many broiler chickens have impaired locomotion or are even unable to walk. Here we present the results of a comprehensive survey of commercial flocks which quantifies the risk factors for poor locomotion in broiler chickens. We assessed the walking ability of 51,000 birds, representing 4.8 million birds within 176 flocks. We also obtained information on approximately 150 different management factors associated with each flock. At a mean age of 40 days, over 27.6% of birds in our study showed poor locomotion and 3.3% were almost unable to walk. The high prevalence of poor locomotion occurred despite culling policies designed to remove severely lame birds from flocks. We show that the primary risk factors associated with impaired locomotion and poor leg health are those specifically associated with rate of growth. Factors significantly associated with high gait score included the age of the bird (older birds), visit (second visit to same flock), bird genotype,[a] not feeding whole wheat, a shorter dark period during the day, higher stocking density at the time of assessment, no use of antibiotic, and the use of intact feed pellets. The welfare implications are profound. Worldwide approximately 2×10^{10} broilers are reared within similar husbandry systems. We identify a range of management factors that could be altered to reduce leg health problems, but implementation of these changes would be likely to reduce growth rate and production. A debate on the sustainability of current practice in the production of this important food source is required.

Introduction

Due to their short reproductive cycle and their worldwide popularity as a food, poultry represent the most highly selected livestock. Selection of broiler chickens (chickens grown for their meat) has been primarily directed at economic traits which have reduced costs of production.[1-3] Throughout the world the majority of broilers are reared using very similar, modern, intensive systems of production where birds are confined for their lifetime within high density housing[4] and reared from hatch to slaughter weight within approximately 40 days. However, there is evidence that in optimising traits for production the resulting birds, whilst producing meat at a low cost, have a reduced viability and reduced welfare[5-7] with poor walking ability, or locomotion, a primary concern.

Previous research has highlighted associations between management practices and levels of leg disorders. Most attention has focussed on the partially effective practices of reducing feed quantity or the nutrient density of feed (e.g.[8,9]), on providing more than 1-hour of darkness each 24-hour period,[10] and on attempts to increase bird activity.[11,12] There are also known genetic effects, with genotype influences on many traits associated with leg health.[7, 13-15] Despite a large body of work investigating the effects of specific risk factors (reviewed in[16]), there has been little previous work to examine how these practices interact on real commercial farms to determine the overall level of leg disorders in particular flocks. There are also many management practices, with potential influences on leg disorders, which have not been looked at quantitatively. This study was therefore commissioned by the UK Department for Environment, Food and Rural Affairs to investigate the extent of variation of leg disorders within UK flocks and to identify methods by which these disorders could possibly be controlled.

---

a *genotype* Genetic makeup.

MATERIALS AND METHODS

We studied broiler flocks belonging to five major UK producers who together accounted for over 50 per cent of UK production. Two other relatively large companies were invited to participate but declined. We obtained data from each producer in proportion to their respective annual broiler production. Visits were randomised to farm and flock and were made by veterinarians who had completed a five-day training course to evaluate broiler walking ability with a standardised gait scoring method.[7] Eighteen veterinarians with postgraduate qualifications in poultry medicine and production, or in welfare science, acted as flock assessors and were trained to categorise gait scores within a range from 0 (completely normal) to 5 (unable to stand). The scoring system primarily assesses walking ability rather than exhaustion, with assessors trained to identify rolling gaits, limping, jerky and unsteady movements and problems with manoeuvrability. The scoring system is also known to correlate well with other methods of assessing leg disorders that do not involve active movement, such as the latency-to-lie test.[a][17] Throughout the study the uniformity of the assessors' scoring was monitored and by the end of the course, average scores for each category were all within half a score. During the subsequent 18 month study, assessors were sent at approximately six and 12 months, a tape containing new video sequences covering a range of gait scores. The scoring of these tapes was monitored to ensure that the assessors remained in agreement. Reference movies of birds' walking ability for each of the six categories are given in the supplementary information.[b] Each of 176 flocks was visited approximately three days before the flock was depopulated[c] for slaughter and at least 250 birds from each flock were gait-scored from ten pre-selected, randomised sites within a house.

Fifty-seven of the 176 flocks in our study were not 'depopulated' for slaughter simultaneously. Instead, one or more groups of birds were removed sequentially over a period of days or weeks in a process known as 'thinning'. This process involved the removal of a portion of the flock, usually the female birds, to allow the remaining birds more room to grow on to a greater weight. To account for the effects of 'thinning' practices, an additional 30 visits were made as second visits approximately three days prior to a later depopulation of one of the original 176 flocks. The flocks visited a second time were also chosen at random from the initial set of flocks.

---

a *latency-to-lie test* Test measuring how long a chicken will stand when encouraged to do so.

b *supplementary information* The original article provided links to the films, available online at *PLOS ONE*.

c *depopulated* I.e., removed from the flock.

5   A primary aim of the study was to investigate possible risk factors associated with the wide inter-flock variation in leg disorders. Of particular interest were risk factors associated with bird husbandry which could possibly be altered when rearing future flocks. Information on these aspects was obtained for each flock by a direct interview with a farm representative. The same questionnaire was used for each visit and comprised 134 questions initially about the breeding flock that had supplied the farm, the facilities where the eggs had been hatched, the distance and time the chicks had been transported, and hatchery vaccination policies. Information was then obtained about the number, weight, sex and time of chicks placed, and their date of arrival. The largest section of the questionnaire sought information on husbandry practices including stocking density and thinning practices, nutritional information, layout and construction of the house, and background information on health, growth rates, mortality and culling policies. Finally, information about the personnel working with the flock, the farm, biosecurity measures and company policies was obtained. After conducting the direct interview, each veterinary assessor collected direct information relating to air quality, temperature, general cleanliness and feed quality.

STATISTICAL ANALYSIS

Statistical models were built to identify between-flock variables that were associated with the differences in average flock walking ability. The multilevel modelling software package MLwiN v2.01 (http://www.cmm.bristol.ac.uk/MLwiN/index.shtml) was used as it allowed us to create linear models[a] within the hierarchical structure of the data of repeated measurements on flocks and flocks within companies.

RESULTS

The overall results of the survey, showing the distribution of gait scores prior to slaughter, are given in Table 1. The figures in Table 1 were calculated using the gait scores of a flock, weighted by the size of the flock as given by the number of birds placed as chicks. The minimum and maximum values in Table 1 show that there was considerable variation in walking ability between flocks, but overall, 27.6 per cent of birds represented by this survey had a gait score of 3 or above.

---

a   *linear models*   Statistical models that highlight a relationship between a dependent variable and an independent variable, presenting that relationship in a form that can be shown as a straight line on a graph.

TABLE 1.

The estimated percentage of birds in the survey population within each gait score category.

	Gait Score					
	0	1	2	3	4	5
Mean	2.2	26.6	43.5	24.3	3.1	0.2
SD	4.8	21.1	15.9	21.3	7.0	0.5
Min	0.0	0.0	1.6	0.0	0.0	0.0
Max	34.7	82.7	74.6	83.7	45.9	3.2

Mean, SD, minimum and maximum for flocks are shown. The values are calculated from flock averages weighted by birds placed and include first and second visits. Total birds placed n=4,485,962. Total birds gait scored – 206 flocks x minimum of 250 birds per flock – n is approximately 51,000.

Table 2 shows the percentage of birds within each gait score category broken down by the five companies and by the first and second visits. There was a large amount of variation in the distribution of gait score in flocks between the different companies. Company 4 notably produced only 8.5 per cent of birds with gait scores of 3 and above at the first visit compared with 22.7 to 29.7 per cent for the other companies, and only 0.6 per cent of birds with gait scores of 4 and above compared with 1.3 to 4.2 per cent for the other companies. Table 2 also shows the deterioration of gait over time for companies 1, 4 and 5 where a minimum of four second visits were made, the figures for the second visit to company 3 representing only one flock. [...]

TABLE 2.

The percentage of birds in each gait score category by producer and by first and second visit.

	Company	Gait Score						Mean GS	Birds Placed (n)	Flocks (n)
		0	1	2	3	4	5			
First	1	2.9	27.0	47.4	20.4	2.1	0.2	1.92	1,484,392	71
Visit	2	1.0	21.9	49.1	25.3	2.2	0.5	2.07	191,295	10
	3	1.0	21.3	48.0	28.4	1.2	0.1	2.08	486,258	20
	4	3.7	44.1	43.6	7.9	0.5	0.1	1.58	773,145	26
	5	1.5	29.0	41.2	24.2	3.9	0.3	2.01	1,225,925	49
Second	1	2.2	8.2	40.3	40.6	8.0	0.7	2.46	206,360	11
Visit	3	1.5	53.2	43.1	2.2	0.0	0.0	1.46	34,000	1
	4	3.5	26.6	37.5	31.1	1.2	0.1	2.00	119,150	4
	5	0.3	4.2	23.7	59.4	11.9	0.4	2.80	329,037	14

The table also shows the mean gait score and the number of birds represented in terms of birds placed.

The mean, minimum and maximum of the predictor variables are shown in Table 4, including all those variables which were centred.

TABLE 4.
Mean, minimum and maximum values of the continuous predictor variables in the model of average flock gait score.

Variable	Mean	Min	Max
Age assessed (day)	39.8	28	56
Breed A (% in flock)	85.6	0	100
Dietary wheat (wk 3) %	9.2	0	30
Average dark (hr/day)	2.9	0	8.5
Stocking density (kg/m^2)	31.3	15.9	44.8

When all other variables were held constant, there was a seasonal pattern to the average gait score of the flocks ([...] Figure 1) with the lowest (best) gait scores occurring in March and the highest (worst) in September. The age at which the birds were assessed was important in determining gait score, with every extra day, across the range of 28 to 56 days, leading to an average daily deterioration in score of 0.048. Although each flock was visited close to slaughter when gait is known to be poorest, within the survey as a whole we were able to evaluate the effect of age on locomotion problems throughout the growth period because of the wide range of age at slaughter. A post-thinning

FIGURE 1.
The modelled seasonal change in average flock gait score.

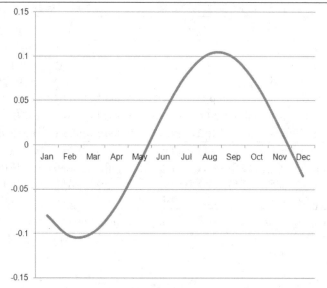

visit was associated with an increased average gait score of 0.25 over and above that due to the age at which the birds were assessed, probably due to the effect of the stress of the first thin, and/or the preponderance of larger, faster-growing male birds remaining in the flock after thinning.

We found that a number of fundamental husbandry practices were significantly associated with average flock gait score and these are detailed below.

A major influence was bird genotype. Broilers worldwide are predominantly of two types, from either one of two major international breeder companies, labelled here A and B. Birds from both genotypes are sometimes reared together within one flock. For every percentage increase in Breed A birds in a flock, from between 0 to 100 per cent, there was a 0.0024 improvement in flock gait score.

Whole wheat is sometimes fed to broilers as part of their diet, predominantly to improve digestive function. For every percentage increase in dietary wheat fed, from 0 to 30 per cent, as measured during their third week of life, there was a 0.017 per cent improvement in flock gait score.

Broilers are reared under a wide variety of artificial lighting regimes. For every 1 hour increase in the daily period of darkness, across the range of 0 to 8.5 hours, there was a 0.079 improvement in flock gait score.

There has been debate about the importance of stocking density as an influence on bird welfare and locomotion.[18] Within limits, putting as many birds in a house as possible for each rearing cycle will improve profitability. For every 1 kg/m^2 increase in stocking density as measured at the time of the flock assessment, across a range from 15.9 to 44.8 kg/m^2, there was a 0.013 deterioration in flock gait score.

Antibiotics are routinely used during different stages of broiler rearing and their use can be quite difficult to quantify accurately. In our study farmers were simply asked if a flock had received antibiotic. A reply of 'yes' in this context meant that a flock had received an extra antibiotic treatment in addition to that which would be part of normal rearing practice. For the flocks for which the farmer had answered 'yes' flock gait score was reduced by 0.17.

Broiler feed is pelleted to minimise wastage and to increase the amount of feed that a bird can consume within a given time. In our study, veterinary assessors made a subjective judgement of whether the quality of the pelleting was good or poor (dusty/broken). For flocks with poor pelleting, gait score was on average 0.15 improved. [...]

Discussion

For comparison with previous surveys we draw attention to the prevalence of birds in our survey that had gait scores of 3 or above. This cut-off point is important because there is evidence from studies of the effects of analgesic drugs that birds in these categories can be in pain.[19-22] Other surveys have reported between 14.1% and 30.1% of birds with gait scores of 3 or above in different European countries, although it is not always clear whether weighted or unweighted estimates have been used.[7,23,24] Our survey presents a conservative estimate of UK prevalence of leg disorders in meat chickens because the national proportion of birds slaughtered at second, and subsequent thinnings, when gait score tends to deteriorate, is higher than in our sample, and because we present results only from companies that volunteered to participate.

All companies had a policy of culling broilers and some farms separately identified "leg culls" conducted because of leg disorders. If a flock within the survey were to be heavily culled because of leg disorders it would be expected that the overall flock gait score would be improved. However, we found no association between the flock average gait score and the percentage of birds culled as "leg culls." However, the lack of a relationship may reflect the difficulties farmers have in recording these data in a standardised manner across the survey as a whole. Companies themselves report that these records are inconsistent between farms.

Despite examining a fuller range of husbandry and management practices than previous surveys, we did not identify many novel or previously unreported risks. An effect of season has been noted before in the US, where a higher percentage of leg abnormalities was reported in the summer.[25] The strong genotype effect that we found, confirms the important genetic component to leg disorders, and many of the husbandry effects detected most likely alter levels of leg disorder through direct or indirect effects on growth rate. Thus, we consider the effect of feeding whole wheat is probably due to the slower rate of digestion for whole wheat resulting in reduced growth rate, whilst providing broken or dusty pellets, rather than whole pellets, probably reduced overall consumption rates.

Birds reduce or cease their feed intake during periods of darkness, with associated reductions in growth rate. This is the most likely explanation for the beneficial effect of longer dark periods on the prevalence of leg disorders. The detrimental effect of higher stocking densities may be more complex, reflecting not only a lack of room available for birds to move and exercise, but also the extra environmental loading from increased biomass (e.g. additional ammonia and litter moisture).[18] Finally, the improvement in leg health observed when antibiotics were used, was probably due to a reduction in the infectious disorders which can cause some types of leg problem.

The study indicates that modern husbandry and genotypes, biased towards economics of production, have been detrimental to poultry welfare in compromising the ability of chickens to walk. However, we demonstrate that within the current framework there is variation in the magnitude of the problem between different flocks, and so some scope to improve walking ability through alterations in husbandry practice. Work needs to be carried out on the predictability of these risks, and the economics of improved welfare practices, for them to gain industry acceptance. An informed balance could then be drawn between profitability and our moral obligation to maintain good standards of animal welfare. The agreement, in May 2007, within the EU of new regulations governing the conditions under which broilers may be reared[26] is a recognition of the problems associated with modern broiler production and is an attempt at a first step towards remedying the situation. The new measures will include the introduction of a maximum stocking density limit, data collection and scientific monitoring of impacts on chicken welfare. The new Directive will not come into force until 2010 but it will prevent farms from stocking birds at densities over 39 kg/m^2 in subsequent flocks where mortality levels in the past seven flocks have exceeded a set level. The animal welfare implications of monitoring mortality and culling rates are potentially complex. In the short-term, the new Directive could lead to less rigorous culling of birds with leg problems and thereby increase suffering. However, in the longer term, the Directive could act as a stimulus to breeding companies to produce more robust genotypes, with a reduced susceptibility to leg disorders.

Research shows that consumers currently know little about how broiler chickens are reared but can be shocked when presented with information about current commercial practices.[27] Since the sustainability of intensive broiler production depends on continued consumer acceptance of the farming practices involved, the broiler industry will need to work with the scientific community to develop more robust and healthier genotypes and to ensure that optimal husbandry and management practices are fully implemented. [...]

(2007)

Acknowledgments: We are most grateful to the broiler companies and farmers who helped with the work and to the members of the expert steering committee, drawn from government, the broiler industry and academia, who gave their time to guide and support the study.

Author Contributions: Conceived and designed the experiments: DP CN TK SK SB AB LG. Performed the experiments: TK SK SH SB AB SP. Analyzed the data: DP CN TK SK SH SB LG. Wrote the paper: DP CN TK SK SH LG.

Funding: This study was carried out with funding from the UK Department of Environment, Food and Rural Affairs (Grant AW0230) with whom the design and execution were agreed. When reviewing the paper DEFRA have suggested modifications to the terminology used to describe broiler lameness within the paper which the authors have accepted.

Competing interests: The authors have declared that no competing interests exist.

REFERENCES

1. Emmans GC, Kyriazakis I (2000) Issues arising from genetic selection for growth and body composition characteristics in poultry and pigs. Occ Publi Br Soc Anim Sc 27: 39.
2. Farm Animal Welfare Council (1992) Report on the Welfare of Broiler Chickens: PB 0910. Tolworth, UK: FAWC.
3. Compassion in World Farming (2005) The welfare of Broiler Chickens in the European Union. CIWF Report, Petersfield, UK. ISBN 1 900156 35 0. Available: http://www.ciwf.org.uk/publications/reports/Welfare_of_Broiler_Chickens_in_the_EU_2005.pdf.
4. FAO–PPLPI Research Report. Industrial Livestock Production and Global Health Risks (June 2007). Available: http://www.fao.org/ag/AGAinfo/projects/en/pplpi/docarc/rep-hpai_industrialisationrisks.pdf.
5. Bessei W (2006) Welfare of broilers: a review. World Poultry Sci J 62: 455–466.
6. Butterworth A, Weeks CA, Crea PR, Kestin SC (2002) Dehydration and lameness in a broiler flock. Animal Welfare 11: 89–94.
7. Kestin SC, Knowles TG, Tinch AE, Gregory NE (1992) The prevalence of leg weakness in broiler chickens assessed by gait scoring and its relationship to genotype. Vet Rec 131: 190–194.
8. Su G, Sorensen P, Kestin SC (1999) Meal feeding is more effective than early feed restriction at reducing the prevalence of leg weakness in broiler chickens. Poult Sci 78: 949–955.
9. Brickett KE, Dahiya JP, Classen HL, Annett CB, Gomis S (2007) The impact of nutrient density, feed form, and photoperiod on the walking ability and skeletal quality of broiler chickens. Poult Sci 86: 2117–2125.
10. Moller AP, Sanotra GS, Vestergaard KS (1999) Developmental instability and light regime in chickens. Appl Anim Behav Sci 62: 57–71.
11. Classen HL, Riddell C (1989) Photoperiodic effects on performance and leg abnormalities in broiler chickens. Poult Sci 68: 873–879.
12. Reiter K (2006) Behaviour and welfare of broiler chicken. Archiv. fur Geflugelkunde 70: 208–215.
13. Kestin SC, Su G, Sorensen P (1999) Different commercial broiler crosses have different susceptibilities to leg weakness. Poult Sci 78: 1085–1090.
14. Kestin SC, Gorden S, Su G, Sorensen P (2001) Relationships in broiler chickens between lameness, live weight, growth rate and age. Vet Rec 148: 195–197.
15. Reiter K, Kutritz B (2001) Behaviour and leg weakness in different broiler breeds. Archiv fur Geflugelkunde 65: 137–141.
16. Bradshaw RH, Kirkden RD, Broom DM (2002) A review of the aetiology and pathology of leg weakness in broilers in relation to welfare. Avian and Poultry Biol Rev 13: 45–103.
17. Weeks CA, Knowles TG, Gordon RG, Kerr AE, Peyton ST, et al. (2002) A novel technique for objectively assessing lameness in broiler chickens. Vet Rec 151: 762–764.
18. Dawkins MS, Donnelly CA, Jones TA (2004) Chicken welfare is influenced more by housing conditions than by stocking density. Nature 427: 342–344.

19. Weeks CA, Danbury TD, Davies HC, Hunt P, Kestin SC (2000) The behaviour of broiler chickens and its modification by lameness. Appl Anim Behav Sci 67: 111–125.

20. Scientific Committee on Animal Health and Animal Welfare (2000) The Welfare of Chickens Kept for Meat Production (Broilers). European Commission. Available: http://ec.europa.eu/food/fs/sc/scah/out39_en.pdf.

21. Danbury TC, Weeks CA, Chambers JP, Waterman-Pearson AE, Kestin SC (2000) Self selection of the analgesic drug carprofen by lame broiler chickens. Vet Rec 146: 307–311.

22. McGeown D, Danbury TC, Waterman-Pearson AE, Kestin SC (1999) Effect of carprofen on lameness in broiler chickens. Vet Rec 144: 668–671.

23. Sanotra GS, Lund JD, Ersboll AK, Petersen JS, Vestergaard KS (2001) Monitoring leg problems in broilers: a survey of commercial broiler production in Denmark. World Poultry Sci J 57: 55–69.

24. Sanotra GS, Berg C, Lund JD (2003) A comparison between leg problems in Danish and Swedish broiler production. Animal Welfare 12: 677–683.

25. Laster CP, Hoerr FJ, Bilgili SF, Kincaid SA (1999) Effects of dietary roxarsone supplementation, lighting program, and season on the incidence of leg abnormalities in broiler chickens. Poult Sci 78: 197–203.

26. EU Press Release IP/07/630. Brussels 7th May 2007. Available: http://europa.eu/rapid/pressReleasesAction.doreferenceIP/07/630formatHTMLaged0languageEN.

27. Hall C, Sandilands V (2007) Public attitudes to the welfare of broiler chickens. Animal Welfare 16: 499–512.

Questions:

1. What do you take to be the meaning of the Abstract's statement that "The welfare implications are profound"? Does the article explore these implications? If so, what are they?

2. Provide a paraphrase of the following sentence from the introduction, and comment on its significance as a statement of intent:

   > However, there is evidence that in optimising traits for production the resulting birds, whilst producing meat at a low cost, have a reduced viability and reduced welfare,[5-7] with poor walking ability, or locomotion, a primary concern.

3. Why does the paper include financial concerns such as cost and profitability for producers as one of its main areas of concern?

4. The first two sentences of "Materials and Methods" concern the companies included in the study. In what way(s) might the results have been different had the two companies mentioned in the second sentence participated in the study? In what ways might the results have been different if farms that use non-intensive methods had been included?

5. In what way(s) do you think Company 4's approach might differ from that of the other companies? Why do you think the article does not specify individual company policies or practices?

6. What does it mean that "the authors have declared that no competing interests exist"?

7. Why was gait measurement chosen as the assessment technique? Can you think of other assessment possibilities?

8. In their results, the authors discuss "stocking density" and refer to it as being in "a range from 15.9 to 44.8 kg/m². " How many individual birds might that represent in a square meter (think of the weight of an average roasting chicken)? What would be the purpose(s) of measuring density in terms of kilograms per square meter instead of birds per square meter?

9. The authors report that "broilers are reared under a wide variety of artificial lighting regimes." What do they report as being the range between the operations that give the chickens the most darkness and those that give them the least? Why would some operations keep the lights on in their broiler sheds except for one hour a day?

10. The authors report in their results that antibiotics "are routinely used during different stages of broiler rearing." What sorts of animal husbandry problems does such a routine address?

11. Re-read the third-to-last paragraph of the "Discussion" section. What does the paragraph suggest happens to the chickens' excrement in these operations?

12. Read the last paragraph again. What primary audience is the author addressing here? How can you tell?

13. The authors refer in their final paragraph to "optimal husbandry and management practices." What might "optimal" mean in this context?

thirty-one

EUROPEAN FOOD SAFETY AUTHORITY (EFSA)
PANEL ON PLANT PROTECTION PRODUCTS AND THEIR
RESIDUES (PPR)

from Scientific Opinion on the Science behind the Development of a Risk Assessment of Plant Protection Products on Bees (*Apis mellifera, Bombus* spp. and solitary bees)

This abstract and summary are taken from a statement of scientific opinion, issued by a European Food Safety Authority panel, regarding the effects of pesticides on bees. The document provides an overview of the existing scientific research on the risk pesticides pose for bees, as well as recommendations for future research.

❧

ABSTRACT

The PPR Panel was asked to deliver a scientific opinion on the science behind the development of a risk assessment of plant protection products on bees (*Apis mellifera, Bombus* spp.[a] and solitary bees). Specific protection goals options were suggested based on the ecosystem services approach.[b] The different routes of exposure were analysed in detail for different categories of bees. The existing test guidelines were evaluated and suggestions for improvement and further research needs were listed. A simple prioritisation tool to assess cumulative effects of single pesticides using mortality data is

---

a   *spp.*   Species (plural).
b   *ecosystem services approach*   Policy-making approach that evaluates nature in terms of the benefits it provides for human beings.

suggested. Effects from repeated and simultaneous exposure and synergism[a] are discussed. Proposals for separate risk assessment schemes, one for honey bees and one for bumble bees and solitary bees, were developed.

Summary

Following a request from the European Commission, the Scientific Panel on Plant Protection Products and their Residues (PPR Panel) of EFSA was asked to deliver a scientific opinion on the science behind the development of a risk assessment of Plant Protection Products on bees (*Apis mellifera*, *Bombus* spp. and solitary bees). The opinion will be the scientific basis for the development of a Guidance Document which should provide guidance for notifiers and authorities in the context of the review of Plant Protection Products (PPPs) and their active substances under Regulation (EC) 1107/2009.[b]

For the development of robust and efficient environmental risk assessment procedures, it is crucial to know what to protect, where to protect it and over what time period. Specific protection goals based on ecosystem services were suggested according to the methodology outlined in the Scientific Opinion of EFSA (2010). Pollination, hive products (for honey-bees only) and biodiversity (specifically addressed under genetic resources and cultural services) were identified as relevant ecosystem services. It was suggested to define the attributes to protect as survival and development of colonies and effects on larvae and honey bee behaviour as listed in regulation (EC) No 1107/2009. In addition, abundance/biomass[c] and reproduction were also suggested because of their importance for the development and long-term survival of colonies. The magnitude of effects was defined as negligible if the natural background mortality compared to controls was not exceeded. Further work is needed to give recommendations on the deviation from the controls up to which an effect is still considered negligible. The current methods of field testing would need major improvements in order to detect for example an increase in daily mortality of foragers by 10% with high statistical power. Based on expert judgement it was considered that a small effect could be tolerated for a few days without putting the survival of a hive at risk. Further research (modelling) is proposed to clarify this question and to revise the proposal for the magnitude of effects and the temporal scale of effects. The

a *synergism* I.e., compound effects from more than one substance, which may differ from the effects of each substance added together.

b *Regulation (EC) 1107/2009* European Union regulation document regarding "the placing of plant protection products on the market."

c *abundance/biomass* Measures of a species' presence in a given area, in terms of number of organisms or mass of organisms, respectively.

current risk assessment for honey bees relies on an Hazard Quotient (HQ) approach (application rate/LD50)[a] in lower tiers and on semi-field and field tests in higher tiers. It is particularly difficult to ascertain whether a specific exposure percentile is achieved in field studies. Decisions need to be taken on how conservative the exposure estimate should be and what percentage of exposure situations should be covered in the risk assessment. It is recommended to design a flow chart for checking whether exposure in the semi-field or field study was indeed higher than that corresponding to the desired percentile. Factors that may be included are: the crop and its developmental stage, the dosage, measures ensuring that bees are coming into contact with the compound/formulation, weather conditions, and for instance the generation of guttation droplets[b] by the crop. The final decision on protection goals needs to be taken by risk managers. There is a trade-off between plant protection and the protection of bees. The effects on pollinators need to be weighted against increase in crop yields due to better protection of crops against pests.

Residues in different environmental matrices and bee products were combined with estimates of exposure of different categories of bees. Highest concentrations of residues were found after spray treatments in pollen and nectar. Residues in guttation droplets showed a wide variability due to the number of parameters known to influence guttation production (environmental conditions, crop type, growth stage, etc.). A potentially high exposure was highlighted for bees in some crops (e.g. maize). Exposure to dust drift from sowing treated seeds was identified as a relevant exposure route. The exposure of different categories of bees from different sources and for different application techniques suggests that the potential risk from oral uptake was highest for forager bees, winter bees and larvae. The exposure of nurse bees occurs via a combination of pollen and nectar, of larvae by contact to wax and foragers, drones, queens and swarms intercepting droplets and vapour by contact and inhalation.

Worker bees, queens and larvae of bumble bees and adult females and larvae of solitary bees were considered to be the categories that are most exposed via oral uptake. Larvae of solitary bees consume large mass provisions with unprocessed pollen; thus, compared with honey bee larvae, they are more exposed to residues in pollen. Moreover, bumble bees and solitary bees may be exposed to a larger extent via contact with nesting material (soil

---

a *Hazard Quotient ... LD50*   Approach using the ratio of the application rate of the pesticide to the LD50 (the dose that would kill 50% of a particular population) to determine whether the level of contamination is acceptable.

b *guttation droplets*   Droplets of liquid secreted by plants.

or plants) compared to honey bees, suggesting the need for a separate risk assessment for bumble bees and solitary bees.

5   For the ranking of bees, the inclusion of multiple exposures with appropriate weights would need to be done with a modelling or scenario-based approach that was not available in the current assessment. It was therefore recommended that the categories of bees which represent the worst-case exposure scenarios through multiple exposures are further assessed (e.g. honey bee nurses) and that those categories which highlighted potential but unknown exposures through consumption of water and inhalation of vapour in/out field are further analysed with more studies. Further research is recommended on the testing of the presence and fate of residues (e.g. in bee relevant matrices and in-hive following spray and dust applications) and on the development of reliable exposure models.

The overview of the available studies on sub-lethal doses and long-term effects of pesticides on bees highlighted gaps in knowledge and research needs in the following areas: more toxicological studies to be performed in bees for a wider range of pesticides on both adults and larvae including sub-lethal endpoints, also including contact and inhalation routes of exposure. Few studies were conducted with non-*Apis* bees, considering endpoints such as fecundity (e.g. drones production in *Bombus* and cell production rate in solitary bees), larvae mortality rate, adult longevity and foraging behaviour. The use of micro-colonies in bumble bees appears to be well-suited to measure lethal and sub-lethal effects of pesticides with low doses and long-term effects.

Because of the specific toxicokinetic[a] profile of bees compared with other insects, it is recognised that toxicokinetic data can provide useful information on the potential biological persistence of a pesticide which, in some cases, could have effects after continuous exposure that may be more marked compared with their short-term effects. The integration of toxicokinetic knowledge and low (sub-lethal) dose effects generated from laboratory and field studies in the hazard identification and hazard characterisation of pesticides in *Apis* and non-*Apis* bees can provide a better understanding of short-term and long-term effects. It is therefore concluded that the conventional regulatory tests based on acute toxicity (48 to 96 h) are likely to be unsuited to assess the risks of long-term exposures to pesticides.

A testing protocol and mathematic model, based on Haber's law,[b] have been developed as a simple prioritisation tool to investigate the potential

---

a   *toxicokinetic*   Relating to the absorption and processing of chemicals by human or animal bodies.

b   *Haber's law*   Rule suggesting that, for many toxins in some circumstances, a long period of exposure to a relatively small amount of toxin can produce the same effect as a short period of effects after repeated exposure to single pesticides using mortality data. However, a number of assumptions inherent to the model raise uncertainties. The protocol and model needs further validation in the laboratory and to be tested for sub-lethal endpoints in adult and bee larvae. Finally, combining basic toxicokinetic data for an active substance and its metabolites,[a] such as the half life,[b] will also provide more precise estimates on the potential of bioaccumulation. In the case of potential persistence of the active ingredient, half life of the parent compound and its metabolites should be determined in larvae, newly emerged bees and foragers.

The working group identified the need for improvement of existing laboratory, semi-field and field testing and areas for further research. Several exposure routes of pesticides are not evaluated in laboratory conditions, such as the intermittent and prolonged exposures of adult bees, exposure through inhalation and the exposure of larvae. Likewise, the effects of sub-lethal doses of pesticides are not fully covered in the conventional standard tests.

Sub-lethal effects should be taken into account and observed in laboratory studies. Potential laboratory methods to investigate sub-lethal effects would be testing of *Bombus* microcolonies to investigate effects on reproduction, proboscis extension reflex (PER) test for neurotoxic effects and homing behaviour for effects on foraging, including orientation. Further research is needed in order to integrate the results of these studies in the risk assessment scheme.

Semi-field testing appears to be a useful option of higher tier testing. Nevertheless, weaknesses have been identified for each of the test guidelines e.g. the limited size of crop area, the impossibility to evaluate all the possible exposure routes of the systemic compounds used as seed- and soil-treatments (SSST), the limited potential to extrapolate the findings on larger colony sizes used in field studies or the relatively short timescale (one brood cycle).

The guideline for field testing (EPPO 170) (4)[c] has several major weaknesses (e.g. the small size of the colonies, the very small distance between the hives and the treated field, the very low surface of the test field), leading to uncertainties concerning the real exposures of the honey bees. The guideline is better suited to the assessment of spray products than of seed- and soil-

---

exposure to a relatively large amount.

a *metabolites* Substances produced when the active substance is metabolized.

b *half life* I.e., amount of time it takes for the quantity of the substance in an organism to reduce by 50%.

c *(EPPO 170) (4)* European and Mediterranean Plant Protection Organization bulletin "Side-effects on honeybees."

treatments. Points for research and improvement of methods used in field testing are highlighted (e.g. methods for detection of mortality).

The available protocols for testing of solitary bees are suitable to study the oral and contact toxicity in adults and larvae for several species of solitary bees (*Megachile rotundata*, *Osmia* spp.) but they need to be ring tested. More studies are necessary to compare the susceptibility of honey bees with other non-*Apis* species in order to see to which extent honey bee endpoints also cover non-*Apis* bees.

Future research is recommended to improve laboratory, semi-field and field tests (e.g. extrapolation of the endpoints in first tier to the colony/forager effects, extrapolation of the toxicity between dust and spray, extrapolation of laboratory based *Bombus* micro colonies to *Apis* and solitary bees).

15   Pesticides are often applied in tank mixes (2 to 9 active ingredients at the same time) and in addition non target organisms will be exposed to mixtures of compounds following sequential applications to crops. There is a consensus in the field of mixture toxicology that the customary chemical-by-chemical approach to risk assessment is too simplistic. At low levels of exposure concentration, addition has been observed more often than synergistic or antagonistic[a] effects for mixtures of pesticides with a common mode of action and independent action (response addition)[b] has been observed for compounds with a different mode of action. In some cases synergistic and antagonistic effects have also been observed.

Honey bees and hymenoptera[c] are known to have a specific metabolic profile with the lowest number of copies of detoxification enzymes within the insect kingdom. A number of studies have shown synergistic effects of pesticides and active substances applied in hives as medical treatments against *Varroa* mites in honey bees, for which toxicokinetic interactions were most commonly involved. There is also a growing body of evidence of interaction between honey bee disease (fungi, bacteria and viruses) and pesticides. Currently, full dose responses for synergistic effects between potential inhibitors and different classes of pesticides are rarely available for either lethal effects or sub-lethal effects in bees so that predictions of the magnitude of these interactions at realistic exposure levels cannot be performed. However, there is evidence that where realistic exposure levels have been investigated, de-

---

a   *addition* Cumulative effect in which the effects of each substance are added together (as opposed to interacting and altering each other); *antagonistic* With one or more substances impeding the effects of one or more others.

b   *independent action (response addition)* Combined effect in which the effects of one substance are not directly influenced by the effects of another.

c   *hymenoptera* Order of insects to which bees belong.

viations from concentration addition, such as synergy, is rarely more than a factor of 2 to 3. Such deviations have been observed for mixtures containing small numbers of chemicals and decreases as the complexity of the mixture increases.

In the case of synergism which can be predicted based on the mode of action of the chemical classes involved (e.g. EBI[a] fungicides and insecticides), and in the absence of existing data on toxicity of the mixture, it is recommended to design full dose-response studies in adult bees and larvae for mixtures of potential synergists. Further work is also required to identify the molecular basis of interactions between environmentally realistic exposure to pesticides and the range of honey bee diseases (fungi, bacteria and viruses) to determine whether and how these may be included in risk assessment.

Separate risk assessment schemes are proposed, one for honey bees and one for bumble bees and solitary bees. In the first tier it is suggested to include toxicity testing that covers a longer period of exposure (7 to 10 days) for adult bees as well as larval bees. Both life stages can be exposed for more than one day and this risk was not covered by the standard OECD tests (213 and 214) for oral and contact exposure. Currently there is insufficient evidence that toxicity following extended exposures can be reliably predicted from acute oral LD50 data. It is also proposed to investigate whether there are any indications of cumulative effects for each compound. A new method to detect cumulative toxicity is proposed based on Haber's law. If there is an indication that a compound is a cumulative toxin then this needs further evaluation since the potential effects of prolonged or repeated exposure to low doses may be underestimated. [...]

(2012)

Questions:

1. Why, according to the article, do bumble bees and solitary bees require a separate risk assessment?

2. Find a sentence in this article that has been written in the passive voice; rephrase it using the active voice. Does this represent an improvement? Why or why not?

3. Identify and define the Latin root words that make up the following terms: toxicokinetic, bioaccumulation, sub-lethal.

---

a *EBI* European Bioinformatics Institute.

4. Who is the intended audience for this piece? Identify three ways that the intended audience is reflected in the writing.

5. Rewrite the abstract as though you were presenting the information for a newspaper or popular science magazine.

6. Following the recommendations made in this article to the best of your ability, describe a hypothetical experiment that could be used to test some aspect of the effects of pesticides on bees.

7. This is a summary of a much longer document. What types of information not included here would you expect to find in the full document?

8. This summary was written by the EFSA Panel on Plant Protection Products and their Residues in response to a request from the European Commission. Does knowing about the study's design origins influence your reading of the material? If so, how?

9. What is an "ecosystem services approach"? What are the strengths and weaknesses of such an approach compared to an approach that advocates the preservation of nature for its own sake?

10. Research neonicotinoids. When were neonicotinoids introduced as a commercial insecticide class? When did bee numbers start to rapidly decline? What is your take on whether this temporal correlation is also a causal correlation?

thirty-two

Damian Carrington

Common Pesticides "Can Kill Frogs within an Hour"

In this article from the Guardian, *environmental journalist
Damian Carrington reports on the findings of a study
assessing the effects of common pesticides on frogs.*

❧

Widely used pesticides can kill frogs within an hour, new research has revealed, suggesting the chemicals are playing a significant and previously unknown role in the catastrophic global decline of amphibians.

The scientists behind the study said it was both "astonishing" and "alarming" that common pesticides could be so toxic at the doses approved by regulatory authorities, adding to growing criticism of how pesticides are tested.

"You would not think products registered on the market would have such a toxic effect," said Carsten Brühl, at the University of Koblenz-Landau in Germany. "It is the simplest effect you can think of: you spray the amphibian with the pesticide and it is dead. That should translate into a dramatic effect on populations."

Trenton Garner, an ecologist at the Zoological Society of London, said: "This is a valuable addition to the substantial body of literature detailing how existing standards for the use of agricultural pesticides, herbicides and fertilisers are inadequate for the protection of biodiversity."

Amphibians are the best example of the great extinction of species 5 currently under way, as they are the most threatened and rapidly declining vertebrate group. More than a third of all amphibians are included in the IUCN[a] "red list" of endangered species, with loss of habitat, climate change and disease posing the biggest threats.

Brühl had previously studied how easily frogs can absorb pesticides through their permeable skins, which they can breathe through when under-

---

a *IUCN* The International Union for Conservation of Nature produces the "Red List of Threatened Species," a document evaluating the threat of extinction for a wide range of species.

water. But pesticides are not required to be tested on amphibians, said Brühl: "We could only find one study for one pesticide that was using an exposure likely to occur on farmland."

His team chose widely used fungicides, herbicides and insecticides. The most striking results were for a fungicide called pyraclostrobin, sold as the product Headline by the manufacturer BASF and used on 90 different crops across the world. It killed all the common European frogs used as test animals within an hour when applied at the rate recommended on the label. Other fungicides, herbicides and insecticides also showed acute toxicity, even when applied at just 10% of the label rate, with the insecticide dimethoate, for example, killing 40% of animals within a week.

The study, published on Thursday in Scientific Reports, concluded: "The observation of acute mortality in a vertebrate group caused by commercially available pesticides at recommended field rates is astonishing, since 50 years after the publication of Rachel Carson's *Silent Spring*[a] one would have thought that the development of refined risk-assessment procedures would make such effects virtually impossible."

A BASF spokesman disputed the findings: "This study was performed under laboratory 'worst-case' conditions. Under normal agricultural conditions amphibians are not exposed to such pesticide concentrations. According to our knowledge, no significant impact on amphibian populations has been reported despite the widespread and global use of the fungicide pyraclostrobin."

10   Brühl said the method, a single spray directly on to the frogs, sometimes at just 10% of the label rate, was a "realistic worst-case" scenario. He added that in the field, multiple sprays of a variety of pesticides was likely and that chemicals might run off into ponds where frogs lived.

Sandra Bell, Friends of the Earth's nature campaigner, said: "From frogs to bees, there is mounting evidence that the pesticide bombardment of our farmland is having a major impact on our precious wildlife. Strong action is urgently needed to get farmers off the chemical treadmill.

"As well as banning the most toxic products, governments must set clear targets for reducing all pesticides and ensure farmers have safe and thoroughly tested alternatives."

Earlier this month, the world's most widely used insecticide was for the first time officially labelled an "unacceptable" danger to bees feeding on flowering crops, by the European Food Safety Agency. The agency had previously stated that current "simplistic" regulations contained "major weaknesses."

---

a   *Rachel Carson's Silent Spring*   Widely influential 1962 book that increased public awareness of the dangers of pollution, especially of the insecticide DDT.

"There is an urgency to address [the amphibian issue] as pesticides will be applied again soon because it's spring, and that's when we have all these migrations to ponds," said Brühl.

"We don't have any data from the wild about dead frogs because no one is looking for them – and if you don't look, you don't find. But the pesticides are very widely used and so have the potential to have a significant effect on populations."

(2013)

Questions:

1. How would you go about finding out what (if any) restrictions apply to the use of pyraclostrobin where you live?

2. Argue for or against the proposition that pyraclostrobin should be banned in North America. Can you definitively support your argument using information from the article, or do you need to do outside research? If so, where would you find such information?

3. Pesticides are not the only threat to world frog populations. Through research, identify three other such threats. Write a paragraph describing one of them.

4. List the people whom Carrington quotes directly in this article. Which of these sources strike you as being most reliable? Least reliable? Why?

5. Parse the claims made by the BASF spokesperson in paragraph 9 according to the word choice. What does "According to our knowledge, no significant impact on amphibian populations has been reported despite the widespread and global use of the fungicide pyraclostrobin" actually mean?

6. How are pesticides applied to fields? In what ways does the testing procedure used in the experiment differ from the way frogs would likely be exposed to pesticides in the environment? Do you think these differences suggest that amphibians would be more or less affected by pesticides in real-world use than they were in the experiment? Why?

7. This article summarizes the findings of an article from *Nature* titled "Terrestrial pesticide exposure of amphibians: An underestimated cause of global decline?" by Carsten A. Brühl, Thomas Schmidt, Silvia Pieper, and Annika Alscher. Find that article online; read it and answer the following questions:

   a. Compare the tone of each article, and explain how the tone reflects the audience for which each piece was written. Then, identify five words or phrases appearing in the *Nature* article that would be out of place in the *Guardian* article.

b. What kinds of information were included in the *Nature* article but not included in the *Guardian* article? Why do you think this information might have been left out?

c. Identify one or two pieces of information from the *Nature* article that might be of interest to layperson readers but are not included in the *Guardian* article. In the style of the *Guardian* article, write a brief paragraph on this information.

8. According to the description in Brühl et al.'s article, this experiment involved 150 frogs. Do you think such experimentation on live animals is ethical? What considerations do you feel need to be taken into account in this instance?

thirty-three

Tim Fox and Ceng Fimeche

Executive Summary of *Global Food: Waste Not, Want Not*

This summary outlines a report by the Institution of Mechanical Engineers that analyzes many factors contributing to food waste as well as potential solutions in the face of an ever-growing global population. The report's findings regarding effective land, water, and energy usage lead to recommendations for reducing waste in these categories.

℮

Feeding the 9 Billion: The Tragedy of Waste

By 2075, the United Nations' mid-range projection for global population growth predicts that human numbers will peak at about 9.5 billion people. This means that there could be an extra three billion mouths to feed by the end of the century, a period in which substantial changes are anticipated in the wealth, calorific intake and dietary preferences of people in developing countries across the world.

Such a projection presents mankind with wide-ranging social, economic, environmental and political issues that need to be addressed today to ensure a sustainable future for all. One key issue is how to produce more food in a world of finite resources.

Today, we produce about four billion metric tonnes of food per annum. Yet due to poor practices in harvesting, storage and transportation, as well as market and consumer wastage, it is estimated that 30–50% (or 1.2–2 billion tonnes) of all food produced never reaches a human stomach. Furthermore, this figure does not reflect the fact that large amounts of land, energy, fertilisers and water have also been lost in the production of foodstuffs which simply end up as waste. This level of wastage is a tragedy that cannot continue if we are to succeed in the challenge of sustainably meeting our future food demands.

WHERE FOOD WASTE HAPPENS

In 2010, the Institution of Mechanical Engineers identified three principal emerging population groups across the world, based on characteristics associated with their current and projected stage of economic development.

- Fully developed, mature, post-industrial societies, such as those in Europe, characterised by stable or declining populations which are increasing in age.
- Late-stage developing nations that are currently industrialising rapidly, for example China, which will experience decelerating rates of population growth, coupled with increasing affluence and age profile.
- Newly developing countries that are beginning to industrialise, primarily in Africa, with high to very high population growth rates (typically doubling or tripling their populations by 2050), and characterised by a predominantly young age profile.

Each group over the coming decades will need to address different issues surrounding food production, storage and transportation, as well as consumer expectations, if we are to continue to feed all our people.

Third World and Developing Nations

In less-developed countries, such as those of sub-Saharan Africa and South-East Asia, wastage tends to occur primarily at the farmer-producer end of the supply chain. Inefficient harvesting, inadequate local transportation and poor infrastructure mean that produce is frequently handled inappropriately and stored under unsuitable farm site conditions.

As the development level of a country increases, so the food loss problem generally moves further up the supply chain with deficiencies in regional and national infrastructure having the largest impact. In South-East Asian countries for example, losses of rice can range from 37% to 80% of total production depending on development stage, which amounts to total wastage in the region of about 180 million tonnes annually. In China, a country experiencing rapid development, the rice loss figure is about 45%, whereas in less-developed Vietnam, rice losses between the field and the table can amount to 80% of production.

Developed Nations

In mature, fully developed countries such as the UK, more-efficient farming practices and better transport, storage and processing facilities ensure that a larger proportion of the food produced reaches markets and consumers. However, characteristics associated with modern consumer culture mean produce is often wasted through retail and customer behaviour.

Major supermarkets, in meeting consumer expectations, will often reject entire crops of perfectly edible fruit and vegetables at the farm because they do not meet exacting marketing standards for their physical characteristics, such as size and appearance. For example, up to 30% of the UK's vegetable crop is never harvested as a result of such practices. Globally, retailers generate 1.6 million tonnes of food waste annually in this way.

Of the produce that does appear in the supermarket, commonly used sales promotions frequently encourage customers to purchase excessive quantities which, in the case of perishable foodstuffs, inevitably generates wastage in the home. Overall between 30% and 50% of what has been bought in developed countries is thrown away by the purchaser.

Controlling and reducing the level of wastage is frequently beyond the capability of the individual farmer, distributor or consumer, since it depends on market philosophies, security of energy supply, quality of roads and the presence of transport hubs. These are all related more to societal, political and economic norms, as well as better-engineered infrastructure, rather than to agriculture. In most cases the sustainable solutions needed to reduce waste are well known. The challenge is transferring this know-how to where it is needed, and creating the political and social environment which encourages both transfer and adoption of these ideas to take place.

BETTER USE OF OUR FINITE RESOURCES

Wasting food means losing not only life-supporting nutrition but also precious resources, including land, water and energy. These losses will be exacerbated by future population growth and dietary trends that are seeing a shift away from grain-based foods and towards consumption of animal products. As nations become more affluent in the coming decades through development, per capita calorific intake from meat consumption is set to rise 40% by mid-century. These products require significantly more resource to produce. As a global society therefore, tackling food waste will help contribute towards addressing a number of key resource issues:

Effective Land Usage

Over the last five decades, improved farming techniques and technologies have helped to significantly increase crop yields along with a 12% expansion of farmed land use. However, with global food production already utilising about 4.9Gha[a] of the 10Gha usable land surface available, a further increase in farming area without impacting unfavourably on what remains of the world's natural ecosystems appears unlikely. The challenge is that an increase in animal-based production will require greater land and resource requirement, as livestock farming demands extensive land use. One hectare of land can, for example, produce rice or potatoes for 19–22 people per annum. The same area will produce enough lamb or beef for only one or two people. Considerable tensions are likely to emerge, as the need for food competes with demands for ecosystem preservation and biomass production as a renewable energy source.

Water Usage

Over the past century, fresh water abstraction for human use has increased at more than double the rate of population growth. Currently about 3.8 trillion m³ [b] of water is used by humans per annum. About 70% of this is consumed by the global agriculture sector, and the level of use will continue to rise over the coming decades. Indeed, depending on how food is produced and the validity of forecasts for demographic trends, the demand for water in food production could reach 10–13 trillion m³ annually by mid-century. This is 2.5 to 3.5 times greater than the total human use of fresh water today.

Better irrigation can dramatically improve crop yield and about 40% of the world's food supply is currently derived from irrigated land. However, water used in irrigation is often sourced unsustainably, through boreholes sunk into poorly managed aquifers.[c] In some cases government development programmes and international aid interventions exacerbate this problem. In addition, we continue to use wasteful systems, such as flood or overhead spray, which are difficult to control and lose much of the water to evaporation. Although the drip or trickle irrigation methods are more expensive to install, they can be as much as 33% more efficient in water use as well as being able to carry fertilisers directly to the root.

In processing of foods after the agricultural stage, there are large additional uses of water that need to be tackled in a world of growing demand. This is particularly crucial in the case of meat production, where beef uses about 50 times more water than vegetables. In the future, more effective washing techniques, management procedures, and recycling and purification of water will be needed to reduce wastage.

---

a *Gha* Gigahectares; each gigahectare is one million hectares.

b *m³* Cubic meters.

c *boreholes* Human-made shafts drilled into the ground; *aquifers* Deposits of underground rock that contain water.

Energy Usage

Energy is an essential resource across the entire food production cycle, with 20 estimates showing an average of 7–10 calories of input being required in the production of one calorie of food. This varies dramatically depending on crop, from three calories for plant crops to 35 calories in the production of beef. Since much of this energy comes from the utilisation of fossil fuels, wastage of food potentially contributes to unnecessary global warming as well as inefficient resource utilisation.

In the modern industrialised agricultural process – which developing nations are moving towards in order to increase future yields – energy usage in the making and application of agrochemicals such as fertilisers and pesticides represents the single biggest component. Wheat production takes 50% of its energy input for these two items alone. Indeed, on a global scale, fertiliser manufacturing consumes about 3–5% of the world's annual natural gas supply. With production anticipated to increase by 25% between now and 2030, sustainable energy sourcing will become an increasingly major issue. Energy to power machinery, both on the farm and in the storage and processing facilities, together with the direct use of fuel in field mechanisation and produce transportation, adds to the energy total, which currently represents about 3.1% of annual global energy consumption.

RECOMMENDATIONS

Rising population combined with improved nutrition standards and shifting dietary preferences will exert pressure for increases in global food supply.

Engineers, scientists and agriculturalists have the knowledge, tools and systems that will assist in achieving productivity increases. However, pressure will grow on finite resources of land, energy and water. Although increasing yields in hungry countries is an appropriate response to an emerging food crisis, to ensure we can sustainably meet the food needs of over three billion extra people on the planet by 2075, the Institution of Mechanical Engineers calls for initiatives to be taken to reduce the substantial quantity of food wasted annually around the world. The potential to provide 60–100% more food by simply eliminating losses, while simultaneously freeing up land, energy and water resources for other uses, is an opportunity that should not be ignored. Factors affecting waste relate to engineered infrastructure, economic activity, vocational training, knowledge transfer, culture and politics. In order to begin tackling the challenge, the Institution recommends that:

1. The UN Food and Agriculture Organisation (FAO) works with the international engineering community to ensure governments of developed

nations put in place programmes that transfer engineering knowledge, design know-how, and suitable technology to newly developing countries. This will help improve produce handling in the harvest, and immediate post-harvest stages of food production.

25 2. Governments of rapidly developing countries incorporate waste minimisation thinking into the transport infrastructure and storage facilities currently being planned, engineered and built.

3. Governments in developed nations devise and implement policy that changes consumer expectations. These should discourage retailers from wasteful practices that lead to the rejection of food on the basis of cosmetic characteristics, and losses in the home due to excessive purchasing by consumers.

(2013)

Questions:

1. This text organizes its information according to distinctions between "fully developed, mature, post-industrial societies," "late-stage developing nations," and "newly developing countries that are beginning to industrialise." Can you think of another way to organize the various types of food production and distribution problem than by tiers of countries?

2. Look up the nineteenth-century concept of racial hierarchy (or its twentieth-century versions). Based on your research, what sorts of historical problems might be embedded in terms used in this article, such as "fully, developed, mature, post-industrial" and "Third World"? What does the use of this language say about the likely intended audience for this report?

3. Can you think of parallels in your daily experience with such phrasing as "up to 30%" or "37% to 80%"? Do you think the use of such language helps or hinders the report in making its case? Why?

4. This summary emphasizes the inefficiency of meat production but does not recommend that countries should attempt to reduce meat consumption. Why do you think this suggestion was not included? Should it have been?

5. "Over the last five decades, improved farming techniques and technologies have helped to significantly increase crop yields along with a 12% expansion of farmed land use." What does this sentence mean? Are any parts of it controversial? Do any parts of it lack clarity?

6. Do any of the recommendations at the end of this article surprise you? Do you agree with them? Why or why not?

7. This report covers food, land, water, fertilizers, and energy. Are all of these items equally important and necessary? Why or why not?

8. Consider the first of the three recommendations made at the end of this text. Alexis de Greiff and Mauricio Nieto's article "What We Still Do Not Know about South-North Technoscientific Exchange" (included in this anthology) discusses the North's imposition of its own version of science onto the South. Is this recommendation an example of such an imposition? Why or why not?

9. How could recommendation three be implemented? Give some examples. How effective do you think the recommended plan would be? Why?

10. Using the information in the article, explain how current energy use in food production is inefficient. How do you think the situation could be improved?

11. Assess the structural framework of this text and consider the following questions in your answer. Can you point to a thesis statement? Are claims well supported with evidence? Are the transitions smooth? Would you have done anything differently?

12. Look up a discussion about manufactured fertilizers and pesticides. What are the health implications of these "more efficient" means to food production?

13. Is it fair to say that this report looks only at the implications of commercial agriculture? Why or why not? How do you think the report would change if it included the local food movement and/or subsistence farming?

VI

Engineering and Technology: Risk Platforms

thirty-four

ALEXANDRE ERLER

In Vitro Meat, New Technologies, and the "Yuck Factor"

In this entry from the blog Practical Ethics, *researcher
Alexandre Erler discusses the "yuck factor" as an obstacle
to the widespread acceptance of in vitro meat and argues
that people would be much more disgusted by factory farmed
meat if they fully comprehended how it was produced.*

❧

In vitro[a] meat, recently discussed on this blog by Julian Savulescu,[b] is gradually becoming a reality. It holds great promise, notably considering that billons of animals are slaughtered for food every year, often after spending miserable lives in factory farms, and that the current production of meat contributes significantly to the emission of greenhouse gases. In spite of those facts, it seems highly unlikely that most meat-eaters will agree to give up meat anytime soon (though the success of the "meat-free Mondays" initiative in a number of different places should be saluted), yet they might well prove more willing to switch from traditionally produced meat to in vitro meat, if the latter were as healthy (or even healthier), reasonably priced, and tasted the same as the former.

Discussions of in vitro meat in the media most often cite the so-called "yuck factor" as a major obstacle to its general acceptance: i.e. the instinctive revulsion that many people feel at the idea of eating "unnatural" meat grown in a petri dish. I am inclined to be cautiously optimistic about the prospects of overcoming that obstacle: "unnatural" meat substitutes have already become popular among vegetarians, and some meat-eaters do consume them as well occasionally. Although in vitro meat should bear even more of an uncanny resemblance to the real thing than those substitutes (which might be why

---

a   *in vitro*   Latin for "In glass," this term refers to growth outside of a living body, such as in a test tube.

b   *Julian Savulescu*   Philosopher and bioethicist.

some people are revulsed by the idea), I would expect it to find success if issues of health and taste can be adequately dealt with. Now what if the yuck factor were to prove more of an issue than I anticipate? I believe the following points deserve to be emphasized:

1) *Our negative gut reactions often do not track any facts that should be granted any significance for our decision-making.*

I have said in a previous blog entry that our gut feelings of revulsion towards certain behaviors do sometimes seem to have normative significance, and that we should not dismiss such reactions out of hand as irrelevant. Nevertheless, the revulsion elicited by the arrival of certain new technologies is often irrelevant from a normative point of view, and this very much applies in the case of in vitro meat. Of course, if someone's instinctive reticence towards such meat were grounded e.g. in the concern that it might be unsafe, it would be reasonable. But if such concerns could be addressed, any revulsion one might still feel at the idea of eating something that had been produced in the lab and was therefore "unnatural" would not be tracking anything of any significance. In vitro meat would share those properties with many other inventions that are highly desirable and beneficial, such as life-saving medicines – one would just expect it to taste much better! Of course, medicines usually do not purport to look or taste like anything that exists "naturally," but this again is of no significance whatever – chocolate eggs for instance do have such a purpose, yet this is hardly a reason to find them repugnant.

5   2) *Some of our negative gut reactions have been distorted by powerful economic and social forces.*

I am inclined to think that if people were made keenly aware of the production process on which the meat that they consume depends, their gut feelings would lead them to favor in vitro meat over real meat. The yuck factor is currently biased against in vitro meat mostly because modern society, influenced by the economic interests of the meat industry, encourages an almost complete disconnection in people's minds between the poor treatment and killing of animals, and the sight of meat on their plate. For most meat-eaters (and this was my case when I used to eat meat), a steak or chicken breast fillet is simply something that somehow appears, as if by magic, on a supermarket shelf, ready to be cooked and enjoyed. Yet suppose that when eating out, people had to choose between two types of restaurant, each having various TV screens distributed throughout the premises. The first type serves traditional meat and, accordingly, its TV screens are continually showing the various stages of the process through which that meat was produced. The second restaurant only serves in vitro meat, and its screens show how it was grown in the lab. My guess is that very few people would choose to eat in the first type of restaurant. They would simply feel too uncomfortable and disgusted at e.g. the sight of pigs being confined to crowded, insalubrious warehouses, before being taken to the slaughterhouse to be stunned (sometimes improperly) and bled to death, one after the other. On the other hand, watching scientists manipulating muscle cells in a petri dish (even if they sometimes had to throw away one of their preparations after making a mistake) would, I think, be much less likely to make diners want to run out of the restaurant, or lose their appetite.

These two points also apply to other cases where the yuck factor tends to prevent many people from embracing a new technology that promises great benefits (examples include therapeutic cloning).

So if you are a meat-eater who finds the idea of eating lab-grown meat repulsive, remember that (a) you are perfectly entitled to be concerned about how healthy this meat will be and whether it will be tasty enough, but if your feelings persist even after these concerns have been resolved, then they don't deserve to play any role in your decision-making; (b) if you hadn't been taught to focus exclusively on the pleasurable qualities of meat and to keep the various dark aspects of the meat industry out of your mind, it is most likely that your gut feelings would tell you to prefer in vitro meat to conventional meat.

(2012, revised 2013)

Questions:

1. Define the term "yuck factor" as it is used in this article. Why might the media call this an "instinctive revulsion"?

2. What do you know about factory farming? Do you agree that "if people were made keenly aware of the production process on which the meat that they consume depends, their gut feelings would lead them to favor in vitro meat over real meat"?

3. Consider the restaurant experiment mentioned here. Do you agree with the article's statement that few people would eat at the first type of restaurant? Why or why not?

4. Consider other food items that we eat, such as chocolate bars, soda, and fast food. If stores and restaurants were to play videos of these food items being created, would only a few people continue to consume these items after seeing the videos?

5. Reflect upon the article's writing style and rhetorical approach. How much factual information is there?

6. Is the fact that the author identifies as a vegetarian detrimental to this argument? Why or why not?

7. What is the tone of the article, and how does it affect your reading and response?

8. Do you eat meat? Are you at all interested in eating in vitro meat? Take a poll of the people sitting next to you in class. How many are vegetarians and how many are meat eaters? Does being in either group correlate positively to a specific perception of in vitro meat?

9. The article states that "modern society, influenced by the economic interests of the meat industry, encourages an almost complete disconnection in people's minds between the poor treatment and killing of animals, and the sight of meat on their plate." Can you think of any specific examples of how such a disconnection takes place?

10. Do you think it is possible for people to retrain their minds to remove or redirect their "yuck factor"? If so, how?

11. Why do you think new technologies often elicit an ethically irrelevant gut revulsion?

12. Does this article define its use of "normative"? What would be an example of a gut reaction to a food item that had normative significance? How – if at all – is it possible to determine which gut reactions have normative significance and which do not?

13. According to the article, why would one be unjustified in finding in vitro meat revolting if it proved to be a healthy and tasty alternative to factory farming? Can you come up with a counter-argument?

thirty-five

William F. Baker, D. Stanton Korista, and Lawrence C. Novak

from Engineering the World's Tallest – Burj Dubai

Burj Dubai, renamed Burj Khalifa since the publication of this article, was the world's tallest human-made structure when it was completed in 2010. This article describes the innovations that made the building possible.

℮

Abstract

The goal of the Burj Dubai Tower is not simply to be the world's highest building; it is to embody the world's highest aspirations. The superstructure is currently under construction and as of fall 2007 has reached over 150 stories. The final height of the building is a "well-guarded secret." The height of the multi-use skyscraper will "comfortably" exceed the current record holder, the 509 meter (1671 ft) tall Taipei 101. The 280,000 m² (3,000,000 ft²) reinforced concrete multi-use Burj Dubai tower is utilized for retail, a Giorgio Armani Hotel, residential and office.

As with all super-tall projects, difficult structural engineering problems needed to be addressed and resolved. This paper presents the structural system for the Burj Dubai Tower.

Structural System Description

Designers purposely shaped the structural concrete Burj Dubai – "Y" shaped in plan – to reduce the wind forces on the tower, as well as to keep the structure simple and foster constructability. The structural system can be described as a "buttressed" core (Figur[e] 1 [...]). Each wing, with its own high performance concrete corridor walls and perimeter columns, buttresses[a] the others via a six-sided central core, or hexagonal hub. The result

---

a *buttresses* I.e., supports.

is a tower that is extremely stiff laterally and torsionally.[a] SOM[b] applied a rigorous geometry to the tower that aligned all the common central core, wall, and column elements.

Each tier of the building sets back in a spiral stepping pattern up the building. The setbacks are organized with the Tower's grid, such that the building stepping is accomplished by aligning columns above with walls below to provide a smooth load path.[c] This allows the construction to proceed without the normal difficulties associated with column transfers.

The setbacks are organized such that the Tower's width changes at each setback. The advantage of the stepping and shaping is to "confuse the wind." The wind vortices never get organized because at each new tier the wind encounters a different building shape.

The Tower and Podium structures are currently under construction [...] and the project is scheduled for topping out in 2008.

Definition of World's Tallest

5   From the outset, it has been intended that the Burj Dubai be the World's Tallest Building. The official arbiter of height is the Council on Tall Buildings

FIGURE 1.
Typical Floor Plan.

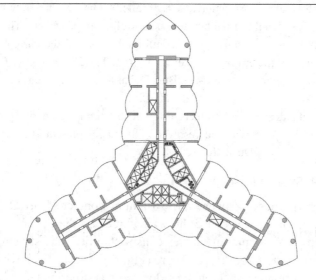

a   *stiff ... torsionally*   I.e., resistant to pressure from the side and to twisting.
b   *SOM*   Skidmore, Owings & Merrill, the firm that developed the structural design.
c   *load path*   Transfer of force from one component of the structure to another.

and Urban Habitat (CTBUH) founded at Lehigh University in Bethlehem, Pennsylvania, and currently housed at the Illinois Institute of Chicago in Chicago, Illinois. The CTBUH measures the height of buildings using four categories (measured from sidewalk at the main entrance). The categories and current record holders are as follows:

1. Highest Occupied Floor:
   Taipei 101   439m
2. Top of Roof:
   Taipei 101   449m
3. Top of Structure:
   Taipei 101   509m
4. Top of Pinnacle, Mast, Antenna or Flagpole:
   Sears Tower   527m

Although not considered to be a "building" the Tallest Freestanding Structure is:

CN Tower   553m

Although the final height of the Tower is a well-guarded secret, Burj Dubai will be the tallest by a significant amount in all the above categories.[a]

ARCHITECTURAL DESIGN

The context of Burj Dubai being located in the city of Dubai, UAE,[b] drove the inspiration for the building form to incorporate cultural and historical [elements] particular to the region. The influences of the Middle Eastern domes and pointed arches in traditional buildings, spiral imagery in Middle Eastern architecture, resulted in the tri-axial geometry of the Burj Dubai and the tower's spiral reduction with height. [...]

The Y-shaped plan is ideal for residential and hotel usage, with the wings allowing maximum outward views and inward natural light.

STRUCTURAL ANALYSIS AND DESIGN

The center hexagonal reinforced concrete core walls provide the torsional re-   10
sistance of the structure similar to a closed tube or axle. The center hexagonal

---

a   *tallest by ... above categories*   At 828m high upon its completion, the tower, as predicted, surpassed the previous record holders in all of these categories.
b   *UAE*   United Arab Emirates.

walls are buttressed by the wing walls and hammer head walls which behave as the webs and flanges[a] of a beam to resist the wind shears and movements. Outriggers[b] at the mechanical floors allow the columns to participate in the lateral load resistance of the structure; hence, all of the vertical concrete is utilized to support both gravity and lateral loads. The wall concrete specified strengths ranged from C80 to C60 cube strength[c] and utilized Portland cement[d] and fly ash.[e] Local aggregates were utilized for the concrete mix design. The C80 concrete for the lower portion of the structure had a specified

Figure 4.
Construction Photo.

a *webs and flanges* I.e., connecting and supporting elements.
b *Outriggers* Supports extending outward from the building.
c *cube strength* Load per unit area at which a cube of Portland cement fails when tested in a standardized manner.
d *Portland cement* Commonly used water-resistant cement.
e *fly ash* Residue generated when coal is burned, often used in combination with Portland cement.

Young's Elastic Modulus[a] of 43,800 N/mm² (6,350ksi) at 90 days. The wall and column sizes were optimized using virtual work/LaGrange multiplier methodology[b] which results in a very efficient structure (Baker et al., 2000).

The wall thicknesses and column sizes were fine-tuned to reduce the effects of creep[c] and shrinkage[d] on the individual elements which compose the structure. To reduce the effects of differential column shortening, due to creep, between the perimeter columns and interior walls, the perimeter columns were sized such that the self-weight gravity stress on the perimeter columns matched the stress on the interior corridor walls. The five (5) sets of outriggers, distributed up the building, tie all the vertical load carrying elements together, further ensuring uniform gravity stresses; hence, reducing differential creep movements. Since the shrinkage of concrete occurs more quickly in thinner walls or columns, the perimeter column thickness of 600mm (24") matched the typical corridor wall thickness (similar volume to surface ratios) [...] to ensure the columns and walls will generally shorten at the same rate due to concrete shrinkage. [...]

Foundations and Site Conditions

The Tower foundations consist of a pile supported raft.[e] The solid reinforced concrete raft is 3.7 meters (12 ft) thick and was poured utilizing C50 (cube strength) self consolidating concrete (SCC). The raft was constructed in four (4) separate pours (three wings and the center core). Each raft pour occurred over at least a 24 hour period. Reinforcement was typically at 300mm spacing in the raft, and arranged such that every 10th bar in each direction was omitted, resulting in a series of "pour enhancement strips" throughout the raft at which 600mm x 600mm openings at regular intervals facilitated access and concrete placement. [...]

The groundwater in which the Burj Dubai substructure is constructed is particularly severe, with chloride concentrations of up to 4.5%, and sulphates of up to 0.6%. The chloride and sulfate concentrations found in the ground water are even higher than the concentrations in sea water. Due to the aggressive conditions present due to the extremely corrosive groundwater, a

---

a *Young's Elastic Modulus* Stiffness of a material, indicated by a ratio of stress to strain. It is measured in units of pressure: here, either milli-pascals (N/mm²) or kilopounds of force per square inch (ksi).

b *LaGrange multiplier methodology* Mathematical strategy, used in this case to determine the most efficient measurements for the building.

c *creep* Gradual deformation of a structure due to stress.

d *shrinkage* Decrease in volume of concrete due to drying and chemical changes over time.

e *raft* Type of foundation designed to prevent settlement.

rigorous program of measures was required to ensure the durability of the foundations. Measures implemented include specialized waterproofing systems, increased concrete cover,[a] the addition of corrosion inhibitors to the concrete mix, stringent crack control design criteria and an impressed current cathodic[b] protection system utilizing titanium mesh. [...] A controlled permeability formwork liner was utilized for the Tower raft which results in a higher strength/lower permeable concrete cover to the rebar. Furthermore, a specially designed concrete mix was formulated to resist attack from the ground water. The concrete mix for the piles was a 60 MPa[c] mix based on a triple blend with 25% fly ash, 7% silica fume,[d] and a water to cement ratio of 0.32. The concrete was also designed as fully self consolidating concrete, incorporating a viscosity modifying admixture with a slump flow[e] of 675 +/- 75mm to limit the possibility of defects during construction.

WIND ENGINEERING

For a building of this height and slenderness, wind forces and the resulting motions in the upper levels become dominant factors in the structural design. An extensive program of wind tunnel tests and other studies were undertaken under the direction of Dr. Peter Irwin of Rowan Williams Davies and Irwin Inc.'s (RWDI) boundary layer wind tunnels in Guelph, Ontario. [...] The wind tunnel program included rigid-model force balance tests,[f] a full multi degree of freedom aeroelastic model studies,[g] measurements of localized pressures, pedestrian wind environment studies and wind climatic studies. Wind tunnel models account for the cross wind effects of wind induced vortex shedding on the building. [...] The aeroelastic and force balance studies used models mostly at 1:500 scale. The RWDI wind engineering was peer reviewed by Dr. Nick Isyumov of the University of Western Ontario Boundary Layer Wind Tunnel Laboratory.

15   In addition to the structural loading tests, the Burj Dubai tower was studied by RWDI for cladding,[h] pedestrian level, and stack effect[i] (Irwin et al., 2006).

---

a   *concrete cover*   Distance between the embedded reinforcement and surface of the concrete.
b   *cathodic*   Using polarized electrodes.
c   *MPa*   Milli-pascals.
d   *silica fume*   Substance that is commonly added to Portland cement to increase its strength and resistance to corrosion.
e   *slump flow*   Measure of consistency.
f   *rigid-model ... tests*   Tests which estimate the overall effects of wind forces by measuring the movements at the base of a lightweight model.
g   *aeroelastic model studies*   Studies mapping deformity due to aerodynamic forces.
h   *cladding*   Exterior covering.
i   *stack effect*   Movement of air into and out of the building.

To determine the wind loading on the main structure wind tunnel tests were undertaken early in the design using the high-frequency-force-balance technique.[a] The wind tunnel data were then combined with the dynamic properties of the tower in order to compute the tower's dynamic response and the overall effective wind force distributions at full scale. For the Burj Dubai the results of the force balance tests were used as early input for the structural design and detailed shape of the Tower and allowed parametric studies to be undertaken on the effects of varying the tower's stiffness and mass distribution. [...]

Several rounds of force balance tests were undertaken as the geometry of the tower evolved and was refined. The three wings set back in a clockwise sequence with the A wing setting back first. After each round of wind tunnel testing, the data was analyzed and the building was reshaped to minimize wind effects and accommodate unrelated changes in the Client's program. In general, the number and spacing of the set backs changed as did the shape of wings. This process resulted in a substantial reduction of wind forces on the tower by "confusing" the wind (Figur[e] 16 [...]) by encouraging disorganized vortex shedding over the height of the tower.

LONG-TERM AND CONSTRUCTION SEQUENCE ANALYSIS

Historically, engineers have typically determined the behavior of concrete structures using linear-elastic finite element analysis[b] and/or summations of vertical column loads. As building height increases, the results of such conventional analysis may increasingly diverge from actual behavior. Long-term, time-dependant deformations in response to construction sequence, creep, and shrinkage can cause redistribution of forces and gravity induced sidesway that would not be detected by conventional methods. When the time-dependant effects of construction, creep, shrinkage, variation of concrete stiffness with time, sequential loading and foundation settlements are not considered, the predicted forces and deflections may be inaccurate. To account for these time-dependant concrete effects in the Burj Dubai Tower structure, a comprehensive construction sequence analysis incorporating the effects of creep and shrinkage was utilized to study the time-dependant behavior of the structure (Baker et al., 2007). [...]

---

a *high ... technique* Wind-tunnel method involving a model mounted on a highly sensitive balance.

b *linear ... analysis* Form of structural analysis used to approximate the ways a building will deform over time due to internal stresses.

CONSTRUCTION

The Burj Dubai utilizes the latest advancements in construction techniques and material technology. The walls are formed using Doka's SKE 100 automatic self-climbing formwork system. [...] The circular nose columns are formed with steel forms, and the floor slabs are poured on MevaDec formwork. Wall reinforcement is prefabricated on the ground in 8m sections to allow for fast placement.

20   The construction sequence for the structure has the central core and slabs being cast first, in three sections; the wing walls and slabs follow behind; and the wing nose columns and slabs follow behind these. [...] Concrete is

FIGURE 16.
Tower Massing.

distributed to each wing utilizing concrete booms which are attached to the jump form system. Due to the limitations of conventional surveying techniques, a special GPS monitoring system has been developed to monitor the verticality of the structure. The construction survey work is being supervised by Mr. Doug Hayes, Chief Surveyor for the Burj Dubai Tower, with the Samsung BeSix Arabtec Joint Venture.

CONCLUSION

When completed, the Burj Dubai Tower will be the world's tallest structure. The architects and engineers worked hand in hand to develop the building form and the structural system, resulting in a tower which efficiently manages its response to the wind, while maintaining the integrity of the design concept.

It represents a significant achievement in terms of utilizing the latest design, materials, and construction technology and methods, in order to provide an efficient, rational structure to rise to heights never before seen.

(2008)

Project Team
Owner: Emaar Properties PJSC
Project Manager: Turner Construction International
Architect/Structural Engineers/MEP Engineers: Skidmore, Owings & Merrill LLP
Adopting Architect & Engineer/Field Supervision: Hyder Consulting Ltd.
Independent Verification and Testing Agency: GHD Global Party Ltd.
General Contractor: Samsung/BeSix/Arabtec
Foundation Contractor: NASA Multiplex

REFERENCES[a]

Baker, Korista, Novak, Pawlikowski & Young (2007), *"Creep & Shrinkage and the Design of Supertall Buildings – A Case Study: The Burj Dubai Tower"*, ACI SP-246: Structural Implications of Shrinkage and Creep of Concrete.

Baker, Novak, Sinn & Viise (2000), *"Structural Optimization of 2000-Foot Tall 7 South Dearborn Building"*, Proceedings of the ACSE Structures Congress 2000 – Advanced Technology in Structural Engineering and 14th Analysis & Computational Conference.

Irwin, Baker, Korista, Weismantle & Novak (2006), *"The Burj Dubai Tower: Wind Tunnel Testing of Cladding and Pedestrian Level"*, Structure Magazine, published by NCSEA, November 2006, pp 47-50.

Questions:

1. In part, the Burj Khalifa tower is presented as remarkable because it has broken so many world records (tallest structure, tallest freestanding

---

a   References have been excerpted to show only those cited in the included material.

structure, highest occupied floor). Why is record-breaking considered admirable? Should it be?

2. The construction of the tower cost about U.S. $1.5 billion. Do you think the result was worth the cost? Why or why not?

3. According to the article, "difficult structural engineering problems needed to be addressed and resolved" in designing the tower. What are some of the problems the article discusses? Is the article clear and specific about how these problems were overcome?

4. Underline all the loaded language in this piece. Do you expect a peer-reviewed engineering article to speak about needing to "confuse the wind," for instance?

5. Is an engineer likely to need to be told that the LaGrange multiplier methodology "results in a very efficient structure"? What specific information included in the article do you think an engineer might be interested in? Who do you think the article's intended readers might be?

6. The journal in which this article was published, *Structural Design of Tall and Special Buildings*, offers a user-pay open-access at a cost of $3,000 per article (otherwise the article is not available free-of-charge online). Identify some of the implications of this charge.

7. Wiley publishes *Structural Design of Tall and Special Buildings*, the journal in which this article appears, on behalf of the Los Angeles Tall Buildings Structural Design Council, described on its website as "a nonprofit organization whose members are those individuals who have demonstrated exceptional professional accomplishments in the structural design of tall buildings" and "a liaison organization to the Council on Tall Buildings and Urban Habitat, an international nonprofit organization of architects, engineers, planners, developers, and other professionals involved in the design, construction and operation of tall buildings." You can find the editorial board here: http://onlinelibrary.wiley.com/ journal/10.1002/%28ISSN%291541-7808/homepage/EditorialBoard. html. Do you consider this an academic peer-reviewed journal? Why or why not?

8. How does the article balance the need to mention engineering information with the need to protect proprietary information? Which need dominates?

9. The Burj Khalifa is featured in *Mission Impossible – Ghost Protocol*, and some BASE jumpers have used the building for jumps. What might the engineering company or the building's owners gain from being embedded in popular culture in this way?

thirty-six

JIMMY LEE SHREEVE

Mile-High Tower Wars: How Tall Is Too Tall?

This article, written for The Independent, *provides an overview of some of the engineering challenges associated with the creation of mega-skyscrapers and highlights divided opinions regarding the value of such structures.*

❦

Prince Charles famously doesn't care for skyscrapers. He sees them as a vain attempt to assert masculinity, like a rock star with a cucumber down his trousers – or, as he puts it: "Trying to make them ever taller than the other person's building is surely taking the commercial macho into the realms of adolescent lunacy."

Phallic icons or not, we're about to move into a new era of mega-tall buildings that will put structures like Chicago's Sears Tower, at 442 metres (1,450 feet) tall, and Taiwan's Taipei 101, at 508m, in the shade.

George Efstathiou, a managing partner at the architectural firm Skidmore, Owings & Merrill, declares that "the age of the super-skyscrapers is starting again." Considering the scale of the structures on the way, he could well be right.

The structure set to beat them all was announced at the end of March. The Mile High Tower, to be built in a "mini city" near the Red Sea port of Jeddah in Saudi Arabia, will be about 1,600 metres tall – seven times the height of the Canary Wharf tower in London Docklands, or four Empire State buildings on top of each other.

The tower is the brainchild of Prince al-Walid bin Talal, a member of 5 the Saudi royal family and the owner of London's Savoy hotel. According to Forbes magazine, he's worth about £11bn and is the 19th-richest person in the world. He needs to be: the estimated cost of the Mile High Tower – which, it is reported, will be built by the British engineering firm Hyder Consulting – is £5bn.

The building will pivot on hi-tech wizardry. A giant computer-controlled damper (shock absorber) will stretch down several floors to counter the nausea-inducing sway caused by the wind. And two mini towers, attached by sky bridges, will flank the building's base, further improving its stability.

You'd imagine that the logistics and know-how needed to build this monolith would be enormous. They are, says Ron Klemencic, president of Magnusson Klemencic Associates, an engineering firm specialising in high-rise constructions. But such structures are now relatively easy to erect. "Structural engineering-wise, it's not even difficult," Klemencic says. Stronger concrete and steel and advances in designing building frames, he says, allow for the safe development of mega-skyscrapers.

The current record-holder as tallest man-made structure is still a construction site. The Burj Dubai[a] in the United Arab Emirates edged past Taiwan's Taipei 101, reaching 509m in July last year. Sources say the tower – being built by Skidmore, Owings & Merrill – will reach just over 800m when it is completed by the end of next year. The spire will be visible from 95km (60 miles) away – and the views will naturally be stunning.

The quantity of materials involved in the building's construction is mind-boggling. For example, the 31,000 tons of steel rods used to reinforce the structure would, if laid end to end, stretch one-quarter of the way round the world. When the tower is finished, 230,000 cubic metres of concrete will have been used. It will weigh about 500,000 tons.

10   When structures are so tall, it isn't just materials you have to worry about; it's the tools needed to build them. The tower cranes used to build conventional high-rise buildings, which are fixed to the ground, can carry enormous weight – but for mega-skyscrapers, helicopters are needed to lift materials to the higher levels.

Numerous high-profile skyscrapers are now being built around the world, including New York's Freedom Tower on the World Trade Centre site. But none will exceed 700m in height.

The building set to replace Southwark Towers in London – the Shard London Bridge (also known as the "shard of glass") – will be a comparatively puny 310m tall and have 72 floors. When it is completed (estimates say in 2011[b]) it will be Britain's tallest building. The 50-storey Canary Wharf tower at One Canada Square is the current record-holder at 250m.

---

a   *Burj Dubai*   The tower was officially completed and given the name Burj Khalifa in 2010 (see "Engineering the World's Tallest – Burj Dubai" in this anthology).

b   *estimates say in 2011*   The Shard was completed in 2012.

Several other high-rises are proposed for the capital. But the Mayor of London, Boris Johnson, is critical of tall buildings and has vowed to strengthen regulations protecting views of historic buildings such as St Paul's Cathedral and the Palace of Westminster. By world standards, of course, what Johnson considers tall isn't at all.

But how high can skyscrapers be? Is the sky the limit? David Scott, chairman of the Council on Tall Buildings and Urban Habitats (CTBUH) and principal at the New York-based engineering firm Arup, doesn't believe there is an absolute limit. "Mount Everest is essentially a pile of stone," he says. "You don't need a lot of technology to create it. Just a lot of money."

Engineers and architects have always speculated about how tall skyscrapers could be. Frank Lloyd Wright designed a mile-high tower, the Illinois, to be built in Chicago. It was proposed in 1956. Most experts agree that the technology was there to build it at the time, but not the investment. 15

But money isn't the only limiting factor. Elevator technology lags behind building technology, and one obvious issue is lift[a] cables; if they had to raise a lift one mile, they would be far too heavy. In Burj Dubai, no elevator goes all the way from the ground to the top.

There's also we frail humans to consider. If an express elevator – at speeds up to 25mph – went from the ground floor to the top, we could pass out due to changes in the air pressure.

Then there's the problem of building movement. Most skyscrapers can sway a few metres in the wind without tumbling down, but the people inside might feel uncomfortable, if not downright nauseous, especially on higher floors. So the tallest structures need sturdy central cores to anchor them. The Empire State Building in New York, and other skyscrapers of that time, had steel beams wrapped around their elevator shafts.

The Burj Dubai has a "buttressed core," or concrete hub, with three wings spreading out to form a kind of tripod. When the wind blows against two of the wings, the third supports them. Taipei 101 has a 730-ton pendulum in the top of the building; the giant ball swings against the movement to keep the upper floors steady.

More difficult are natural disasters, particularly earthquakes that could topple a skyscraper. In areas of high seismic activity, such as California, strict building-codes are in place to protect against earth tremors. The most common defence is to use mass dampers – essentially seismic shock absorbers – made of giant springs or hydraulic systems that move in the opposite direction to the earthquake's oscillations. 20

---

a *lift* Elevator.

And today, there's the threat of terrorism. Tall buildings are a target because they are iconic, and because they contain so many people. Lessons have been learnt from the attacks on the World Trade Center, which led to the collapse of the towers. More internal supports make collapse far less likely. Stairwells are made wider to aid emergency evacuation, and ventilation systems force smoke out rather than letting it spread upwards, as if in a chimney.

But the fact is that there are hazards everywhere in life. And most experts agree that the benefits of skyscrapers outweigh the downsides. Bill Baker, chief structural engineer at Skidmore, Owings & Merrill, believes that building high is not only safe, but necessary to stop cities devouring green-belt land.[a] "Urban density is good," he says. "Everyone uses public transportation, people walk to lunch. Look at the Sears Tower in Chicago. It has 4.4 million square feet – that's 100 acres on one city block."

David Scott of the CTBUH agrees. "If you look at the holistic impact of tall buildings on urban living, they offer many benefits," he says. "Not only do they reduce things like car ownership, but they can use waste heat in winter to warm them, and they have the potential to generate more open spaces at ground level, as well as reducing suburban sprawl. In the case of London, it can either expand outwards or upwards."

But George Efstathiou, the architect, perhaps has the real reason for our love of skyscrapers: "Tall buildings are a matter of ego. Tall buildings are a sign of success."

TALL STOREYS

25
- The 60-storey Woolworth Building in New York, built in 1913, required 17 million bricks, 7,500 tons of terra cotta, 53,000 pounds of bronze and iron hardware and 87 miles of electric wiring.
- The Empire State Building in New York required 60 miles of water pipe and 3,500 miles of telephone and telegraph wire. Rain and snow can sometimes be seen rising instead of falling because of the wind patterns around the building.
- The World Trade Center in New York (built in 1970–71), used enough concrete to lay a pavement 5ft wide from New York City to Washington, DC, a distance of 204 miles. The electrical wiring for the twin towers would have reached from New York to Mexico (about 1,500 miles). More than a million cubic yards of earth and rock were excavated to make way

---

a *green-belt land* Undeveloped land, often supporting wild plant and animal life or used agriculturally, typically surrounding urban areas.

for the World Trade Center. The material went into the Hudson River, creating 23.5 acres of new land.

- The Sears Tower in Chicago contains enough concrete to build an eight-lane highway five miles long. If laid out flat on the ground, the black "skin" of the tower would cover an area of 28 acres.

(2008)

Questions:

1. Does the opening reference to Prince Charles immediately make you want to read this article? Why do you think the author uses it? What is the article's tone in this section?

2. What is a "phallic icon"? Why is the use of the term meant to be humorous? Locate three other examples of humour.

3. Has the "Mile High Tower" been built?

4. What sorts of comparisons does the article use to try to help you understand the height of the building? Do they work? What sort of audience do you think they are intended for?

5. What does this article seem to admire about the building process?

6. What sorts of justifications does the article offer for building super-skyscrapers? Do you find them convincing? Why or why not?

7. Pick one quotation given in the article in support of these buildings and argue against it using an analysis of the logic in the quotation.

8. The article lists all sorts of reasons not to build such a building (or "difficulties" in the process). Which ones do you think are the most significant? Why?

9. Rewrite one of the article's paragraphs in a scientific writing style.

10. Do you feel this article is more or less well-supported than Baker, Korista, and Novak's (included in this anthology)? Why or why not?

thirty-seven

ETHAN KROSS, PHILIPPE VERDUYN, EMRE DEMIRALP,
JIYOUNG PARK, DAVID SEUNGJAE LEE, NATALIE LIN,
HOLLY SHABLACK, JOHN JONIDES, AND OSCAR YBARRA

from Facebook Use Predicts Declines in Subjective Well-Being in Young Adults

The results of this study on Facebook use and its relationship to subjective well-being suggest that Facebook negatively influences how young adults feel in the present moment, and how they perceive their lives overall.

❦

ABSTRACT

Over 500 million people interact daily with Facebook. Yet, whether Facebook use influences subjective well-being over time is unknown. We addressed this issue using experience-sampling, the most reliable method for measuring in-vivo[a] behavior and psychological experience. We text-messaged people five times per day for two-weeks to examine how Facebook use influences the two components of subjective well-being: how people feel moment-to-moment and how satisfied they are with their lives. Our results indicate that Facebook use predicts negative shifts on both of these variables over time. The more people used Facebook at one time point, the worse they felt the next time we text-messaged them; the more they used Facebook over two-weeks, the more their life satisfaction levels declined over time. Interacting with other people "directly" did not predict these negative outcomes. They were also not moderated by the size of people's Facebook networks, their perceived supportiveness, motivation for using Facebook, gender,

---

a *experience-sampling* Method of gathering experimental data by requiring participants to provide regular updates about their behavior and feelings; *in-vivo* Occurring within a complete, live organism.

loneliness, self-esteem, or depression. On the surface, Facebook provides an invaluable resource for fulfilling the basic human need for social connection. Rather than enhancing well-being, however, these findings suggest that Facebook may undermine it.

INTRODUCTION

Online social networks are rapidly changing the way human beings interact. Over a billion people belong to Facebook, the world's largest online social network, and over half of them log in daily.[1] Yet, no research has examined how interacting with Facebook influences subjective well-being over time. Indeed, a recent article that examined every peer-reviewed publication and conference proceeding on Facebook between 1/2005 and 1/2012 (412 in total) did not reveal a single study that examined how using this technology influences subjective well-being over time.[2, Supporting Information 1]

Subjective well-being is one of the most highly studied variables in the behavioral sciences. Although significant in its own right, it also predicts a range of consequential benefits including enhanced health and longevity.[3-5] Given the frequency of Facebook usage, identifying how interacting with this technology influences subjective well-being represents a basic research challenge that has important practical implications.

This issue is particularly vexing because prior research provides mixed clues about how Facebook use should influence subjective well-being. Whereas some cross-sectional research reveals positive associations between online social network use (in particular Facebook) and well-being,[6] other work reveals the opposite.[7,8] Still other work suggests that the relationship between Facebook use and well-being may be more nuanced and potentially influenced by multiple factors including number of Facebook friends, perceived supportiveness of one's online network, depressive symptomatology, loneliness, and self-esteem.[9,10,11]

So, how does Facebook usage influence subjective well-being over time? The cross-sectional approach[a] used in previous studies makes it impossible to know. We addressed this issue by using experience-sampling, the most reliable method for measuring in-vivo behavior and psychological experience over time.[12] We text-messaged participants five times per day for 14-days. Each text-message contained a link to an online survey, which participants completed using their smartphones. We performed lagged analyses[b] on par-

---

a *cross-sectional approach* Approach to gathering experimental data by observing an entire group at a specific time.

b *lagged analyses* Identification of patterns in data collected over time.

ticipants' responses, as well as their answers to the Satisfaction With Life Questionnaire (SWLS),[13] which they completed before and immediately following the 14-day experience-sampling period, to examine how interacting with Facebook influences the two components of subjective well-being: how people feel ("affective" well-being) and how satisfied they are with their lives ("cognitive" well-being).[14,15] This approach allowed us to take advantage of the relative timing of participants' natural Facebook behavior and psychological states to draw inferences about their likely causal sequence.[16-19]

METHODS

Participants

5   Eighty-two people (M_{age} = 19.52, SD_{age} = 2.17;[a] 53 females; 60.5% European American, 28.4% Asian, 6.2% African American, and 4.9% other) were recruited for a study on Facebook through flyers posted around Ann Arbor, Michigan. Participants needed a Facebook account and a touch-screen smartphone to qualify for the study. They received $20 and were entered into a raffle to receive an iPad2 for participating.

Ethics Statement

The University of Michigan Institutional Review Board approved this study. Informed written consent was obtained from all participants prior to participation.

Materials and Procedure

PHASE 1

Participants completed a set of questionnaires, which included the SWLS (M = 4.96, SD = 1.17), Beck Depression Inventory[20] (M = 9.02, SD = 7.20), the Rosenberg Self-Esteem Scale[21] (M = 30.40, SD = 4.96), and the Social Provision Scale[22] (M = 3.55, SD = .34), which we modified to assess perceptions of Facebook support. We also assessed participants' motivation for using Facebook by asking them to indicate whether they use Facebook "to keep in touch with friends (98% answered yes)," "to find new friends (23% answered yes)," "to share good things with friends (78% answered yes)," "to share bad things with friends (36% answered yes)," "to obtain new information (62% answered yes)," or "other: please explain (17% answered yes)." Examples of other reasons included chatting with others, keeping in touch with family, and facilitating schoolwork and business.

---

a M Mean; SD Standard Deviation, a number indicating the extent of difference within a group.

PHASE 2

Participants were text-messaged 5 times per day between 10am and midnight over 14-days. Text-messages occurred at random times within 168-minute windows per day. Each text-message contained a link to an online survey, which asked participants to answer five questions using a slider scale: (1) How do you feel right now? (*very positive* [0] to *very negative* [100]; $M =$ 37.47, $SD = 25.88$); (2) How worried are you right now? (*not at all* [0] to *a lot* [100]; $M = 44.04$, $SD = 30.42$); (3) How lonely do you feel right now? (*not at all* [0] to *a lot* [100]; $M = 27.61$, $SD = 26.13$); (4) How much have you used Facebook since the last time we asked? (*not at all* [0] to *a lot* [100]; $M = 33.90$, $SD = 30.48$); (5) How much have you interacted with other people "directly" since the last time we asked? (*not at all* [0] to *a lot* [100]; $M = 64.26$, $SD = 31.11$). When the protocol for answering these questions was explained, interacting with other people "directly" was defined as face-to-face or phone interactions. An experimenter carefully walked participants through this protocol to ensure that they understood how to answer each question and fulfill the study requirements.

Participants always answered the affect question first. Next the worry and loneliness questions were presented in random order. The Facebook use and direct social interaction questions were always administered last, again in random order. Our analyses focused primarily on affect (rather than worry and loneliness) because this affect question is the way "affective well-being" is typically operationalized.

PHASE 3

Participants returned to the laboratory following Phase 2 to complete another set of questionnaires, which included the SWLS ($M = 5.13$, $SD = 1.26$) and the Revised UCLA Loneliness Scale[23] ($M = 1.69$, $SD = .46$). Participants' number of Facebook friends ($M = 664.25$, $SD = 383.64$) was also recorded during this session from participants' Facebook accounts.Supporting Information 2

RESULTS

Attrition and compliance

Three participants did not complete the study. As the methods section notes, participants received a text message directing them to complete a block of five questions once every 168 minutes on average (the text message was delivered randomly within this 168-minute window). A response to any question within a block was considered "compliant" if it was answered *before* participants received a subsequent text-message directing them to complete the next block of questions. Participants responded to an average of 83.6% of text-messages (range: 18.6%–100%). Following prior research,[24] we pruned the data[a] by excluding all of the data from two participants who responded to <33% of the texts, resulting in 4,589 total observations. The results did not change substantively when additional cutoff rates were used.

Analyses overview

We examined the relationship between Facebook use and affect using multilevel analyses to account for the nested data structure.[b] Specifically, we examined whether T_2[c] affect (i.e., How do you feel *right now?*) was predicted by T_{1-2} Facebook use (i.e., How much have you used Facebook *since the last time we asked?*), controlling for T_1 affect at level-1 of the model (between-day lags were excluded). Note that although this analysis assesses Facebook use at T_2, the question refers to usage between T_1 and T_2 (hence the notation T_{1-2}). This analysis allowed us to explore whether Facebook use during the time period separating T_1 and T_2 predicted changes in affect over this time span. [...]

The relationship between mean Facebook use and life satisfaction was assessed using OLS regressions[d] because these data were not nested. Both unstandardized (B) and standardized (β) OLS regression coefficients[e] are reported.Supporting Information 3

Facebook use and well-being

AFFECTIVE WELL-BEING

We examined whether people's tendency to interact with Facebook during the time period separating two text messages influenced how they felt at T_2, controlling for how they felt at T_1. Nested time-lag analyses indicated that the more people used Facebook the worse they subsequently felt, $B = .08$, $\chi^2 = 28.90$, $p<.0001$[f][....] The reverse pathway (T_1 Affect predicting T_{1-2}

---

a *pruned the data* Removed unnecessary data.

b *multilevel analyses* Methods of analyzing data which recognize that there are multiple sources which may explain variations in data. Multilevel analyses are used specifically in research where data may be classified and organized at several levels; *nested data structure* Describes data that is obtained from multiple observations of individuals in particular groups, e.g. students in a particular class, or data that is obtained through repeated observation of the same individual over time.

c T_2 Time 2.

d *OLS regressions* Ordinary Least Squares regressions, models that calculate the relationship between a dependent variable and an independent variable in order to estimate the boundaries of the variables.

e *OLS regression coefficients* Numerical indications of the relationship between two variables.

f χ^2 Chi squared, used in statistical tests to determine the probability that a set of data reflects a significant relationship between variables; *p* Probability, here an indication of the likelihood

Facebook use, controlling for T_{0-1} Facebook use) was not significant, $B = -.005$, $\chi^2 = .05$, $p = .82$, indicating that people do not use Facebook more or less depending on how they feel.[Supporting Information 4,5] [...]

COGNITIVE WELL-BEING

To examine how Facebook use influenced "cognitive well-being," we ana- 15
lyzed whether people's average Facebook use over the 14-day period predicted their life satisfaction at the end of the study, controlling for baseline
life satisfaction and average emotion levels over the 14-day period. The more
participants used Facebook, the more their life satisfaction levels declined
over time, $B = -.012$, $\beta = -.124$, $t(73) = -2.39$,[a] $p = .02$[....]

ALTERNATIVE EXPLANATIONS

An alternative explanation for these results is that any form of social interaction undermines well-being. Because we also asked people to indicate how
frequently they interacted with other people "directly" since the last time we
text messaged them, we were able to test this idea. Specifically, we repeated
each of the aforementioned analyses substituting "direct" social interaction
for Facebook use. In contrast to Facebook use, "direct" social interaction did
not predict changes in cognitive well-being, $B = -.006$, $\beta = -.059$, $t(73) =
1.04$, $p = .30$, and predicted *increases* (not decreases) in affective well-being,
$B = -.15$, $\chi^2 = 65.30$, $p<.0001$. Controlling for direct social interaction did
not substantively alter the significant relationship between Facebook use and
affective well-being, $B = .05$, $\chi^2 = 10.78$, $p<.01$.

Another alternative explanation for these results is that people use
Facebook when they feel bad (i.e., when they are bored, lonely, worried or
otherwise distressed), and feeling bad leads to declines in well-being rather
than Facebook use per se. The analyses we reported earlier partially address
this issue by demonstrating that affect does not predict changes in Facebook
use over time and Facebook use continues to significantly predict declines
in life satisfaction over time when controlling for affect. However, because
participants also rated how lonely and worried they felt each time we text
messaged them, we were able to test this proposal further.

We first examined whether worry or loneliness predicted changes in Facebook use over time (i.e., T_1 worry [or T_1 loneliness] predicting T_{1-2} Facebook
use, controlling for T_{0-1} Facebook use). Worry did not predict changes in

---

of getting the same experimental results as the ones observed if there were no relationship
between the variables being studied.

a   *t*   Variable used in *t*-tests to calculate the significance of the differences between two sets of
data.

Facebook use, $B = .04$, $\chi^2 = 2.37$, $p = .12$, but loneliness did, $B = .07$, $\chi^2 = 8.54$, $p<.01$. The more lonely people felt at one time point, the more people used Facebook over time. Given this significant relationship, we next examined whether controlling for loneliness renders the relationship between Facebook use and changes in affective and cognitive well-being non-significant – what one would predict if Facebook use is a proxy for loneliness. This was not the case. Facebook use continued to predict declines in affective well-being, $B = .08$, $\chi^2 = 27.87$, $p<.0001$, and cognitive well-being, $B = -.012$, $\beta = -.126$, $t(72) = 2.34$, $p = .02$, when loneliness was controlled for in each analysis. Neither worry nor loneliness interacted significantly with Facebook use to predict changes in affective or cognitive well-being ($ps>.44$).

MODERATION

Next, we examined whether a number of theoretically relevant individual-difference variables[a] including participants' number of Facebook Friends, their perceptions of their Facebook network support, depressive symptoms, loneliness, gender, self-esteem, time of study participation, and motivation for using Facebook (e.g., to find new friends, to share good or bad things, to obtain new information) interacted with Facebook use to predict changes in affective or cognitive well-being.[Supporting Information 6] In no case did we observe any significant interactions ($ps>.16$).

EXPLORATORY ANALYSES

20 Although we did not have *a priori* predictions[b] about whether Facebook use and direct social contact would interact to predict changes in affective and cognitive well-being, we nevertheless explored this issue in our final set of analyses. The results of these analyses indicated that Facebook use and direct social contact interacted significantly to predict changes in affective well-being, $B = .002$, $\chi^2 = 19.55$, $p <.0001$, but not changes in cognitive well-being, $B = .000$, $\beta = .129$, $t(71)=.39$, $p = .70$. To understand the meaning of the former interaction, we performed simple slope analyses.[c] These analyses indicated that the relationship between Facebook use and declines in affective well-being increased linearly with direct social contact. Specifically, whereas Facebook use did not predict significant declines in affective well-being when participants experienced low levels of direct social contact (i.e., 1 standard

---

a   *individual-difference variables*   Variables indicating characteristics that individual participants already possess, which may affect study results but are not controlled by the study.

b   *a priori predictions*   Predictions made before research began.

c   *slope analyses*   Analyses showing the incline of the line that would be formed if a set of data were depicted in a graph.

deviation below the sample mean for direct social contact; $B = .00$, $\chi^2 = .04$, $p = .84$), it did predict significant declines in well-being when participants experienced moderate levels of direct social contact (i.e., at the sample mean for direct social contact; $B = .05$, $\chi^2 = 11.21$, $p<.001$) and high levels of direct social contact (i.e., 1 standard deviation above the sample mean for direct social contact; $B = .10$, $\chi^2 = 28.82$, $p<.0001$).

DISCUSSION

Within a relatively short timespan, Facebook has revolutionized the way people interact. Yet, whether using Facebook predicts changes in subjective well-being over time is unknown. We addressed this issue by performing lagged analyses on experience sampled data, an approach that allowed us to take advantage of the relative timing of participants' naturally occurring be-haviors and psychological states to draw inferences about their likely causal sequence.[17,18] These analyses indicated that Facebook use predicts declines in the two components of subjective well-being: how people feel moment to moment and how satisfied they are with their lives.

Critically, we found no evidence to support two plausible alternative interpretations of these results. First, interacting with other people "directly" did not predict declines in well-being. In fact, direct social network interac-tions led people to feel *better* over time. This suggests that Facebook use may constitute a unique form of social network interaction that predicts impoverished well-being. Second, multiple types of evidence indicated that it was not the case that Facebook use led to declines in well-being because people are more likely to use Facebook when they feel bad – neither affect nor worry predicted Facebook use and Facebook use continued to predict significant declines in well-being when controlling for loneliness (which did predict increases in Facebook use and reductions in emotional well-being).

Would engaging in any solitary activity similarly predict declines in well-being? We suspect that they would not because people often derive pleasure from engaging in some solitary activities (e.g., exercising, reading). Support-ing this view, a number of recent studies indicate that people's *perceptions* of social isolation (i.e., how lonely they feel) – a variable that we assessed in this study, which did not influence our results – are a more powerful determinant of well-being than *objective* social isolation.[25] A related question concerns whether engaging in any Internet activity (e.g., email, web surfing) would likewise predict well-being declines. Here too prior research suggests that it would not. A number of studies indicate that whether interacting with the Internet predicts changes in well-being depends on how you use it (i.e., what sites you visit) and who you interact with.[26]

Future research

Although these findings raise numerous future research questions, four stand out as most pressing. First, do these findings generalize? We concentrated on young adults in this study because they represent a core Facebook user demographic. However, examining whether these findings generalize to additional age groups is important. Future research should also examine whether these findings generalize to other online social networks. As a recent review of the Facebook literature indicated[2] "[different online social networks] have varied histories and are associated with different patterns of use, user characteristics, and social functions" (p. 205). Therefore, it is possible that the current findings may not neatly generalize to other online social networks.

25   Second, what mechanisms underlie the deleterious effects of Facebook usage on well-being? Some researchers have speculated that online social networking may interfere with physical activity, which has cognitive and emotional replenishing effects[27] or trigger damaging social comparisons.[8, 28] The latter idea is particularly interesting in light of the significant interaction we observed between direct social contact and Facebook use in this study – i.e., the more people interacted with other people directly, the more strongly Facebook use predicted declines in their affective well-being. If harmful social comparisons explain how Facebook use predicts declines in affective well-being, it is possible that interacting with other people directly either enhances the frequency of such comparisons or magnifies their emotional impact. Examining whether these or other mechanisms explain the relationship between Facebook usage and well-being is important both from a basic science and practical perspective.

Finally, although the analytic approach we used in this study is useful for drawing inferences about the likely causal ordering of associations between naturally occurring variables, experiments that manipulate Facebook use in daily life are needed to corroborate these findings and establish definitive causal relations. Though potentially challenging to perform – Facebook use prevalence, its centrality to young adult daily social interactions, and addictive properties may make it a difficult intervention target – such studies are important for extending this work and informing future interventions.

Caveats

Two caveats are in order before concluding. First, although we observed statistically significant associations between Facebook usage and well-being, the sizes of these effects were relatively "small." This should not, however, undermine their practical significance.[29] Subjective well-being is a multiply determined outcome – it is unrealistic to expect any single factor to powerfully influence it. Moreover, in addition to being consequential in its own right, subjective well-being predicts an array of mental and physical health consequences. Therefore, identifying any factor that systematically influences it is important, especially when that factor is likely to accumulate over time among large numbers of people. Facebook usage would seem to fit both of these criteria.

Second, some research suggests that asking people to indicate how good or bad they feel using a single bipolar scale, as we did in this study, can obscure interesting differences regarding whether a variable leads people to feel less positive, more negative or both less positive and more negative. Future research should administer two unipolar affect questions to assess positive and negative affect separately to address this issue.

CONCLUDING COMMENT

The human need for social connection is well established, as are the benefits that people derive from such connections.[30-34] On the surface, Facebook provides an invaluable resource for fulfilling such needs by allowing people to instantly connect. Rather than enhancing well-being, as frequent interactions with supportive "offline" social networks powerfully do, the current findings demonstrate that interacting with Facebook may predict the opposite result for young adults – it may undermine it.

(2013)

Acknowledgments: We thank Emily Kean for her assistance running the study and Ozlem Ayduk and Phoebe Ellsworth for their feedback.

Author Contributions: Conceived and designed the experiments: EK ED JP DSL NL JJ OY. Performed the experiments: HS NL. Analyzed the data: PV ED. Wrote the paper: EK ED PV JJ OY. Discussed the results and commented on the manuscript: EK PV ED JP DSL NL HS JJ OY.

Supporting Information:
1: We do not imply that no longitudinal research on Facebook has been performed. Rather, no published work that we are aware of has examined how Facebook influences subjective well-being over time (i.e., how people feel and their life satisfaction).
2: Additional measures were administered during Phases 1 and 2 for other purposes. The measures reported in the MS are those that were theoretically motivated.
3: Raw data are available upon request for replication purposes.
4: We also examined whether T_{0-1} (rather than T_{1-2}) Facebook use influences T_2 affect, controlling for T_1 affect. Nested time-lagged analyses indicated that this was also true, $B = .03$, $\chi^2 = 4.67$, $p = .03$.
5: Some research suggests that affect fluctuates throughout the day. Replicating this work, time of day was related to affective well-being such that people reported feeling better as the day progressed

(B = –1.06, χ^2 = 21.49, p < .0001). Controlling for time of day did not, however, substantively influence any of the results.

6: 98% of participants reported using Facebook to "keep in touch with friends." Therefore, we did not test for moderation with this variable.

REFERENCES

1. Facebook_Information (2012) Facebook Newsroom Website. Available: http://newsroom. fb.com/content/default.aspx?NewsAreaId=22. Accessed 2012 April 23.

2. Wilson RE, Gosling SD, Graham LT (2012) A Review of Facebook Research in the Social Sciences. Perspect Psychol Sci 7: 203–220.

3. Steptoe A, Wardle J (2011) Positive affect measured using ecological momentary assessment and survival in older men and women. Proc Natl Acad Sci USA 108: 18244–18248. doi:10.1073/pnas.1110892108.

4. Boehm JK, Peterson C, Kivimaki M, Kubzansky L (2011) A prospective study of positive psychological well-being and coronary heart disease. Health Psychol 30: 259–267. doi:10.1037/a0023124.

5. Diener E (2011) Happy people live longer: Subjective well-being contributes to health and longevity. Appl Psychol Health Well Being 3: 1–43. doi:10.1111/j.1758-0854.2010.01045.x.

6. Valenzuela S, Park N, Kee KF (2009) Is There Social Capital in a Social Network Site?: Facebook Use and College Students' Life Satisfaction, Trust, and Participation. J Comput Mediat Commun 14: 875–901. doi:10.1111/j.1083-6101.2009.01474.x.

7. Huang C (2010) Internet use and psychological well-being: A meta-analysis. Cyberpsychol Behav Soc Netw 13: 241–248.

8. Chou H, Edge N (2012) 'They are happier and having better lives than I am': The impact of using Facebook on perceptions of others' lives. Cyberpsychol Behav Soc Netw 15: 117–120. doi:10.1089/cyber.2011.0324.

9. Forest AL, Wood JV (2012) When Social Networking Is Not Working: Individuals With Low Self-Esteem Recognize but Do Not Reap the Benefits of Self-Disclosure on Facebook. Psychol Sci 23: 295–302. doi:10.1177/0956797611429709.

10. Manago AM, Taylor T, Greenfield PM (2012) Me and my 400 friends: The anatomy of college students' Facebook networks, their communication patterns, and well-being. Dev Psychol 48: 369–380. doi:10.1037/a0026338.

11. Kim J, LaRose R, Peng W (2009) Loneliness as the cause and the effect of problematic Internet use: the relationship between Internet use and psychological well-being. Cyberpsychology & behavior: the impact of the Internet, multimedia and virtual reality on behavior and society 12: 451–455. doi:10.1089/cpb.2008.0327.

12. Kahneman D, Krueger AB, Schkade DA, Schwarz N, Stone AA (2004) A survey method for characterizing daily life experience: The day reconstruction method. Science 306: 1776–1780. doi:10.1126/science.1103572.

13. Diener E, Emmons RA, Larsen RJ, Griffin S (1985) The Satisfaction with Life Scale. J Pers Assess 49: 71–74. doi:10.1207/s15327752jpa4901_13.

14. Kahneman D, Deaton A (2010) High income improves evaluation of life but not emotional well-being. Proc Natl Acad Sci USA 107: 16489–16493. doi:10.1073/pnas.1011492107.

15. Diener E (1984) Subjective Well-Being. Psychol Bull 95: 542–575. doi:10.1037/0033-2909.95.3.542.

16. Hofmann W, Vohs KD, Baumeister RF (2012) What people desire, feel conflicted about, and try to resist in everyday life. Psychol Sci doi:10.1177/0956797612437426.

17. Bolger N, Davis A, Rafaeli E (2003) Diary methods: Capturing life as it is lived. Annu Rev Psychol 54: 579–616. doi:10.1146/annurev.psych.54.101601.145030.

18. Adam EK, Hawkley LC, Kudielka BM, Cacioppo JT (2006) Day-to-day dynamics of experience–cortisol associations in a population-based sample of older adults. Proc Natl Acad Sci USA 103: 17058–17063. doi:10.1073/pnas.0605053103.

19. Killingsworth MA, Gilbert DT (2010) A Wandering Mind Is an Unhappy Mind. Science 330: 932–932. doi:10.1126/science.1192439.

20. Beck AT, Steer RA, Brown GK (1996) BDI-II Manual San Antonio: Harcourt Brace & Company.

21. Rosenberg M (1965) Society and the adolescent self-image. Princeton: Princeton University Press.

22. Cutrona CE (1989) Ratings of social support by adolescents and adult informants: Degree of correspondence and prediction of depressive symptoms. Journal of Personality and Social Psychology 57: 723–730. doi:10.1037/0022-3514.57.4.723.

23. Russell D, Peplau LA, Cutrona CE (1980) The revised UCLA Loneliness Scale: Concurrent and discriminant validity evidence. J Pers Soc Psychol 39: 472–480. doi:10.1037/0022-3514.39.3.472.

24. Moberly NJ, Watkins ER (2008) Ruminative self-focus, negative life events, and negative affect. Behav Res Ther 46: 1034–1039. doi:10.1016/j.brat.2008.06.004.

25. Cacioppo JT, Hawkley LC, Norman GJ, Berntson GG (2011) Social isolation. Ann N Y Acad Sci 1231: 17–22. doi:10.1111/j.1749-6632.2011.06028.x.

26. Bessiére K, Kiesler S, Kraut R, Boneva BS (2008) Effects of Internet use and social resources on changes in depression. Information, Communication, and Society 11: 47–70.

27. Kaplan S, Berman MG (2010) Directed Attention as a Common Resource for Executive Functioning and Self-Regulation. Perspect Psychol Sci 5: 43–57. doi:10.1177/1745691609356784.

28. Haferkamp N, Kramer NC (2011) Social Comparison 2.0: Examining the Effects of Online Profiles on Social-Networking Sites. Cyberpsychol Behav Soc Netw 14: 309–314. doi:10.1089/cyber.2010.0120.

29. Prentice DA, Miller DT (1992) When small effects are impressive. Psychological Bulletin 112: 160–164. doi:10.1037/0033-2909.112.1.160.

30. Baumeister RF, Leary MR (1995) The need to belong: desire for interpersonal attachments as a fundamental human motivation. Psychol Bull 117: 497–529. doi:10.1037/0033-2909.117.3.497.

31. Kross E, Berman MG, Mischel W, Smith EE, Wager TD (2011) Social rejection shares somatosensory representations with physical pain. Proc Natl Acad Sci USA 108: 6270–6275. doi:10.1073/pnas.1102693108.

32. Eisenberger NI, Cole SW (2012) Social neuroscience and health: neurophysiological mechanisms linking social ties with physical health. Nat Neurosci 15: 669–674. doi:10.1038/nn.3086.

33. House JS, Landis KR, Umberson D (1988) Social relationships and health. Science 241: 540–545. doi:10.1126/science.3399889.

34. Ybarra O, Burnstein E, Winkielman P, Keller MC, Chan E, et al. (2008) Mental exercising through simple socializing: Social interaction promotes general cognitive functioning. Pers Soc Psychol Bull Pers Soc Psychol Bull 34: 248–259. doi:10.1177/0146167207310454.

Questions:

1. This article claims that, "Given the frequency of Facebook usage, identifying how interacting with this technology influences subjective well-being represents a basic research challenge that has important practical implications." What is the research challenge being identified? Name some of the "practical implications."

2. The article states that "Subjective well-being is one of the most highly studied variables in the behavioral sciences." Why do you think the authors might have chosen this variable to examine in their study? What reasons were given for their choice in the article? What potential drawbacks to this choice can you identify?

3. Look online for a definition of "informed consent." Why is informed consent necessary for psychological studies?

4. What graphic format would you choose to present the data recorded in Phase 2 of the experiment? Why?

5. The article suggests that Facebook "undermines" well-being. What does this mean? What criteria were used to measure well-being?

6. Why do you think the authors might have chosen to text-message study participants to gather data? What are some benefits and drawbacks of using text-messaging this way? Can you think of another way they could have gathered this data?

7. What is the difference between "affective" and "cognitive" well-being, according to the article? Do you think this experiment adequately addresses both components of subjective well-being? Why or why not?

8. What element of social media would you like to see studied? Why?

9. The article considers and rejects alternative explanations of the study data. Do you think all of these alternative explanations are adequately addressed? Why or why not?

10. Find a media article or video report that discusses this research study on Facebook use. Does the media report accurately reflect the results of the study? What are the similarities and/or differences?

11. Describe the writing style of the conclusion to the paper. What sort of diction, sentence length, syntax, voice, or tone does the conclusion use? Why do you think it does so?

VII

Human Gender: Some Cultural and Laboratory Perspectives

thirty-eight

EMILY MARTIN

The Egg and the Sperm: How Science Has Constructed a Romance Based on Stereotypical Male-Female Roles

This 1991 article by anthropologist Emily Martin examines the ways in which scientific writing about human reproduction reflects gender stereotypes. Western science, she argues, has portrayed the story of male reproductive biology as one of "remarkable feats" and that of female reproductive biology as one of "passivity and decay."

❧

The theory of the human body is always a part of a world-picture.... The theory of the human body is always a part of a fantasy.

[James Hillman, *The Myth of Analysis*][1]

As an anthropologist, I am intrigued by the possibility that culture shapes how biological scientists describe what they discover about the natural world. If this were so, we would be learning about more than the natural world in high school biology class; we would be learning about cultural beliefs and practices as if they were part of nature. In the course of my research I realized that the picture of egg and sperm drawn in popular as well as scientific accounts of reproductive biology relies on stereotypes central to our cultural definitions of male and female. The stereotypes imply not only that female biological processes are less worthy than their male counterparts but also that women are less worthy than men. Part of my goal in writing this article is to shine a bright light on the gender stereotypes hidden within the scientific language of biology. Exposed in such a light, I hope they will lose much of their power to harm us.

EGG AND SPERM: A SCIENTIFIC FAIRY TALE

At a fundamental level, all major scientific textbooks depict male and female reproductive organs as systems for the production of valuable substances, such as eggs and sperm.[2] In the case of women, the monthly cycle is described as being designed to produce eggs and prepare a suitable place for them to be fertilized and grown – all to the end of making babies. But the enthusiasm ends there. By extolling the female cycle as a productive enterprise, menstruation must necessarily be viewed as a failure. Medical texts describe menstruation as the "debris" of the uterine lining, the result of necrosis, or death of tissue. The descriptions imply that a system has gone awry, making products of no use, not to specification, unsalable, wasted, scrap. An illustration in a widely used medical text shows menstruation as a chaotic disintegration of form, complementing the many texts that describe it as "ceasing," "dying," "losing," "denuding," "expelling."[3]

Male reproductive physiology is evaluated quite differently. One of the texts that sees menstruation as failed production employs a sort of breathless prose when it describes the maturation of sperm: "The mechanisms which guide the remarkable cellular transformation from spermatid to mature sperm remain uncertain.... Perhaps the most amazing characteristic of spermato-genesis is its sheer magnitude: the normal human male may manufacture several hundred million sperm per day."[4] In the classic text *Medical Physiology*, edited by Vernon Mountcastle, the male/female, productive/destructive comparison is more explicit: "Whereas the female *sheds* only a single gamete each month, the seminiferous tubules *produce* hundreds of millions of sperm each day" (emphasis mine).[5] The female author of another text marvels at the length of the microscopic seminiferous tubules, which, if uncoiled and placed end to end, "would span almost one-third of a mile!" She writes, "In an adult male these structures produce millions of sperm cells each day." Later she asks, "How is this feat accomplished?"[6] None of these texts expresses such intense enthusiasm for any female processes. It is surely no accident that the "remarkable" process of making sperm involves precisely what, in the medical view, menstruation does not: production of something deemed valuable.[7]

One could argue that menstruation and spermatogenesis are not analogous processes and, therefore, should not be expected to elicit the same kind of response. The proper female analogy to spermatogenesis, biologically, is ovulation. Yet ovulation does not merit enthusiasm in these texts either. Textbook descriptions stress that all of the ovarian follicles containing ova are already present at birth. Far from being *produced*, as sperm are, they merely sit on the shelf, slowly degenerating and aging like overstocked inventory: "At birth, normal human ovaries contain an estimated one million follicles [each], and no new ones appear after birth. Thus, in marked contrast to the male, the newborn female already has all the germ cells she will ever have. Only a few, perhaps 400, are destined to reach full maturity during her active productive life. All the others degenerate at some point in their development so that few, if any remain by the time she reaches menopause at approximately 50 years of age."[8] Note the "marked contrast" that this description sets up between male and female, who has stockpiled germ cells by birth and is faced with their degeneration.

Nor are the female organs spared such vivid descriptions. One scientist writes in a newspaper article that a woman's ovaries become old and worn out from ripening eggs every month, even though the woman herself is still relatively young: "When you look through a laparoscope ... at an ovary that has been through hundreds of cycles, even in a superbly healthy American female, you see a scarred, battered organ."[9]

To avoid the negative connotations that some people associate with the female reproductive system, scientists could begin to describe male and female processes as homologous. They might credit females with "producing" mature ova one at a time, as they're needed each month, and describe males as having to face problems of degenerating germ cells. This degeneration would occur throughout life among spermatogonia, the undifferentiated germ cells in the testes that are the long-lived, dormant precursors of sperm.

But the texts have an almost dogged insistence on casting female processes in a negative light. The texts celebrate sperm production because it is continuous from puberty to senescence, while they portray egg production as inferior because it is finished at birth. This makes the female seem unproductive, but some texts will also insist that it is she who is wasteful.[10] In a section heading for *Molecular Biology of the Cell*, a best-selling text, we are told that "Oogenesis is wasteful." The text goes on to emphasize that of the seven million oogonia, or egg germ cells, in the female embryo, most degenerate in the ovary. Of those that do go on to become oocytes, or eggs, many also degenerate, so that at birth only two million eggs remain in the ovaries. Degeneration continues throughout a woman's life: by puberty 300,000 eggs remain, and only a few are present by menopause. "During the 40 or so years of a woman's reproductive life only 400 to 500 eggs will have been released," the authors write. "All the rest will have degenerated. It is still a mystery why so many eggs are formed only to die in the ovaries."[11]

The real mystery is why the male's vast production of sperm is not seen as wasteful.[12] Assuming that a man "produces" 100 million (10^8) sperm per day (a conservative estimate) during an average reproductive life of sixty years, he would produce well over two trillion sperm in his lifetime. Assuming that a woman "ripens" one egg per lunar month, or thirteen per year, over the course of her forty-year reproductive life, she would total five hundred eggs in her lifetime. But the word "waste" implies an excess, too much produced. Assuming two or three offspring, for every baby a woman produces, she wastes only around two hundred eggs. For every baby a man produces, he wastes more than one trillion (10^{12}) sperm.

How is it that positive images are denied to the bodies of women? A look at language – in this case, scientific language – provides the first clue. Take the egg and the sperm.[13] It is remarkable how "femininely" the egg behaves and how "masculinely" the sperm.[14] The egg is seen as large and passive.[15] It does not *move* or *journey*, but passively "is transported," "is swept,"[16] or even "drifts"[17] along the fallopian tube. In utter contrast, sperm are small, "streamlined,"[18] and invariably active. They "deliver" their genes to the egg, "activate the developmental program of the egg,"[19] and have a "velocity" that is often remarked upon.[20] Their tails are "strong" and efficiently powered.[21] Together with the forces of ejaculation, they can "propel the semen into the deepest recesses of the vagina."[22] For this they need "energy," "fuel,"[23] so that with a "whiplashlike motion and strong lurches"[24] they can "burrow through the egg coat"[25] and "penetrate" it.[26]

10 At its extreme, the age-old relationship of the egg and the sperm takes on a royal or religious patina. The egg coat, its protective barrier, is sometimes called its "vestments," a term usually reserved for sacred, religious dress. The egg is said to have a "corona,"[27] a crown, and to be accompanied by "attendant cells."[28] It is holy, set apart and above, the queen to the sperm's king. The egg is also passive, which means it must depend on sperm for rescue. Gerald Schatten and Helen Schatten liken the egg's role to that of Sleeping Beauty: "a dormant bride awaiting her mate's magic kiss, which instills the spirit that brings her to life."[29] Sperm, by contrast, have a "mission,"[30] which is to "move through the female genital tract in quest of the ovum."[31] One popular account has it that the sperm carry out a "perilous journey" into the "warm darkness," where some fall away "exhausted." "Survivors" "assault" the egg, the successful candidates "surrounding the prize."[32] Part of the urgency of this journey, in more scientific terms, is that "once released from the supportive environment of the ovary, an egg will die within hours unless rescued by a sperm."[33] The wording stresses the fragility and dependency of the egg, even though the same text acknowledges elsewhere that sperm also live for only a few hours.[34]

In 1948, in a book remarkable for its early insights into these matters, Ruth Herschberger argued that female reproductive organs are seen as biologically interdependent, while male organs are viewed as autonomous, operating independently and in isolation:

> At present the functional is stressed only in connection with women: it is in
> them that ovaries, tubes, uterus, and vagina have endless interdependence.
> In the male, reproduction would seem to involve "organs" only.

Yet the sperm, just as much as the egg, is dependent on a great many related processes. There are secretions which mitigate the urine in the urethra before ejaculation, to protect the sperm. There is the reflex shutting off of the bladder connection, the provision of prostatic secretions, and various types of muscular propulsion. The sperm is no more independent of its milieu than the egg, and yet from a wish that it were, biologists have lent their support to the notion that the human female, beginning with the egg, is congenitally more dependent than the male.[35]

Bringing out another aspect of the sperm's autonomy, an article in the journal *Cell* has the sperm making an "existential decision" to penetrate the egg: "Sperm are cells with a limited behavioral repertoire, one that is directed toward fertilizing eggs. To execute the decision to abandon the haploid state, sperm swim to an egg and there acquire the ability to effect membrane fusion."[36] Is this a corporate manager's version of the sperm's activities – "executing decisions" while fraught with dismay over difficult options that bring with them very high risk?

There is another way that sperm, despite their small size, can be made   15 to loom in importance over the egg. In a collection of scientific papers, an electron micrograph of an enormous egg and tiny sperm is titled "A Portrait of the Sperm."[37] This is a little like showing a photo of a dog and calling it a picture of the fleas. Granted, microscopic sperm are harder to photograph than eggs, which are just large enough to see with the naked eye. But surely the use of the term "portrait," a word associated with the powerful and wealthy, is significant. Eggs have only micrographs or pictures, not portraits.

One depiction of sperm as weak and timid, instead of strong and powerful – the only such representation in western civilization, so far as I know – occurs in Woody Allen's movie *Everything You Always Wanted To Know About Sex* *But Were Afraid to Ask*. Allen, playing the part of an apprehensive sperm inside a man's testicles, is scared of the man's approaching orgasm. He is reluctant to launch himself into the darkness, afraid of contraceptive devices, afraid of winding up on the ceiling if the man masturbates.

The more common picture – egg as damsel in distress, shielded only by her sacred garments; sperm as heroic warrior to the rescue – cannot be proved to be dictated by the biology of these events. While the "facts" of biology may not *always* be constructed in cultural terms, I would argue that in this case they are. The degree of metaphorical content in these descriptions, the extent to which differences between egg and sperm are emphasized, and the parallels between cultural stereotypes of male and female behavior and the character of egg and sperm all point to this conclusion.

NEW RESEARCH, OLD IMAGERY

As new understandings of egg and sperm emerge, textbook gender imagery is being revised. But the new research, far from escaping the stereotypical representations of egg and sperm, simply replicates elements of textbook gender imagery in a different form. The persistence of this imagery calls to mind what Ludwig Fleck termed "the self-contained" nature of scientific thought. As he described it, "the interaction between what is already known, what remains to be learned, and those who are to apprehend it, go to ensure harmony within the system. But at the same time they also preserve the harmony of illusions, which is quite secure within the confines of a given thought style."[38] We need to understand the way in which the cultural content in scientific descriptions changes as biological discoveries unfold, and whether that cultural content is solidly entrenched or easily changed.

In all of the texts quoted above, sperm are described as penetrating the egg, and specific substances on a sperm's head are described as binding to the egg. Recently, this description of events was rewritten in a biophysics lab at Johns Hopkins University – transforming the egg from the passive to the active party.[39]

20   Prior to this research, it was thought that the zona, the inner vestments of the egg, formed an impenetrable barrier. Sperm overcame the barrier by mechanically burrowing through, thrashing their tails and slowly working their way along. Later research showed that the sperm released digestive enzymes that chemically broke down the zona; thus, scientists presumed that the sperm used mechanical and chemical means to get through to the egg.

In this recent investigation, the researchers began to ask questions about the mechanical force of the sperm's tail. (The lab's goal was to develop a contraceptive that worked topically on sperm.) They discovered, to their great surprise, that the forward thrust of sperm is extremely weak, which contradicts the assumption that sperm are forceful penetrators.[40] Rather than thrusting forward, the sperm's head was now seen to move mostly back and forth. The sideways motion of the sperm's tail makes the head move sideways with a force that is ten times stronger than its forward movement. So even if the overall force of the sperm were strong enough to mechanically break the zona, most of its force would be directed sideways rather than forward. In fact, its strongest tendency, by tenfold, is to *escape* by attempting to pry itself off the egg. Sperm, then, must be exceptionally efficient at escaping from any cell surface they contact. And the surface of the egg must be designed to trap the sperm and prevent their escape. Otherwise, few if any sperm would reach the egg.

The researchers at Johns Hopkins concluded that the sperm and egg stick together because of adhesive molecules on the surfaces of each. The egg traps the sperm and adheres to it so tightly that the sperm's head is forced to lie flat against the surface of the zona, a little bit, they told me, "like Br'er Rabbit getting more and more stuck to tar baby the more he wriggles." The trapped sperm continues to wiggle ineffectually side to side. The mechanical force of its tail is so weak that a sperm cannot break even one chemical bond. This is where the digestive enzymes released by the sperm come in. If they start to soften the zona just at the tip of the sperm and the sides remain stuck, then the weak, flailing sperm can get oriented in the right direction and make it through the zona – provided that its bonds to the zona dissolve as it moves in.

Although this new version of the saga of the egg and the sperm broke through cultural expectations, the researchers who made the discovery continued to write papers and abstracts as if the sperm were the active party who attacks, binds, penetrates, and enters the egg. The only difference was that the sperm were now seen as performing these actions weakly.[41] Not until August 1987, more than three years after the findings described above, did these researchers reconceptualize the process to give the egg a more active role. They began to describe the zona as an aggressive sperm catcher, covered with adhesive molecules that can capture a sperm with a single bond and clasp it to the zona's surface.[42] In the words of their published account: "The innermost vestment, the *zona pellucida*, is a glycoprotein shell, which captures and tethers the sperm before they penetrate it.... The sperm is captured at the initial contact between the sperm tip and the *zona*.... Since the thrust [of the sperm] is much smaller than the force needed to break a single affinity bond, the first bond made upon the tip-first meeting of the sperm and *zona* can result in the capture of the sperm."[43]

Experiments in another lab reveal similar patterns of data interpretation. Gerald Schatten and Helen Schatten set out to show that, contrary to conventional wisdom, the "egg is not merely a large, yolk-filled sphere into which the sperm burrows to endow new life. Rather, recent research suggests the almost heretical view that sperm and egg are mutually active partners."[44] This sounds like a departure from the stereotypical textbook view, but further reading reveals Schatten and Schatten's conformity to the aggressive-sperm metaphor. They describe how "the sperm and egg first touch when, from the tip of the sperm's triangular head, a long, thin filament shoots out and harpoons the egg." Then we learn that "remarkably, the harpoon is not so much fired as assembled at great speed, molecule by molecule, from a pool of protein stored in a specialized region called the acrosome. The filament may grow as much as twenty times longer than the sperm head itself before its tip reaches the egg and sticks."[45] Why not call this "making a bridge" or "throwing out a line" rather than firing a harpoon? Harpoons pierce prey and injure or kill them, while this filament only sticks. And why not focus, as the Hopkins lab did, on the stickiness of the egg, rather than the stickiness of the sperm?[46] Later in the article, the Schattens replicate the common view of the sperm's perilous journey into the warm darkness of the vagina, this time for the purpose of explaining its journey into the egg itself: "[The sperm] still has an arduous journey ahead. It must penetrate farther into the egg's huge sphere of cytoplasm and somehow locate the nucleus, so that the two cells' chromosomes can fuse. The sperm dives down into the cytoplasm, its tail beating. But it is soon interrupted by the sudden and swift migration of the egg nucleus, which rushes toward the sperm with a velocity triple that of the movement of chromosomes during cell division, crossing the entire egg in about a minute."[47]

25   Like Schatten and Schatten and the biophysicists at Johns Hopkins, another researcher has recently made discoveries that seem to point to a more interactive view of the relationship of egg and sperm. This work, which Paul Wassarman conducted on the sperm and eggs of mice, focuses on identifying the specific molecules in the egg coat (the zona pellucida) that are involved in egg-sperm interaction. At first glance, his descriptions seem to fit the model of an egalitarian relationship. Male and female gametes "recognize one another," and "interactions ... take place between sperm and egg."[48] But the article in *Scientific American* in which those descriptions appear begins with a vignette that presages the dominant motif of their presentation: "It has been more than a century since Hermann Fol, a Swiss zoologist, peered into his microscope and became the first person to see a sperm penetrate an egg, fertilize it and form the first cell of a new embryo."[49] This portrayal of the sperm as the active party – the one that *penetrates* and *fertilizes* the egg and *produces* the embryo – is not cited as an example of an earlier, now out-moded view. In fact, the author reiterates the point later in the article: "Many sperm can bind to and penetrate the zona pellucida, or outer coat, of an unfertilized mouse egg, but only one sperm will eventually fuse with the thin plasma membrane surrounding the egg proper (*inner sphere*), *fertilizing the egg and giving rise to a new embryo.*"[50]

The imagery of sperm as aggressor is particularly startling in this case: the main discovery being reported is isolation of a particular molecule *on the egg coat* that plays an important role in fertilization! Wassarman's choice of language sustains the picture. He calls the molecule that has been isolated, ZP3, a "sperm receptor." By allocating the passive, waiting role to the egg, Wassarman can continue to describe the sperm as the actor, the one that makes it all happen: "The basic process begins when many sperm first attach loosely and then bind tenaciously to receptors on the surface of the egg's thick outer coat, the zona pellucida. Each sperm, which has a large number of egg-binding proteins on its surface, binds to many sperm receptors on the egg. More specifically, a site on each of the egg-binding proteins fits a complementary site on a sperm receptor, much as a key fits a lock."[51] With the sperm designated as the "key" and the egg the "lock," it is obvious which one acts and which one is acted upon. Could this imagery not be reversed, letting the sperm (the lock) wait until the egg produces the key? Or could we speak of two halves of a locket matching, and regard the matching itself as the action that initiates the fertilization?

It is as if Wassarman were determined to make the egg the receiving partner. Usually in biological research, the protein member of the pair of binding molecules is called the receptor, and physically it has a pocket in it rather like a lock. As the diagrams that illustrate Wassarman's article show, the molecules on the sperm are proteins and have "pockets." The small, mobile molecules that fit into these pockets are called ligands. As shown in the diagrams, ZP3 on the egg is a polymer of "keys"; many small knobs stick out. Typically, molecules in the sperm would be called receptors and molecules on the egg would be called ligands. But Wassarman chose to name ZP3 on the egg the receptor and to create a new term, "the egg-binding protein," for the molecule on the sperm that otherwise would have been called the receptor.[52]

Wassarman does credit the egg coat with having more functions than those of a sperm receptor. While he notes that "the zona pellucida has at times been viewed by investigators as a nuisance, a barrier to sperm and hence an impediment to fertilization," his new research reveals that the egg coat "serves as a sophisticated biological security system that screens incoming sperm, selects only those compatible with fertilization and development, prepares sperm for fusion with the egg and later protects the resulting embryo from polyspermy [a lethal condition caused by fusion of more than one sperm with a single egg]."[53] Although this description gives the egg an active role, that role is drawn in stereotypically feminine terms. The egg *selects* an appropriate mate, *prepares* him for fusion, and then *protects* the resulting offspring from harm. This is courtship and mating behavior as seen through the eyes of a sociobiologist: woman as the hard-to-get prize, who, following union with the chosen one, becomes woman as servant and mother.

And Wassarman does not quit there. In a review article for *Science*, he outlines the "chronology of fertilization."[54] Near the end of the article are two subject headings. One is "Sperm Penetration," in which Wassarman describes how the chemical dissolving of the zona pellucida combines with the "substantial propulsive force generated by sperm." The next heading is "Sperm-Egg Fusion." This section details what happens inside the zona after a sperm "penetrates" it. Sperm "can make contact with, adhere to, and fuse with (that is, fertilize) an egg."[55] Wassarman's word choice, again, is astonishingly skewed in favor of the sperm's activity, for in the next breath he says that sperm *lose* all motility upon fusion with the egg's surface. In mouse and sea urchin eggs, the sperm enters at the *egg's* volition, according to Wassarman's description: "Once fused with egg plasma membrane [the surface of the egg], how does a sperm enter the egg? The surface of both mouse and sea urchin eggs is covered with thousands of plasma membrane-bound projections, called microvilli [tiny 'hairs']. Evidence in sea urchins suggests that, after membrane fusion, a group of elongated microvilli cluster tightly around and interdigitate over the sperm head. As these microvilli are resorbed, the sperm is drawn into the egg. Therefore, sperm motility, which ceases at the time of fusion in both sea urchins and mice, is not required for sperm entry."[56] The section called "Sperm Penetration" more logically would be followed by a section called "The Egg Envelopes," rather than "Sperm-Egg Fusion." This would give a parallel – and more accurate – sense that both the egg and the sperm initiate action.

30   Another way that Wassarman makes less of the egg's activity is by describing components of the egg but referring to the sperm as a whole entity. Deborah Gordon has described such an approach as "atomism" ("the part is independent of and primordial to the whole") and identified it as one of the "tenacious assumptions" of Western science and medicine.[57] Wassarman employs atomism to his advantage. When he refers to processes going on within sperm, he consistently returns to descriptions that remind us from whence these activities came: they are part of sperm that penetrate an egg or generate propulsive force. When he refers to processes going on within eggs, he stops there. As a result, any active role he grants them appears to be assigned to the parts of the egg, and not to the egg itself. In the quote above, it is the microvilli that actively cluster around the sperm. In another example, "the driving force for engulfment of a fused sperm comes from a region of cytoplasm just beneath an egg's plasma membrane."[58]

SOCIAL IMPLICATIONS

All three of these revisionist accounts of egg and sperm cannot seem to escape the hierarchical imagery of older accounts. Even though each new account gives the egg a larger and more active role, taken together they bring into play another cultural stereotype: woman as a dangerous and aggressive threat. In the Johns Hopkins lab's revised model, the egg ends up as the female aggressor who "captures and tethers" the sperm with her sticky zona, rather like a spider lying in wait in her web.[59] The Schatten lab has the egg's nucleus "interrupt" the sperm's dive with a "sudden and swift" rush by which she "clasps the sperm and guides its nucleus to the center."[60] Wassarman's description of the surface of the egg "covered with thousands of plasma membrane-bound projections, called microvilli" that reach out and clasp the sperm adds to the spiderlike imagery.[61]

These images grant the egg an active role but at the cost of appearing disturbingly aggressive. Images of woman as dangerous and aggressive, the femme fatale who victimizes men, are widespread in Western literature and culture.[62] More specific is the connection of spider imagery with the idea of an engulfing, devouring mother.[63] New data did not lead scientists to eliminate gender stereotypes in their descriptions of egg and sperm. Instead, scientists simply began to describe egg and sperm in different, but no less damaging, terms.

Can we envision a less stereotypical view? Biology itself provides another model that could be applied to the egg and the sperm. The cybernetic model – with its feedback loops, flexible adaptation to change, coordination of the parts within a whole, evolution over time, and changing response to the environment – is common in genetics, endocrinology, and ecology and has a growing influence in medicine in general.[64] This model has the potential to shift our imagery from the negative, in which the female reproductive system is castigated both for not producing eggs after birth and for producing (and thus wasting) too many eggs overall, to something more positive. The female reproductive system could be seen as responding to the environment (pregnancy or menopause), adjusting to monthly changes (menstruation), and flexibly changing from reproductivity after puberty to nonreproductivity later in life. The sperm and egg's interaction could also be described in cybernetic terms. J.F. Hartman's research in reproductive biology demonstrated fifteen years ago that if an egg is killed by being pricked with a needle, live sperm cannot get through the zona.[65] Clearly, this evidence shows that the egg and sperm do interact on more mutual terms, making biology's refusal to portray them that way all the more disturbing.

We would do well to be aware, however, that cybernetic imagery is hardly neutral. In the past, cybernetic models have played an important part in the imposition of social control. These models inherently provide a way of thinking about a "field" of interacting components. Once the field can be seen, it can become the object of new forms of knowledge, which in turn can allow new forms of social control to be exerted over the components of the field. During the 1950s, for example, medicine began to recognize the psychosocial *environment* of the patient: the patient's family and its psychodynamics. Professions such as social work began to focus on this new environment, and the resulting knowledge became one way to further control the patient. Patients began to be seen not as isolated, individual bodies, but as psychosocial entities located in an "ecological" system: management of "the patient's psychology was a new entrée to patient control."[66]

35   The models that biologists use to describe their data can have important social effects. During the nineteenth century, the social and natural sciences strongly influenced each other: the social ideas of Malthus about how to avoid the natural increase of the poor inspired Darwin's *Origin of Species*.[67] Once the *Origin* stood as a description of the natural world, complete with competition and market struggles, it could be reimported into social science as social Darwinism, in order to justify the social order of the time. What we are seeing now is similar: the importation of cultural ideas about passive females and heroic males into the "personalities" of gametes. This amounts to the "implanting of social imagery on representations of nature so as to lay a firm basis for reimporting exactly that same imagery as natural explanations of social phenomena."[68]

Further research would show us exactly what social effects are being wrought from the biological imagery of egg and sperm. At the very least, the imagery keeps alive some of the hoariest old stereotypes about weak damsels in distress and their strong male rescuers. That these stereotypes are now being written in at the level of the cell constitutes a powerful move to make them seem so natural as to be beyond alteration.

The stereotypical imagery might also encourage people to imagine that what results from the interaction of egg and sperm – a fertilized egg – is the result of deliberate "human" action at the cellular level. Whatever the intentions of the human couple, in this microscopic "culture" a cellular "bride" (or femme fatale) and a cellular "groom" (her victim) make a cellular baby. Rosalind Petchesky points out that through visual representations such as sonograms, we are given "*images* of younger and younger, and tinier and tinier, fetuses being 'saved.'" This leads to "the point of visibility being 'pushed back' *indefinitely*."[69] Endowing egg and sperm with intentional action, a key aspect of personhood in our culture, lays the foundation for the point of viability being pushed back to the moment of fertilization. This will likely lead to greater acceptance of technological developments and new forms of scrutiny and manipulation, for the benefit of these inner "persons": court-ordered restrictions on a pregnant woman's activities in order to protect her fetus, fetal surgery, amniocentesis, and rescinding of abortion rights, to name but a few examples.[70]

Even if we succeed in substituting more egalitarian, interactive metaphors to describe the activities of egg and sperm, and manage to avoid the pitfalls of cybernetic models, we would still be guilty of endowing cellular entities with personhood. More crucial, then, than what *kinds* of personalities we bestow on cells is the very fact that we are doing it at all. This process could ultimately have the most disturbing social consequences.

One clear feminist challenge is to wake up sleeping metaphors in science, particularly those involved in descriptions of the egg and the sperm. Although the literary convention is to call such metaphors "dead," they are not so much dead as sleeping, hidden within the scientific content of texts – and all the more powerful for it.[71] Waking up such metaphors, by becoming aware of when we are projecting cultural imagery onto what we study, will improve our ability to investigate and understand nature. Waking up such metaphors, by becoming aware of their implications, will rob them of their power to naturalize our social conventions about gender.

(1991)

ENDNOTES

1. James Hillman, *The Myth of Analysis* (Evanston, Ill.: Northwestern University Press, 1972), 220.

2. The textbooks I consulted are the main ones used in classes for undergraduate premedical students or medical students (or those held on reserve in the library for these classes) during the past few years at Johns Hopkins University. These texts are widely used at other universities in the country as well.

3. Arthur C. Guyton, *Physiology of the Human Body*, 6th ed. (Philadelphia: Saunders College Publishing, 1984), 624.

4. Arthur J. Vander, James H. Sherman, and Dorothy S. Luciano, *Human Physiology: The Mechanisms of Body Function*, 3rd ed. (New York: McGraw Hill, 1980), 483-84.

5. Vernon B. Mountcastle, *Medical Physiology*, 14th ed. (London: Mosby, 1980), 2:1624.

6. Eldra Pearl Solomon, *Human Anatomy and Physiology* (New York: CBS College Publishing, 1983), 678.

7. For elaboration, see Emily Martin, *The Woman in the Body: A Cultural Analysis of Reproduction* (Boston: Beacon, 1987), 27-53.

8. Vander, Sherman, and Luciano, 568.

9. Melvin Konner, "Childbearing and Age," *New York Times Magazine* (December 27, 1987), 22-23, esp. 22.

10. I have found but one exception to the opinion that the female is wasteful: "Smallpox being the nasty disease it is, one might expect nature to have designed antibody molecules with combining sites that specifically recognize the epitopes on smallpox virus. Nature differs from technology, however: it thinks nothing of wastefulness. (For example, rather than improving the chance that a spermatozoon will meet an egg cell, nature finds it easier to produce millions of spermatozoa.)" (Niels Kaj Jerne, "The Immune System," *Scientific American* 229, no. 1 [July 1973]: 53). Thanks to a *Signs* reviewer for bringing this reference to my attention.

11. Bruce Alberts et al., *Molecular Biology of the Cell* (New York: Garland, 1983), 795.

12. In her essay "Have Only Men Evolved?" (in *Discovering Reality: Feminist Perspectives on Epistemology, Metaphysics, Methodology, and Philosophy of Science*, ed. Sandra Harding and Merrill B. Hintikka [Dordrecht: Reidel, 1983], 45-69, esp. 60-61), Ruth Hubbard points out that sociobiologists have said the female invests more energy than the male in the production of her large gametes, claiming that this explains why the female provides parental care. Hubbard questions whether it "really takes more 'energy' to generate the one or relatively few eggs than the large excess of sperms required to achieve fertilization." For further critique of how the greater size of eggs is interpreted in sociobiology, see Donna Haraway, "Investment Strategies for the Evolving Portfolio of Primate Females," in *Body/Politics*, ed. Mary Jacobus, Evelyn Fox Keller, and Sally Shuttleworth (New York: Routledge, 1990), 155-56.

13. The sources I used for this article provide compelling information on interactions among sperm. Lack of space prevents me from taking up this theme here, but the elements include competition, hierarchy, and sacrifice. For a newspaper report, see Malcolm W. Browne, "Some Thoughts on Self Sacrifice," *New York Times* (July 5, 1988), C6. For a literary rendition, see John Barth, "Night-Sea Journey," in his *Lost in the Funhouse* (Garden City, NY: Doubleday, 1968), 3-13.

14. See Carol Delaney, "The Meaning of Paternity and the Virgin Birth Debate," *Man* 21, no. 3 (September 1986): 494-513. She discusses the difference between this scientific view that women contribute genetic material to the fetus and the claim of long-standing Western folk theories that the origin and identity of the fetus comes from the male, as in the metaphor of planting a seed in soil.

15. For a suggested direct link between human behavior and purportedly passive eggs and active sperm, see Erik H. Erikson, "Inner and Outer Space: Reflections on Womanhood," *Daedalus* 93, no. 2 (Spring 1964): 582-606, esp. 591.

16. Guyton (n. 3 above), 619; and Mountcastle (n. 5 above), 1609.

17. Jonathan Miller and David Pelham, *The Facts of Life* (New York: Viking Penguin, 1984), 5.

18. Alberts et al., 796.

19. Ibid., 796.

20. See, e.g., William F. Ganong, *Review of Medical Physiology*, 7th ed. (Los Altos, Calif.: Lange Medical Publications, 1975), 322.

21. Alberts et al. (n. 11 above), 796.

22. Guyton, 615.

23. Solomon (n. 6 above), 683.

24. Vander, Sherman, and Luciano (n. 4 above), 4th ed. (1985), 580.

25. Alberts et al., 796.

26. All biology texts quoted above use the word "penetrate."

27. Solomon, 700.

28. A. Beldecos et al., "The Importance of Feminist Critique for Contemporary Cell Biology," *Hypatia* 3, no. 1 (Spring 1988): 61-76.

29. Gerald Schatten and Helen Schatten, "The Energetic Egg," *Medical World News* 23 (January 23, 1984): 51-53, esp. 51.

30. Alberts et al., 796.

31. Guyton (n. 3 above), 613.

32. Miller and Pelham (n. 17 above), 7.

33. Alberts et al. (n. 11 above), 804.

34. Ibid., 801.

35. Ruth Herschberger, *Adam's Rib* (New York: Pelligrini & Cudaby, 1948), esp. 84. I am indebted to Ruth Hubbard for telling me about Herschberger's work, although at a point when this paper was already in draft form.

36. Bennett M. Shapiro. "The Existential Decision of a Sperm," *Cell* 49, no. 3 (May 1987): 293-94, esp. 293.

37. Lennart Nilsson, "A Portrait of the Sperm," in *The Functional Anatomy of the Spermatozoan*, ed. Bjorn A. Afzelius (New York: Pergamon, 1975), 79-82.

38. Ludwig Fleck, *Genesis and Development of a Scientific Fact*, ed. Thaddeus J. Trenn and Robert K. Merton (Chicago: University of Chicago Press, 1979), 38.

39. Jay M. Baltz carried out the research I describe when he was a graduate student in the Thomas C. Jenkins Department of Biophysics at Johns Hopkins University.

40. Far less is known about the physiology of sperm than comparable female substances, which some feminists claim is no accident. Greater scientific scrutiny of female reproduction has long enabled the burden of birth control to be placed on women. In this case, the researchers' discovery did not depend on development of any new technology: The experiments made use of glass pipettes, a manometer, and a simple microscope, all of which have been available for more than one hundred years.

41. Jay Baltz and Richard A. Cone, "What Force Is Needed to Tether a Sperm?" (abstract for Society for the Study of Reproduction, 1985), and "Flagellar Torque on the Head Determines the Force Needed to Tether a Sperm" (abstract for Biophysical Society, 1986).

42. Jay M. Baltz, David F. Katz, and Richard A. Cone, "The Mechanics of the Sperm-Egg Interaction at the Zona Pellucida," *Biophysical Journal* 54, no. 4 (October 1988): 643-54. Lab members were somewhat familiar with work on metaphors in the biology of female reproduction. Richard Cone, who runs the lab, is my husband, and he talked with them about my earlier research on the subject from time to time. Even though my current research focuses on biological imagery and I heard about the lab's work from my husband every day, I myself did not recognize the role of imagery in the sperm research until many weeks after the period of research and writing I describe. Therefore, I assume that any awareness the lab members may have had about how underlying metaphor might be guiding this particular research was fairly inchoate.

43. Ibid., 643, 650.

44. Schatten and Schatten (n. 29 above), 51.

45. Ibid., 52.

46. Surprisingly, in an article intended for a general audience, the authors do not point out that these are sea urchin sperm and note that human sperm do not shoot out filaments at all.

47. Schatten and Schatten, 53.

48. Paul M. Wassarman, "Fertilization in Mammals," *Scientific American* 259, no. 6 (December 1988): 78-84, esp. 78, 84.

49. Ibid., 78.

50. Ibid., 79.

51. Ibid., 78.

52. Since receptor molecules are relatively *immotile* and the ligands that bind to them relatively *motile*, one might imagine the egg being called the receptor and the sperm the ligand. But the molecules in question on egg and sperm are immotile molecules. It is the sperm as a *cell* that has motility, and the egg as a cell that has relative immotility.

53. Wassarman, 78-79.

54. Paul M. Wassarman, "The Biology and Chemistry of Fertilization," *Science* 235, no. 4788 (January 30, 1987): 553-60, esp. 554.

55. Ibid., 557.

56. Ibid., 557-58. This finding throws into question Schatten and Schatten's description (n. 29 above) of the sperm, its tail beating, diving down into the egg.

57. Deborah R. Gordon, "Tenacious Assumptions in Western Medicine," in *Bio-medicine Examined*, ed. Margaret Lock and Deborah Gordon (Dordrecht: Kluwer, 1988), 19-56, esp. 26.

58. Wassarman, "The Biology and Chemistry of Fertilization," 558.

59. Baltz, Katz, and Cone (n. 42 above), 643, 650.

60. Schatten and Schatten, 53.

61. Wassarman, "The Biology and Chemistry of Fertilization," 557.

62. Mary Ellman, *Thinking about Women* (New York: Harcourt Brace Jovanovich, 1968), 140; Nina Auerbach, *Woman and the Demon* (Cambridge, Mass.: Harvard University Press, 1982), esp. 186.

63. Kenneth Alan Adams, "Arachnophobia: Love American Style," *Journal of Psychoanalytic Anthropology* 4, no. 2 (1981): 157-97.

64. William Ray Arney and Bernard Bergen, *Medicine and the Management of Living* (Chicago: University of Chicago Press, 1984).

65. J.F. Hartman, R.B. Gwatkin, and C.F. Hutchison, "Early Contact Interactions between Mammalian Gametes In Vitro," *Proceedings of the National Academy of Sciences* (US) 69, no. 10 (1972): 2767-69.

66. Arney and Bergen, 68.

67. Ruth Hubbard, "Have Only Men Evolved?" (n. 12 above), 51-52.

68. David Harvey, personal communication, November 1989.

69. Rosalind Petchesky, "Fetal Images: The Power of Visual Culture in the Politics of Reproduction," *Feminist Studies* 13, no. 2 (Summer 1987): 263-92, esp. 272.

70. Rita Arditti, Renate Klein, and Shelley Minden, *Test-Tube Women* (London: Pandora, 1984); Ellen Goodman, "Whose Right to Life?" *Baltimore Sun* (November 17, 1987); Tamar Lewin, "Courts Acting to Force Care of the Unborn," *New York Times* (November 23, 1987), A1 and B10; Susan Irwin and Brigitte Jordan, "Knowledge, Practice, and Power: Court Ordered Cesarean Sections," *Medical Anthropology Quarterly* 1, no. 3 (September 1987): 319-34.

71. Thanks to Elizabeth Fee and David Spain, who in February 1989 and April 1989, respectively, made points related to this.

Questions:

1. Why, according to the article, is it harmful when human reproduction is described in language that reflects gender stereotypes? What practical problems arise from this metaphoric conceptualization?

2. Summarize the discoveries of the Johns Hopkins researchers (not named in the body of the article, but identified in the references as J.M. Baltz, David F. Catz, and Richard A. Cone) who carried out the work Martin discusses in paragraphs 19-23.

3. Discuss the organization of Martin's essay. In particular, outline the overall progression in the ideas she presents.

4. Are there any instances where the use of metaphors Martin presents as biased might be justified by the biological facts? If so, why are they justified?

5. The article begins with a quotation from James Hillman's *The Myth of Analysis*. What is the significance of the quotation in the context of the article?

6. Consider the kind of references that Martin cites. Roughly, what percentage are biology science sources? University textbooks? Works of feminist analysis and/or anthropology? Compare this set of citations to one from one of the PLOS One articles in this reader. Categorize the differences. Do they influence your interpretations of either article?

7. What is "'atomism'" (paragraph 30)? What does the article suggest is wrong with an atomistic approach?

8. In paragraphs 37-38, Martin suggests that the metaphors used to describe human reproduction might encourage people to unconsciously imagine the egg and sperm as having "personhood." Discuss the implications of this.

9. What effect might the use of the active or passive voice have in scientific research that describes gendered processes or behaviors?

10. What does Martin's use of her own experience add to the article?

11. Martin writes that "as new understandings of egg and sperm emerge, textbook gender imagery is being revised." Martin's article was published in 1991; find a more recent article or textbook chapter on the subject of human reproduction. In what ways, if any, does the imagery differ from that discussed by Martin?

12. Martin says that "the 'facts' of biology may not *always* be constructed in cultural terms." Choose another article in this book that addresses biology (e.g., the articles by Wakefield et al. or Knowles et al.). Does the article use culturally biased metaphors? If so, describe the biases.

13. Research the ZP3 protein (discussed in paragraphs 26-27 of this article) and write a paragraph describing its function; in doing so, try to avoid using gender-stereotyped metaphors. How successfully are you able to avoid gender stereotypes? What parts of your description are particularly effective or ineffective in this regard?

thirty-nine

CORRINE A. MOSS-RACUSIN, JOHN F. DOVIDIO,
VICTORIA L. BRESCOLL, MARK J. GRAHAM,
AND JO HANDELSMAN

from Science Faculty's Subtle Gender Biases Favor Male Students

*This 2012 study examining gender inequalities and bias in academic
science found that both male and female science faculty members
exhibited a gender bias against women when considering
applications for a managerial position in a laboratory.*

❧

ABSTRACT

Despite efforts to recruit and retain more women, a stark gender disparity persists within academic science. Abundant research has demonstrated
gender bias in many demographic groups, but has yet to experimentally
investigate whether science faculty exhibit a bias against female students
that could contribute to the gender disparity in academic science. In a
randomized double-blind study[a] (n = 127), science faculty from research-
intensive universities rated the application materials of a student – who was
randomly assigned either a male or female name – for a laboratory manager position. Faculty participants rated the male applicant as significantly
more competent and hireable than the (identical) female applicant. These
participants also selected a higher starting salary and offered more career
mentoring to the male applicant. The gender of the faculty participants did
not affect responses, such that female and male faculty were equally likely
to exhibit bias against the female student. Mediation analyses indicated that
the female student was less likely to be hired because she was viewed as
less competent. We also assessed faculty participants' preexisting subtle

---

a *randomized ... study* Study in which participants are randomly assigned to participate in the
competing variables in the study, and neither researchers nor participants are aware of any
information about the study that could influence its outcome.

bias against women using a standard instrument and found that preexisting subtle bias against women played a moderating role, such that subtle bias against women was associated with less support for the female student, but was unrelated to reactions to the male student. These results suggest that interventions addressing faculty gender bias might advance the goal of increasing the participation of women in science.

A 2012 report from the President's Council of Advisors on Science and Technology[a] indicates that training scientists and engineers at current rates will result in a deficit of 1,000,000 workers to meet United States workforce demands over the next decade.[1] To help close this formidable gap, the report calls for the increased training and retention of women, who are starkly underrepresented within many fields of science, especially among the professoriate.[b] [2-4] Although the proportion of science degrees granted to women has increased,[5] there is a persistent disparity between the number of women receiving PhDs and those hired as junior faculty.[1-4] This gap suggests that the problem will not resolve itself solely by more generations of women moving through the academic pipeline but that instead, women's advancement within academic science may be actively impeded.

With evidence suggesting that biological sex differences in inherent aptitude for math and science are small or nonexistent,[6-8] the efforts of many researchers and academic leaders to identify causes of the science gender disparity have focused instead on the life choices that may compete with women's pursuit of the most demanding positions. Some research suggests that these lifestyle choices (whether free or constrained) likely contribute to the gender imbalance,[9-11] but because the majority of these studies are correlational,[c] whether lifestyle factors are solely or primarily responsible remains unclear. Still, some researchers have argued that women's preference for nonscience disciplines and their tendency to take on a disproportionate amount of child- and family-care are the primary causes of the gender disparity in science,[9-11] and that it "is not caused by discrimination in these domains."[10] This assertion has received substantial attention and generated significant debate among the scientific community, leading some to conclude that gender discrimination indeed does not exist nor contribute to the gender disparity within academic science (e.g., refs. 12 and 13).

---

a *President's ... Technology*   American council providing advice and education to the President on matters concerning science and technology.

b *among the professoriate*   Among university professors.

c *correlational*   Identifying a connection but not determining the nature of the connection (including, for example, whether the connection is causal).

Despite this controversy, experimental research testing for the presence and magnitude of gender discrimination in the biological and physical sciences has yet to be conducted. Although acknowledging that various lifestyle choices likely contribute to the gender imbalance in science,[9-11] the present research is unique in investigating whether faculty gender bias exists within academic biological and physical sciences, and whether it might exert an independent effect on the gender disparity as students progress through the pipeline to careers in science. Specifically, the present experiment examined whether, given an equally qualified male and female student, science faculty members would show preferential evaluation and treatment of the male student to work in their laboratory. Although the correlational and related laboratory studies discussed below suggest that such bias is likely (contrary to previous arguments),[9-11] we know of no previous experiments that have tested for faculty bias against female students within academic science.

If faculty express gender biases, we are not suggesting that these biases are intentional or stem from a conscious desire to impede the progress of women in science. Past studies indicate that people's behavior is shaped by implicit or unintended biases, stemming from repeated exposure to pervasive cultural stereotypes[14] that portray women as less competent but simultaneously emphasize their warmth and likeability compared with men.[15] Despite significant decreases in overt sexism over the last few decades (particularly among highly educated people),[16] these subtle gender biases are often still held by even the most egalitarian individuals,[17] and are exhibited by both men and women.[18] Given this body of work, we expected that female faculty would be just as likely as male faculty to express an unintended bias against female undergraduate science students. The fact that these prevalent biases often remain undetected highlights the need for an experimental investigation to determine whether they may be present within academic science and, if so, raise awareness of their potential impact.

5   Whether these gender biases operate in academic sciences remains an open question. On the one hand, although considerable research demonstrates gender bias in a variety of other domains,[19-23] science faculty members may not exhibit this bias because they have been rigorously trained to be objective. On the other hand, research demonstrates that people who value their objectivity and fairness are paradoxically particularly likely to fall prey to biases, in part because they are not on guard against subtle bias.[24,25] Thus, by investigating whether science faculty exhibit a bias that could contribute to the gender disparity within the fields of science, technology, engineering, and mathematics (in which objectivity is emphasized), the current study addressed

critical theoretical and practical gaps in that it provided an experimental test of faculty discrimination against female students within academic science.

A number of lines of research suggest that such discrimination is likely. Science is robustly male gender-typed,[26,27] resources are inequitably distributed among men and women in many academic science settings,[28] some undergraduate women perceive unequal treatment of the genders within science fields,[29] and nonexperimental evidence suggests that gender bias is present in other fields.[19] Some experimental evidence suggests that even though evaluators report liking women more than men,[15] they judge women as less competent than men even when they have identical backgrounds.[20] However, these studies used undergraduate students as participants (rather than experienced faculty members), and focused on performance domains outside of academic science, such as completing perceptual tasks,[21] writing nonscience articles,[22] and being evaluated for a corporate managerial position.[23]

Thus, whether aspiring women scientists encounter discrimination from faculty members remains unknown. The formative predoctoral years[a] are a critical window, because students' experiences at this juncture shape both their beliefs about their own abilities and subsequent persistence in science.[30,31] Therefore, we selected this career stage as the focus of the present study because it represents an opportunity to address issues that manifest immediately and also resurface much later, potentially contributing to the persistent faculty gender disparity.[32,33]

CURRENT STUDY

In addition to determining whether faculty expressed a bias against female students, we also sought to identify the processes contributing to this bias. To do so, we investigated whether faculty members' perceptions of student competence would help to explain why they would be less likely to hire a female (relative to an identical male) student for a laboratory manager position. Additionally, we examined the role of faculty members' preexisting subtle bias against women. We reasoned that pervasive cultural messages regarding women's lack of competence in science could lead faculty members to hold gender-biased attitudes that might subtly affect their support for female (but not male) science students. These generalized, subtly biased attitudes toward women could impel faculty to judge equivalent students differently as a function of their gender.

The present study sought to test for differences in faculty perceptions and treatment of equally qualified men and women pursuing careers in science

---

a *predoctoral years*   Years of study a student completes before entering a PhD program.

and, if such a bias were discovered, reveal its mechanisms and consequences within academic science. We focused on hiring for a laboratory manager position as the primary dependent variable of interest because it functions as a professional launching pad for subsequent opportunities. As secondary measures, which are related to hiring, we assessed: (i) perceived student competence; (ii) salary offers, which reflect the extent to which a student is valued for these competitive positions; and (iii) the extent to which the student was viewed as deserving of faculty mentoring.

10   Our hypotheses were that: Science faculty's perceptions and treatment of students would reveal a gender bias favoring male students in perceptions of competence and hireability, salary conferral, and willingness to mentor (hypothesis A); Faculty gender would not influence this gender bias (hypothesis B); Hiring discrimination against the female student would be mediated (i.e., explained) by faculty perceptions that a female student is less competent than an identical male student (hypothesis C); and participants' preexisting subtle bias against women would moderate (i.e., impact) results, such that subtle bias against women would be negatively related to evaluations of the female student, but unrelated to evaluations of the male student (hypothesis D).

Results

A broad, nationwide sample of biology, chemistry, and physics professors ($n = 127$) evaluated the application materials of an undergraduate science student who had ostensibly applied for a science laboratory manager position. All participants received the same materials, which were randomly assigned either the name of a male ($n = 63$) or a female ($n = 64$) student; student gender was thus the only variable that differed between conditions. Using previously validated scales, participants rated the student's competence and hireability, as well as the amount of salary and amount of mentoring they would offer the student. Faculty participants believed that their feedback would be shared with the student they had rated (see *Materials and Methods* for details).

Student Gender Differences

The competence, hireability, salary conferral, and mentoring scales were each submitted to a two (student gender; male, female) × two (faculty gender; male, female) between-subjects ANOVA.[a] In each case, the effect of student gender was significant (all $P < 0.01$), whereas the effect of faculty participant gender and their interaction was not (all $P > 0.19$). Tests of simple effects (all

---

a *ANOVA* Analysis of variance, a method of determining whether there are significant differences between groups of data.

$d > 0.60$) indicated that faculty participants viewed the female student as less competent [t(125) = 3.89, $P < 0.001$] and less hireable [t(125) = 4.22, $P < 0.001$] than the identical male student ([...]Table 1). Faculty participants also offered less career mentoring to the female student than to the male student [t(125) = 3.77, $P < 0.001$]. The mean starting salary offered the female student, $26,507.94, was significantly lower than that of $30,238.10 to the male student [t(124) = 3.42, $P < 0.01$] (Fig. 2). These results support hypothesis A.

TABLE 1.
Means for student competence, hireability, mentoring and salary conferral by student gender condition and faculty gender.

Variable	Compe-tence	Hire-ability	Mentor-ing	Salary
Male target student				
Male faculty Mean	4.01$_a$	3.74$_a$	4.74$_a$	30,520.83$_a$
SD	(0.92)	(1.24)	(1.11)	(5,764.86)
Female faculty Mean	4.1$_a$	3.92$_a$	4.73$_a$	29,333.33$_a$
SD	(1.19)	(1.27)	(1.31)	(4,952.15)
Female target student				
Male faculty Mean	3.33$_b$	2.96$_b$	4.00$_b$	27,111.11$_b$
SD	(1.07)	(1.13)	(1.21)	(6,948.58)
Female faculty Mean	3.32$_b$	2.84$_b$	3.91$_b$	25,000.00$_b$
SD	(1.10)	(0.84)	(0.91)	(7,965.56)
d	0.71	0.75	0.67	0.60

Scales for competence, hireability, and mentoring range from 1 to 7, with higher numbers reflecting a greater extent of each variable. The scale for salary conferral ranges from $15,000 to $50,000. Means with different subscripts within each row differ significantly ($P < 0.05$). Effect sizes (Cohen's d) represent target student gender differences (no faculty gender differences were significant, all $P > 0.14$). Positive effect sizes favor male students. Conventional small, medium, and large effect sizes for d are 0.20, 0.50, and 0.80, respectively (51). $n_{\text{male student condition}} = 63$, $n_{\text{female student condition}} = 64$. ***$P < 0.001$.

In support of hypothesis B, faculty gender did not affect bias (Table 1). Tests of simple effects[a] (all $d < 0.33$) indicated that female faculty participants did not rate the female student as more competent [t(62) = 0.06, $P = 0.95$] or

---

a *Tests of simple effects* Tests that follow up any significant interaction or relationship concerning a variable in a study. Such tests investigate the degree to which a second, related variable has a significant effect on the main variable being studied.

hireable [t(62) = 0.41, P = 0.69] than did male faculty. Female faculty also did not offer more mentoring [t(62) = 0.29, P = 0.77] or a higher salary [t(61) = 1.14, P = 0.26] to the female student than did their male colleagues. In addition, faculty participants' scientific field, age, and tenure status had no effect (all P > 0.53). Thus, the bias appears pervasive among faculty and is not limited to a certain demographic subgroup. [...]

FIG. 2.

Salary conferral by student gender condition (collapsed across faculty gender).

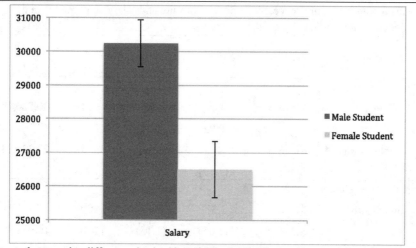

The student gender difference is significant (P < 0.01). The scale ranges from $15,000 to $50,000. Error bars represent SEs. $n_{\text{male student condition}}$ = 63, $n_{\text{female student condition}}$ = 64.

DISCUSSION

The present study is unique in investigating subtle gender bias on the part of faculty in the biological and physical sciences. It therefore informs the debate on possible causes of the gender disparity in academic science by providing unique experimental evidence that science faculty of both genders exhibit bias against female undergraduates. As a controlled experiment, it fills a critical gap in the existing literature, which consisted only of experiments in other domains (with undergraduate students as participants) and correlational data that could not conclusively rule out the influence of other variables.

15   Our results revealed that both male and female faculty judged a female student to be less competent and less worthy of being hired than an identical male student, and also offered her a smaller starting salary and less career mentoring. Although the differences in ratings may be perceived as modest, the effect sizes were all moderate to large (d = 0.60–0.75). Thus, the current

results suggest that subtle gender bias is important to address because it could translate into large real-world disadvantages in the judgment and treatment of female science students.[39] Moreover, our mediation findings shed light on the processes responsible for this bias, suggesting that the female student was less likely to be hired than the male student because she was perceived as less competent. Additionally, moderation results indicated that faculty participants' preexisting subtle bias against women undermined their perceptions and treatment of the female (but not the male) student, further suggesting that chronic subtle biases may harm women within academic science. Use of a randomized controlled design and established practices from audit study methodology support the ecological validity and educational implications of our findings (see *SI Materials and Methods*[a]).

It is noteworthy that female faculty members were just as likely as their male colleagues to favor the male student. The fact that faculty members' bias was independent of their gender, scientific discipline, age, and tenure status suggests that it is likely unintentional, generated from widespread cultural stereotypes rather than a conscious intention to harm women.[17] Additionally, the fact that faculty participants reported liking the female more than the male student further underscores the point that our results likely do not reflect faculty members' overt hostility toward women. Instead, despite expressing warmth toward emerging female scientists, faculty members of both genders appear to be affected by enduring cultural stereotypes about women's lack of science competence that translate into biases in student evaluation and mentoring.

Our careful selection of expert participants revealed gender discrimination among existing science faculty members who interact with students on a regular basis (see *SI Materials and Methods: Subjects and Recruitment Strategy*). This method allowed for a high degree of ecological validity[b] and generalizability relative to an approach using nonexpert participants, such as other undergraduates or lay people unfamiliar with laboratory manager job requirements and academic science mentoring (i.e., the participants in much psychological research on gender discrimination). The results presented here reinforce those of Steinpreis, Anders, and Ritzke,[40] the only other experiment we know of that recruited faculty participants. Because this previous experiment also indicated bias within academic science, its results raised serious concerns about the potential for faculty bias within the biological and physi-

---

a   *SI Materials and Methods*   The original article linked to a document containing supporting information: http://www.pnas.org/content/109/41/16474/suppl/DCSupplemental.

b   *ecological validity*   An experiment is "ecologically valid" when the circumstances created in the experiment closely match a set of real-world circumstances.

cal sciences, casting further doubt on assertions (based on correlational data) that such biases do not exist.[9-11] In the Steinpreis et al. experiment, psychologists were more likely to hire a psychology faculty job applicant when the applicant's curriculum vitae was assigned a male (rather than female) name.[40] This previous work invited a study that would extend the finding to faculty in the biological and physical sciences and to reactions to undergraduates, whose competence was not already fairly established by accomplishments associated with the advanced career status of the faculty target group of the previous study. By providing this unique investigation of faculty bias against female students in biological and physical sciences, the present study extends past work to a critical early career stage, and to fields where women's underrepresentation remains stark.[2-4]

Indeed, our findings raise concerns about the extent to which negative predoctoral experiences may shape women's subsequent decisions about persistence and career specialization. Following conventions established in classic experimental studies to create enough ambiguity to leave room for potentially biased responses,[20,23] the student applicants in the present research were described as qualified to succeed in academic science (i.e., having coauthored a publication after obtaining 2 y of research experience), but not irrefutably excellent. As such, they represented a majority of aspiring scientists, and were precisely the type of students most affected by faculty judgments and mentoring (see *SI: Materials and Methods* for more discussion). Our results raise the possibility that not only do such women encounter biased judgments of their competence and hireability, but also receive less faculty encouragement and financial rewards than identical male counterparts. Because most students depend on feedback from their environments to calibrate their own worth,[41] faculty's assessments of students' competence likely contribute to students' self-efficacy and goal setting as scientists, which may influence decisions much later in their careers. Likewise, inasmuch as the advice and mentoring that students receive affect their ambitions and choices, it is significant that the faculty in this study were less inclined to mentor women than men. This finding raises the possibility that women may opt out of academic science careers in part because of diminished competence judgments, rewards, and mentoring received in the early years of the careers. In sum, the predoctoral years represent a window during which students' experiences of faculty bias or encouragement are particularly likely to shape their persistence in academic science.[30-33] Thus, the present study not only fills an important gap in the research literature, but also has critical implications for pressing social and educational issues associated with the gender disparity in science.

If women's decisions to leave science fields when or before they reach the faculty level are influenced by unequal treatment by undergraduate advisors, then existing efforts to create more flexible work settings[42] or increase women's identification with science[27] may not fully alleviate a critical underlying problem. Our results suggest that academic policies and mentoring interventions targeting undergraduate advisors could contribute to reducing the gender disparity. Future research should evaluate the efficacy of educating faculty and students about the existence and impact of bias within academia, an approach that has reduced racial bias among students.[43] Educational efforts might address research on factors that attenuate gender bias in real-world settings, such as increasing women's self-monitoring.[44] Our results also point to the importance of establishing objective, transparent student evaluation and admissions criteria to guard against observers' tendency to unintentionally use different standards when assessing women relative to men.[45,46] Without such actions, faculty bias against female undergraduates may continue to undermine meritocratic[a] advancement, to the detriment of research and education.

CONCLUSIONS

The dearth of women within academic science reflects a significant wasted   20
opportunity to benefit from the capabilities of our best potential scientists, whether male or female. Although women have begun to enter some science fields in greater numbers,[5] their mere increased presence is not evidence of the absence of bias. Rather, some women may persist in academic science despite the damaging effects of unintended gender bias on the part of faculty. Similarly, it is not yet possible to conclude that the preferences for other fields and lifestyle choices[9-11] that lead many women to leave academic science (even after obtaining advanced degrees) are not themselves influenced by experiences of bias, at least to some degree. To the extent that faculty gender bias impedes women's full participation in science, it may undercut not only academic meritocracy, but also the expansion of the scientific workforce needed for the next decade's advancement of national competitiveness.[1]

MATERIALS AND METHODS

Participants

We recruited faculty participants from Biology, Chemistry, and Physics departments at three public and three private large, geographically diverse

---

a *meritocratic* According to merit.

research-intensive universities in the United States, strategically selected for their representative characteristics (see *SI Materials and Methods* for more information on department selection). The demographics of the 127 respondents corresponded to both the averages for the selected departments and faculty at all United States research-intensive institutions, meeting the criteria for generalizability even from nonrandom samples (see *SI Materials and Methods* for more information on recruitment strategy and participant characteristics). Indeed, we were particularly careful to obtain a sample representative of the underlying population, because many past studies have demonstrated that when this is the case, respondents and nonrespondents typically do not differ on demographic characteristics and responses to focal variables.[47]

Additionally, in keeping with recommended practices, we conducted an a priori power analysis[a] before beginning data collection to determine the optimal sample size needed to detect effects without biasing results toward obtaining significance (*SI Materials and Methods: Recruitment Strategy*).[48] Thus, although our sample size may appear small to some readers, it is important to note that we obtained the necessary power and representativeness to generalize from our results while purposefully avoiding an unnecessarily large sample that could have biased our results toward a false-positive type I[b] error.[48]

Procedure

Participants were asked to provide feedback on the materials of an undergraduate science student who stated their intention to go on to graduate school, and who had recently applied for a science laboratory manager position. Of importance, participants believed they were evaluating a real student who would subsequently receive the faculty participants' ratings as feedback to help their career development (see *SI Materials and Methods* for more information, and Fig. S1 for the full text of the cover story). Thus, the faculty participants' ratings were associated with definite consequences.

Following established practices, the laboratory manager application was designed to reflect high but slightly ambiguous competence, allowing for variability in participant responses.[20,23] In addition, a promising but still-nascent[c] applicant is precisely the type of student whose persistence in academic sci-

---

a   *a priori power analysis*   Analysis carried out before research begins to determine the sample size required to produce a meaningful result.

b   *false-positive type I*   Statistical term referring to a finding that indicates a relationship between two measured phenomena when in fact no such relationship exists.

c   *nascent*   Newly developing.

ence is most likely to be affected by faculty support or discouragement,[30-33] rendering faculty reactions to such a student of particular interest for the present purposes. The materials were developed in consultation with a panel of academic science researchers (who had extensive experience hiring and supervising student research assistants) to ensure that they would be perceived as realistic (*SI Materials and Methods*). Results of a funneled debriefing[a] [49] indicated that this was successful; no participant reported suspicions that the target was not an actual student who would receive their evaluation.

Participants were randomly assigned to one of two student gender conditions: application materials were attributed to either a male student (John, $n = 63$), or a female student (Jennifer, $n = 64$), two names that have been pretested as equivalent in likability and recognizeability.[50] Thus, each participant saw only one set of materials, from either the male or female applicant (see Fig. S2 for the full text of the laboratory manager application and *SI Materials and Methods* for more information on all materials). Because all other information was held constant between conditions, any differences in participants' responses are attributable to the gender of the student. Using validated scales, participants rated student competence, their own likelihood of hiring the student, selected an annual starting salary for the student, indicated how much career mentoring they would provide to such a student, and completed the Modern Sexism Scale.

(2012)

Acknowledgments

We thank faculty members from six anonymous universities for their involvement as participants; and Jessamina Blum, John Crosnick, Jennifer Frederick, Jaime Napier, Jojanneke van der Toorn, Tiffany Tsang, Tessa West, James Young, and two anonymous reviewers for valuable input. This research was supported by a grant from the Howard Hughes Medical Institute Professors Program (to J.H.).

References

1. President's Council of Advisors on Science and Technology (2012). *Engage to excel: Producing one million additional college graduates with degrees in science, technology, engineering, and mathematics.* Available at http://www.whitehouse.gov/sites/default/files/microsites/ostp/pcast-engage-to-excel-final_feb.pdf. (Accessed February 13, 2012).
2. Handelsman J, et al. (2005) Careers in science. More women in science. *Science* **309**: 1190–1191.
3. United States National Academy of Sciences (2007) *Beyond Bias and Barriers: Fulfilling the Potential of Women in Academic Science and Engineering* (National Academies, Washington, DC).
4. National Science Foundation (2009) *Women, Minorities, and Persons with Disabilities in Science and Engineering* (National Science Foundation, Arlington).

---

a *funneled debriefing* Post-experiment interview that begins with very open-ended questions and progresses toward narrower questions.

5. Bell N (2010) *Graduate Enrollment and Degrees: 1999 to 2009* (Council of Graduate Schools, Washington, DC).

6. Halpern DF, et al. (2007) The science of gender differences in science and mathematics. *Psychol Sci* **8**(1):1–51.

7. Hyde JS, Linn MC (2006) Diversity. Gender similarities in mathematics and science. *Science* **314**:599–600.

8. Spelke ES (2005) Sex differences in intrinsic aptitude for mathematics and science?: A critical review. *Am Psychol* **60**:950–958.

9. Ceci SJ, Williams WM (2010) Gender differences in math-intensive fields. *Curr Dir Psychol Sci* **19**:275–279.

10. Ceci SJ, Williams WM (2011) Understanding current causes of women's underrepresentation in science. *Proc Natl Acad Sci* USA **108**:3157–3162.

11. Ceci SJ, et al. (2011) Do subtle cues about belongingness constrain women's career choices? *Psychol Inq* **22**:255–258.

12. Berezow AB (2011) *Gender discrimination in science is a myth.* Available at http://www.nationalreview.com/articles/256816/gender-discrimination-science-myth-alex-bberezow?pg=2. (Accessed June 6, 2012).

13. Dickey Zakaib G (2011) Science gender gap probed. *Nature* **470**:153.

14. Devine PG (1989) Stereotypes and prejudice: Their automatic and controlled components. *J Pers Soc Psychol* **56**(1):5–18.

15. Eagly AH, Mladinic A (1994) Are people prejudiced against women? Some answers from research on attitudes, gender stereotypes, and judgments of competence. *Eur Rev Soc Psychol* **5**(1):1–35.

16. Spence JT, Hahn ED (1997) The attitudes toward women scale and attitude change in college students. *Psychol Women Q* **21**(1):17–34.

17. Dovidio JF, Gaertner SL (2004) in *Advances in Experimental Social Psychology*, ed Zanna MP (Elsevier, New York), pp 1–51.

18. Nosek BA, Banaji M, Greenwald AG (2002) Harvesting implicit group attitudes and beliefs from a demonstration web site. *Group Dyn* **6**(1):101–115.

19. Goldin C, Rouse C (2000) Orchestrating impartiality: The impact of "blind" auditions on female musicians. *Am Econ Rev* **90**:715–741.

20. Foschi M (2000) Double standards for competence: Theory and research. *Annu Rev Sociol* **26**(1):21–42.

21. Foschi M (1996) Double standards in the evaluation of men and women. *Soc Psychol Q* **59**:237–254.

22. Goldberg P (1968) Are women prejudiced against women? *Transaction* **5**(5):28–30.

23. Heilman ME, Wallen AS, Fuchs D, Tamkins MM (2004) Penalties for success: Reactions to women who succeed at male gender-typed tasks. *J Appl Psychol* **89**:416–427.

24. Monin B, Miller DT (2001) Moral credentials and the expression of prejudice. *J Pers Soc Psychol* **81**:33–43.

25. Uhlmann EL, Cohen GL (2007) "I think it, therefore it's true": Effects of self perceived objectivity on hiring discrimination. *Organ Behav Hum Dec* **104**:207–223.

26. Nosek BA, Banaji MR, Greenwald AG (2002) Math = male, me = female, therefore math not = me. *J Pers Soc Psychol* **83**:44–59.

27. Dasgupta N (2011) Ingroup experts and peers as social vaccines who inoculate the self-concept: The stereotype inoculation model. *Psychol Inq* **22**:231–246.

28. The Massachusetts Institute of Technology (1999) *A study on the status of women faculty in science at MIT.* Available at http://web.mit.edu/fnl/women/women.html#The%20Study (Accessed February 13, 2012).

29. Steele J, James JB, Barnett RC (2002) Learning in a man's world: Examining the perceptions of undergraduate women in male-dominated academic areas. *Psychol Women Q* **26**(1):46–50.

30. Lent RW, et al. (2001) The role of contextual supports and barriers in the choice of math/science educational options: A test of social cognitive hypotheses. *J Couns Psychol* **48**:474–483.

31. Seymour E, Hewitt NM (1996) *Talking about Leaving: Why Undergraduates Leave the Sciences* (Westview, Boulder).

32. Gasiewski JA, et al. (2012) From gatekeeping to engagement: A multicontextual, mixed method study of student academic engagement in introductory STEM courses. *Res Higher Educ* **53**:229–261.

33. Byars-Winston A, Gutierrez B, Topp S, Carnes M (2011) Integrating theory and practice to increase scientific workforce diversity: A framework for career development in graduate research training. *CBE Life Sci Educ* **10**:357–367.

34. Campbell DT, Fiske DW (1959) Convergent and discriminant validation by the multitrait-multimethod matrix. *Psychol Bull* **56**:81–105.

35. Robins RW, Hendin HM, Trzesniewski KH (2001) Measuring global self esteem: Construct validation of a single-item measure and the Rosenberg Self-Esteem Scale. *Pers Soc Psychol Bull* **27**(2):151–161.

36. Moss-Racusin CA, Rudman LA (2010) Disruptions in women's self-promotion: The backlash avoidance model. *Psychol Women Q* **34**:186–202.

37. Rudman LA, Moss-Racusin CA, Glick P, Phelan JE (2012) in *Advances in Experimental Social Psychology*, eds Devine P, Plant A (Elsevier, New York), pp 167–227.

38. Swim JK, Aikin KJ, Hall WS, Hunter BA (1995) Sexism and racism: Old-fashioned and modern prejudices. *J Pers Soc Psychol* **68**:199–214.

39. Martell RF, Lane DM, Emrich C (1996) Male-female differences: A computer simulation. *Am Psychol* **51**(2):157–158.

40. Steinpreis RE, Anders KA, Ritzke D (1999) The impact of gender on the review of the curricula vitae of job applicants and tenure candidates: A national empirical study. *Sex Roles* **41**:509–528.

41. Bandura A (1982) Self-efficacy mechanism in human agency. *Am Psychol* **37**:122–147.

42. Scandura TA, Lankau MJ (1997) Relationships of gender, family responsibility and flexible work hours to organizational commitment and job satisfaction. *J Organ Behav* **18**:377–391.

43. Rudman LA, Ashmore RD, Gary ML (2001) "Unlearning" automatic biases: The malleability of implicit prejudice and stereotypes. *J Pers Soc Psychol* **81**:856–868.

44. O'Neill OA, O'Reilly CA (2011) Reducing the backlash effect: Self-monitoring and women's promotions. *J Occup Organ Psychol* **84**:825–832.

45. Uhlmann EL, Cohen GL (2005) Constructed criteria: Redefining merit to justify discrimination. *Psychol Sci* **16**:474–480.

46. Phelan JE, Moss-Racusin CA, Rudman LA (2008) Competent yet out in the cold: Shifting criteria for hiring reflects backlash toward agentic women. *Psychol Women Q* **32**:406–413.

47. Holbrook AL, Krosnick JA, Pfent A (2007) in *Advances in Telephone Survey Methodology*, eds Lepkowski JM, et al. (John Wiley & Sons, Hoboken), pp 499–528.

48. Simmons JP, Nelson LD, Simonsohn U (2011) False-positive psychology: Undisclosed flexibility in data collection and analysis allows presenting anything as significant. *Psychol Sci* **22**:1359–1366.

49. Bargh JA, Chartrand TL (2000) in *Handbook of Research Methods in Social and Personality Psychology*, eds Reis HT, Judd CM (Cambridge Univ Press, New York), pp 253–285.

50. Brescoll VL, Uhlmann EL (2008) Can angry women get ahead? Status conferral, gender, and workplace emotion expression. *Psychol Sci* **19**:268–275.

51. Cohen J (1998) *Statistical Power Analysis for the Behavioral Sciences* (Erlbaum, Hillsdale), 2nd Ed.

Questions:

1. What are the article's main findings? How surprising do you find them? Why?

2. Consider the article's suggestion that the President's Council of Advisors on Science and Technology "indicates" there will be a "deficit of 1,000,000 workers ... over the next decade." Does the use of the number as a justification for the article's research seem relevant? Why or why not?

3. What arguments can you think of as to why women should be more effectively integrated into the sciences and technology?

4. Note one place in this article where a citation has been given without sufficient explanation of what is being cited and/or how this information fits into the argument. Explain what type of information is missing from the example you found.

5. This article's "Materials and Methods" section comes in an unusual place. Why do you think it is situated at the end?

6. Do you think the argument regarding "pervasive cultural stereotypes" needs to be included here (i.e., is this an obvious argument)? Why or why not? For whom is this section included (who is the intended audience for the specific section)?

7. In your own words, describe the approach and results of the research conducted by Steinpreis et al.

8. Six of the article's fifty-one references are to work done by one or more of the article's authors. Interpret this finding. What sorts of things might it mean?

9. How might one test for the likeability of a job applicant?

10. What ethical issues are associated with misleading the study's participants as to the purpose of the study and their role in it? Do you think concerns about those ethical issues are overridden by the importance of the science?

11. Why would comparing the extent of male and female participants' bias show that the bias was not intentional? Suggest an alternative approach for a study to assess whether such a bias was intentional or not intentional.

12. Pick out two examples of loaded language in this article. Is the loading justified, in your opinion, by the seriousness of the claim?

13. Consider the hypothesis of "academic meritocracy." What is it? How do you think research jobs are allocated at your institution?

forty

David Reilly

from Gender, Culture, and Sex-Typed Cognitive Abilities

This article by David Reilly, a graduate student in applied psychology, uses data from international student assessments to identify national and international gender disparities in science, math, and reading literacy. The article examines the relationships between social factors, such as gender inequality and economic prosperity, and gender gaps in academic achievement.

❧

ABSTRACT

Although gender differences in cognitive abilities are frequently reported, the magnitude of these differences and whether they hold practical significance in the educational outcomes of boys and girls is highly debated. Furthermore, when gender gaps in reading, mathematics and science literacy are reported they are often attributed to innate, biological differences rather than social and cultural factors. Cross-cultural evidence may contribute to this debate, and this study reports national gender differences in reading, mathematics and science literacy from 65 nations participating in the 2009 round of the Programme for International Student Assessment (PISA). Consistently across all nations, girls outperform boys in reading literacy, $d = -.44$.[a] Boys outperform girls in mathematics in the USA, $d = .22$ and across OECD[b] nations, $d = .13$. For science literacy, while the USA showed the largest gender difference across all OECD nations, $d = .14$, gender differences across OECD nations were non-significant, and a small female advantage was found for non-OECD nations, $d = -.09$. Across all three domains, these differences were more pronounced at both tails of the distribu-

a *d* Cohen's d, a variable indicating the extent of difference between two groups of data.
b *OECD* Organisation for Economic Co-operation and Development. Most countries in the OECD are wealthy and industrially developed.

tion for low- and high-achievers. Considerable cross-cultural variability was also observed, and national gender differences were correlated with gender equity measures, economic prosperity, and Hofstede's cultural dimension of *power distance*. Educational and societal implications of such gender gaps are addressed, as well as the mechanisms by which gender differences in cognitive abilities are culturally mediated.

INTRODUCTION

Rightly or wrongly, the topic of gender differences in cognitive abilities appears perennial, holding curiosity not only for social scientists but also for the general public and media.[1-4] Intelligence is multifaceted,[5-10] and comprises a range of culturally-valued cognitive abilities. While there is almost unanimous consensus that men and women do not differ in general intelligence,[11-14] there are several domains where either males or females as a group may show an advantage, such as *visuospatial*[15-16] and *verbal* abilities[17-18] respectively. However, gender differences in *quantitative abilities*,[19] such as science and mathematics, remain contentious. Researchers are divided between arguing for small but still influential differences in quantitative reasoning,[9-11] and claiming that any observed differences in maths are so small, in fact, that they can be categorised as "trivial."[12-14]

A key limitation of research in this area is that it is largely US-centric, and does not speak to gender differences between males and females raised under different social and educational environments in other cultures. Additional lines of evidence are required, and one such source is international testing of students. Secondly, research primarily focuses on mean gender differences, and fails to address gender differences in the tails of distributions which Hyde, et al.[20] argues may forecast the underrepresentation of women in the science, technology, engineering and mathematics (STEM) related professions.

To this aim, I present findings from the 2009 OECD Programme for International Student Assessment (PISA), which to my knowledge has not yet been widely discussed in psychology journals. This information provides a snapshot of current gender differences and similarities in reading, mathematics and science across 65 nations. It also highlights the wide degree of cultural variation between nations, and examines the role that social and environmental factors play in the development of gender differences. Before reviewing the PISA findings, I will briefly discuss the advantages that national and cross-national testing have to offer the debate on the nature of gender differences in cognitive abilities.

ADVANTAGES OF NATIONALLY REPRESENTATIVE SAMPLES FOR ASSESSING GENDER DIFFERENCES

Large national and international samples can provide a 'yardstick' estimate of gender differences within a given region, at a given point in time. By drawing from a broad population of students, national and international testing provide us with stronger evidence for gender similarities or differences than could be found from smaller, more selective samples. It is common practice for gender difference studies to use convenience samples drawn from psychology student subject pools,[21] as well as from groups of high performing students such as gifted and talented programmes[22] – conclusions drawn from such samples may not be generalizable to wider populations. There is evidence to suggest that the performance of males is more widely distributed, with greater numbers of high and low achievers.[23] This has been termed the *greater male variability hypothesis*,[10,15-16] and presents a problem for researchers recruiting from only high achievers – even though mean differences between males and females may be equal, if the distribution of male scores is wider than females, males will be overrepresented as high-achievers in a selective sample. This may lead to the erroneous conclusion that gender differences exist in the population of males and females.

A good example of this in practice comes in the form of the Scholastic Assessment Test (SAT) used for assessing suitability of students for college entry within the United States. Males consistently outperform females on the mathematical component.[22,24-25] Gender differences in SAT-M[a] are extremely robust across decades [...]. On the basis of this evidence alone, one might erroneously conclude that the gender gap in mathematics is pervasive unless consideration is given to the demographics of the sample. Students considering college admission are motivated to undertake the SAT, and this is largely a self-selected sample that may differ on important characteristics such as socioeconomic status, and general ability level. Additionally many more girls sit the SAT than boys,[24-26] reflecting the higher admission rate of women in college.[27] Thus the sample of males is more selective, while the sample of females is more general. One cannot rule out the possibility that the male sample includes a greater proportion of high achieving students and that the female sample may have included students of more mediocre mathematical ability, lowering mean performance. [...]

Additionally, many cognitive abilities show an interaction between gender and socioeconomic status.[1,25-28] Studies that selectively recruit from college subject pools in medium- to high-socioeconomic status regions would

---

a   *SAT-M*   Mathematical component of the SAT.

therefore be more likely to find gender differences than those recruiting from lower socioeconomic regions, as there will be greater differentiation between high and low ability levels. Likewise, samples drawing from a college pool may find greater gender differences than if they were recruited from a high school sample, or from the general population. Potentially, this could give a distorted picture of actual gender gaps when generalising from these selective samples to the wider population of males and females.

[In contrast, l]arge national samples allow researchers to investigate objectively the existence and magnitude of gender differences or similarities. We can be more confident that any observed differences are reflective of what we would find in the general population of boys and girls, and are not simply due to sampling bias. [...]

GENDER DIFFERENCES IN MATHEMATICS AND SCIENCE WITHIN THE UNITED STATES

For the United States, one such program is the National Assessment of Educational Progress (NAEP), a federal assessment of educational achievement. The NAEP is conducted for all states within the United States and since participation is both comprehensive and not self-selected, is ideally suited to answering the question of whether males and females differ in mathematical ability (a type of quantitative reasoning). Hyde[20] and colleagues examined gender differences between boys and girls in mathematics from grades 2 through 11, drawing on a sample of students from ten states which amounted to a sample of over seven million students. Hyde, et al.[20] reported an effect size for gender differences in each grade that approached zero, and categorised differences between males and females as "trivial."[29] [...]

INTERNATIONAL SAMPLING OF SCIENCE AND MATHEMATICAL ABILITY

Another source of evidence for evaluating claims of gender differences comes from international testing of students' educational attainments as part of the OECD's Programme for International Assessment (PISA). Beginning in 2000 and conducted every three years, participating nations assess the educational attainment of students using a standardized exam that allows their performance to be compared globally. PISA aims to assess the educational progress of students as they reach the end of compulsory education, at age 15, across three skill areas: these being reading literacy, mathematical literacy, and science literacy. Samples are stratified random probability samples, selected from a range of public and private institutions across geographical regions, and weighted so as to be nationally representative.[31] This overcomes the selection-bias of tests such as the SAT-M,[24-26] as well as providing a more valid assessment of the general population of boys and girls at that age than could be found in college-bound students. [...]

CROSS-NATIONAL VARIATION IN COGNITIVE ABILITIES

Cross-national variation in the magnitude of gender differences can provide useful information about the environmental conditions that foster, or inhibit, gender differences in domains such as mathematics. While gender differences in mathematics are frequently found at a national level, they are not found universally across *all* nations.[32] Social roles for women vary greatly from culture to culture, with some cultures promoting higher standards of gender equality and access to education than others.[33] Even those nations that have progressive attitudes towards women may still have strongly-held cultural stereotypes that narrowly constrain them.[34-38] Cultural stereotypes that girls and women are less able than boys and men in mathematics and science still endure,[39-40] and these stereotypes have damaging consequences for the self-efficacy of young girls.[41]

 Cross-cultural comparisons of the performance of males and females might help answer some theoretical questions about the *origins* of any observed gender differences. When we see consistent gender differences across many or all nations, and when they are large enough in magnitude to have a practical impact on the educational and occupational aspirations of boys and girls, then we might reasonably conclude some systematic process is responsible – be this biological or institutional. When we see changes in the *magnitude* and the *direction* of gender differences, as is the case for science performance reported below, then we might reasonably conclude that either cultural or environmental influences are strong moderators in the development of cognitive ability – gender differences are not an inevitable consequence of biology. Finally, if we were to see more *similarities* than differences in the performance of boys and girls, then this would also be useful information for shaping public policy and educational practices such as continuing support for coeducation.[42]

 [...] Gender differences in mathematics [are] smaller in more gender-equal nations than in less-equal nations. Though the precise mechanism by which this occurs is unclear, these findings have been replicated by a number of researchers.[31-32,43] This suggests that two factors influencing the cognitive abilities of women are the gender stereotypes that a culture holds, and the gender-roles for women in a society.[29-32] This has been referred to in the literature as the *gender stratification* hypothesis,[33,43] which argues that gender differences are more pronounced when the roles of men and women are tightly controlled into separate spheres and duties.[35,37,44-45]

Mathematics is not the only cognitive domain where we see an influence of gender-equality and gender stereotypes on cognitive performance. The female advantage in reading and language, while universal, also differs in magnitude between nations. Guiso, et al.[32] examined data from the PISA 2003 round of testing, replicating the finding of Baker and Jones for mathematics as well as finding an association between gender equity and the gender gap in reading. Although this might be expected given that correlations between mathematics performance and reading overlap, the direction of the association differed. Instead of finding reduced gender differences in reading for countries fostering greater gender-equality, the gender gap between boys and girls actually *increased*. One possibility for this seemingly paradoxical finding is that whatever natural advantage girls may have for reading is suppressed in more restrictive countries, but that under favorable conditions is allowed to flourish to its full potential. However, further replication of these findings with subsequent waves of testing is required to determine whether this association is stable across time.

Programme for International Student Assessment (PISA) 2009

[...] The aim of this study was to explore sociocultural factors that promote, or inhibit, the development of gender gaps in highly sex-typed academic domains of reading, mathematics and science.[46] It presents findings from international assessment of student abilities as part of the Programme for International Student Assessment (PISA), conducted by the Organisation for Economic Co-operation and Development (OECD). The study uses data from the most recent round of testing to calculate national and international gender gaps in reading, mathematics, and science literacy.

15   In addition to presenting data on national gender differences, it uses meta-analytic techniques to calculate global gender differences to examine evidence for Hyde's *gender similarities* hypothesis,[47] which posits there are no meaningful gender differences in cognitive performance. The study also seeks to replicate the findings of past researchers for the *gender stratification* hypothesis,[27,38,43-44] using several measures of gender equity and occupational segregation. A number of other sociocultural constructs are also examined to determine the extent to which gender differences are culturally mediated by factors other than biology.

One hypothesised influence [on results] is the economic prosperity of a nation,[39-41] which reflects two mechanisms. Firstly, greater economic prosperity allows for a greater proportion of national resources to be spent on education, resulting in a higher quality of education and emphasis on skills such as mathematics and science. Secondly, skills in these technical areas are in greater demand, and represent a pathway to a higher standard of living. This may result in greater competition for these occupations, and such competition may not always be helpful to the career aspirations of women wishing to enter male-dominated fields. While increases in gender equity are strongly associated with economic prosperity (and hence should be associated with smaller gender gaps), these may be partially offset by increased occupational stratification and stronger cultural stereotypes associating maths and science with gender roles.[27,32-33,44-45] Thus increased gender differences are not purely the result of increased spending on education and also reflect social processes.

A second mechanism by which gender differences may be culturally mediated is through the attitudes, values, and beliefs of a nation. While beliefs about the role of women in society vary considerably from nation to nation, there are few instruments available that have wide global coverage of gender stereotypes and attitudes towards women.[38, 48-49] One of the most widely used cultural instruments is Hofstede's[50] five cultural dimensions. One of these is theoretically relevant to cultural mediation of gender differences in cognitive ability, the dimension of power distance.

The dimension *power distance* describes the ways in which societies address the issue of human inequality, and the ways in which social groups are segregated.[50] In a lower power distance culture, there are reduced distinctions between social classes, between employees and employers, between students and teachers, and between genders. Higher power distance cultures have greater social division, and a compensatory strategy for those who are lower in power is to acquire culturally valued skills through education. Girls may have increased motivation to learn maths and science and pursue higher status occupations as a way of overcoming social inequity.

HYPOTHESES

Based on prior research and theoretical perspectives, it was hypothesised that:
1. Gender differences in the domains of mathematics and science would be found for the United States, and these would be larger than those reported by Hyde.[51] These would reflect gender stereotypes associating these domains with masculinity and males.[39] However gender differences cross-culturally would be much smaller, in partial support of a global gender similarities hypothesis.
2. Gender differences in reading performance in favor of girls would be found in reading for the United States and cross-culturally, reflecting an inherent biological disposition that is only weakly influenced by cultural environment.

3. Measures of national gender equity would be associated with smaller gender gaps in mathematics and science, in support of the gender stratification hypothesis. Furthermore, increased gender equity would be weakly associated with wider reading gaps in favor of girls.

4. Economic prosperity would be associated with wider gender gaps in mathematics and science than in less prosperous nations, reflecting increased spending on education, increased demand for these skills, and heightened competition by males. Such competition may not be helpful to the career aspirations of women, but will not influence reading performance which is less malleable to social and cultural influences.

5. Countries that score highly on Hofstede's power distance dimension have greater segregation and foster inequalities, particularly for women. A compensatory strategy for women is to acquire culturally valued skills such as science and mathematics. High power distance nations would be associated with smaller gender gaps or a slight female advantage in these domains. Boys may have increased motivation to develop reading and writing proficiency in high power distance cultures, resulting in smaller gender gaps for reading literacy.

Methods

Participants

25 Performance data for students accessed under PISA is offered as a publicly accessible archive for researchers. Additionally, aggregate national performance profiles are published as separate male and female subgroups,[31] which were used for analysis. PISA 2009 included 34 OECD countries, as well as 31 additional partner nations. This amounts to a total participant size of 480,405 students (50.6% female) drawn from across 65 nations. This represents the most recent round of testing, as well as providing performance data for a broader range of nations than earlier PISA assessments.

Analysis

National performance profiles in reading, mathematics and science literacy were obtained from OECD,[31] which reports the assessment of boys and girls separately. Because of the large sample sizes involved in national testing, even slight or trivial differences between boys and girls may be deemed *statistically* significant, even though it may have no *practical* significance. For this reason, an effect size is presented in the form of Cohen's *d*, the mean standardized difference. This allows the reader to draw his or her own conclusions as to the practical significance of reported gender differences.

The computation is calculated as the mean difference between male and female scores, divided by the pooled within-gender standard deviation. By convention, female scores are subtracted from male scores, so that a positive d indicates higher scores for males while a negative d reflects higher scores for females. This convention is observed for readability reasons only, and the interested reader may choose to rephrase the equations so that male scores are subtracted from female scores simply by inverting the sign of any effect size given.

Conventional criteria for labelling effect sizes as "small", "medium", or "large" have many limitations and should be used with great caution.[52-53] Cohen[53] offered a rule of thumb that an effect size of $d \leq .20$ could be considered a "small" effect for the purpose of estimating statistical power, and that many legitimate psychological phenomena studied are in fact small effects. The label of small is perhaps an unfortunate one as some researchers have mistakenly taken small to be of no practical significance, a practice Rosenthal and Rubin[54] caution against. However Hyde, et al.[20] have argued that effect sizes as small as $d = .04$ should be regarded as trivial, a cut-off which seems sound practice. Hyde[47] has also suggested that $d \leq .10$ should be actually be regarded "as close to zero" (p.581), a cut-off which is overly conservative and dismisses what are legitimate, albeit very small, between-group differences. Accordingly, Cohen's conventions for labelling are followed for reporting. Additionally, gender differences are presented using Rosenthal and Rubin's[54-55] Binomial Effect Size Display (BESD) which presents results in a metric that represents effect size in a format suitable for interpretation by non-statisticians.[56]

In order to test the gender similarities hypothesis, national gender gaps in reading, mathematics, and science were combined using meta-analysis. Comprehensive Meta Analysis (CMA) V2 software was used for the calculation of statistics.[57] A random-effects model was chosen[58] due to the high degree of cross-cultural variability, which would make a fixed-effects model[a] unsuitable.[56,59] Such a method is more conservative in estimating error terms and produces wider confidence intervals, giving us greater assurance that the true effect size falls within this range.

Favreau[60] argues against the use of null hypothesis[b] testing for evaluating claims of gender difference because it may be overly sensitive, and does not present a clear picture of how differences are distributed across groups. Ac-

---

a   *random-effects model*   Statistical approach in which variables can have any of a wide or infinite range of values;   *fixed effects model*   Statistical approach in which variables are assumed to have a small range of possible values.

b   *null hypothesis*   Hypothesis that there is no meaningful relationship between variables. Statistical analyses can show that the relationship between variables is significant by disproving the null hypothesis for that set of variables.

cordingly, data is presented showing high and low-achievers, as well as effect sizes. Even when a mean gender difference may be regarded as 'small' by Cohen's[53] conventions, or 'trivial' by Hyde,[47] a more pronounced difference may be found at the tails of a distribution in high and low-achieving students, resulting in quite disparate educational outcomes.

Moderation effects of sociocultural factors were examined to test the gender stratification hypothesis for national gender gaps using correlational analysis. Although past researchers[32, 61] have examined the gender stratification hypothesis for mathematics and reading, exploration of the relationship with science has gone largely untested. Multiple measures of gender equity were used, as each instrument operationalises the construct of gender equity differently, and prior research has shown that they vary in their predictive validity for educational and social outcomes. Other moderators tested include economic prosperity, as measured by GDP,[a] and Hofstede's power distance dimension.

GENDER GAP INDEX

For comparability with Guiso, et al.'s findings, the Gender Gap Index (GGI) produced by the World Economic Forum was selected as one measure of gender equity.[62] Data for the calendar year of PISA testing was used. This measure assesses four areas: economic participation, educational attainment, political empowerment, and health and survival. While the first three are theoretical relevant to the gender stratification hypothesis, health and survival (which measures differences in male and female life expectancy, as well as sex ratio) may reflect other – largely biological – factors, thus lowering predictive validity of this measure. An additional criticism of this measure is that the economic participation component emphasises male to female participation across various sectors, but gives less emphasis to income disparities.

RELATIVE STATUS OF WOMEN

As an alternative conceptualisation of gender equity, the Relative Status of Women (RSW) measures gender differences across educational attainment, life expectancy, and women's share of income.[63] This reflects a stronger economic and educational component in estimation of gender stratification, with wage inequality playing a greater weighting.

WOMEN IN RESEARCH

Else-Quest, et al.[61] argued that domain-specific indicators of gender equity may play an important role in the development of gender differences, with those related to gender stratification in educational outcomes showing strong

---

a GDP Gross domestic product, the total value of products and services produced by a country, usually within a given year.

predictive validity. One such marker is the relative share of research positions held by women. Data for this measure was obtained from the UNESCO Institute for Statistics, and supplemented by data from the National Science Foundation and Statistics Canada. Data was selected for the calendar year 2009 when possible, or earlier if not available. Women's relative share of research positions was available for forty-one nations.

GROSS DOMESTIC PRODUCT (GDP)

Economic data was obtained from the World Economic Outlook database 35 produced by the International Monetary Fund. Archived information for the calendar year 2009 was obtained for sixty-one nations.

HOFSTEDE'S POWER DISTANCE INDEX

National power distance scores are published in Hofstede's text *"Culture's consequences,"*[50] which ranks nations across this dimension. Data was unavailable however for many of the non-OECD partner nations, and several European countries, and was supplemented by national profiles published online (http://geert-hofstede.com). This provided coverage of fifty-two nations. [...]

RESULTS

Although assessing qualitatively different abilities, there was a strong overlap between national gender differences in reading, mathematics and science. The quantitative abilities of mathematics and science showed the greatest overlap. [...] Tables 3 and 4 present national sample size and calculated effect sizes across the three domains for OECD and partner nations respectively.

Reading Literacy

Table 5 presents summary statistics for reading achievement. Within the United States, girls outperformed boys in overall reading, Cohen's $d = -.26$ which is just over a quarter of a standard deviation. By comparison, the OECD gender difference in reading was larger, $d = -.42$. Examining performance data for the US sample further, boys were overrepresented at the lowest level of reading proficiency, with approximately 4.5 boys to every girl. When we consider the vocational and economic outcomes associated with poor literacy, such a large disparity is alarming. Such findings are consistent with previous findings on reading literacy assessed by PISA[32] and gender differences in the prevalence of reading difficulties.[64-65] When we look at students attaining the highest level of reading proficiency (Level *6*), the trend is reversed with over twice the number of girls than boys achieving the highest standard. Thus boys are overrepresented at the lower end of the spectrum, while girls are overrepresented at the highest end.

TABLE 3.

National Gender Differences in Reading, Mathematics, and Science Literacy for Countries within the OECD.

Country	Sample size		Effect sizes (Cohen's d)		
	Males	Females	Reading	Mathematics	Science
Australia	7020	7231	-0.37	0.11	-0.01
Austria	3252	3338	-0.41	0.20	0.08
Belgium	4345	4156	-0.27	0.21	0.06
Canada	11431	11776	-0.38	0.14	0.05
Chile	2870	2799	-0.27	0.26	0.11
Czech Republic	3115	2949	-0.53	0.05	-0.05
Denmark	2886	3038	-0.34	0.19	0.13
Estonia	2430	2297	-0.53	0.11	-0.01
Finland	2856	2954	-0.64	0.03	-0.17
France	2087	2211	-0.38	0.16	0.03
Germany	2545	2434	-0.42	0.16	0.05
Greece	2412	2557	-0.50	0.15	-0.11
Hungary	2294	2311	-0.42	0.13	0.00
Iceland	1792	1854	-0.46	0.04	0.02
Ireland	1973	1964	-0.41	0.09	-0.03
Israel	2648	3113	-0.38	0.08	-0.03
Italy	15696	15209	-0.48	0.16	-0.02
Japan	3126	2962	-0.39	0.10	-0.12
Korea	2590	2399	-0.45	0.04	-0.03
Luxembourg	2319	2303	-0.38	0.20	0.07
Mexico	18209	20041	-0.29	0.17	0.08
Netherlands	2348	2412	-0.27	0.19	0.04
New Zealand	2396	2247	-0.44	0.08	-0.06
Norway	2375	2285	-0.52	0.06	-0.04
Poland	2243	2474	-0.56	0.04	-0.07
Portugal	3020	3278	-0.44	0.13	-0.04
Slovak Republic	2238	2317	-0.57	0.03	-0.01
Slovenia	3333	2822	-0.60	0.01	-0.15
Spain	13141	12746	-0.33	0.21	0.08
Sweden	2311	2256	-0.46	-0.02	-0.04
Switzerland	6020	5790	-0.42	0.20	0.08
Turkey	2551	2445	-0.52	0.12	-0.15
United Kingdom	6062	6117	-0.26	0.23	0.10
United States	2687	2546	-0.26	0.22	0.14

Note: Significant gender differences are highlighted in bold.

TABLE 4.
National Gender Differences in Reading, Mathematics, and Science Literacy for PISA Partner Countries.

Country	Sample size		Effect sizes (Cohen's d)		
	Males	Females	Reading	Mathematics	Science
Albania	2321	2275	-0.62	**-0.12**	**-0.33**
Argentina	2183	2591	**-0.34**	**0.11**	**-0.08**
Azerbaijan	2443	2248	**-0.31**	**0.13**	**-0.10**
Brazil	9101	11026	**-0.30**	**0.19**	**0.04**
Bulgaria	2231	2276	**-0.54**	-0.04	**-0.19**
Colombia	3711	4210	**-0.11**	**0.43**	**0.26**
Croatia	2653	2341	**-0.58**	**0.12**	**-0.10**
Dubai (UAE)	5554	5313	**-0.47**	0.02	**-0.26**
Hong Kong-China	2257	2280	**-0.39**	**0.15**	0.03
Indonesia	2534	2602	**-0.55**	-0.02	**-0.13**
Jordan	3120	3366	**-0.63**	-0.01	**-0.10**
Kazakhstan	2723	2689	**-0.47**	-0.01	**-0.10**
Kyrgyzstan	2381	2605	**-0.54**	**-0.07**	**-0.24**
Latvia	2175	2327	**-0.59**	0.02	**-0.09**
Liechtenstein*	181	148	**-0.39**	**0.28**	0.18
Lithuania	2287	2241	**-0.68**	**-0.07**	**-0.20**
Macao-China	3011	2941	**-0.45**	**0.13**	-0.03
Montenegro	2443	2382	**-0.57**	**0.14**	**-0.14**
Panama	1936	2033	**-0.33**	**0.06**	-0.02
Peru	3000	2985	**-0.23**	**0.20**	**0.05**
Qatar	4510	4568	**-0.44**	**-0.05**	**-0.25**
Romania	2378	2398	**-0.47**	0.04	**-0.13**
Russian Federation	2623	2685	**-0.50**	0.03	-0.03
Serbia	2680	2843	**-0.47**	**0.13**	-0.01
Shanghai-China	2528	2587	**-0.50**	-0.01	-0.01
Singapore	2626	2657	**-0.32**	**0.05**	-0.01
Chinese Taipei	2911	2920	**-0.43**	**0.05**	**-0.16**
Thailand	2681	3544	**-0.52**	**0.05**	**-0.16**
Trinidad and Tobago	2283	2495	**-0.51**	**-0.08**	**-0.17**
Tunisia	2359	2596	**-0.37**	**0.16**	0.01
Uruguay	2810	3147	**-0.42**	**0.13**	-0.01

Note: Significant gender differences are highlighted in bold.
*Although effect sizes are large, caution must be taken interpreting due to small sample size.

TABLE 5.
Reading Ability for Girls and Boys for the USA and OECD nations.

	Girls	Boys	Standard Deviation	Effect Size (d)
United States	513	488	(97)	-.26
OECD Average	513	474	(93)	-.42
% students at lowest ability level, USA	0.2%	0.9%	4.5 boys : 1 girl	
% at highest ability level, USA	2.1%	0.9%	2.4 girls : 1 boy	

Overall, across all sixty-five nations the gender difference in reading literacy favored girls, $d = -.44$ [95%CI $= -.41, -.46$], $Zma = -31.04, p<.001$,[a] with a similar gender difference also being found for OECD nations only as a group. Additionally, statistically significant gender differences in reading favoring girls were found in *every* nation surveyed, and have since the first assessment in 2000.[66] These effect sizes ranged from $-.11$ to $-.68$, from a small- to a medium-sized difference in reading literacy.

40   To investigate the gender stratification hypothesis, I examined correlations between gender equity and the gender gap in reading. Partial support was found for the gender stratification hypothesis. National scores on the Relative Status of Women (RSW) measure were negatively correlated with reading, $r = -.33$,[b] $p = .018$, such that increased gender equality was associated with larger reading gaps favoring females. Additionally, the educational measure of women in research (WIR) was associated with larger reading gaps, $r = -.38, p = .016$. Surprisingly though, there was no association between the gender gap index (GGI) and reading ability, $r = .01$. Examination of the scatterplot showed no discernible pattern, and the result was not driven by outliers.

Stronger support for the gender stratification hypothesis was found when examining gender differences in the percentage of students attaining the highest level of reading. Improvements in national gender equity was associated with a wider gender gap in high achieving girls, RSW, $r = -.32, p = .021$; GGI, $r = -.41, p = .002$, which is consistent with the findings of Guiso, et al.[32] Somewhat surprisingly, however, the educational measure of gender equity showed a strong positive association, with increases in the percentage of women in research associated with smaller gender gaps, $r = .57, p<.001$.

a   *CI* Confidence interval, an indicator of how precisely the given statistics reflect the data; *p* Probability, here an indication of the likelihood of getting the same experimental results as the ones observed if there were no relationship between the variables being studied.

b   *r* Refers to the Pearson product-moment correlation coefficient, which indicates how close the relationship between the variables is to that of a linear equation.

While the role of women in higher education may make a contribution to the mean performance of girls and boys in *basic* reading literacy, it may be the case that for high-achieving reading comprehension skills, boys and girls benefit equally from female role-models in higher learning.

No association between GDP and gender differences in reading was found, $r = .04$, consistent with predictions. However, a strong association with economic prosperity was found for reading high achievers, $r = -.43$, $p<.001$ with a greater ratio of female to male high achievers as GDP increased. This suggests an interaction between gender and GDP, with girls benefiting more from economic prosperity than boys. Furthermore, while no association was found between power distance and mean reading literacy scores of boys and girls, a strong positive association with the gender gap in high achievers was found as hypothesized, $r = .40$, $p = .003$ with gender ratios approaching more equal representation as power distance increased. Cultural mediation through economic prosperity and power distance was not found for mean male and female performance, only for gender ratios in high achievement.

Mathematics Literacy

Table 6 presents summary statistics for mathematics literacy. Within the United States, boys scored higher on mathematical literacy than girls, $d = .22$ which is a small but non-trivial effect size. Additionally, the size of the gender differences was almost twice that of the OECD average. This is in contrast to previous studies examining national mathematics performance by Hyde, et al.[20] which had found a gender gap that approached zero. At the lower end of ability level for the US sample, the difference in prevalence between girls and boys was extremely slight; however at the highest ability levels there were just over twice as many boys than girls reaching this proficiency level.

TABLE 6.
Mean Mathematical Ability for Girls and Boys for the USA and OECD nations.

	Girls	Boys	Standard Deviation	Effect Size (d)
United States	477	497	(91)	.22
OECD Average	490	501	(92)	.12
% students at lowest ability level, USA	9.5%	6.8%	1.40 girls : 1 boy	
% at highest ability level, USA	1.2%	2.5%	2.12 boys : 1 girl	

As the distribution of gender differences differed somewhat between OECD and partner nations, they are reported separately. Overall, across all 34 OECD nations, there was a significant gender difference favoring males on mathematical literacy, Cohen's $d = .13$ 95%CI [.11,.15], $Zma = 11.22$, $p<.001$. While this is a small effect size, it does exceed the criteria set forth by Hyde and Linn[51] for trivial gender differences. Gender differences across PISA partner nations also favored males, Cohen's $d = .07$ 95%CI [.02,.11], $Zma = 3.10$, $p = .001$ although this difference was somewhat smaller.

45   While statistically significant differences were found in most countries, they showed considerable variability ranging from $d = -.12$ to $d = .43$[...]. For many nations the gender gap is negligible, while others show small to medium sized differences. Additionally the direction of the gender gap was sometimes reversed, with girls outperforming boys in many nations. Under different social and educational environments, a gender advantage supporting either males or females emerges. This would be inconsistent with Hyde's[47] *gender similarities* hypothesis; rather, gender differences or similarities in mathematics are strongly mediated by cultural factors.

To explore the *gender stratification* hypothesis, correlations between gender equity measures and the gender gap in maths were examined. As hypothesized there was a strong negative relationship between the educational measure of women in research and the gender gap in mathematics, $r = -.38$, $p = .014$. Greater representation of women in research was associated with smaller gender gaps or a female advantage, consistent with the findings of Else-Quest, et al.[61] However, only a weak association was found between gender equity measure of RSW, $r = -.14$, and no association was found between GGI and maths, in contrast to the findings of Guiso, et al.[32]

Since the PISA 2009 dataset includes a much broader range of partner nations than was examined by Guiso, et al.,[32] the strength of the gender equity association may have been obscured by additional noise reflecting developed/developing nationhood. When restricting analysis to OECD nations only, the hypothesized gender equity association was found for the relative status of women (RSW) measure, $r = -.42$, $p = .020$, as well as a weak association with GGI, $r = -.21$ that fell short of statistical significance. While gender equity plays an important role in the development of gender differences in mathematical literacy for developed nations, it may be the case that there are more proximate needs for girls in developing nations (such as access to schooling, parental support, freedom from work and home duties) that these gender equity measures do not assess.

A similar pattern of associations was found for gender differences in high achieving mathematics students across all nations. There was a strong association between women in research educational measure, $r = -.63$, $p<.001$, with increased representation of women in research positions associated with a smaller gender difference in high achievers approaching zero [...]. However no association was found between the gender gap in high achievers and other gender equity measures, nor was this found when restricting to OECD nations only.

Support was also found for the economic prosperity hypothesis. Mean gender differences in mathematics literacy were larger in more economically prosperous nations, $r = .31$, $p = .015$. This relationship was stronger for high achievement, $r = .53$, $p<.001$ with a greater number of males attaining this level of proficiency.

Examining the relationship between Hofstede's power distance cultural dimension and mathematics literacy, support was also found for cultural mediation. There was a strong negative relationship between power distance and mean gender differences in mathematics, $r = -.28$, $p = .044$, as well as for gender ratios in high achievement, $r = -.33$, $p = .019$. Gender differences were smaller in nations with greater tolerance for inequality, suggesting a compensatory strategy to acquire culturally and economically valued skills in mathematics.

Science Literacy

TABLE 7.
US National Science performance for girls and boys, including high and low achievers.

	Girls	Boys	Standard Deviation	Effect Size (d)
United States	495	509	(98)	.14
OECD Average	501	501	(94)	.00
% students at lowest ability level, USA	4.6%	3.8%	1.20 girls : 1 boy	
% at highest ability level, USA	1.0%	1.5%	1.52 boys : 1 girl	

Table 7 presents summary statistics for science literacy achievement scores. For the United States, a gender difference of $d = .14$ was found. Furthermore, the United States showed the largest gender difference across *all* OECD countries. Although statistically significant, the difference between the average boy and girl is small, but neither is it of a trivial magnitude either. Boys in the US scored higher than boys internationally, while girls scored lower than their international peers. Additionally, at both ends of the ability level spectrum, gender differences were more pronounced – there are approximately 1.5 boys to every girl achieving the highest level of science proficiency. Thus while the mean difference between males and females may be "small" by Cohen's[53] effect size conventions, it may have more of an impact than one might assume from that label.

In contrast to US performance, across OECD countries there was no difference between boys and girls, $d = .00$ 95%CI $[-.03,.03]$, $Zma = 0.10$, $p = .919$. However there was a large degree of cultural variability, with gender differences favoring both boys and girls. Indeed, statistically significant differences in favor of boys were only found in nine countries, and only three were higher than the 'close-to-zero' criterion suggested by Hyde of $d < .10$. Figure 4 shows mean standardized effect sizes (Cohen's d) for gender-gaps in science across OECD nations. Gender similarities, rather than differences, were the norm which is consistent with the findings of Hyde and Linn.[51] Somewhat surprisingly, there were also five nations where girls *outperformed* boys to a statistically significant degree (the largest being Finland, $d = -.17$).

FIGURE 4.
Distribution of effect sizes for gender differences in science literacy across OECD nations.

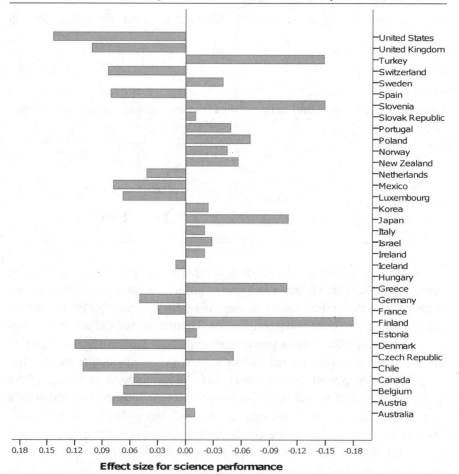

One of the advantages of cross-cultural comparisons in national testing is that it highlights just how powerfully cultural and environmental influences can be in either promoting – or inhibiting – the cognitive development and learning of a child.

A markedly different picture of gender differences in science can be found across the 31 non-OECD nations. In general, females scored higher in science literacy than males across most nations. Overall, across non-OECD nations surveyed there was a statistically significant difference in science literacy favoring girls, $d = -.09$ 95%CI [$-.14, -.04$], $Zma = -3.44$, $p = .001$. For some nations, the gender difference was trivial or favored boys, but these were the exception; this is in contrast to the gender similarities in science noted above for OECD nations.

When both OECD and non-OECD nations were combined, there was a statistically significant difference in favor of girls, $d = -.04$ [95%CI$-.070, -.013$] $Zma = -2.84$, $p = .005$. This effect size would fall into the trivial size by Hyde's[67] conventions, but a focus on the combined sample overlooks the pattern of gender differences at a national level where girls show small but meaningful gains over boys in science literacy across large parts of the world. Given that women are underrepresented in science, particularly in the United States[68] such findings call into question the validity of cultural stereotypes that associate science with masculinity,[69] and highlight the need for further efforts at challenging these damaging cultural stereotypes.

Examining mean gender differences in science literacy, partial support 55 for the gender stratification hypothesis was found. There was a strong correlation between national GGI scores and science, $r = .29$, $p = .035$, with greater gender equity associated with smaller gender gaps approaching zero. However, only a weak non-significant association was found for the RSW, $r = .14$.

Additionally, there was a strong negative correlation between the percentage of women in research and gender gaps in science, $r = -.39$, $p = .011$, with increased representation of women being associated with a stronger female advantage over males in science. Thus increased gender equity was associated with more equal science performance, but this was offset by higher female performance as the share of women in research positions increased.

Only weak support for the gender stratification hypothesis was found for gender differences in high achievement in science. Increased gender equity as measured by the percentage of researchers who are women was associated with smaller gender gaps in the number of high achievers, $r = -.57$, $p < .001$ [...]. While positive female role models are certainly important for challenging gender stereotypes about women in science generally, they may be even more so for encouraging young women to excel in science and pursue it as a career path. In contrast to this finding, there was no association between the relative status of women measure, $r = .12$ and a slight positive correlation with gender equity as measured by the GGI, $r = .29$, $p = .029$, with increased gender equity associated with more male high achievers than female which is contrary to predictions. This anomalous association may be at least partly explained by the underlying construct measured by the GGI. It incorporates a strong economic component in its formula, with a correlation of $r = .43$ between national GGI scores and economic productivity as measured by GDP. When controlling for economic productivity, the association between GGI and science high achievers becomes non-significant, $r = .12$, $p = .373$.

Strong support was also found for cultural mediation of gender differences in science. Positive relationships were observed between GDP and gender differences in mean science scores, $r = .42$, $p = .001$, as well as for gender ratios in high achievement, $r = .27$, $p = .036$, as hypothesised. In contrast, a negative relationship was found between the power-distance dimension and mean gender differences in science, $r = -.39$, $p = .005$ with gender differences favoring girls in high power-distance nations. This effect was even stronger for gender ratios in high achievement, $r = -.45$, $p = .001$.

DISCUSSION

Does the size of gender differences in reading, mathematics, and science from PISA assessment merit further research into the social and cultural factors that promote, or inhibit, differential educational outcomes for boys and girls? Evidence presented for the United States shows that there are meaningful gender gaps across all three domains. Furthermore, they are larger than those found in most OECD nations placing the US among the highest gender gaps in mathematics and science in the developed world, but somewhat smaller than other nations in reading literacy. However, quite different patterns are found when examining gender gaps globally. US performance is reviewed first, followed by a discussion of cross-cultural evidence.

Reading Literacy

60 While a small-to-medium sized gender difference in reading was found for US students $d = -.26$, this was comparatively smaller than that found in other OECD nations. However, gender differences were strikingly different at both tails of the distribution, with boys overrepresented in the lowest level of reading proficiency and girls overrepresented in the highest. PISA sampling allows for exclusion of students with limited language proficiency, so it is likely that this result reflects poorer reading ability generally rather than male

overrepresentation in reading difficulties students. This pattern is consistent with existing research on gender ratios for reading difficulties.[64-65]

Cross-culturally, a medium sized gender difference ($d = -.44$) was found for reading literacy, which would be inconsistent with Hyde's gender similarities hypothesis.[47] Expressed in the BESD format, the likelihood of being average or higher in reading ability increases from 39% for boys to 61% for girls. Reading performance was higher for girls than boys across *every* nation, but also showed considerable between-nation variation. Though the direction of gender differences would be consistent with a biological explanation, it appears at least partially malleable by social and cultural factors. While there was no support for cultural mediation through economic prosperity and power distance in mean gender differences, contrary to predictions associations were found for high achievers in reading literacy.

It has been a common research finding that boys are generally poorer readers and writers than girls,[70] and considerable effort has been made to address the gender gap over recent decades with focus on early identification and intervention for reading difficulties. Basic literacy is an essential life skill for all children, and for full participation as a citizen. While much attention is given to the issue of math and science gender gaps, gender gaps in reading are in fact much larger and favor girls at both tails of the distribution. While gender gaps in reading literacy for the USA were smaller than those found internationally, the need for further progress remains. Enrolments of women outnumber men in college, with higher female GPA and completion rates than their male peers.[16,65-66] Raising the educational aspirations of boys who experience difficulties in reading literacy, and continuing support for early intervention is critical as a matter of gender equity.

Mathematics Literacy

Gender differences in mathematics literacy were comparatively larger for the United States than those found across other OECD nations. These findings are consistent with student test data reported by Hedges and Nowell,[23] as well as findings from PISA 2003[32,61] that a small gender difference in mathematics exists, but is also inconsistent with findings of no difference reported by Hyde and colleagues using data from the NAEP.[20] How are we to reconcile this discrepancy?

[...P]roblem-solving for complex and novel mathematics tasks show a small to medium sized male advantage,[30] and PISA assessment of mathematical literacy is somewhat different to that of the NAEP. This may allow for greater differentiation between high and low ability students if a ceiling-effect is present, and may provide a more thorough test of the gender similarities hypothesis. It may well be the case that gender differences in basic mathematical literacy are trivial in size,[71] but that gender differences can be found in more complex tasks[30] requiring more than just curriculum knowledge.

65   Gender differences were observed for US performance, $d = .22$, which is small in size by Cohen's[53] conventions and non-trivial by Hyde's[47] criteria. When expressed in the BESD format, the likelihood of being average or higher in mathematics increases from 44.5% for girls to 55.5% for boys. One should be careful not to make too much, or too little, of this gender difference. As Hyde[47] points out, the degree of overlap between male and female performance is large for effect sizes in the small range, with many girls performing at or above the male average in mathematics. This perspective does not diminish the observation that a gender gap exists. As can be seen from the cross-cultural evaluation of mathematics, gender gaps in mathematics are not an inevitability, with many countries in fact showing higher female performance.

This difference is most apparent when examining student attainment of the highest proficiency level in mathematics, with double the amount of boys than girls reaching this stage. Benbow[22] argued that gender differences in high-achievement for mathematics could be at least partially explained by greater male variability and a combination of biological and environmental factors. It is likely that greater male variability explains at least part of the gender difference in high achievement, but that sociocultural factors also play a role in the development of mathematics at the extreme tails of the distribution. While general proficiency in mathematics is an important life goal for *all* students, attainment of an advanced level of mathematics is an important prerequisite for pursuing more technical degrees in STEM-related fields.[72] [...] At least for students within the USA, gender differences in mean and high achievement for mathematics have not been eliminated, and highlight the need for further progress.

While cross-culturally, gender differences favored males across OECD and partner nations, the magnitude of this difference ($d = .13$) was also small in size and subject to wide cultural variation. The likelihood of being average or higher in mathematical ability increases from 46.7% for girls to 53.2% for boys, a small but non-trivial difference. Unlike reading literacy, there were a number of countries which had non-significant gender differences, which would be inconsistent with strong biological differences between boys and girls in mathematical reasoning.[11,15,65-66] It may be the case that whatever slight advantage boys have is magnified by social and cultural reinforcement, to produce gender differences in some countries but that other nations raise girls and boys to equivalent performance. [...]

Science Literacy

While the effect size for gender differences in science literacy for the USA was relatively small compared to that of reading and mathematics, it stands out as the largest effect size across all OECD nations, $d = .14$. This is a small effect size, but also not a trivial one by Hyde's[47] conventions. Represented in the BESD format, the likelihood of being average or higher in science literacy increases from 46.5% for girls to 53.5% for boys. Additionally, boys were slightly overrepresented in attaining the highest level of science proficiency, but not to the same degree as for mathematics. Of all the domains assessed, science literacy appears to be the most variable cross-culturally, with many countries showing no difference whatsoever, and many showing a female advantage. This is a promising sign, and a benchmark to which the USA can aspire. This pattern of results was consistent with the gender similarities hypothesis.

Gender Stratification Hypothesis

In order to test the gender stratification hypothesis, this study examined the relationship between national measures of gender equity and gender gaps in reading, mathematics and science literacy. While some support for the gender stratification hypothesis was found, the predictive validity of gender equity measures varied across instruments and domains. In particular, relationships between the Gender Gap Index instrument were often weak, and in the case of science literacy high achievers in a direction contrary to hypotheses. This failure to support the gender stratification hypothesis using all gender equity measures should not be interpreted as a refutation of the hypothesis, but means that one should evaluate the hypothesis carefully. Each instrument taps different aspects of the underlying gender equity construct, and it is likely that some elements of equity have greater bearing on educational outcomes than others. A consistent finding across all three domains, and across both mean performance and high achievers, was that the relative share of women in research accurately predicted the presence or absence of gender differences. However, composite measures of gender equity showed weaker or inconsistent associations.

It may be the case that measures more closely related to education, such 70 as gender differences in relative share of research and science positions, may more accurately measure the underlying social and cultural conditions that foster or inhibit the development of gender differences in reading, mathematics and science literacy. None of the instruments directly measure attitudes towards women in STEM-related fields, or gender stereotypes about the relative abilities of males and females.[69,84] Instead, the composite measures relate to the role of women in society in general, which may lack the specificity required to consistently predict gender differences in learning outcomes. Although increased gender equity generally may be associated with the presence or absence of gender gaps in reading, mathematics and science, it may not be the direct cause.

The relative share of women employed in scientific research may be more directly related to societal attitudes about the role of women in technical fields, and to gender stereotypes about the capabilities of males and females in sex-typed achievement domains. Girls growing up in a society that praises the scientific and technical achievements of men but lacks equivalent female role models may perceive that women are less capable in this area, or that their skills are not culturally valued. They may instead be motivated to develop other talents, such as high proficiency in language, and to pursue careers in less-segregated professions. Conversely, if girls grow up in a social environment where they see progression into further education and specialisation in STEM-related fields is not only possible but also commonplace, they may be more motivated to acquire and master mathematics and science skills. In such a culture, encouragement from parents and teachers may be higher, and they may show greater confidence and improved self-efficacy in these domains than children from other cultures. While mean gender differences are smaller (or favor females) in such nations, this also translates to increased female representation in high achievers as well. This provides for stronger support of the gender stratification hypothesis.

Economic Prosperity

Mean gender differences were larger for mathematics and science in economically prosperous nations as hypothesised but were largely unrelated to reading literacy. This likely reflects both increased educational spending for economically prosperous nations, as well as increased emphasis being placed on mathematics and science skills. Student achievement in less prosperous nations may be more homogenous with smaller gender differences, and there may be a reduced focus on teaching of these skills. It may also be the case that there is greater competition by males to achieve in these masculine sex-typed domains. These associations were also found for gender ratios in high achievement. Additionally, gender ratios for high achievers in reading literacy were also related to economic prosperity, which was unexpected.

Power Distance

Hofstede[50] argued that cultures differed in their tolerance for inequality, with some cultures observing social class distinctions more strongly than others. Such cultures may place greater emphasis on social roles and stratification, but one way of overcoming inequity is the pursuit of culturally valued skills and traits. As a compensatory strategy, girls may seek out higher social status positions by obtaining education in mathematics and science, and this may help to explain the female advantage for science observed for non-OECD nations. As hypothesised, these associations were found for mean gender differences in mathematics and science as well as for gender ratios of high achievers. Lesser support was found for cultural mediation in reading literacy, with no association for mean gender differences but a positive association for gender ratios in high achievement.

Social Implications

The question of whether gender differences exist in cognitive abilities has important implications for parents, educators, and policy-makers.[20,47,72, 82-83] Yet great caution must be taken when interpreting empirical evidence – Hyde[47] raises a legitimate concern that inflated claims of wide gender difference might contribute to increased gender segregation in education and the workforce, and that the potential of girls may be overlooked by parents and teachers.[78-82] This study finds evidence of gender similarities rather than differences cross-culturally but also that meaningful gender gaps in maths and science remain and are related to cultural factors. [...]

SUMMARY

Evidence from national testing for the United States shows that there are meaningful gender gaps to be addressed in academic achievement across reading, mathematical and science literacy. Furthermore, these are larger than that found cross-culturally, where evidence for the gender similarities hypothesis is stronger. Globally, there is a small gender difference in mathematics literacy favoring males, and a small difference in science literacy favoring girls in non-OECD nations. However, a consistent finding for reading literacy is that girls outperform boys both in mean differences overall and gender ratios in attaining high reading achievement. Correlational analyses show that economic prosperity, gender equity, and the dimension of power distance are good predictors of global gender differences in cognitive abilities.

(2012)

Author Contributions
Analyzed the data: DR. Wrote the paper: DR.

REFERENCES

1. Eagly AH (1995) The science and politics of comparing women and men. American Psychologist 50: 145–158.
2. Halpern DF, Lamay M (2000) The smarter sex: A critical review of sex differences in intelligence. Educational Psychology Review 12: 229–246.

3. Caplan PJ, Caplan JB (1997) Do sex-related cognitive differences exist, and why do people seek them out? In: Caplan PJ, Crawford M, Hyde JS, Richardson JTE, editors. pp. 52–80. New York: Oxford University Press.

4. Reilly D (2012) Exploring the science behind sex and gender differences in cognitive abilities. Sex Roles. doi:10.1007/s11199-012-0134-6.

5. Gardner H (1985) Frames of mind: The theory of multiple intelligences. Basic Books.

6. Horn J (1994) Theory of fluid and crystallized intelligence. In: Sternberg RJ, editor. pp. 443–451. New York: Macmillan.

7. Gardner H (1999) Intelligence reframed: Multiple intelligences for the 21st century. Basic Books. (AZ).

8. Neisser U (1967) Cognitive psychology. New York: Appleton-Century-Crofts.

9. Carroll JB (1997) The three-stratum theory of cognitive abilities. In: Flanagan DP, Harrison PL, editors. pp. 69–76. New York: Guilford.

10. Sternberg RJ (1994) Encyclopedia of human intelligence. New York: Macmillan.

11. Neisser UC, Boodoo G, Bouchard TJ Jr, Boykin AW, Brody N, et al. (1996) Intelligence: Knowns and unknowns. American Psychologist 51: 77–101.

12. Hyde JS, Lindberg SM (2007) Facts and assumptions about the nature of gender differences and the implications for gender equity. In: Klein SS, editor. pp. 19–32. Mahwah, NJ: Lawrence Erlbaum Associates.

13. Halpern DF (2007) Science, sex, and good sense: why women are underrepresented in some areas of science and math. In: Ceci SJ, editor. pp. 121–130. Washington, D.C.: American Psychological Association.

14. Jensen AR (1998) The g factor: the science of mental ability. London: Praeger.

15. Voyer D, Voyer S, Bryden MP (1995) Magnitude of sex differences in spatial abilities: A meta-analysis and consideration of critical variables. Psychological Bulletin 117: 250–270.

16. Linn MC, Petersen AC (1985) Emergence and characterization of sex differences in spatial ability: A meta-analysis. Child Development 56: 1479–1498.

17. Hyde JS, Linn MC (1988) Gender differences in verbal ability: A meta-analysis. Psychological Bulletin 104: 53–69.

18. Halpern DF (2000) Sex differences in cognitive abilities. Mahwah, NJ: Erlbaum.

19. Maccoby EE, Jacklin CN (1974) The psychology of sex differences. Stanford: Stanford University Press.

20. Hyde JS, Lindberg SM, Linn MC, Ellis AB, Williams CC (2008) Gender similarities characterize math performance. Science 321: 494–495.

21. Sears DO (1986) College sophomores in the laboratory: Influences of a narrow data base on social psychology's view of human nature. Journal of Personality and Social Psychology 51: 515–530.

22. Benbow CP (1988) Sex differences in mathematical reasoning ability in intellectually talented preadolescents: Their nature, effects, and possible causes. Behavioral and Brain Sciences 11: 169–232.

23. Hedges LV, Nowell A (1995) Sex differences in mental test scores, variability, and numbers of high-scoring individuals. Science 269: 41–45.

24. Halpern DF (2011) Sex differences in cognitive abilities. Mahwah, NJ: Erlbaum.

25. Gallagher AM, Kaufman JC, editors (2005) Gender differences in mathematics. New York: Cambridge University Press.

26. Spelke ES (2005) Sex differences in intrinsic aptitude for mathematics and science?: a critical review. American Psychologist 60: 950–958.

27. Alon S, Gelbgiser D (2011) The female advantage in college academic achievements and horizontal sex segregation. Social Science Research 40: 107–119.

28. Henry PJ (2008) College sophomores in the laboratory redux: Influences of a narrow data base on social psychology's view of the nature of prejudice. Psychological Inquiry 19: 49–71.

29. Hyde JS, Mertz JE (2009) Gender, culture, and mathematics performance. Proceedings of the National Academy of Sciences 106: 8801–8807.

30. Hyde JS, Fennema E, Lamon SJ (1990) Gender differences in mathematics performance: A meta-analysis. Psychological Bulletin 107: 139–155.

31. OECD (2010) PISA 2009 Results: What students know and can do – Student performance in reading, mathematics and science. (Volume I).

32. Guiso L, Monte F, Sapienza P, Zingales L (2008) Culture, gender, and math. Science 320: 1164–1165.

33. Riegle-Crumb C (2005) The cross-national context of the gender gap in math and science. In: Hedges LV, Schneider B, editors. pp. 227–243. New York, NY: Russell Sage Foundation.

34. Eccles JS (1987) Gender roles and women's achievement-related decisions. Psychology of Women Quarterly 11: 135–172.

35. Eagly AH (1987) Sex differences in social behavior: A social-role interpretation. Hillsdale: Erlbaum.

36. Eagly AH (1987) The interpretation of sex differences in social behavior. In: Eagly AH, editor. London: Lawrence Erlbaum Associates.

37. Eagly AH, Mitchell AA (2004) Social role theory of sex differences and similarities: Implications for the sociopolitical attitudes of men and women. In: Paludi MA, editor. Praeger guide to the psychology of gender. Westport, Conn.: Praeger Publishers. 183–206.

38. Best DL, Williams JE (1997) Sex, gender, and culture. In: Berry JW, Segall MH, Kagitcibasi C, editors. pp. 163–212. Needleham Heights, MA: Allyn and Bacon.

39. Halpern DF, Straight CA, Stephenson CL (2011) Beliefs about cognitive gender differences: Accurate for direction, underestimated for size. Sex Roles 64: 336–347.

40. Eccles JS, Jacobs JE, Harold RD (1990) Gender role stereotypes, expectancy effects, and parents' socialization of gender differences. Journal of Social Issues 46: 183–201.

41. Steele CM (1997) A threat in the air: How stereotypes shape intellectual identity and performance. American Psychologist 52: 613–629.

42. Halpern DF, Eliot L, Bigler RS, Fabes RA, Hanish LD, et al. (2011) The pseudoscience of single-sex schooling. Science 333: 1706–1707.

43. Baker DP, Jones DP (1993) Creating gender equality: Cross-national gender stratification and mathematical performance. Sociology of Education 66: 91–103.

44. Eagly AH, Wood W (1999) The origins of sex differences in human behavior: Evolved dispositions versus social roles. American Psychologist 54: 408–423.

45. Eagly AH, Wood W, Diekman AB (2000) Social role theory of sex differences and similarities: A current appraisal. The developmental social psychology of gender. pp. 123–174.

46. Nash SC (1979) Sex role as mediator of intellectual functioning. In: Wittig MA, Petersen AC, editors. pp. 263–302. New York: Academic Press.

47. Hyde JS (2005) The gender similarities hypothesis. American Psychologist 60: 581–592.

48. Williams JE, Best DL (1990) Measuring sex stereotypes: a multination study. Newbury Park: Sage Publications.

49. Triandis HC (2004) The many dimensions of culture. The Academy of Management Executive (1993–2005) 18: 88–93.

50. Hofstede GH (2001) Culture's consequences: Comparing values, behaviors, institutions, and organizations across nations. Thousand Oaks, Calif.: Sage Publications.

51. Hyde JS, Linn MC (2006) Gender similarities in mathematics and science. Science 314: 599–600.

52. Hedges LV (2008) What are effect sizes and why do we need them? Child Development Perspectives 2: 167–171.

53. Cohen J (1988) Statistical power analysis for the behavioral sciences. Hillsdale, NJ: Lawrence Earlbaum Associates.

54. Rosenthal R, Rubin DB (1982) A simple, general purpose display of magnitude of experimental effect. Journal of Educational Psychology 74: 166–169.

55. Rosenthal R, Rubin DB (1982) Further meta-analytic procedures for assessing cognitive gender differences. Journal of Educational Psychology 74: 708–712.

56. Rosenthal R, DiMatteo MR (2001) Meta-analysis: Recent developments in quantitative methods for literature reviews. Annual Review of Psychology 52: 59–82.

57. Borenstein M, Rothstein HR (1999) Comprehensive meta-analysis: A computer program for research synthesis. Englewood, NJ: BioStat.

58. Borenstein M, Hedges LV, Higgins JPT, Rothstein HR (2009) Introduction to Meta-Analysis. West Sussex. UK: John Wiley & Sons, Ltd.

59. Field AP (2001) Meta-analysis of correlation coefficients: A Monte Carlo comparison of fixed- and random-effects methods. Psychological methods 6: 161–180.

60. Favreau OE (1997) Sex and gender comparisons: Does null hypothesis testing create a false dichotomy? Feminism & Psychology 7: 63–81.

61. Else-Quest NM, Hyde JS, Linn MC (2010) Cross-national patterns of gender differences in mathematics: A meta-analysis. Psychological Bulletin 136: 103–127.

62. World Economic Forum (2009) The Global Gender Gap Report 2009.

63. Dijkstra AG, Hanmer LC (2000) Measuring socio-economic gender inequality: Toward an alternative to the UNDP Gender-Related Development Index. Feminist Economics 6: 41–75.

64. Hawke JL, Olson RK, Willcut EG, Wadsworth SJ, DeFries JC (2009) Gender ratios for reading difficulties. Dyslexia 15: 239–242.

65. Shaywitz SE, Escobar MD, Shaywitz BA, Fletcher JM, Makuch R (1992) Evidence that dyslexia may represent the lower tail of a normal distribution of reading ability. New England Journal of Medicine 326: 145–150.

66. OECD (2010) PISA 2009 Results: Learning trends: Changes in student performance since 2000. (Volume V).

67. Hyde JS (2007) New directions in the study of gender similarities and differences. Current Directions in Psychological Science 16: 259–263.

68. National Science Foundation (2011) Women, minorities, and persons with disabilities in science and engineering: 2011. Arlington, VA: National Science Foundation.

69. Nosek BA, Banaji MR, Greenwald AG (2002) Math = male, me = female, therefore math≠ me. Journal of Personality and Social Psychology 83: 44–59.

70. Dwyer CA (1973) Sex differences in reading: An evaluation and a critique of current theories. Review of Educational Research 43: 455–467.

71. Hyde JS, Fennema E, Ryan M, Frost LA, Hopp C (1990) Gender comparisons of mathematics attitudes and affect. Psychology of Women Quarterly 14: 299–324.

72. Hanson SL, Schaub M, Baker DP (1996) Gender stratification in the science pipeline: A comparative analysis of seven countries. Gender and Society 10: 271–290.

73. Wai J, Lubinski D, Benbow CP (2009) Spatial ability for STEM domains: Aligning over 50 years of cumulative psychological knowledge solidifies its importance. Journal of Educational Psychology 101: 817–835

74. Nuttall RL, Casey MB, Pezaris E (2005) Spatial ability as a mediator of gender differences on mathematics tests: A biological-environmental framework. In: Gallagher AM, Kaufman JC, editors. pp. 121–142. Cambridge, UK: Cambridge University Pres.

75. Casey MB, Nuttall R, Pezaris E, Benbow CP (1995) The influence of spatial ability on gender differences in mathematics college entrance test scores across diverse samples. Developmental Psychology 31: 697–705.

76. Geary DC, Saults SJ, Liu F, Hoard MK (2000) Sex differences in spatial cognition, computational fluency, and arithmetical reasoning. Journal of Experimental Child Psychology 77: 337–353.

77. Lippa R, Collaer M, Peters M (2010) Sex differences in mental rotation and line-angle judgments are positively associated with gender equality and economic development across 53 nations. Archives of Sexual Behavior 39: 990–997.

78. Eccles JS, Blumenfeld P (1985) Classroom experiences and student gender: Are there differences and do they matter? In: Wilkinson L, Marrett CB, editors. pp. 79–114. Orlando, FL: Academic Press.

79. Eccles JS, Wigfield A, Harold RD, Blumenfeld P (1993) Age and gender differences in children's self- and task perceptions during elementary school. Child Development 64: 830–847.

80. Frome PM, Eccles JS (1998) Parents' influence on children's achievement-related perceptions. Journal of Personality and Social Psychology 74: 435.

81. Furnham A, Hosoe T, Tang T (2001) Male hubris and female humility? A cross-cultural study of ratings of self, parental, and sibling multiple intelligence in America, Britain, and Japan. Intelligence 30: 101–115.

82. Furnham A, Petrides K (2004) Parental estimates of five types of intelligence. Australian Journal of Psychology 56: 10–17.

83. Furnham A, Reeves E, Budhani S (2002) Parents think their sons are brighter than their daughters: Sex differences in parental self-estimations and estimations of their children's multiple intelligences. The Journal of Genetic Psychology 163: 24–39.

84. Nosek BA, Smyth FL, Sriram N, Lindner NM, Devos T, et al. (2009) National differences in gender–science stereotypes predict national sex differences in science and math achievement. Proceedings of the National Academy of Sciences 106: 10593–10597.

85. Newcombe NS, Ambady N, Eccles JS, Gomez L, Klahr D, et al. (2009) Psychology's role in mathematics and science education. American Psychologist 64: 538.

Questions:

1. Consider the section titled "Advantages of Nationally Representative Samples for Assessing Gender Differences." Why do you think this passage discusses sampling bias in the way it does?

2. Assess the strength of the article's structural framework. Point to the thesis, topic sentences, transitions, and guiding sub-headings. Does their presence work to make the article more clearly understandable? What elements work best?

3. List the article's hypotheses. Which of them are found to be borne out by the study data?

4. How does the article define "high power distance"? Does the "high power distance" labeling make sense to you? Are the countries clearly identified as to their high/low power distance status? Identify one of the

countries in either Table 3 or Table 4 as high power distance and justify your choice.

5. Consider the difference between citations which are included as though they made a point definitively, without any discussion provided, and citations which frame the inclusion of the material. For instance, is "Since the scores of males are more variable[12,18-19]" adequately convincing, especially for those who might not agree with the article's basic premise? Compare it to the description of Hyde et al. under the "Gender Differences in Mathematics and Science within the United States" heading. Find another example of each type of citation, and discuss the merits of the latter type of presentation.

6. What audience(s) is the article aimed at? What groups are the most likely audiences? Does the variation in audiences explain the ways the article presents its statistical choices? Why might the article emphasize the US as much as it does, given that the author is a postgraduate in Australia?

7. What are the article's major findings? Were any of the findings surprising to you? Does the article present any of its findings as being unexpected?

8. What do you make of the difference between math and science literacy rates for females in the data? What explanations can you think of for the fact that the science literacy rates are more variable than the math rates? What do you make of the difference in males' reading literacy across various countries?

9. Explain how to read Tables 6 and 7. Do you need more information in order to do so?

10. Identify 6 major grammar errors in this document. Class at least one of the types of errors that recur. To what extent does the inclusion of errors such as these affect your perception of the article? How might the Open Access platform lend itself to such errors? How might they be effectively addressed?

forty-one

Amy B. Wisniewski, Claude J. Migeon, Heino F. L. Meyer-Bahlburg, John P. Gearhart, Gary D. Berkovitz, Terry R. Brown, and John Money

Complete Androgen Insensitivity Syndrome: Long-Term Medical, Surgical, and Psychosexual Outcome

People with complete androgen insensitivity syndrome have a genetic mutation, on the X chromosome of an XY chromosome set, that means their bodies do not respond to male sex hormones. This scholarly article sets out to examine the physical characteristics and medical and psychosexual experiences of 14 participants with CAIS.

℘

ABSTRACT

Controversy concerning the most appropriate treatment guidelines for intersex[a] children currently exists. This is due to a lack of long-term information regarding medical, surgical, and psychosexual outcome in affected adults. We have assessed by questionnaire and medical examination the physical and psychosexual status of 14 women with documented complete androgen[b] insensitivity syndrome (CAIS). We have also determined participant knowledge of CAIS as well as opinion of medical and surgical treatment. As a whole, secondary sexual development of these women was satisfactory, as judged by both participants and physicians. In general, most women were satisfied with their psychosexual development and sexual function. Factors reported to contribute to dissatisfaction were sexual abuse in one case and marked obesity in another. All of the women who participated were satisfied

---

a *intersex* Possessing both male and female visible and/or genetic sex characteristics.

b *androgen* Refers to testosterone and other steroid hormones that are typically found in men. Androgens are responsible for stimulating the growth of male secondary sex attributes, such as facial hair.

with having been raised as females, and none desired a gender reassignment. Although not perfect, the medical, surgical, and psychosexual outcomes for women with CAIS were satisfactory; however, specific ways for improving long-term treatment of this population were identified.

The relative contributions of variables such as prenatal hormones and social rearing to psychosexual development have been studied in a variety of intersex populations.[1,2,3,4,5,6] These investigations have led to the wide acceptance of a multivariate conceptualization of gender development that emphasizes the importance of both nature and nurture.[7,8,9]

Recent reports of psychosexual outcome in two penis ablatio[a] patients[10,11] have led to reconsideration of sex assignment practices for intersexuals. One of the issues raised by these reports is the possibility that fetal androgen exposure influences the brains of such patients, resulting in male-typical psychosexual development. However, reports of psychosexual development in penis ablatio patients are conflicting, and to date it remains unclear how treatment of intersex patients should be revised.[12]

If androgens alone are important for male psychosexual development, then a group of intersex patients with complete androgen insensitivity syndrome (CAIS) would not be expected to exhibit a male bias, as they have a complete end-organ resistance to androgenic effects. However, other variables related to CAIS, such as the presence of a Y-chromosome, testes and shallow vagina, have been suggested to pose obstacles to healthy psychosexual development in this group.[13] Additionally, CAIS women provide an ideal opportunity to investigate potential influences of estrogens on gender development in 46,XY[b] individuals who are unresponsive to androgens, yet respond to estrogens.[14]

SUBJECTS AND METHODS

The research reported here was approved by the Joint Committee of Clinical Investigations of The Johns Hopkins University School of Medicine, The Johns Hopkins Hospital (Baltimore, MD). Written, informed consent was obtained from all subjects before participation. Participants were asked to complete a written questionnaire before their physical examinations. During the physical examination, participants were asked to confirm their original questionnaire responses and also to elaborate on responses that were unclear or for which the participant desired further discussion.

---

a *ablatio* Surgical removal.

b *46,XY* Term designating the number and type of chromosomes that are normal for a human male.

Diagnostic criteria for CAIS subjects

CAIS diagnosis was based on the following: 1) presence of testes along with 5 normal female external genitalia in a 46,XY individual, 2) identification of an androgen receptor[a] (AR) gene mutation, 3) spontaneous feminization (but with no menses) at puberty before gonadectomy with no virilization[b] despite normal or high male levels of testosterone, and 4) markedly decreased or absent postpubertal axillary[c] and pubic hair growth. The identification of an AR gene mutation is necessary for accurate diagnosis of AIS, as other syndromes of 46,XY intersex phenotypically resemble and can be reported erroneously as AIS.[15] This is mostly true for partial forms of AIS.

Subjects

The total population of adult women with CAIS recorded in The Johns Hopkins Pediatric Endocrinology[d] Clinic at the time of study consisted of 20 individuals, among whom 4 were not located, 1 did not respond to a study invitation, and 1 chose not to participate. The 14 women with CAIS who participated ranged in age from late 20s to mid-60s (mean age, 45 yr). All had been previously examined at the Johns Hopkins Pediatric Endocrinology Clinic. Nonparticipants did not differ from participants in terms of age or medical/surgical treatment received. To maintain participant anonymity age is presented in categories of 5 yr at the time of participation. Five of the 14 participants had been included in previous psychological studies when they were adolescents or young adults.[16,17,18]

The specific AR gene mutations identified in subjects varied, as expected in light of the large number of AR gene lesions known to be associated with CAIS.[19]

The only variance in the physical characteristics related to CAIS was with respect to adult sexual hair. Two women (14%) had absolutely no pubic or axillary hair, whereas the remaining 12 participants (86%) had no axillary or facial hair, but had minimal amounts of fine, silky pubic hair (early Tanner stage[e] 2; Table 1).

---

a *androgen receptor* Protein that attaches itself to androgens, aiding in the development of male physical characteristics.

b *feminization* Development of female attributes in a male body; *gonadectomy* Surgical procedure in which one or both ovaries or testes are removed, and the production of hormones in the body is reduced as a result; *virilization* Development of male physical attributes.

c *axillary* Refers to the underarm area.

d *Endocrinology* Study of the endocrine glands, hormones, and their effects on the body.

e *Tanner stage* The Tanner scale indicates stages of physical sexual development; there are 5 stages in total.

Table 1.
Information regarding race, age, androgen receptor gene mutation, ages at surgeries, and appearance of genitalia for CAIS subjects at time of participation

ID no.	Race	Age (yr)	Androgen receptor gene mutation	Adult sexual hair	Age (yr) at gonad-ectomy	Age (yr) at vagino-plasty	Length of clitoris (cm)	Depth of vagina (cm)
1	Caucasian	25–30	Exon 5	Minimal	14	None	1	12
2	Caucasian	30–35	Exon 1	Minimal	2	None	1.5	8.5
3	Caucasian	36–40	Exon 3	None	16	None	1	8
4	African American	36–40	Exon 3	None	19	None	1	4.5
5	Caucasian	36–40	Deletion	Minimal	17	17, 35	0.75	14
6	African American	36–40	Exon 5	Minimal	18	None	1	13
7	African American	36–40	Exon 5	Minimal	15	None	1	13
8	Caucasian	41–45	Deletion	Minimal	18	None	1	6
9	Caucasian	41–45	Deletion	Minimal	17	None	0.2	7
10	African American	46–50	Exon 5	Minimal	16	16	1	8
11	Caucasian	46–50	Exon 7	Minimal	16	16	1	7
12	Caucasian	51–55	Not available	Minimal	13	21	0.1	4.5
13	Caucasian	56–60	Exon 7	Minimal	20	20	0.1	8
14	Caucasian	61–65	Exon 7	Minimal	21	21	1	10

Physical measurements

Physical measurements at birth. Appearance of external genitalia and weight at birth were determined from medical records. Birth weight was considered in light of previous reports of reduced birth weight associated with AIS.[20]

Physical measurements in adulthood. These were obtained during an examination that took place in the Pediatric Clinical Research Unit at The Johns Hopkins Hospital as well as from medical records. Adult height was calculated as a percentile extrapolated from a growth chart for women (Serono, Norwell, MA). Weight at time of participation was determined in relation to medium build statistical norms published by Metropolitan Life Insurance Co. for women (www.indexmedico.com). Weight was documented according to the number of kilograms below or above the mean value for healthy women in relation to height. This mode of data presentation was elected to maintain the anonymity requested by participants.

The cosmetic appearance of external genitalia and breast development were evaluated independently and then agreed upon by an experienced pediatric endocrinologist (C.J.M.[a]) and a reconstructive surgeon (J.P.G.). Vaginal depth was measured with graduated vaginal dilators and reported in centimeters. Clitoral length was measured with a tape measure and reported in centimeters. Breast size was also measured with a tape measure and reported in centimeters, both horizontally and vertically for each breast, as conventionally performed by pediatric endocrinologists. Ages at gonadectomy and vaginoplasty[b] were obtained from medical records. The number of surgical procedures as well as type of vaginoplasty procedure performed were obtained from medical records and were verified during physical examination.

Participants were asked first with an open-ended question and then as a review of the major body systems if they experienced medical conditions other than CAIS, to determine whether women with this syndrome were at an increased risk for additional problems. Finally, participants were asked during the physical exam to indicate their level of compliance with estrogen therapy after gonadectomy.

Psychosexual assessment

Sexual function. Participants were asked about the overall adequacy of their genitalia for sexual functioning, the estimated strength of their libido, and their ability to experience orgasms in a written questionnaire and again during the physical examination. Subjects also reported their degree of overall satisfaction with sexual functioning.

Body image. Participants were asked about their level of satisfaction with their physical appearance in a written questionnaire and again during the physical examination. When dissatisfied with their physical appearance, participants were asked which physical characteristics contributed to this dissatisfaction.

Self-perceived femininity and masculinity. Participants were asked in a written questionnaire, how masculine and feminine they considered themselves during their childhood, adolescence, and adulthood. This section of the questionnaire was based on the Sexual Behavior Assessment Schedule–adult version (Meyer-Bahlburg, H.F.L., and A.A. Ehrhardt, unpublished).

Sexual orientation. Participants were asked in a written questionnaire if they were sexually attracted to, fantasized about, or participated in sexual

---

a *C.J.M.* Claude J. Migeon, one of the paper's co-authors. Throughout this article, the co-author responsible for a particular portion of the research is identified by her or his initials in parentheses.

b *vaginoplasty* Surgical procedure in which a vagina is repaired or created.

activity with men (female heterosexual orientation), women (female homosexual orientation), or both (female bisexual orientation) during adolescence and adulthood. This section of the questionnaire was adapted from the Sexual Behavior Assessment Schedule–adult version (Meyer-Bahlburg, H.F.L., and A.A. Ehrhardt, unpublished).

Marriage and motherhood. Frequencies of marriage and motherhood by adoption were asked for all participants in a written questionnaire.

Satisfaction with sex of rearing. Participants reported in a written questionnaire and during the physical examination on their degree of satisfaction with being a woman. Furthermore, participants were asked if at any point in their lives they had questioned their female gender or considered changing sex.

Participants' opinions concerning timing of surgical treatment/third sex. Participants were asked their opinion regarding appropriate timing of gonadectomy and vaginoplasty in a written questionnaire and during the physical examination. Although the question of optimal timing for surgical treatment of intersex patients has focused on the reconstruction of ambiguous external genitalia, surgical modification of gonadal and posterior vaginal status in CAIS women also leads to permanent consequences and therefore should be considered in this patient group.

Participants were asked their opinion regarding the categorization of intersex children as a "third gender" in opposition to the more traditional categories of male or female. Although this third gender classification usually applies to individuals with ambiguous external genitalia, it can be extended to CAIS women who possess a Y-chromosome and testes. Participants were also asked if they agreed with the concept of recognizing a third intersex category within our society as an alternative to treating intersex patients as either males or females. This question was included due to suggestions that intersex patients should not be raised according to strictly male or female categories.[21,22]

Long-term psychological treatment

Participants reported in a written questionnaire and during physical examination whether they had ever received counseling concerning their condition.

Knowledge of medical history

Participants reported their level of satisfaction regarding their knowledge of AIS in a written questionnaire. Participants were also evaluated by a pediatric endocrinologist (C.J.M.) and psychologist (A.B.W.) to determine their level of understanding of AIS at the time of study participation. Specifically documented was participants' knowledge of their gonadal development, karyotype,[a] and importance of estrogen therapy as presented in our Patients' Guide to Syndromes of Abnormal Sex Differentiation.[23]

Statistical analysis

Due to the unique nature of CAIS, it is impossible to identify an appropriate control group. Additionally, the low frequency of this syndrome in the general population led to data presentation in the form of descriptive statistics and raw data.

RESULTS

Physical measurements at birth

Mean birth weight for participants was 3.5 kg (range, 2.26–4.1 kg), which did not differ from that of the general population (3.4 kg; range, 2.5–4.6 kg).[24] External genitalia was completely female with no abnormalities at birth for all participants.

Physical measurements in adulthood

Adult height of eight subjects (57%) fell at or exceeded the 90th percentile [25] of the range of control adult females. Adult height of the remaining women (43%) fell between the 50th and 75th percentiles. Seven participants (50%) were within ±10 kg of their ideal weight range. The remaining seven participants exceeded their ideal weight by 15 kg or more, three of whom exceeded their ideal by 80 kg or more.

Most gonadectomies and vaginoplasties were performed on participants during their adolescence or adulthood (Table 1). Eight participants did not require vaginoplasty for peno-vaginal intercourse (one was homosexual), and one woman required vaginoplasty twice. Vaginoplasty consisted of a McIndoe partial thickness skin graft in all cases. Both examining physicians rated the appearance of the external genitalia as good for all women. Average clitoral length was 0.83 cm (range, 0.1–1.5 cm); no participant exhibited clitoral enlargement, and none required clitoroplasty.[b] The eight women who did not experience vaginoplasty had an average vaginal length of 9 cm (range, 4.5–13 cm). The six women with a history of vaginoplasty had an average vaginal length of 8.6 cm (range, 4.5–14 cm). These measures are consistent with previous measures of mean vaginal length ranging from 7–11

---

a *karyotype* Chromosomal characteristics.
b *clitoroplasty* Surgical procedure in which a clitoris is repaired or created.

cm.[25,26,27] All women had breast development, with wide variability in breast size (range, 16 × 14 to 41 × 31 cm).

Regarding the occurrence of other medical conditions among CAIS women, obesity (43%) and bone loss (43%) were both reported most frequently.

The degree of compliance with estrogen replacement therapy was determined for all participants based on information obtained from written questionnaire responses, medical charts, and discussion during the physical examination. Nine participants reported that they were compliant, whereas the remaining five women reported they had not taken estrogen replacement for most of their adult lives after gonadectomy.

Psychosexual assessment

Sexual function. Eleven of the women with CAIS studied (78%) reported satisfaction with their genitalia in terms of sexual functioning, whereas 3 others (22%) reported dissatisfaction. Ten women (71%) reported satisfaction with their sexual function overall, and four women (29%) were dissatisfied.

Despite their complete end-organ resistance to androgens, the majority of CAIS women (71%) reported a self-estimated libido of average strength or stronger, and 10 of the 13 who responded (77%) reported an ability to experience orgasms.

Body image. When provided with three response options (mainly satisfied, somewhat dissatisfied, mainly dissatisfied) regarding degree of satisfaction with physical appearance, eight women (57%) reported they were mainly satisfied, five (36%) reported being somewhat dissatisfied, and one woman reported mainly dissatisfaction with her physical appearance. Of the six women who reported being either somewhat dissatisfied or mainly dissatisfied with their physical appearance, five reported inadequate body hair, five reported looking younger than actual age, and three reported obesity as factors that contributed to this dissatisfaction.

Self-perceived femininity and masculinity. Women with CAIS overwhelmingly reported a high degree of femininity along with a low degree of masculinity throughout development. Mean femininity rating (not feminine = 1 to highly feminine = 5) for subjects during childhood was 4.4 (range, 2–5), during adolescence was 4.2 (range, 3–5) and during adulthood was 4.6 (range, 4–5). Mean masculinity rating (not masculine = 1 to highly masculine = 5) for subjects during childhood was 1.4 (range, 1–4), during adolescence was 1.4 (range, 1–3), and during adulthood was 1.1 (range, 1–2).

Sexual orientation. A large majority of CAIS women reported female het-erosexual orientation in terms of sexual attraction, fantasies, and experience during both adolescence (100%) and adulthood (93%). The one participant who reported homosexual attraction, fantasies, and experiences indicated that a lesbian orientation applied only to her adulthood. Clearly, in this case the development of female homosexuality was not associated with androgen exposure.

Marriage and motherhood. Seven women (50%) were married at the time of participation, and the mean age at first marriage was 27 yr (range, 16–38 yr). Five of these and one unmarried woman (43%) were mothers through adoption. Of the seven participants who were not married, one was engaged, one was homosexual, one was not interested in marriage, and the remaining four women expressed a desire for marriage to a man but had not yet met a satisfactory partner.

Satisfaction with sex of rearing. CAIS women unanimously reported satisfac-tion with being a woman (100%). Two participants (18%) questioned their physical status as women, but none reported a desire to change sex to that of a man.

Participants' opinions of timing of surgical treatment/third sex. The majority of women who responded indicated the most appropriate timing for their sur-gical procedures was during adolescence or adulthood (8 of 10 respondents, or 80%). Two women reported the most appropriate timing for surgery was during infancy.

The vast majority of participants (81% of the 11 who responded) did not approve of rearing intersex children according to a third gender.

Long-term psychological treatment

The majority of CAIS women studied received some form of counseling (83%) at various ages and for various lengths of time (1–15 yr) concerning aspects of their syndrome.

Knowledge of medical history

Eight CAIS women (57%) exhibited no understanding of CAIS (*i.e.* were unaware of their karyotype, gonadal characteristics, or the importance of es-trogen replacement) at the time of participation, and only 64% indicated that they were satisfied with their level of knowledge regarding their condition. It could not be determined if participants ever received information about their syndrome or if they were previously informed but did not recall the information. Nevertheless, this finding emphasizes the importance of following CAIS patients into adulthood and subsequently offering them adult education about their syndrome.

Discussion

40 To determine the natural history of CAIS and long-term outcome of treatment intervention, it is necessary to follow affected individuals over time. Previous knowledge of CAIS has been based on studies of relatively young patients. Such young CAIS women overwhelmingly report contentment with being female, a desire for marriage and motherhood, sexual attraction and practice exclusively oriented toward men, as well as the ability to experience orgasms.[16,17,18,28] However, it is likely that older women have the benefit of greater experience with their gender and sexuality compared to younger women and girls. The present study extends our knowledge of CAIS to an older cohort of women.

Furthermore, quality of sexual function and appearance of genitalia and secondary sexual characteristics were considered in conjunction with our psychosexual evaluation. Previous psychosexual studies of intersex patients have not consistently assessed outcome of medical and surgical treatment in terms of sexual function and appearance. It is likely that factors such as an individual's cosmetic appearance and quality of sexual function influence gender and sexuality, and therefore should be considered in psychosexual outcome research concerning intersex populations.

Physical

CAIS babies appeared no different from unaffected infants in terms of birth weight or genital appearance. Consistent with previous reports, CAIS women tend to be tall[29,30] and experience normal feminization of secondary sexual characteristics, with the exception of lacking female-typical amounts of axillary and pubic hair in adulthood.[31] Interestingly, 12 participants had a minimal amount of fine, soft pubic hair but no axillary hair. This is consistent with previous observations in this group of vellus down on the body that is not androgen dependent.[32] Aside from lacking sexual hair, the external genitalia of all participants appeared normal. Six women (43%) had undergone vaginoplasty, but none required or had clitoroplasty.

Psychosexual

Most CAIS women were satisfied with their sexual functioning. However, three (21%) were dissatisfied. Of these, one did not receive any genital reconstructive surgery despite a shallow vaginal depth and thinks she would be unable to participate in peno-vaginal intercourse. Although this woman has never attempted peno-vaginal intercourse, she has discovered great satisfaction from homosexual sexual activity. The remaining two had undergone vaginoplasties that resulted in sufficient vaginal length for peno-vaginal intercourse. Their dissatisfaction may be related to other variables. One woman was a victim of sexual abuse early in life and is presently in poor health resulting from substance abuse; the other reported severe dissatisfaction with her body image.

Libido and ability to experience orgasms were not a problem for the CAIS women in this study. This illustrates that although androgens may contribute to libido and orgasmic potential in non-CAIS women,[33,34,35] libido and orgasm can be experienced by women who exhibit complete end-organ insensitivity to androgens. Furthermore, there were no differences in self-reported libido based on compliance with estrogen therapy. Vaginal lubrication is another characteristic of sexual functioning thought to be related to estrogen levels. No participant reported difficulty with vaginal lubrication on the written questionnaire or during the physical examination.

Despite some of the unhappiness with physical attributes, the majority of women with CAIS were mainly satisfied with their physical appearance. 45

CAIS women overwhelmingly perceive themselves as highly feminine and not masculine throughout development. Additionally, CAIS women largely report their sexual attraction, fantasies, and experiences were best described as female heterosexual. Concerning the one woman who reported heterosexual attraction and fantasies in adolescence followed by homosexual thoughts and actions in adulthood, perhaps a short vagina coupled with fear of vaginoplasty contributed to this change. Several women were married and/or mothers. All participants reported being mainly satisfied when asked their degree of satisfaction in being a woman. Of the two women who stated that they questioned their physical status as women, one responded that this was due to her inability to menstruate and become pregnant, and the other reported that this was in response to media articles she encountered regarding intersexuality.

Treatment

Several issues of long-term health status are of concern in our CAIS participants. First, long-term compliance with estrogen replacement is less than optimal in this group. Second, older CAIS women as a group are obese. However, obesity among women with CAIS mirrors rates observed in the population of American women at large and does not appear to be directly related to this condition.[36,37] Lastly, women with CAIS appear at to be at risk for bone-related disease. It is unclear at this time, however, if this is a result of androgen insensitivity *per se*, a consequence of inadequate estrogen replacement, or both.[38]

Most women (80% who responded) sought psychological counseling at some point in development, as indicated by the written questionnaire. The majority of participants did not believe a third gender category was appropriate for intersex patients. These points stress the importance of providing well rounded care to these patients that includes counseling services as well as medical and surgical care.

The great majority of these CAIS women (78%) reported that the most appropriate timing of gonadectomy and vaginoplasty procedures was during adolescence or adulthood. More than half (64%) did not fully understand their diagnosis in adulthood, and the majority indicated a desire to better understand their condition. However, all women stressed the importance of confidentiality regarding their condition. These results support the concerns of CAIS patient advocacy groups regarding the disclosure of medical information to patients and the postponement of vaginoplasty to late adolescence or adulthood for CAIS patients.[8,39]

50   Complete insensitivity to androgen action is clearly an extreme in the spectrum of congenital malformations of sex organs. Additional studies of long-range outcome of subjects with partial AIS and other conditions associated with ambiguous genital development are in progress.

(2000)

This work was supported by a grant from The Genentech Foundation for Growth and Development (98–33C to C.J.M.), NIH National Research Service Award F32HD08544 (to A.B.W.), and by NIH, National Center for Research Resources, General Clinical Research Center Grant RR-00052.

References

1. **Money JM, Hampson JG, Hampson JL.** 1957 Imprinting and the establishment of gender role. *Arch Neurol Psychol.* **77**:333–336.

2. **Money J, Dalery J.** 1976 Iatrogenic homosexuality: Gender identity in seven 46,XX chromosomal females with hyperadrenocortical hermaphroditism born with a penis, three reared as boys, four reared as girls. *J Homosexuality.* **1**:357–371.

3. **Money J, Schwartz M, Lewis VG.** 1984 Adult erotosexual status and fetal hormonal masculinization and demasculinization: 46,XX congenital virilizing adrenal hyperplasia and 46,XY androgen-insensitivity syndrome compared. *Psychoneuroendocrinology.* **9**:405–414.

4. **Money J, Lehne GK, Pierre-Jerome F.** 1985 Micropenis: gender, erotosexual coping strategy and behavioral health in nine pediatric cases followed to adulthood. *Comp Psychol.* **26**:29–42.

5. **Money J, Devore H, Norman BF.** 1986 Gender identity and gender transposition: longitudinal outcome study of 32 male hermaphrodites assigned as girls. *J Sex Mar Ther.* **12**:165–181.

6. **Money J, Norman BF.** 1987 Gender identity and gender transposition: longitudinal outcome study of 24 male hermaphrodites assigned as boys. *J Sex Mar Ther.* **13**:75–79.

7. **Money J.** 1994 Hormones, hormonal anomalies, and psychologic health care. In: Kappy MS, Blizzard RM, Migeon CJ, eds. Wilkins' the diagnosis and treatment of endocrine disorders in childhood and adolescence, 4th Ed. Springfield: Thomas; 1141–1178.

8. **Money J.** 1994 Sex errors of the body and related syndromes: a guide to counseling children, adolescents and their families, 2nd Ed. Baltimore: Brookes.

9. **Money J, Ehrhardt A.** 1972 Man and woman, boy and girl: differentiation and dimorphism of gender identity from conception to maturity. Baltimore: Johns Hopkins University Press.

10. **Bradley SJ, Oliver GD, Chernick AB, Zucker KJ.** 1982 Experiment of nurture: ablatio penis at 2 months, sex reassignment at 7 months, and a psychosexual follow-up in young adulthood. Pediatrics. 102:E91–E95 (http://www.pediatrics.org/cgi/content/full/102/1/e9).

11. **Diamond M, Sigmundson HK.** 1997 Sex reassignment at birth: long-term review and clinical implications. *Arch Pediatr Adolesc Med.* **151**:298–304.

12. **Zucker KJ.** 2000 Intersexuality, and gender identity differentiation. *Annu Rev Sex Res.***10**:1–69.

13. **Migeon CJ, Wisniewski AB.** 1998 Sex differentiation: from genes to gender. *Horm Res.* **50**:245–251.

14. **Meyer-Bahlburg HFL.** 1997 The role of prenatal estrogens in sexual orientation. In: Ellis L, Ebertz L, eds. Sexual orientation: toward biological understanding. Westport: Praeger.

15. **Boehmer ALM, Brinkmann AO, Sandkuijl LA, et al.** 17β-Hydroxysteroid dehydrogenase 3 deficiency: diagnosis, phenotypic variability, population genetics, and world-wide distribution of ancient and *de novo* mutations. J Clin Endocrinol Metab. In press.

16. **Money J, Ehrhardt AA, Masica DN.** 1968 Fetal feminization induced by androgen insensitivity in the testicular feminizing syndrome: effect on marriage and maternalism. *Johns Hopkins Med J.* **123**:105–114.

17. **Masica DN, Money J, Ehrhardt AA.** 1971 Fetal feminization and female gender identity in the testicular feminizing syndrome of androgen insensitivity. *Arch Sex Behav.* **1**:131–142.

18. **Lewis VG, Money J.** 1983 Gender-identity/role: G-I/R. A. XY (androgen-insensitive) syndrome and XX (Rokitansky) syndrome of vaginal atresia. In: Dennerstein L, Burrows G, eds. Handbook of psychosomatic obstetrics and gynaecology. Amsterdam: Elsevier.

19. **Gottlieb B, Trifiro M, Lumbroso R, Pinsky L.** 1997 The androgen receptor gene mutation database. *Nucleic Acids Res.* **25**:158–162 (www.mcgill.ca/androgendb).

20. **de Zegher F, Francois I, Ibanez L.** 1999 Pediatric endocrinopathies related to reduced fetal growth. *Growth Genet Horm.* **15**:1–6.

21. **Fausto-Sterling A.** 1993 The five sexes: why male, and female are not enough. *The Sciences.* **33**:20–25.

22. **Diamond M, HK Sigmundson.** 1997 Management of intersexuality: guidelines for dealing with persons with ambiguous genitalia. *Arch Pediatr Adolesc Med.* **151**:1046–1050.

23. **Migeon CJ, Wisniewski AB, Gearhart JP.** 1999 Syndromes of abnormal sex differentiation: a guide for patients and their families, http://www.med.jhu.edu/pedendo/intersex/.

24. **Kaplan SL.** 1987 Growth: normal and abnormal. In: Rudolph AM, Hoffman JIE, eds. Pediatrics, 18th Ed. Norwalk: Appleton & Lange; 75–91.

25. **Dickinson RL.** 1949 The vagina. In: Human sex anatomy, 2nd Ed. Baltimore: Williams & Wilkins; 34–35.

26. **Masters WH, Johnson VE.** 1966 The vagina. In: Human sexual response. Boston: Little Brown; 73–75.

27. **Given FT, Muhlendorf IK, Browning GM.** 1993 Vaginal length and sexual function after colopexy for complete uterovaginal eversion. *Am J Obstet Gynecol.* **169**:284–287.

28. **Slijper FME, Drop SLS, Molenaar JC, de Muinck Keizer-Schrama MPF.** 1998 Long-term psychological evaluation of intersex children. *Arch Sex Behav.* **27**:125–144.

29. **Polani PE.** 1970 Hormonal, and clinical aspects of hermaphroditism and the testicular feminizing syndrome in man. Phil Trans R. *Soc.* **259**:187–204.

30. **Migeon CJ, Brown TR, Fichman KR.** 1981 Androgen insensitivity syndrome. In: Josso N, ed. The intersex child. Pediatric and adolescent endocrinology. Basel: Karger; 171–202.

31. **Migeon CJM, Berkovitz G, Brown TR.** 1994 Sexual differentiation and ambiguity. In: Kappy MS, Blizzard RM, Migeon CJ, eds. Wilkins' the diagnosis and treatment of endocrine disorders in childhood and adolescence, 4th Ed. Springfield; Thomas; 573–715.

32. **Quigley CA, De Bellis A, Marschke KB, El-Awady MK, Wilson EM, French FS.** 1995 Androgen receptor defects: historical, clinical and molecular perspectives. *Endocr Rev.* **16**:271–321.

33. **Money J.** 1965 Influence of hormones on sexual behavior. *Annu Rev Med.* **16**:67–82.

34. **Sherwin BB.** 1988 A comparative analysis of the role of androgen in human male, and female sexual behavior. *Psychobiology.* **16**:416–425.

35. **Wiebke A, Callies F, Van Vlijmen JC, et al.** 1999 Dehydroepiandrosterone replacement in women with adrenal insufficiency. *N Engl J Med.* **341**:1013–1020.

36. **Calle EE, Thun MJ, Petrelli JM, Rodriguez C, Heath Jr CW.** 1999 Body-mass index and mortality in a prospective cohort of U.S. adults. *N Engl J Med.* **341**:1097–1105.

37. **Sichieri R, Everhart JE, Hubbard VS.** 1991 Relative weight classifications in the assessment of underweight and overweight in the United States. *Int J Obesity.* **16**:303–312.

38. **Marcus R, Leary D, Schneider DL, Shane E, Favus M, Quigley CA.** 2000 The contributions of testosterone to skeletal development and maintenance: lessons from the androgen insensitivity syndrome. *J Clin Endocrinol Metab.* **85**:1032–1037.

39. **Medhelp.** 1999. http://www.medhelp.org/www/ais.

Questions:

1. In what ways might this article suffer from confirmation bias?

2. Assess the scales used in this study. Do you think a scale of perceived masculinity/perceived femininity is useful in a twenty-first-century context? Why or why not?

3. Can you find the "Sexual Behavior Assessment Schedule – adult version" online? Can you find it using your academic library or your public library? Is an unpublished schedule a clinically useful tool? Why or why not?

4. Look at Table 1. Based on the contents and organization of the table, what do you think the researchers considered most important? In what ways are the participants similar? In what ways are they different?

5. Consider the impact on study results of conducting vaginal and breast measurements during an assessment of "satisfaction" with gender assignment.

6. What are the problems with using case reports and small, non-random, non-anonymous surveys or interviews as the basis of a scientific report?

7. Over a third of these participants did not take estrogen therapy. Based on the information given in the article, can you tell if these are the same people who did not have vaginoplasty? What, if any, types of connections does Table 1 allow you to make? What does it privilege as linked information?

8. None of these participants were raised as intersex children, but the investigators report on their approval/disapproval of "rearing intersex children according to a third gender." Can you tell what question or questions were asked to elicit this response? In a consideration of whether intersex children should be raised according to a third gender, how much weight should be given to the opinions the participants expressed in this study?

9. Should a child have to be raised as either male or female? Why or why not?

10. Consider the implications of the following description: "Concerning the one woman who reported heterosexual attraction and fantasies in adolescence followed by homosexual thoughts and actions in adulthood, perhaps a short vagina coupled with fear of vaginoplasty contributed to this change." What does it tell you about the beliefs of the researchers?

11. Research John Money, one of the researchers involved in this study, with particular attention to the controversy surrounding his treatment of David Reimer. Identify ways in which "Complete Androgen Insensitivity Syndrome" is both responding to that controversy and aimed more specifically at showing that his techniques are appropriate for "intersex" individuals (if not for sex reassignment to another sex).

Appendix A:
How to Use Sources[a]

DOCUMENTATION AND RESEARCH

This final section looks at something that concerns you as a student in a very vital way – and that is how to get other people's scientific information into your own paper without misrepresenting or plagiarizing it. There are a variety of reasons for including the results of research in your own writing. Outside sources can help support or clarify your points, or they can serve as counter-arguments against which to demonstrate the strengths of your arguments. Sources are also useful for showing where your paper's argument is located in the wider scientific conversation among writers studying the same subject. At the very least, including source material shows that you are acquainted with the latest thinking on your research subject.

In incorporating research into your writing, you must make sure to document your sources accurately and completely. This is a service to your reader, who might like to investigate your topic by looking up your paper's sources directly; every academic citation system gives readers all the information they need to access original source material. But it is also critical that there be complete clarity about which parts of a work are yours and which parts come from other sources. Anything less is dishonest, as it engages in a kind of cheating known as plagiarism.

Plagiarism

Most people understand that taking someone else's writing and passing it off as one's own is intellectual thievery. But it is important to be aware that you may commit plagiarism even if you do not use precisely the same words another person wrote in precisely the same order. For instance, here is an actual example of plagiarism. *Globe and Mail* newspaper columnist Margaret Wente[b] borrowed material for one of her columns from a number of works,

---

a   This material has been adapted from the "Documentation and Research" section of *The Broadview Pocket Guide to Citation and Documentation*," by Maureen Okun; these pages are copyright © Maureen Okun 2013 and Broadview Press 2014, with revisions by Catherine Nelson-McDermott.

b   Note that if this is the first use of these writers' names in one of your sentences (i.e., not the parenthetical reference), MLA citation format requires full names. However, the science-writing formats most frequently chosen (APA, Chicago) do not usually require this courtesy.

including an article by Dan Gardner that had appeared the previous year in another newspaper (the *Ottawa Citizen*) and a book by Robert Paarlberg called *Starved for Science* (which was the subject of Gardner's article). The similarities were brought to light by media commentator Carol Wainio, who presented a series of parallel passages, including the following, on her blog *Media Culpa* (the fonts are Wainio's – simple bold is for direct copying; the bold + italics is for "near copying"):

> Gardner: ***Many NGOs working in Africa in the area of development and the environment have been advocating against the modernization of traditional farming practices***, Paarlberg says. "**They believe that traditional farming in Africa incorporates indigenous knowledge that shouldn't be replaced by science-based knowledge introduced from the outside.** They encourage Africa to stay away from fertilizers, and be certified as organic instead. And in the case of genetic engineering, they warn African governments against making these technologies available to farmers."

> Wente: ***Yet, many NGOs working in Africa have tenaciously fought the modernization of traditional farming practices.*** **They believe traditional farming in Africa incorporates indigenous knowledge that shouldn't be replaced by science-based knowledge introduced from the outside.** As Prof. Paarlberg writes, "They encourage African farmers to stay away from fertilizers and be certified organic instead. And they warn African governments to stay away from genetic engineering."

Wente does not always use exactly the same words as her sources, but no one reading the passages can doubt that one writer is appropriating the phrasings of the others. Additionally, where Wente *does* quote Paarlberg directly, the quotation is lifted from Gardner's article and should be identified as such.

The penalties for such practices are not trivial; Wente was publicly reprimanded by her employer, and the CBC radio program *Q* removed her from its media panel. Other reporters have been, justifiably, fired under similar circumstances. At most colleges and universities, students are likely to receive a zero if they are caught plagiarizing – and they may be expelled from the institution. It's important to be aware, too, that penalties for plagiarism make no allowance for intent; it is no defense that a writer took someone else's words "by mistake" rather than intentionally.

How, then, can you be sure to avoid plagiarism? First of all, be extremely careful in your note-taking, so as to make it impossible to imagine, a few days later, that words you have jotted down from somewhere else are your own. This is why notes need to be in separate file or book from your own ideas. (In her *Globe and Mail* column responding to the plagiarism charges, Wente, in fact, claimed that she had accidentally mixed a quotation into her own ideas.) If your note-taking is reliable, then you will know which words need to be credited. One way to rewrite the passage above would simply be to remove the material taken from Gardner and to credit Paarlberg by quoting him directly, if you were able to access his book and could do so:"As Robert Paarlberg has argued in his book *Starved for Science*, many NGOs 'believe that traditional farming in Africa incorporates indigenous knowledge that shouldn't be replaced by science-based knowledge introduced from the outside.'" You would, of course, look up and provide the page number as well.

You may notice that the quoted material is a statement of opinion rather than fact – controversial views are being given, but without any evidence provided to back them up – so a careful reader would wonder whether NGOs are really as anti-science as the quotation suggests, or whether the writer hasn't done enough research on the debate. If you were to make an assertion like this in a paper of your own it would not be enough just to quote Paarlberg; you would need to do much more research and find information to support or deny your claim. If you are including quotations in an essay, the best sources to quote are not necessarily those which express opinions that mirror the ones you are putting forward. In a case such as this, for example, the argument would have been much more persuasive if Wente had quoted an official statement from one of the NGOs she was attacking. If her article had quoted a source making this specific case against "science-based knowledge" and then argued directly against that source's argument, Wente's own position would have been strengthened. Quoting many such sources would provide proof that the article's characterization of the position of NGOs was factually accurate.

Whenever you do quote someone else, it's important to cite the source. But do you need a citation for everything that did not come from your own knowledge? Not necessarily. Citations are usually unnecessary when you are touching on common knowledge (provided it is, in fact, common knowledge, and provided your instructor has not asked you to do otherwise). If you refer to the chemical composition of water, or the date when penicillin was discovered, you are unlikely to need to provide any citation, even if you used a source to find the information, since such facts are generally available and uncontroversial. (Make sure, however, to check any "common knowledge" with several reputable sources; if your information is incorrect, it reflects poorly on you, especially if you have not cited your source.) If you have any doubts about whether something is common knowledge or not, cite it; overcautiousness is not a serious problem, but plagiarism always is.

Citation and Documentation Styles

Citing sources is fundamental to writing a good research paper, but no matter how diligent you are in making your acknowledgments, your paper will not be taken seriously unless its documentation is formatted according to an appropriate and accepted referencing style. Each academic discipline has adopted a particular system of referencing as its standard, and students and researchers writing in that discipline are expected to follow these guidelines. *The Broadview Guide to Citation and Documentation* outlines the four most common systems.[a] Almost all of the humanities use the documentation guidelines created by the Modern Language Association (MLA). The discipline of history, which might be said to be on the boundary between the humanities and the social sciences, tends, in North America, to prefer the documentation guidelines of the *Chicago Manual of Style* (Chicago Style). Some authorities in the applied sciences (such as engineering) also prefer Chicago Style. The social and some health sciences typically follow the style rules of the American Psychological Association (APA), while other sciences commonly use the referencing systems of the Council of Science Editors (CSE). Each of these styles is different, and sometimes an individual journal will use a "house" modification of one of the styles. In order to be considered a responsible researcher and writer, you will need to carefully follow the specific format required for your discipline or intended journal. Each set of citation guidelines uses specific formulae for types of entries, so the fact that there are different guidelines is not as problematic as it might at first seem. Conveniently, some word processors and other sources will put your paper's references into an MLA, APA, or Chicago style so long as you provide the necessary information.

Incorporating Sources

There are three main ways of working source material into a paper: summaries, paraphrases, and direct quotations. The following examples show you how each of the three approaches can go wrong, and how to fix the resulting accidental plagiarism. The passage used in the following examples is from Terrence W. Deacon's *The Symbolic Species: The Co-Evolution of Language and the Brain*, page 102. Note that these examples, and the ones that follow them, use various in-text parenthetical citation styles (mostly MLA and APA or Chicago), so be sure not to use them as examples of how to do MLA,

---

a   Purdue University's Online Writing Lab is also a respected source for learning how to do postsecondary writing correctly, and you can visit its website for more information on how to ethically paraphrase and summarize. OWL also has up-to-date information on the various citation styles.

APA, Chicago, or CSE. Nonetheless, whatever style your discipline uses, the requirements for correct incorporation of sources do not change.

> *original source* Over the last few decades language researchers seem to have reached a consensus that language is an innate ability, and that only a significant contribution from innate knowledge can explain our ability to learn such a complex communication system. Without question, children enter the world predisposed to learn human languages. All normal children, raised in normal social environments, inevitably learn their local language, whereas other species, even when raised and taught in this same environment, do not. This demonstrates that human brains come into the world specially equipped for this function.

SUMMARIZING

A summary presents the main ideas of a passage in a compressed form. To create an honest and competent summary, whether of a passage or of an entire book, you must not only represent the source accurately but also use your own wording. And, of course, you will need to include a citation. It is a common misconception that only quotations need to be acknowledged in the body of an essay. However, without a citation, even a fairly worded summary or paraphrase is an act of plagiarism. The first example below is plagiarized on two counts: it borrows wording (underlined) from the source, and it has no parenthetical citation.

> *incorrect* Researchers agree that language learning is innate, and that only innate knowledge can explain how we are able to learn a system of communication that is so complex. Normal children raised in normal ways will always learn their local language, whereas other species do not, even when taught human language and exposed to the same environment.

To avoid plagiarism, the summary needs to be revised to avoid the wording of the source passage; it also needs to identify the source. The next example includes a signal phrase identifying the author ("As Terrence W. Deacon notes") and a parenthetical citation indicating the page number.

> *correct* As Terrence W. Deacon notes, there is now wide agreement among linguists that the ease with which human children acquire their native tongues, under the conditions of a normal childhood, demonstrates an inborn capacity for language that

is not shared by any other animals, not even those who are reared in comparable ways and given human language training (102).

PARAPHRASING

A paraphrase rephrases information. Whereas a summary is a shorter version of an original passage, a paraphrase tends to be about the same length. However, a paraphrase, like a summary, must reflect the source accurately without using the source's wording, and it must include a citation. The example below, even though it is properly cited, is a failed attempt to paraphrase the first sentence of the Deacon passage. It is something called a "near quote," which is a type of (usually unintended) plagiarism. The phrases and words copied from the original have been underlined so that you can see this passage the way your instructor will. This is a very common error (indeed, some students are taught to do this in high school), but it is a clear example of plagiarism.

> *incorrect*   Researchers in <u>language</u> have come to <u>a consensus</u> in the past <u>few decades</u> that the acquisition of language is <u>innate</u>; such <u>contributions</u> <u>from knowledge</u> <u>contribute significantly</u> to <u>our ability</u> to master <u>such a complex system</u> of <u>communication</u> (Deacon 102).

Simply substituting synonyms for the words and phrases used in the source, however, is not enough to avoid plagiarism. The next example shows you how following the original passage's sentence structure is still a form of plagiarism, even if you use different words. In the second paragraph, the italicized material in parentheses shows the phrasing taken directly from Deacon's original text.

> *incorrect*   Recently, linguists appear to have come to an agreement that speaking is an in-born skill, and that nothing but a substantial input from in-born cognition can account for the human capacity to acquire such a complicated means of expression (Deacon 102).

> Recently (*over the last few decades*), linguists (*language researchers*) appear to have come to an agreement (*seem to have reached a consensus*) that speaking is an in-born skill (*that language is an innate ability*), and that nothing but a substantial input (*and that only a significant contribution*) from

in-born cognition (*from innate knowledge*) can account for the human capacity (*can explain our ability*) to acquire such a complicated means of expression (*to learn such a complex communication system*) (Deacon 102).

Now here is a good paraphrase of the passage's opening sentence; this paraphrase captures the sense of the original without echoing the details and shape of its language.

> *correct*   Many linguists now believe that children are born with the ability to learn language; in fact, these social scientists agree that the human capacity to acquire such a difficult skill cannot easily be accounted for in any other way (Deacon 102).

SITUATING CITATIONS

In referencing other writers' work, it is essential that you make clear to the reader two fundamental things:

- what the actual ideas in the original material are, and
- what the relationship is between that material and the current writer's work.

It's more common than one might think for writers to use citations in such a way that the cited research's conclusions are entirely unclear. Below, for example, is a sentence from the "Discussion" section of Jeff Downing's "Non-invasive Assessment of Stress in Commercial Housing Systems" (included in this anthology).

> *incorrect*   Social interactions can be stressful for laying hens (Craig and Guhl, 1969; Hughes et al., 1997; Bilick and Keeling, 2000; Keeling et al., 2003).

On its own, this statement implies that social interaction in general is the problem leading to stress in hens. But the cited studies do not make this point; in fact the study by Boris Bilcík (not "Bilick," as Downing has it) and Linda J. Keeling was, according to its title, designed to study "pecking in laying hens and the effect of group size." Bilcík and Keeling's conclusion was not that social interaction in general can be stressful, but that their study's results showed "some evidence of an increasing frequency of aggressive pecks with increasing group size" (55). If a great many hens are packed tightly together, in other words, they will tend to display more aggression toward one another. So here is a revised version that acknowledges that meaning.

> *correct*   Research has pointed to some important conclusions regarding social interaction and stress among laying hens. One important study has linked larger flock size with increased aggression (Bilcík and Keeling, 2000).

In the revised version, the information is much more specific – and much clearer.

It's also very common for writers to fail to make clear what the relationship is between material they are citing and the ideas they are putting forward themselves. The reader must be able to see clearly whether the research being cited supports the writer's argument (and if so, how) or disagrees with it (and, if so, why the writer's argument differs).

> *incorrect*   Probiotics have developed a reputation for preventing or treating gastroenterological troubles. Looking at studies on gut results, one can find good examples and bad examples in which design is so variable that a comparative analysis cannot be performed (Floch and Montrose, 2005).

In the above example, it is not easy to tell where the writer's argument ends and the cited article's material starts. Also, the writer's own point is not clear. Is this just a restatement of Floch and Montrose's argument? Here is one way to fix the material:

> *correct*   Probiotics have developed a reputation for preventing or treating gastroenterological troubles. However, while the numbers of investigations of probiotics have been swiftly increasing, the studies are justifiably seen as too varied in their design to allow for overall assessment of the effectiveness of this treatment. In their 2005 study of articles in reliable journals, for instance, Floch and Montrose found that, for pediatric studies, "the number of patients is too small, and there are too many dose and probiotic organism variables to perform a meta-analysis."

Now, the passage has been altered to show what was argued in the original source and how the cited material relates to the writer's point. Words and phrases such as "however," "justifiably," and "for instance" help to demonstrate how the ideas in the passage fit together. The use of quotation helps to clarify what material belongs to the article. A paraphrase such as the following would also clearly distinguish what material belongs to Floch and Montrose:

correct   In their 2005 study of articles in reliable journals, for instance, Floch and Montrose found that, in pediatric studies, there were too few participants and the study regimens were too different to allow them to do a meta-analysis.

CHECKING THE ORIGINAL RESEARCH BEFORE YOU CITE

One might think that facts are facts, and that if the results of research are reported on or summarized in a reputable publication, the results are indeed just as presented. But that's often not the case. There doesn't even need to be any intention to misrepresent the data for misleading information to appear; confusion can occur in a variety of ways. Let's look at a real-life example. On March 14, 2013, in *Psychology Today*'s online "Memory Medic" section, William Klemm published an article entitled "What Learning Cursive Does for Your Brain." Klemm's article argues that cursive writing improves one's cognitive skills, and that it should therefore be taught more widely in schools. In the middle of the article, two sentences specifically concerned with cursive writing (as opposed to printing) are followed by this sentence:

> Other research highlights the hand's unique relationship with the brain when it comes to composing thoughts and ideas. Virginia Berninger, a professor at the University of Washington, reported her study of children in grades two, four, and six that revealed they wrote more words, faster, and expressed more ideas when writing essays by hand versus with a keyboard.

It is natural for the reader to assume that Berninger's study concerned cursive writing. It also seems reasonable to think that one could use this material as a reference in writing something such as the following sentence:

> A *Psychology Today* article has reported on research showing that children using cursive writing in grades two, four, and six write more words, write faster, and express more ideas than do children using a keyboard (Klemm).

But let's follow the trail to the evidence itself. Klemm's footnotes cite the following reference:

> Berninger, V. "Evidence-based, Developmentally Appropriate Writing Skills K-5: Teaching Orthographic Loop of Working Memory to Write Letters So Developing Writers Can Spell Words and Express Ideas." Presented at Handwriting in the 21st Century: An Educational Summit, Washington, D.C., January 23, 2012.

The PowerPoint slides from that presentation are readily available online, but they do not provide any details of the research itself. The only reference to the original data in Berninger's presentation is seen in one PowerPoint slide that includes the following statement:

> UW Findings: Both those with and without dysgraphia in grades 2, 4, and 6 wrote more words, wrote words faster, and expressed more ideas in essays by pen than by keyboard; in 4th grade they wrote more complete sentences by pen than keyboard.

Fortunately, the study itself has been widely cited and is not difficult to find using a search engine (though it is behind a paywall, and readers thus need to pay a stiff access fee or have access to a university or corporate account to be able to read the full article). The article by Berninger et al., "Comparison of Pen and Keyboard Transcription Modes in Children with and without Learning Disabilities," appeared in *Learning Disability Quarterly* 32.3 (Summer 2009): 123–41.

Does the study indeed report what it is said to report? Yes and no. The reported results do in fact appear in the study. But so do contrasting results, mentioned neither by Berninger in her 2012 set of PowerPoint slides nor by Klemm in his *Psychology Today* article. Though the subjects in Berninger's study wrote more words by hand than by keyboard when given ten minutes to write an essay, when they were given no time limit and asked to write a sentence, both the Grade 4 and the Grade 6 subjects wrote significantly *fewer* words by pen than by keyboard.

The most important thing to note about this research paper, though, is that it did *not* compare cursive writing with keyboard writing. Berninger et al. are quite specific on this point:

> For these sentence-level and text-level writing tasks, we did not specify whether children had to use manuscript or cursive letters. That is because past research ... has shown that both typically developing and poor writers use manuscript, cursive, or a mix of manuscript and cursive, and we did not want to disrupt the practiced handwriting format of individual writers. (127)

If we go back to the phrasing used in Klemm's *Psychology Today* piece, we can see that what he wrote was in fact narrowly correct; the phrase "by hand" is included in his reporting on Berninger's research. And that is the phrase that is used in her PowerPoint presentation to the January 23, 2012, conference funded by Zaner Bloser (a company that produces a range of handwriting products). Everything, then, is narrowly correct. No one has egregiously misrepresented anyone else, and everything could be said to be above board, including the role played by Zaner Bloser, which is also mentioned in passing in Klemm's *Psychology Today* article. But this is a good example of why it is important to read carefully, and to check what is said in the actual research, before you cite. While it's unlikely anyone will make students go back to learning cursive on the basis of Klemm's article, you can imagine the consequences if the load-bearing ability of a new formula for concrete were narrowly correctly (but not exactly and fully) reported in an engineering article.

QUOTING DIRECTLY

Direct quotation is much less common in the sciences than in the humanities; the general assumption is that it is the results of scientific research that matter, not the specific way in which the researchers may have discussed those results. On those grounds many science writers refer to a study's results without quoting. But quotation may be useful in scientific writing as well; sometimes there is no need to rephrase everything. And, if you quote from the original, you will reduce the risk of inadvertently misrepresenting what the original says. One sensible way to look at this question is as follows: "Direct quotations are rarely used in scientific writing except for definitions, or when an idea or principle is exceptional, or [when an idea] cannot be expressed any better" ("Quoting").

You should, of course, scan a journal's current issue to see what approach that journal prefers, and you should read through its policies on the actual writing up of your evidence. Where a journal or other science publication routinely encourages minimizing quotations or avoiding them entirely, do so. If your instructor asks you not to use quotations, then obviously you should not use them. But – whether you use quotations or not – always be sure to make it clear how the material you use functions as part of your argument, and never take the material out of context or misrepresent it.

This example shows how a lack of quotation can cause misrepresentation of data.

> *incorrect* Recent research suggests that Facebook use causes people to be unhappy (Kross et al. 2013).

The problem here is that Ethan Kross et al. do not quite say this; they are much more careful in their phrasing, using the verb "predict" rather than the verb "cause" and specifying that their research sample was made up of young people – not people of all ages.

> *correct* Recent research suggests that, "rather than enhancing well-being, ... interacting with Facebook may predict the opposite result for young adults – it may undermine it" (Kross et al. 2013).

Using the exact words that another writer has used can have real advantages. But direct quotation should be used with great care. Quote too frequently, and you risk making your readers wonder why they are not reading your sources instead of your paper. Indeed, some institutions suggest that a paper that is more than 50% quotation is "functionally plagiarized" (meaning it was not written by the author but instead is a collection of source materials).

Your paper should present something you want to say. It should be informed and supported by properly documented sources, but it needs to offer a contribution that is yours alone. To that end, use secondary material to help you build a strong framework of supporting evidence for your work, not to replace it.

Exact and Accurate Quotation

To avoid misrepresenting your sources, be sure to quote accurately; to avoid plagiarism, take care to indicate quotations as quotations, and cite them properly. Below are two problematic quotations. The first does not show which words come directly from the source.

> *incorrect* Terrence W. Deacon maintains that children enter the world predisposed to learn human languages (102).

The second quotation fails to identify the source at all.

> *incorrect* Linguists believe that "children enter the world predisposed to learn human languages."

The next example corrects both problems by naming the source and indicating clearly which words come directly from it.

> *correct* Terrence W. Deacon maintains that "children enter the world predisposed to learn human languages" (102).

Be sure, also, to check your quotations against the originals after you have printed out your paper. Sloppy quotation can be quite noticeable, and it leads readers to wonder if your own research argument can be trusted.

Formatting Quotations

There are two ways to signal an exact borrowing in the North American system:[a] by enclosing it in double quotation marks and by indenting it as a block of text. Which way you should choose depends on the length and genre of the quotation and the style guide you are following.

---

a Other geographical areas use different arrangements, and you should know if the journal you have chosen to submit an article to follows these other arrangements.

SHORT PROSE QUOTATIONS

What counts as a short prose quotation differs among the various reference guides. In APA style, "short" means up to forty words; in Chicago Style, up to one hundred words; in MLA style, up to four lines of your paper; CSE style does not specify a precise number differentiating a short quotation from a long one. All the guides agree, however, that short quotations must be enclosed in double quotation marks, as in the examples below.

Short quotation, full sentence	According to Terrence W. Deacon, linguists agree that a human child's capacity to acquire language is inborn: "Without question, children enter the world predisposed to learn human languages" (102).
Short quotation, partial sentence	According to Terrence W. Deacon, linguists agree that human "children enter the world predisposed to learn human languages" (102).

LONG PROSE QUOTATIONS

Longer prose quotations should be double-spaced and indented, as a block, away from the left margin (the size of the indentation varies according to the style you are using). Do not include quotation marks; the indentation indicates that the words come exactly from the source. Note that indented quotations are often introduced with a full sentence followed by a colon.

correct	Terrence W. Deacon, like most other linguists, believes that human beings are born with a unique cognitive capacity:

> Without question, children enter the world predisposed to learn human languages. All normal children, raised in normal social environments, inevitably learn their local language, whereas other species, even when raised and taught in this same environment, do not. This demonstrates that human brains come into the world specially equipped for this function. (102)

QUOTATIONS WITHIN QUOTATIONS

The usual North American quotation format uses double quotation marks. However, you may sometimes find words already enclosed in double quotation marks within the passage you wish to quote. If your quotation is short, enclose it all in double quotation marks, and change the original quotation marks to single quotation marks.

   correct   Terrence W. Deacon is firm in maintaining that human language differs from other communication systems in kind rather than degree: "Of no other natural form of communication is it legitimate to say that 'language is a more complicated version of that'" (44).

If your quotation is a long quotation, keep the double quotation marks of the original, as this example shows.

   correct   Terrence W. Deacon is firm in maintaining that human language differs from other communication systems in kind rather than degree:

> Of no other natural form of communication is it legitimate to say that "language is a more complicated version of that." It is just as misleading to call other species' communication systems *simple* languages as it is to call them languages. In addition to asserting that a Procrustean mapping of one to the other is possible, the analogy ignores the sophistication and power of animals' non-linguistic communication, whose capabilities may also be without language parallels. (44)

ADDING TO OR DELETING FROM A QUOTATION

While it is important to use the original's exact wording in a quotation, it is accepted use to modify a quotation somewhat, as long as the changes are clearly indicated and they do not distort the meaning of the original. You might, for instance, want to add to a quotation in order to clarify elements that are unclear once the quotation has been removed from its context. Put whatever you add in square brackets, as in the following example:

   correct   Terrence W. Deacon writes that children are born "specially equipped for this [language] function" (102).

If you would like to streamline a quotation by omitting something unnecessary to your point, insert an ellipsis (three spaced dots) to show that you've left material out.

   correct   Terrence W. Deacon talks about "children ... predisposed to learn human languages" (102).

When the omitted material runs over a sentence boundary or constitutes a whole sentence or more, insert a period plus an ellipsis:

> *correct*   Terrence W. Deacon, like most other linguists, believes that human children are born with a unique ability to acquire their native language: "Without question, children enter the world predisposed to learn human languages.... [H]uman brains come into the world specially equipped for this function" (102).

Be sparing in modifying quotations; it is fine to have some altered quotations in a paper, but if you find yourself changing quotations often, or adding to and omitting from one quotation more than once, reconsider quoting at all. A paraphrase or summary might be a more effective choice.

Integrating Quotations

Quotations must be worked smoothly and grammatically into your sentences and paragraphs. Always, of course, mark a quotation as such, but, for the purpose of integrating it into your writing, treat it as if it were simply part of your sentence. The boundary between what you say and what your source says should be grammatically seamless.

> *incorrect*   Terrence W. Deacon points out, "whereas other species, even when raised and taught in this same environment, do not" (102).

> *correct*   According to Terrence W. Deacon, while human children brought up under normal conditions acquire the language they are exposed to, "other species, even when raised and taught in this same environment, do not" (102).

CONTEXTUALIZING QUOTATIONS

Integrating quotations well also means providing a context for them. Don't merely drop them into your paper or string them together like beads on a necklace; make sure to introduce them by noting where the material comes from and how it connects to whatever point you are making. It's important to show 1) where the original material is to be distinguished from your written work, 2) what the original material's actual ideas are, and 3) how the material relates to the rest of your paper.

> *incorrect*   For many years, linguists have studied how human children acquire language: "Without question, children enter the world predisposed to learn human language" (Deacon 102).

correct   Most linguists studying how human children acquire language have come to share a conclusion about how they do acquire it, articulated here by Terrence W. Deacon: "Without question, children enter the world predisposed to learn human language" (102).

incorrect   "Without question, children enter the world predisposed to learn human language" (Deacon 102): "There is ... something special about human brains that enables us to do with ease what no other species can do even minimally without intense effort and remarkably insightful training" (Deacon 103).

correct   Terrence W. Deacon bases his claim that we "enter the world predisposed to learn human language" on the fact that very young humans can "do with ease what no other species can do even minimally without intense effort and remarkably insightful training" (102–03).

SIGNAL PHRASES

To leave no doubt in your reader's mind about which parts of your essay are yours and which come from elsewhere, identify the sources of your summaries, paraphrases, and quotations with signal phrases, as in the following examples.

- As Carter and Rosenthal have demonstrated, ...
- In the words of one researcher, ...
- In his most recent book McGann advances the view that, as he puts it, ...
- As Nussbaum observes, ...
- Kendal suggests that ...
- Freschi and other scholars have rejected these claims, arguing that ...
- Morgan has emphasized this point in her recent research: ...
- As Sayeed puts it, ...
- To be sure, Mtele allows that ...
- In his later novels Hardy takes a bleaker view, frequently suggesting that ...

In order to help establish your paper's credibility, you may also find it useful at times to include information in your signal phrase that shows why readers should take the source seriously, as in the following example.

> In *Silent Spring*, a work that almost single-handedly initiated the North American environmental movement in the 1960s, biologist and conservationist Rachel Carson warned that pesticides such as DDT were overwhelmingly toxic across the whole food chain.

Here, the signal phrase mentions the author's professional credentials; it also points out the importance of her book, which is appropriate to do in the case of a work as famous as *Silent Spring*.

To Sum Up

Plagiarism is one of the worst academic sins and treated as such. In contrast, academic generosity and careful acknowledgment of how you are interacting with the work of other scholars is highly valued. Therefore, you should get all of these citation and source use strategies built into your writing now, so that your work will always show your reader that you are a careful, ethical researcher and writer, just the sort of investigator they can turn to for solidly evidenced, well-written work.

Works Cited

Berninger, Virginia, et al. "Comparison of Pen and Keyboard Transcription Modes in Children with and without Learning Disabilities." *Learning Disability Quarterly* 32.3 (Summer 2009): 123-41. Print.

Berninger, Virginia. "Evidence-based, Developmentally Appropriate Writing Skills K-5: Teaching Orthographic Loop of Working Memory to Write Letters So Developing Writers Can Spell Words and Express Ideas." 2012. *Microsoft Powerpoint* file.

Bilcík, Boris, and Linda J. Keeling. "Relationship between Feather Pecking and Ground Pecking in Laying Hens and the Effect of Group Size." *Applied Animal Behaviour Science* 68.1 (May 2000): 55-66. Web. 12 Dec. 2013.

Deacon, Terrence W. *The Symbolic Species: The Co-Evolution of Language and the Brain.* New York: Norton, 1997. Print.

Klemm, William. "What Learning Cursive Does for Your Brain." *Psychology Today*. Sussex Publishers, 14 Mar. 2013. Web. 12 Dec. 2013.

"Quoting." *Language and Learning Online*. Monash University, 2014. Web. 7 Jan. 2014.

Wainio, Carol. "Margaret Wente: 'A Zero for Plagiarism?'" *Media Culpa*. Blogspot, 18 September 2012. Web. 15 Dec. 2013.

Wente, Margaret. "Columnist Margaret Wente Defends Herself." *Globe and Mail*. Globe and Mail, 25 Sept. 2012. Web. 15 Dec. 2013.

Appendix B:
Writing about Science: A Closer Look

The many academic forms of writing about science include the review article (in which a writer surveys and assesses evidence on a particular topic from various sources); the research or experimental report, usually in the form of an article (in which a researcher writes up the results of an experiment of specific scope – lab reports are good practice for articles of this kind); the conference presentation or a poster board (in which scientists report in a limited or abbreviated fashion about specific research); the abstract (in which scientists provide a brief summary of a report or research paper); and the funding application, generally intended for a national or commercial grant or funding body (which usually includes a clear summary of the proposed research process and a statement of its importance, along with a costing breakdown).

Then there are the multiple forms of writing about science that look to communicate to non-specialist audiences – among them news articles about scientific developments; magazine articles and book-length treatments of scientific subjects for non-specialist readers; and advocacy articles that present arguments on matters involving science and public policy. These forms of writing may be written by the specialists themselves, but often they are written by writers of other sorts – such as academics from other disciplines, journalists, and activists.

IMRAD FORMAT

The peer-reviewed science article is, of course, the gold standard mode of scientific communication. Unlike articles in the humanities, research articles in the sciences are generally organized into specific sections that usually appear in a set order: Introduction, Methods, Results, and Discussion. The opening section of the article, the *Introduction,* tells the reader the purpose and nature of the study. How does it fit in with previous research? What has it been designed to show? What was the hypothesis? It is normal to keep the introduction fairly brief; typically, it explains a bit about the background to the paper and indicates why the research was undertaken. The paper's introduction may, in fact, include a background section or be called *Background.* While it is appropriate to position the paper in the larger context of previous research here, extended discussion of that larger picture is not normally included in the introduction.

The *Methods* section is often very detailed. An important principle is that the paper should provide enough information about how the research was set up and conducted that other researchers can replicate the results. The reader also needs enough information to be able to understand the rationale for each step in the process.

The Methods section in a scientific research paper is always written in the past tense. If you are conducting your own research, however, setting out the details of the process beforehand in the present and/or future tenses will allow you to consider whether the research is being set up in the most unbiased way possible and whether the proposed methodology is best fitted to provide evidence one way or the other as to whether the hypothesis is valid. Similarly, you can give some consideration to whether the most useful statistical methods have been chosen to assess the data.

The *Results* section, as its name suggests, details the results at length. It often includes tables, charts, and graphs as a way of visually presenting these results. The *Discussion* section provides an analysis of the meaning of those results, both in terms of the particular experiment and in terms of past experiments in the same area. Do the results confirm or refute the original hypothesis? What questions are left unanswered? What significance does the data have in the context of other research in this area? Are there reasons to qualify whatever conclusions are drawn on the basis of the research results (e.g., small sample size, confounding elements, short timeframe)? Do the results prompt any recommendations for the future? *Results* and *Discussion* are usually treated as separate sections of a paper; some journals, however, will ask that they be combined into one section of the paper.

Finally, a scientific paper is almost always accompanied by an abstract, which comes before the actual paper but is written after the article has been completed. The *Abstract* summarizes the entire paper, usually in several hundred words or less; it is meant to convey the essence of the research to those who may not have time to read the entire paper, or who may be trying to determine if the entire paper will be of interest to them. Abstracts are often included in searchable databases, precisely so that researchers can determine whether to access the whole paper.

SCIENTIFIC TONE AND STYLISTIC CHOICES

Most contemporary writers in the natural sciences and applied sciences strive for objectivity of tone, while writers in the humanities and in social sciences such as sociology and anthropology may foreground their own subject positions, sometimes adopting a less formal tone. Look, for example, at the way in which anthropologist Emily Martin opens her article on "The Egg and the Sperm" (included in this anthology):

> As an anthropologist, I am intrigued by the possibility that culture shapes how biological scientists describe what they discover about the natural world.... In the course of my research I realized that the picture of egg and sperm drawn in popular as well as scientific accounts of reproductive biology relies on stereotypes central to our cultural definitions of male and female. The stereotypes imply not only that female biological processes are less worthy than their male counterparts but also that women are less worthy than men. Part of my goal in writing this article is to shine a bright light on the gender stereotypes hidden within the scientific language of biology. Exposed in such a light, I hope they will lose their power to harm us.

Martin acknowledges at the outset that she occupies a specific position in relation to her research, and she uses the first person frequently ("I am intrigued," "I realized," "my goal"). Moreover, she acknowledges that her motive for conducting and publishing her research is not merely to expand abstract objective knowledge about the world; she aims not only to shine a bright light on gender stereotypes, but also to reduce "their power to harm us." And yet, at its core, Martin's article is an example of careful scientific inquiry: she conducted a thorough survey of relevant scientific literature, her assessment of it follows scientific standards of objectivity, and she makes a strong argument based on clear and well-documented evidence.

Like Martin, academics writing in natural science disciplines such as biology, chemistry, and physics (as well as those writing in behavioral psychology and in social sciences such as economics and political science) value scientific standards of objectivity and strive to make strong arguments based on strong evidence. The difference between Martin's approach to writing and that which prevails in some other areas of the sciences, then, is largely one of *style* and *tone*, with the natural science disciplines tending towards a more formal and impersonal style than Martin adopts. Compare her discussion of egg and sperm representations with the following behavioral science abstract for "Facebook Use Predicts Declines in Subjective Well-Being in Young Adults" by Ethan Kross et al. (included in this anthology):

> Over 500 million people interact daily with Facebook. Yet, whether Facebook use influences subjective well-being over time is unknown. We addressed this issue using experience-sampling, the most reliable method for measuring in-vivo behavior and psychological experience. We text-messaged

people five times per day for two-weeks to examine how Facebook use influences the two components of subjective well-being: how people feel moment-to-moment and how satisfied they are with their lives. Our results indicate that Facebook use predicts negative shifts on both of these variables over time. The more people used Facebook at one time point, the worse they felt the next time we text-messaged them; the more they used Facebook over two-weeks [sic], the more their life satisfaction levels declined over time. Interacting with other people "directly" did not predict these negative outcomes. They were also not moderated by the size of people's Facebook networks, their perceived supportiveness, motivation for using Facebook, gender, loneliness, self-esteem, or depression. On the surface, Facebook provides an invaluable resource for fulfilling the basic human need for social connection. Rather than enhancing well-being, however, these findings suggest that Facebook may undermine it.

Kross et al. employ terminology specific to the academic discipline in which they are writing (which would be inappropriate jargon in any other context): "subjective well-being," and "experience sampling," for example. More generally, the abstract is much more impersonal in tone than the first paragraph of Martin's paper; the reader is not told of any moment of realization which led to these researchers' work, or of the effects they hope their research will have. Instead, the passage provides a summary of the reasons for the study, the nature of the study, the main results, and some sense of their significance – all in reasonably clear, direct and specific prose.[a]

But what is it about the prose that makes this writing clear to the reader? Notice first that the abstract succinctly sketches in the background and the gap in the research that it aims to fill: "Over 500 million people interact daily with Facebook. Yet, whether Facebook use influences subjective well-being over time is unknown." If the study instead opened with these three sentences, would you want to read it?

Facebook is used every day by people all round the world – overall, a total of more than 500 million people. There have been numerous studies that have examined Facebook use; a number of these have found correlations between Facebook use and lower reported levels of subjective well-being. Those studies, however, have not tracked Facebook use and levels of subjective well-being together over time; doing so may provide a clearer picture

---

a   Even so, you can probably make a good guess at both what type of realization led to the research and what the researchers hope the impact of their research will be. Try rewriting "Facebook Use" using Martin's style; now try rewriting "The Egg and the Sperm" passage using the style of "Facebook Use." Which version of each do you like best and why?

of whether or not Facebook use does indeed negatively influence subjective well-being.

Notice the relative brevity of the sentences used as the original abstract opens – the first is only eight words long. Longer sentences follow; varying your sentence lengths is a good way to help maintain reader interest. The following two sentences from the abstract might easily have been combined into one very long sentence; long paragraphs and long sentences are very common in academic writing. But the information is conveyed more effectively here because the ideas are presented, instead, in two closely related sentences: "Interacting with other people 'directly' did not predict these negative outcomes. They were also not moderated by the size of people's Facebook networks, their perceived supportiveness, motivation for using Facebook, gender, loneliness, self-esteem, or depression."

Expressing ideas concisely and varying sentence length are two ways of making your writing clear and readable. Using parallel or balanced sentence structures is another. Notice how the parallel grammatical structures in the following sentence (*the more ... the worse ...; the more ..., the more...*) emphasize the study's findings: "The more people used Facebook at one time point, the worse they felt the next time we text-messaged them; the more they used Facebook over two-weeks, the more their life satisfaction levels declined over time."

Would you feel drawn to read the article if it began this way?

Subjects who reported high levels of Facebook use at one time point also reported when we next text-messaged them that they felt worse, and this pattern was repeated over the full study, with subjects who used Facebook at high levels over the two-week period reporting declining levels of life satisfaction, a result that was strongest among those who reported the highest extended levels of Facebook use.

Many more words and a difficult sentence structure (successive subordinate clauses tacked onto a compound sentence) make the reader work unnecessarily hard to understand what the writer is saying.

It is often imagined that complex and difficult-to-read sentence structures strike an appropriately academic tone – that direct sentences and rhetorical balance are for journalists or novelists, not students and scholars. If scientific research is important, however, then surely it's worth communicating about that research in clear and readable prose. The next section discusses some additional ways to make your writing as clear as possible.

The First Person and the Active Voice

High school students have often been told they cannot use the first person in "school writing." Similarly, those writing in disciplines such as biology, physics, psychology, and engineering have often been advised to avoid grammatical structures associated with what might seem a subjective approach – most notably, the first person (*I* or *we*). Such advice is often also given to students writing English or history papers, and the reason is much the same: instructors want to discourage students from thinking of the writing of an academic paper as an exercise in expressing one's likes or dislikes. Unsubstantiated opinions are not appreciated in any of the academic disciplines. But notice that how well one supports one's argument bears no necessary relation to whether or not one uses the first person. In the Emily Martin example quoted above, Martin uses the first person extensively, but she doesn't just express an opinion; her study is supported with a good deal of evidence, and the discussion's findings are objective in nature.

Similarly, the abstract of the paper about Facebook by Kross et al. does not achieve its more formal and impersonal tone by avoiding the first person. Quite the contrary: Kross et al. use the first person twice in the first three sentences of the abstract: "We addressed," "We text-messaged."[a] Scientific objectivity, then, is not a matter of avoiding the first person; it is rather a matter of avoiding bias when framing one's research questions, of designing research projects intelligently and ethically, and of interpreting the results in the same fashion.

Discussions of impersonality and objectivity in scientific writing have also often been framed in terms of the question of whether to use the active or the passive voice. For most of the twentieth century, many instructors in the natural sciences tried to train students to use the passive voice[b] – to write sentences like "*It **was decided** that the experiment **would be conducted** in three stages*" and "*These results **will be discussed** from several perspectives*" in order to convey a more impersonal and objective tone. In fact, statements in the active voice (such as "*We **conducted** the experiment in three stages*" and "*We **will discuss** these results from several perspectives*") are not any more or less objective. The reader of an article knows the researchers did

---

a   In the case of group authorship, which is very common in the sciences, the first person plural must, of course, be used.

b   As Randy Moore and others have pointed out, nineteenth-century scientists used the active voice and the first person freely. The active voice and "first-person pronouns such as *I* and *we* began to disappear from scientific writing in the United States in the 1920s" (Moore 388).

the experiment, so using the passive voice does not change the degree of objectivity – though it may add unnecessary words.

While instructors in the natural sciences were encouraging use of the passive voice through most of the twentieth century, instructors in English departments were during the same period strongly *discouraging* use of the passive, in line with the advice of the famous political essayist George Orwell, not to use the passive voice. Indeed, generations of English instructors were reluctant to acknowledge that the passive voice could be useful, even in scientific writing. In the 1990s, Andrea Lunsford and Robert Connors were playing something of a pioneering role in their discipline when they suggested, in their influential handbook, "Much scientific and technical writing uses the passive voice effectively to highlight what is being studied rather than who is doing the studying" (117). More than fifteen years later, a great many fellow writing studies scholars and composition teachers would give the same advice. Interestingly, just as handbook authors had begun to acknowledge that there could be a specific place for the passive voice in scientific writing, the scientific community had begun to swing around to the view that, much as the passive might sometimes have its place, the active voice should be the default writing choice. Here is Randy Moore, writing in *The American Biology Teacher* in 1991:

> The notion that passive voice ensures objectivity is ridiculous[;]... objectivity has nothing to do with one's writing style or with personal pronouns. Objectivity in science results from the choice of subjects, facts that you choose to include or omit, sampling techniques, and how you state your conclusions. Scientific objectivity is a personal trait unrelated to writing. (389)

The 1990s saw heated debates in the pages of certain scientific journals on the matter of whether the active or the passive should be the default. But in the end the decision was clearly that the active voice was the best choice. Nearly every major scientific journal[a] now recommends that its authors use the active voice in most situations – and that they use the first person where appropriate as well. The various *Nature* journals, the American Chemical Society Style Guide, and the American Society of Civil Engineers Style Guide may be taken as representative. The following instructions are from their respective style guides:

---

a   On his blog in 2012, Allen Downey conducted an informal survey and was able to identify only three exceptions: the *ICES Journal of Marine Science*, the *Journal of Animal Ecology*, and *Clinical Oncology and Cancer Research*.

Nature journals prefer authors to write in the active voice ("we performed the experiment ...") as experience has shown that readers find concepts and results to be conveyed more clearly if written directly. ("How to Write a Paper")

Use the active voice when it is less wordy and more direct than the passive.... Use first person when it helps to keep your meaning clear and to express a purpose or a decision. (Coghill and Garson 42-43)

Active Versus Passive Voice
Wherever possible, use active verbs that demonstrate what is being done and who is doing it....

> *Instead of:* Six possible causes of failure were identified in the forensic investigation.
> *Use:* The forensic investigation identified six possible causes of failure. (*ASCE Author's Guide*)

Nonetheless, though the active voice is better as a default choice, in some circumstances it can still be better to use the passive. In some cases, indeed, it is essential. It is hard to imagine, for instance, how the following sentence could be recast in the active voice:

- This compound is made up of three elements.

The passive may also be useful as a means of shifting attention away from the researcher(s) to the experiment or event itself, as in the following example:

- Phenomena of this sort may be seen only during an eclipse.

Problems arise, however, when use of the passive reduces clarity or when the passive is used to disguise responsibility. Consider the following sentence in the passive voice, taken from the opening of the Toby Knowles et al. article on "Leg Disorders in Broiler Chickens" (included in this anthology): "Broiler chickens have been subjected to intense genetic selection." That is to say, they have been bred to grow so fast and become so heavy that they can barely walk. But this has not just happened; humans have done this as part of a broad-based (factory farming) effort to generate more meat, more efficiently, and less expensively. "Intense genetic selection," then, did not just happen to the birds. Here, the article's use of the passive voice functions to de-emphasize responsibility. As a critical reader, you should think skeptically about why that might be.

524 | Appendix B

It is useful to remember that the passive voice is only one of many writing strategies that can be used to disguise or downplay responsibility for an action. Consider this example from Jeff Downing's "Non-invasive Assessment of Stress in Commercial Housing Systems" (included in this anthology): "hens need to deal with [challenges] in their environment.... In any flock there are likely to be some hens that perceive the challenges as more severe than others and have high corticosterone concentrations." In Downing's sentences, the hens are, grammatically, the active parties; they are the subjects of the verbs: "*hens **need** ... hens ... **perceive** the challenges*." But of course the controlling agents here in any sense other than a grammatical one are not the birds; it is humans who have subjected them to these "challenges" – if that euphemism is to be accepted. ("Hardships," "privations," or "cruelties" are some of the other nouns that might be substituted for "challenges.")

It would seem, then, that there is nothing pernicious in the passive voice itself. The passive voice is one means that *may* be used to disguise agency – and it is also a verbal construction that *may* involve unnecessary or inappropriate wordiness. In these cases, to avoid disguising agency or making your writing more difficult to read, it is often better to rephrase.

As a writer and reader, you have a responsibility not only to make yourself aware of the conventions of science writing, but also to pay attention to the ethical and social nuances of communication. Scientific research of many types has implications for the ways in which humans interact with the environment; it has implications for the health of humans and other animals as well. Indeed, there are few scientific topics that do not affect you and the world around you.

Works Cited

ASCE Author's Guide: Writing Style. 2013. American Society of Civil Engineers, 2013. Web. 2 Dec. 2013.

Coghill, A.M., and Lorrin R. Garson, eds. *The ACS Style Guide: Effective Communication of Scientific Information*. 3rd ed. New York: Oxford University Press, 2006. Print.

Downey, Allen. "The Passive Voice Is a Hoax!" *Probably Overthinking It*. Blogspot, 2 Apr. 2012. Web. 7 Dec. 2013.

"How to Write a Paper." *Nature*. Nature Publishing Group, 2013. Web. 2 Dec. 2013.

Lunsford, Angela, and Robert Connors. *The Everyday Writer: A Brief Reference*. Boston: St. Martin's Press, 1997. Print.

Moore, Randy. "How We Write About Biology." *American Biology Teacher* 53.7 (Oct. 1991): 388-89. Print.

Orwell, George. "Politics and the English Language." *Shooting an Elephant and Other Essays*. London: Secker & Warburg, 1950. 84-101. Print.

Permissions Acknowledgments

Luis W. Alvarez, Walter Alvarez, Frank Asaro, and Helen V. Michel. Adapted from "Extraterrestrial Cause for the Cretaceous-Tertiary Extinction." *Science*, Volume 208, Number 4448 (1980): 1095-1108. Reprinted with permission from AAAS and Walter Alvarez.

William F. Baker, D. Stanton Korista, and Lawrence C. Novak. "Burj Dubai: Engineering the World's Tallest Building." *Structural Design of Tall and Special Buildings*, December 1, 2007. Reproduced with permission of John Wiley & Sons Ltd. via Copyright Clearance Center.

Diana Baumrind. "Some Thoughts on the Ethics of Research: After Reading Milgram's 'Behavioral Study of Obedience.'" *American Psychologist* 19.6 (1964): 421-423. Copyright © 1964 by the American Psychological Association. Reproduced with the permission of the APA and Diana Baumrind.

Keith Baverstock and Mauno Rönkkö, 2008. "Epigenetic Regulation of the Mammalian Cell." *PLoS ONE* 3(6): e2290. doi:10.1371/journal.pone.0002290

Damian Carrington. "Common pesticides 'can kill frogs within an hour.'" *The Guardian*, Thursday 24 January 2013. http://www.theguardian.com/environment/2013/jan/24/pesticides-kill-frogs-within-hour. Reprinted with the permission of Guardian News and Media Limited.

George Constantinou; Stavros Melides and Bernadette Modell; Michael Spino and Fernando Tricta; David Nathan and David Weatherall. "The Olivieri Case," (Correspondence). *New England Journal of Medicine* 348: 860-863, February 27 2003. Copyright © 2003 Massachusetts Medical Society. Reprinted with permission from Massachusetts Medical Society.

Alexis De Greiff A. and Mauricio Nieto Olarte. "The Race for Third World Hearts and Minds: The Seduction of Development" and "The Old and New Green 'Revolution'" from "What We Still Do Not Know About South-North Technoscientific Exchange: North-Centrism, Scientific Diffusion, and the Social Studies of Science." *The Historiography of Contemporary Science, Technology and Medicine*, edited by Ronald E. Doel and Thomas Söderqvist. NY: Routledge, 2006. Copyright © 2006, Routledge. Reproduced by permission of Taylor & Francis Books UK.

Brian Deer. "Secrets of the MMR Scare: How the Case against the MMR Vaccine was Fixed." *British Medical Journal* 342 (6 January 2011): 77-82. Reproduced with permission from BMJ Publishing Group Ltd.

Jared Diamond. "Easter's End." *Discover Magazine*, August 1995. Discover Media. All rights reserved. Used by permission and protected by the Copyright Laws of the United States. The printing, copying, redistribution, or retransmission of this Content without express written permission is prohibited.

Joseph Dimow. "Resisting Authority: A Personal Account of the Milgram Obedience Experiments" *Jewish Currents* (January 2004); http://jewishcurrents.org/resisting-authority-15383. Reprinted with the permission of Carl Dimow.

Jeff Downing. Excerpted from "Non-invasive Assessment of Stress in Commercial Housing Systems: A Report for the Australian Egg Corporation Ltd." *AECL Publication No.US108A*, March 2012. © 2012 Australian Egg Corporation Limited. Reprinted with permission.

Alexandre Erler. "In Vitro Meat, New Technologies, and the 'Yuck Factor.'" *Practical Ethics*, March 5, 2012 (http://blog.practicalethics.ox.ac.uk/2012/03/in-vitro-meat-new-technologies-and-the-yuck-factor/); revised by the author, August 2013. Reprinted with the permission of Alexandre Erler.

European Food Safety Authority. EFSA Panel on Plant Protection Products and their Residues (PPR). Scientific Opinion on the Science Behind the Development of a Risk Assessment of Plant Protection Products on Bees (Apis mellifera, Bombus spp. and solitary bees). *EFSA Journal* 2012; 10(5):2668. [275 pp.]. doi:10.2903/j.efsa.2012.2668. Available online: www.efsa.europa.eu/efsajournal. Copyright © European Food Safety Authority (EFSA), Parma, Italy.

Malcolm Gladwell. "None of The Above: What I.Q. Doesn't Tell You About Race." Copyright © 2007 by Malcolm Gladwell. Originally published in *The New Yorker*. Reprinted by permission of the author.

Ben Goldacre. Excerpt from "Marketing" in *Bad Pharma: How Drug Companies Mislead Doctors and Harm Patients*, by Ben Goldacre. Copyright © 2012 by Ben Goldacre. Reprinted by permission of Faber and Faber, Inc., an affiliate of Farrar, Straus and Giroux, LLC. Copyright © 2013 Better Doctors Ltd. Reprinted by permission of McClelland & Stewart.

Ben Goldacre. "Ghostwriters in the Sky," article published online September 18th, 2010 at http://www.badscience.net/2010/09/ghostwriters/. Reproduced by permission of United Agents LLP on behalf of Ben Goldacre.

Stephen Jay Gould. "Critique of The Bell Curve" from *The Mismeasure of Man*, second edition. W.W. Norton, 1996, 1981. Reprinted with the permission of Rhonda Roland Shearer.

Humane Society International. "Beyond Doubt: Intensive Farm Practices Results in Stressed Birds." Humane Society International, March 26, 2012 (http://hsi.org.au/?catID=1107)

Institution of Mechanical Engineers. "Executive Summary," *Global Food: Waste Not, Want Not*. 2013: pp 2-5. http://www.imeche.org/knowledge/themes/environment/global-food. Reprinted with the permission of the Institution of Mechanical Engineers, London.

Intergovernmental Panel on Climate Change. Hockey Stick Graph from *Summary for Policy Makers*, Intergovernmental Panel on Climate Change, 3rd Assessment Report, 2001. Image supplied by Michael Mann.

Toby G. Knowles, Steve C. Kestin, Susan M. Haslam, Steven N. Brown, Laura E. Green, Andrew Butterworth, Stuart J. Pope, Dirk Pfeiffer, Christine J. Nicol, 2008. "Leg Disorders in Broiler Chickens: Prevalence, Risk Factors and Prevention." *PLoS ONE* 3(2): e1545. doi:10.1371/journal.pone.0001545 (excerpt).

Elizabeth Kolbert. "The Sixth Extinction." *The New Yorker Magazine*, May 25, 2009. Reprinted by permission of Elizabeth Kolbert.

Ethan Kross, Philippe Verduyn, Emre Demiralp, Jiyoung Park, David Seungjae Lee, Natalie Lin, Holly Shablack, John Jonides, Oscar Ybarra. "Facebook Use Predicts Declines in Subjective Well-Being in Young Adults." (2013) PLoS ONE 8(8): e69841. doi:10.1371/journal.pone.0069841

Michael Mann. "Myth vs. Fact Regarding the 'Hockey Stick.'" *RealClimate*, 2004. http://www.realclimate.org/index.php/archives/2004/12/myths-vs-fact-regarding-the-hockey-stick/.

Emily Martin. "The Egg and the Sperm: How Science Has Constructed a Romance Based on Stereotypical Male-Female Roles." *Signs: Journal of Woman in Culture and Society*, 16:3 (1991): 485-501. Reprinted with the permission of the publisher, University of Chicago Press.

Stanley Milgram. "Behavioral Study of Obedience," *Journal of Abnormal and Social Psychology* 67.4 (1963). Compliments of Alexandra Milgram.

Stanley Milgram. "Issues in the Study of Obedience: A Reply to Baumrind." *American Psychologist* 19.11 (1964): 848-852. Compliments of Alexandra Milgram.

Corinne A. Moss-Racusin, John F. Dovidio, Victoria L. Brescoll, Mark J. Graham, and Jo Handelsman. "Science Faculty's Subtle Gender Biases Favor Male Students." *Proceedings of the National Academy of Sciences of the United States of America* 2012 109: 16474-16479. http://www.pnas.org/content/109/41/16474.full.pdf+html. Reprinted with the permission of the Proceedings of the National Academy of Sciences.

David G. Nathan, M.D. and David J. Weatherall, M.D. "Academic Freedom in Clinical Research," *New England Journal of Medicine* 347: 1368-1371, October 24 2002. Copyright © 2002 Massachusetts Medical Society. Reprinted with permission from Massachusetts Medical Society.

Ian Nicholson. "'Torture at Yale': Experimental Subjects, Laboratory Torment and the 'Rehabilitation' of Milgram's 'Obedience to Authority.'" *Theory & Psychology* 21.6 (12/2011): 737-761 (excerpt). Reprinted by permission of SAGE.

ODT Maps Inc. "What's Up? South!" World Map, http://odtmaps.com/images/products/WUS-36x56-LT.jpg. Reprinted with the permission of ODT Maps Inc., Amherst, Massachusetts.

Nancy F. Olivieri, Gary M. Brittenham, Christine E. McLaren, Douglas M. Templeton, Ross G. Cameron, Robert A. McClelland, Alastair D. Burt, Kenneth A. Fleming. "Long-Term Safety and Effectiveness of Iron-Chelation Therapy with Deferiprone for Thalassemia Major." *New England Journal of Medicine* 339

(August 13, 1998): 417-423. Copyright © 1998 Massachusetts Medical Society. Reprinted with permission from Massachusetts Medical Society.

Christine Parker. "The Truth about Free Range Eggs Is Tough to Crack." *The Conversation*, 5 August 2013, 2.26pm AEST. http://theconversation.com/the-truth-about-free-range-eggs-is-tough-to-crack-16661. Reprinted with the permission of Christine Parker.

Fred Pearce. "Battle over climate data turned to war between scientists and sceptics." *The Guardian*, Feb 9, 2010. http://www.guardian.co.uk/environment/2010/feb/09/climate-change-data-request-war. Reprinted with the permission of Guardian News and Media Limited.

Fred Pearce. "Climate change debate overheated after sceptic grasped 'hockey stick.'" *The Guardian*, Feb. 9, 2010. http://www.guardian.co.uk/environment/2010/feb/09/hockey-stick-michael-mann-steve-mcintyre. Reprinted with the permission of Guardian News and Media Limited.

David Reilly, 2012. "Gender, Culture, and Sex-Typed Cognitive Abilities." *PLoS ONE* 7(7): e39904. doi:10.1371/journal.pone.0039904

Jimmy Lee Shreeve. "Mile-high tower wars: How tall is too tall?" *The Independent*, June 18 2008 (http://www.independent.co.uk/arts-entertainment/art/features/milehigh-tower-wars-how-tall-is-too-tall-849088.html). Reprinted with the permission of Jimmy Lee Shreeve.

Richard C. Strohman. "Linear Genetics, Non-Linear Epigenetics: Complementary Approach to Understanding Complex Disease." *Integrative Physiological and Behavioral Science* 30.4 (September-December 1995): 273-282. Reproduced with kind permission from Springer Science and Business Media.

F.E. Vera-Badillo, R. Shapiro, A. Ocana, E. Amir and I. F. Tannock. "Bias in reporting of end points of efficacy and toxicity in randomized, clinical trials for women with breast cancer." *Annals of Oncology*. First published online January 9, 2013. Reprinted by permission of Oxford University Press.

A.J. Wakefield, S.H. Murch, A. Anthony, J. Linnell, D.M. Casson, M. Malik, M. Berelowitz, A.P. Dhillon, M.A. Thomson, P. Harvey, A.Valentine, S.E. Davies, J.A. Walker-Smith. "RETRACTED: Ileal-lymphoid-nodular Hyperplasia, Non-specific Colitis, and Pervasive Developmental Disorder in Children." Reprinted from *The Lancet*, Vol. 351, No. 9103 (Feb. 28, 1998): 637-641. Copyright © 1998, with permission from Elsevier. Reproduced with permission of Lancet Publishing Group via Copyright Clearance Center.

Amy B. Wisniewski et al. "Complete Androgen Insensitivity Syndrome: Long-Term Medical, Surgical, and Psychosexual Outcome." *The Journal of Clinical Endocrinology & Metabolism* 85 (August 1, 2000): 2664-2669. Reprinted with the permission of *The Journal of Clinical Endocrinology & Metabolism* via Copyright Clearance Center, Inc.

Author-Title Index

from the publisher

A name never says it all, but the word "broadview" expresses a good deal of the philosophy behind our company. We are open to a broad range of academic approaches and political viewpoints. We pay attention to the broad impact book publishing and book printing has in the wider world; we began using recycled stock more than a decade ago, and for some years now we have used 100% recycled paper for most titles. As a Canadian-based company we naturally publish a number of titles with a Canadian emphasis, but our publishing program overall is internationally oriented and broad-ranging. Our individual titles often appeal to a broad readership too; many are of interest as much to general readers as to academics and students.

Founded in 1985, Broadview remains a fully independent company owned by its shareholders—not an imprint or subsidiary of a larger multinational.

If you would like to find out more about Broadview and about the books we publish, please visit us at **www.broadviewpress.com**. And if you'd like to place an order through the site, we'd like to show our appreciation by extending a special discount to you: by entering the code below you will receive a 20% discount on purchases made through the Broadview website.

Discount code: **broadview20%**

Thank you for choosing Broadview.

Please note: this offer applies only to sales of bound books within the United States or Canada.

The interior of this book is printed on 30% recycled paper.